T0320648

Numerical Analysis for Engineers and Scientists

Striking a balance between theory and practice, this graduate-level text is perfect for students in the applied sciences. The author provides a clear introduction to the classical methods, how they work and why they sometimes fail. Crucially, he also demonstrates how these simple and classical techniques can be combined to address difficult problems. Many worked examples and sample programs are provided to help the reader make practical use of the subject material. Further mathematical background, if required, is summarized in an appendix.

Topics covered include classical methods for linear systems, eigenvalues, interpolation and integration, ODEs and data fitting, and also more modern ideas such as adaptivity and stochastic differential equations.

G. Miller is a Professor in the Department of Chemical Engineering and Materials Science at University of California, Davis.

Numerical Analysis for Engineers and Scientists

G. MILLER

Department of Chemical Engineering and Materials Science
University of California, Davis

CAMBRIDGE
UNIVERSITY PRESS

Shaftesbury Road, Cambridge CB2 8EA, United Kingdom

One Liberty Plaza, 20th Floor, New York, NY 10006, USA

477 Williamstown Road, Port Melbourne, VIC 3207, Australia

314–321, 3rd Floor, Plot 3, Splendor Forum, Jasola District Centre, New Delhi – 110025, India

103 Penang Road, #05–06/07, Visioncrest Commercial, Singapore 238467

Cambridge University Press is part of Cambridge University Press & Assessment, a department of the University of Cambridge.

We share the University's mission to contribute to society through the pursuit of education, learning and research at the highest international levels of excellence.

www.cambridge.org
Information on this title: www.cambridge.org/9781107021082

First published 2014

A catalogue record for this publication is available from the British Library

ISBN   978-1-107-02108-2   Hardback

Contents

Preface   *page* ix

1   **Numerical error**   1
     1.1   Types of error   1
     1.2   Floating point numbers   1
     1.3   Algorithms and error   6
     1.4   Approximation error vs. algorithm error   13
     1.5   An important example   16
     1.6   Backward error   17
     Problems   19

2   **Direct solution of linear systems**   22
     2.1   Gaussian elimination   23
     2.2   Pivot selection   30
     2.3   Errors in Gaussian elimination   36
     2.4   Householder reduction   43
     2.5   Cholesky decomposition   49
     2.6   The residual correction method   52
     Problems   54

3   **Eigenvalues and eigenvectors**   56
     3.1   Gerschgorin's estimate   57
     3.2   The power method   62
     3.3   The QR algorithm   70
     3.4   Singular value decomposition   81
     3.5   Hyman's method   89
     Problems   91

4   **Iterative approaches for linear systems**   93
     4.1   Conjugate gradient   93
     4.2   Relaxation methods   103
     4.3   Jacobi   105
     4.4   Irreducibility   109
     4.5   Gauss–Seidel   111

	4.6	Multigrid	115
		Problems	123
5		**Interpolation**	125
	5.1	Modified Lagrange interpolation and the barycentric form	128
	5.2	Neville's algorithm	129
	5.3	Newton	131
	5.4	Hermite	136
	5.5	Discrete Fourier transform	138
		Problems	148
6		**Iterative methods and the roots of polynomials**	153
	6.1	Convergence and rates	153
	6.2	Bisection	155
	6.3	Regula falsi	157
	6.4	The secant method	159
	6.5	Newton–Raphson	162
	6.6	Roots of a polynomial	168
	6.7	Newton–Raphson on the complex plane	172
	6.8	Bairstow's method	175
	6.9	Improving convergence	179
		Problems	180
7		**Optimization**	182
	7.1	1D: Bracketing	182
	7.2	1D: Refinement by interpolation	183
	7.3	1D: Refinement by golden section search	184
	7.4	**n**D: Variable metric methods	185
	7.5	Linear programming	191
	7.6	Quadratic programming	201
		Problems	211
8		**Data fitting**	213
	8.1	Least squares	213
	8.2	An application to the Taylor series	217
	8.3	Data with experimental error	219
	8.4	Error in $\mathbf{x}$ and $\mathbf{y}$	225
	8.5	Nonlinear least squares	230
	8.6	Fits in other norms	230
	8.7	Splines	235
		Problems	241

9	**Integration**	243
	9.1 Newton–Cotes	243
	9.2 Extrapolation	252
	9.3 Adaptivity	257
	9.4 Gaussian quadrature	259
	9.5 Special cases	271
	Problems	273

10	**Ordinary differential equations**	275
	10.1 Initial value problems I: one-step methods	275
	10.2 Initial value problems II: multistep methods	278
	10.3 Adaptivity	287
	10.4 Boundary value problems	292
	10.5 Stiff systems	298
	Problems	300

11	**Introduction to stochastic ODEs**	302
	11.1 White noise and the Wiener process	303
	11.2 Itô and Stratonovich calculus	306
	11.3 Itô's formula	308
	11.4 The Itô–Taylor series	309
	11.5 Orders of accuracy	311
	11.6 Strong convergence	314
	11.7 Weak convergence	318
	11.8 Modeling	320
	Problems	324

12	**A big integrative example**	326
	12.1 The Schrödinger equation	326
	12.2 Gaussian basis functions	337
	12.3 Results I: H_2	341
	12.4 Angular momentum	342
	12.5 Rys polynomials	345
	12.6 Results II: H_2O	349

Appendix A	**Mathematical background**	353
	A.1 Continuity	353
	A.2 Triangle inequality	354
	A.3 Rolle's theorem	354
	A.4 Mean value theorem	355
	A.5 Geometric series	356
	A.6 Taylor series	357
	A.7 Linear algebra	361
	A.8 Complex numbers	369

Appendix B   Sample codes   371
    B.1   Utility routines   371
    B.2   Gaussian elimination   375
    B.3   Householder reduction   380
    B.4   Cholesky reduction   386
    B.5   The QR method with shifts for symmetric real matrices   388
    B.6   Singular value decomposition   393
    B.7   Conjugate gradient   400
    B.8   Jacobi, Gauss–Seidel, and multigrid   402
    B.9   Cooley–Tukey FFT   405
    B.10   Variable metric methods   408
    B.11   The simplex method for linear programming   413
    B.12   Quadratic programming for convex systems   420
    B.13   Adaptive Simpson's rule integration   426
    B.14   Adaptive Runge–Kutta ODE example   428
    B.15   Adaptive multistep ODE example   430
    B.16   Stochastic integration and testing   434
    B.17   Big example: Hartree–Fock–Roothaan   438

    Solutions   454
    References   555
    Index   567

Preface

This book is an introduction to numerical analysis: the solution of mathematical problems using numerical algorithms. Typically these algorithms are implemented with computers. To solve numerically a problem in science or engineering one is typically faced with four concerns:

1. How can the science/engineering problem be posed as a mathematical problem?
2. How can the mathematical problem be solved at all, using a computer?
3. How can it be solved accurately?
4. How can it be solved quickly?

The first concern comes from the science and engineering disciplines, and is outside the scope of this book. However, there are many practical examples and problems drawn from engineering, chemistry, and economics applications.

The focus of most introductory texts on numerical methods is, appropriately, concern #2, and that is also the main emphasis here. Accordingly, a number of different subjects are described that facilitate solution of a wide array of science and engineering problems.

Accuracy, concern #3, deals with numerical error and approximation error. There is a brief introductory chapter on error that presents the main ideas. Throughout the remainder of this book, algorithm choices and implementation details that affect accuracy are described. For the most part, where a claim of accuracy is made an example is given to illuminate the point, and to show how such claims can be tested.

The speed of computational methods, concern #4, is addressed by emphasizing two aspects of algorithm design that directly impact performance in a desktop environment – rates of convergence and operation count, and by introducing adaptive algorithms which use resources judiciously. Numerous examples are provided to illustrate rates of convergence. Modern high-performance algorithms are concerned also with cache, memory, and communication latency, which are not addressed here.

In some circles there is a tendency, bolstered by Moore's law [166], to suppose that accuracy and speed are not terribly important in algorithm design. In two years, one can likely buy a computer that is twice as fast as the best available today. So, if speed is important it might be best to acquire a faster platform. Likewise, a faster, more capable, computer could employ arbitrarily high precision to overcome any of today's accuracy problems. However, in a given computing environment the fast algorithm will always out-perform the slow ones, so Moore's law does not affect the relative performance.

Similarly, one can always implement a more accurate algorithm with higher precision, and for given precision the more accurate algorithm will always prevail.

The material covered in this book includes representative algorithms that are commonly used. Most are easily implemented or tested with pencil, paper, and a simple hand calculator. It is interesting to note that most of the methods that will be described predate modern computers, so their implementation by hand is not at all unreasonable. In fact, until the 1950s the term "computer" referred to a person whose profession was performing calculations by hand or with slide rules. To emphasize the antiquity of some of these ideas, and to give proper recognition to the pioneers that discovered them, I attempt to provide references to the original works.

Texts on numerical analysis and numerical methods range from very practical to very theoretical, and in this one I hope to strike a balance. On the practical side, there are numerous worked solutions and code examples. The code examples are intended to be a compromise between pseudocode and production code – functional and readable, but not state of the art. I hope the interested reader will see the similarity between the equations in the text and the C++ code to get a better appreciation of the logic, and the accessibility of these methods (i.e., if I can do it, so can you). These codes are available online for download at `http://www.cambridge.org/9781107021082`. On the theoretical side, the mathematical approaches used to derive and explain numerical algorithms are different from those a typical engineering student will have encountered in calculus and analytical partial differential equations. This is both interesting and useful, and I have included some in an informal way. There are no theorems, but the logic is displayed through equations and text. Some of the mathematical background needed to understand these concepts is summarized in an appendix.

This book grew from class notes developed over a dozen years of teaching numerical methods to engineers at both undergraduate and graduate levels. In a 10-week undergraduate course one or two examples from each of the first 10 chapters can be discussed to give an overview of the field and to equip the students with some numerical problem solving skills. In a 20-week graduate course, most of the material can be covered with reasonable depth.

I thank the students of EAD 115 and EAD 210 who all contributed to the development of this book through their engagement and feedback over the years. In particular, I thank Bakytzhan Kallemov for helping to develop the chapter on stochastic methods, and Mehdi Vahab for improving the sample codes. I am especially grateful to my wife Carolyn for her support and encouragement.

1 Numerical error

1.1 Types of error

The term "error" is going to appear throughout this book in different contexts. The varieties of error we will be concerned with are:

- Experimental error. We may wish to calculate some function $y(x_1, \ldots, x_n)$, where the quantities x_i are measured. Any such measurement has associated errors, and they will affect the accuracy of the calculated y.
- Roundoff error. Even if x were measured exactly, odds are it cannot be represented exactly in a digital computer. Consider π, which cannot be represented exactly in decimal form. We can write $\pi \approx 3.1416$, by rounding the exact number to fit 5 decimal figures. Some roundoff error occurs in almost every calculation with real numbers, and controlling how strongly it impacts the final result of a calculation is always an important numerical consideration.
- Approximation error. Sometimes we want one thing but calculate another, intentionally, because the other is easier or has more favorable properties. For example, one might choose to represent a complicated function by its Taylor series. When substituting expressions that are not mathematically identical we introduce approximation error.

Experimental error is largely outside the scope of numerical treatment, and we'll assume here, with few exceptions, that it's just something we have to live with. Experimental error plays an important role in data fitting, which will be described at length in Chapter 8. Sampling error in statistical processes can be thought of as a type of experimental error, and this will be discussed in Chapter 11.

Controlling roundoff error, sometimes by accepting some approximation error, is the main point of this chapter, and it will be a recurring theme throughout this book. J. H. Wilkinson [242] describes error analysis generally, with special emphasis on matrix methods. An excellent and comprehensive modern text is N. J. Higham's [104].

1.2 Floating point numbers

Binary computers represent everything in "bits" – fundamental units of measure that can take on the values 0 or 1 only. A byte is a collection of 8 bits, and real numbers are typically stored in 4 bytes (single precision floating point) or 8 bytes (double precision floating point). The representation of real numbers in floating point format is governed by standards, the current one being IEEE Std. 754-2008 [1].

A decimal number like "0.1" means

$$0 \times 10^0 + 1 \times 10^{-1},$$

and in binary we can represent numbers in a similar fashion. The decimal number "23" has a binary representation

$$1 \times 2^4 + 0 \times 2^3 + 1 \times 2^2 + 1 \times 2^1 + 1 \times 2^0,$$

or 10 111. The decimal number "0.1" has binary representation

$$1 \times 2^{-4} + 1 \times 2^{-5} + 0 \times 2^{-6} + 0 \times 2^{-7} + \cdots,$$

or

$$0.\overline{1100} \times 2^{-3}, \tag{1.1}$$

where the line indicates that the sequence 1100 repeats forever. There exist many numbers like "0.1" that can be represented in a compact fashion in decimal notation, but that cannot be represented in a compact notation in binary.

On a binary computer, the real number "0.1" is represented in single precision as

$$\underbrace{+}_{\text{sign}} \ 0.\underbrace{110\,011\,001\,100\,110\,011\,001\,101}_{\text{binary mantissa}} \ \underbrace{2^{-3}}_{\text{exponent}}, \tag{1.2}$$

with the 32 bits divided as follows:

- 1 sign bit
- 23+1 mantissa bits
- 8 exponent bits.

(The notation of (1.2) may be confusing because the exponent is clearly not written in a binary fashion. Of course it will be on a binary computer, but our concern here is with the mantissa.) The total number of bits in this example is 33, not 32. This is because we enforce (where possible) the following convention: the mantissa will be adjusted so that the first bit after the decimal place is 1. This is called a *normalized representation*. When this convention is respected, there is no need to store this bit – it is implicit.

In double precision, "0.1" is

$$+0.110\,011\,001\,100\,110\,011\,001\,100\,110\,011\,001\,100\,110\,011\,001\,100\,110\,10 \times 2^{-3}, \tag{1.3}$$

with

- 1 sign bit
- 52+1 mantissa bits
- 11 exponent bits.

Neither representation is exact, because the exact solution (1.1) cannot fit in 24 or 53 mantissa places. We will use the symbol t for the number of mantissa places, with $t = 53$ understood unless stated otherwise. Internal machine registers can have still more mantissa bits. For example, the Intel Pentium uses 64 mantissa bits in its internal floating point registers.

When a real number can be represented exactly with a machine floating point convention, it is called *machine representable*. If a real number is not machine representable, it is *rounded* to a close machine representable number. The rounding rule used by default on most modern desktop computers is called *round ties to even*. If the exact number lies exactly between two machine representable numbers, this rounding method chooses the machine representable number with 0 in the least significant bit. Otherwise, rounding chooses the closest machine representable number.

You can see in (1.3) that this rounding took place. The least significant bits of the exact mantissa read ...110011... where the final underlined 1 is in the $2^{-(t+1)}$ place. By the rounding rule, the mantissa changed to ...110 10.

It is of interest to determine the maximum error that can be introduced by rounding. To determine this, consider the two binary numbers

$$x_1 = 0.011$$
$$x_2 = 0.100\overline{1} = 0.101.$$

x_1 is the smallest number that rounds up to 0.10, and x_2 is the largest number that rounds down to 0.10, with $t = 2$ and the round ties to even rule. These numbers both differ from 0.10 by 0.001, or $2^{-(t+1)}$, therefore rounding introduces a mantissa error as large as $2^{-(t+1)}$. Real numbers include also an exponent part, which we should account for. If a number x has a value of d in the exponent,

$$|x - \text{round}(x)| \leq 2^{-(t+1)} \times 2^d. \tag{1.4}$$

But, because of the normalized representation,

$$|x| \geq 0.1_2 \times 2^d = 2^{d-1} \tag{1.5}$$

(the subscript 2 was used to emphasize that 0.1 here is the leading part of the binary mantissa). Combining inequalities (1.4) and (1.5),

$$\frac{|x - \text{round}(x)|}{|x|} \leq 2^{-t} = \begin{cases} 2^{-24} \approx 6.0 \times 10^{-8} & \text{single precision} \\ 2^{-53} \approx 1.1 \times 10^{-16} & \text{double precision.} \end{cases}$$

The expression of this rounding error as a *relative error*, i.e., error divided by value, emphasizes the role of the finite mantissa.

Even if two numbers x and y were represented exactly in normalized binary form, their sum $x + y$ might not be. Likewise, for any function (multiplication, division, sine, cosine, exponentiation, etc.) the result of the operation – if carried out exactly – is unlikely to fit exactly into the finite number of mantissa places available for it. Let's denote $\tilde{x}$ as the rounded version of x, and $\tilde{f}$ the computed (hence rounded) version of function $f(x, y)$. As a best case scenario, we have the situation

$$\tilde{f}(\tilde{x}, \tilde{y}) = \text{round}(f(\tilde{x}, \tilde{y}))$$

$$\tilde{f}(\tilde{x}, \tilde{y}) - f(\tilde{x}, \tilde{y}) = f(\tilde{x}, \tilde{y})\epsilon_f, \quad |\epsilon_f| \le |2^{-t}|,$$

(1.6)

which supposes that the function $f(\tilde{x}, \tilde{y})$ is as accurate as if it were computed exactly, but then rounded to fit the available space. We will make the optimistic assumption (1.6) that all *elementary operations* have an error equivalent to rounding. The number 2^{-t} is called *machine precision*, and will be denoted by symbol ϵ without subscripts.

If x and y are not machine representable, then, for example, $z = x + y$ has errors from the rounding of x, the rounding of y and the rounding of the sum. With z standing for the exact sum, and $\tilde{z}$ standing for the floating point version,

$$\tilde{z} = \text{round}(\text{round}(x) + \text{round}(y))$$

$$= (x(1 + \epsilon_x) + y(1 + \epsilon_y))(1 + \epsilon_+) \quad \text{with all } |\epsilon_i| \le \epsilon \equiv 2^{-t}$$

$$\approx x + x(\epsilon_x + \epsilon_+) + y + y(\epsilon_y + \epsilon_+)$$

(1.7)

$$\frac{\tilde{z} - z}{z} = \frac{\Delta z}{z} \approx \frac{x}{x + y}\epsilon_x + \frac{y}{x + y}\epsilon_y + \epsilon_+,$$

where terms of *order of magnitude* ϵ^2 are ignored.

It is impossible to determine the error from (1.7) alone, because the relative errors ϵ_x, ϵ_y, and ϵ_+ are not known – we only know the upper bound of their magnitude. To use that information, we need to take the absolute value of (1.7) and use the triangle inequality:

$$\frac{|\Delta z|}{|z|} = \left| \frac{x}{x + y}\epsilon_x + \frac{y}{x + y}\epsilon_y + \epsilon_+ \right|$$

$$\le \left| \frac{x}{x + y} \right| |\epsilon_x| + \left| \frac{y}{x + y} \right| |\epsilon_y| + |\epsilon_+|$$

(1.8)

$$\le \left(\left| \frac{x}{x + y} \right| + \left| \frac{y}{x + y} \right| + 1 \right) \epsilon.$$

If $z = x + y$ is genuine addition (i.e., if the signs of x and y are identical), then $0 \le x/(x + y) \le 1$ and $0 \le y/(x + y) \le 1$, and $|x/(x + y)| + |y/(x + y)| = 1$ so the error has an upper bound of $|\Delta z/z| \le 2\epsilon$.

However, if the signs of x and y are different, then this is really subtraction and $|x/(x + y)| \ge 1$. If z is a small difference in relatively large numbers then the factor $|x/(x + y)|$ can be very large, and the calculation $\tilde{z}$ can be correspondingly very inaccurate.

The reason for this is *cancellation*: when two numbers are subtracted, there can be a loss of significant figures. For example,

$$+0.100\,00_{10} \approx +0.110\,011\,001\,100\,110\,011\,001\,101 \times 2^{-3}$$

$$-0.099\,85_{10} \approx -0.110\,011\,000\,111\,111\,000\,101\,000 \times 2^{-3}$$

$$\rule{4cm}{0.4pt}$$

$$+0.000\,15_{10} \approx +0.100\,111\,010\,100\,100\,101\,010\,010 \times 2^{-12} = \text{round}(z),$$

but

$$+0.110\,011\,001\,100\,110\,011\,001\,101 \times 2^{-3}$$

$$-0.110\,011\,000\,111\,111\,000\,101\,000 \times 2^{-3}$$

$$=0.\underline{000\,000\,000}\,100\,111\,010\,100\,101 \times 2^{-3}$$

$$=0.100\,111\,010\,100\,101\,\underline{000\,000\,000} \times 2^{-12} = \tilde{z}.$$

Round(z) and $\tilde{z}$ differ in 10 places: we find a relative error of approximately 2^{-14} or about 1024ϵ. This difference is largely because upon subtraction 9 significant digits of the result were lost (underlined). When the normalized form is reestablished, these lost digits become zeros in the least significant figures.

Note that although there is a large loss of significant figures, this subtraction did not in itself introduce any numerical error [123, p. 12].† The effect of cancellation is to amplify the errors associated with rounding the inputs x and y.

A theoretical upper bound to the error from (1.8) is

$$\left(\frac{0.1}{0.000\,15} + \frac{0.099\,85}{0.000\,15} + 1\right)\epsilon \approx 1333\epsilon,$$

which is pretty close to the observed error.

The error formula (1.7) for addition has a special form that can be generalized to other functions. The idea is *differential error analysis*, and to understand it consider the function

$$y = \cos(x).$$

Accounting for roundoff error, the numerical result $\tilde{y}$ is

$$\tilde{y} = \text{round}(\cos(\tilde{x})) \tag{1.9}$$

and, in contrast to the expansion for addition, the exact value $y = \cos(x)$ does not appear when (1.9) is written

$$\tilde{y} = \cos(x(1 + \epsilon_x))(1 + \epsilon_c). \tag{1.10}$$

However, since ϵ_x is very small compared to 1, we could expand (1.10) in a Taylor series. Keeping only the leading terms,

$$\tilde{y} \approx \cos(x)(1 + \epsilon_c) - x\epsilon_x \sin(x)$$

$$\frac{\Delta y}{y} = \frac{\tilde{y} - y}{y} = -x \tan(x)\epsilon_x + \epsilon_c.$$

† "If p and q are represented exactly in the same conventional floating-point format, and if $1/2 \le p/q \le 2$, then $p - q$ too is representable exactly in the same format, unless $p - q$ suffers exponent underflow."

To generalize this to arbitrary unary and binary functions,

$$\frac{\Delta f(x)}{f(x)} = \left(\frac{x}{f}\frac{df}{dx}\right)\frac{\Delta x}{x} + \epsilon_f$$

$$\frac{\Delta g(x, y)}{g(x, y)} = \left(\frac{x}{g}\frac{dg}{dx}\right)\frac{\Delta x}{x} + \left(\frac{y}{g}\frac{dg}{dy}\right)\frac{\Delta y}{y} + \epsilon_g.$$

The factors

$$\left(\frac{x}{f}\frac{df}{dx}\right)$$

are called *condition numbers* or *amplification factors*. For genuine addition, we saw that these numbers lie between 0 and 1, so they do not cause error to grow. For subtraction, these numbers are greater than 1 in magnitude and they amplify error.

1.3   Algorithms and error

Formulas that are mathematically identical can incur different numerical errors, therefore to assess the numerical error associated with some function it is important to specify completely the algorithm that will be used to evaluate it.

Example 1.1   The variance S^2 of a set of observations $x_1, ..., x_n$ is to be determined (S is the standard deviation). Which of the formulas,

$$S_1^2 = \frac{1}{n-1}\left(\sum_{i=1}^{n} x_i^2 - n\bar{x}^2\right) \qquad (1.11a)$$

$$S_2^2 = \frac{1}{n-1}\sum_{i=1}^{n}(x_i - \bar{x})^2, \qquad (1.11b)$$

with

$$\bar{x} = \frac{1}{n}\sum_{i=1}^{n} x_i,$$

is better from a numerical error point of view? We would like to ask "which formula has the lowest error," but this question cannot be answered. Instead, we can ask and answer "which formula has the lowest maximum error?" This is sometimes worded "which algorithm is more *numerically trustworthy?*"

Solution

First, we note that these expressions are mathematically identical:

$$
S_2^2 = \frac{1}{n-1} \sum_{i=1}^{n} (x_i - \bar{x})^2
$$

$$
= \frac{1}{n-1} \left[\sum_{i=1}^{n} x_i^2 - 2 \sum_{i=1}^{n} x_i \bar{x} + \sum_{i=1}^{n} \bar{x}^2 \right]
$$

$$
= \frac{1}{n-1} \left[\sum_{i=1}^{n} x_i^2 - 2n\bar{x}^2 + n\bar{x}^2 \right]
$$

$$
= \frac{1}{n-1} \left[\sum_{i=1}^{n} x_i^2 - n\bar{x}^2 \right] = S_1^2,
$$

so any numerical differences will be due to the algorithm only.

To answer the question of numerical trustworthiness, we need to formally state the algorithm associated with the mathematical formulas, then analyze these algorithms by tracking the rounding errors.

The first algorithm can be written, for $n = 2$, $\phi_1 = x_1 + x_2$, $\bar{x} = \phi_1/2$, $\phi_2 = x_1^2$, $\phi_3 = x_2^2$, $\phi_4 = \phi_2 + \phi_3$, $\phi_5 = \bar{x}^2$, $\phi_6 = n\phi_5$, $\phi_7 = \phi_4 - \phi_6$, $S_1^2 = \phi_7/(n-1)$. The error associated with this algorithm might be written with differential error analysis

$$
\frac{\Delta x_1}{x_1} = \epsilon_{x_1}
$$

$$
\frac{\Delta x_2}{x_2} = \epsilon_{x_2}
$$

$$
\frac{\Delta \phi_1}{\phi_1} = \frac{x_1}{\phi_1} \frac{\Delta x_1}{x_1} + \frac{x_2}{\phi_1} \frac{\Delta x_2}{x_2} + \epsilon_1
$$

$$
\frac{\Delta \bar{x}}{\bar{x}} = \frac{\Delta \phi_1}{\phi_1}
$$

$$
\frac{\Delta \phi_2}{\phi_2} = 2\frac{\Delta x_1}{x_1} + \epsilon_2
$$

$$
\frac{\Delta \phi_3}{\phi_3} = 2\frac{\Delta x_2}{x_2} + \epsilon_3 \tag{1.12}
$$

$$
\frac{\Delta \phi_4}{\phi_4} = \frac{\phi_2}{\phi_4} \frac{\Delta \phi_2}{\phi_2} + \frac{\phi_3}{\phi_4} \frac{\Delta \phi_3}{\phi_3} + \epsilon_4
$$

$$
\frac{\Delta \phi_5}{\phi_5} = 2\frac{\Delta \bar{x}}{\bar{x}} + \epsilon_5
$$

$$
\frac{\Delta \phi_6}{\phi_6} = \frac{\Delta \phi_5}{\phi_5}
$$

$$
\frac{\Delta \phi_7}{\phi_7} = \frac{\phi_4}{\phi_7} \frac{\Delta \phi_4}{\phi_4} - \frac{\phi_6}{\phi_7} \frac{\Delta \phi_7}{\phi_7} + \epsilon_7
$$

$$
\frac{\Delta S_1^2}{S_1^2} = \frac{\Delta \phi_7}{\phi_7}.
$$

Note that division by 1 carries no error, and division by 2 carries no error either – division or multiplication by any power of 2 will affect the exponent of the number, not its mantissa.

Combining the expressions, and writing the intermediate variables in terms of x_1 and x_2 gives:

$$\frac{\Delta S_1^2}{S_1^2} = \underbrace{\epsilon_7 + \frac{2x_1}{x_1 - x_2}\epsilon_{x_1} - \frac{2x_2}{x_1 - x_2}\epsilon_{x_2}}$$

$$- 2\frac{(x_1 + x_2)^2}{(x_1 - x_2)^2}\epsilon_1 + 2\frac{x_1^2}{(x_1 - x_2)^2}\epsilon_2 \qquad (1.13)$$

$$+ 2\frac{x_2^2}{(x_1 - x_2)^2}\epsilon_3 + 2\frac{x_1^2 + x_2^2}{(x_1 - x_2)^2}\epsilon_4 - \frac{(x_1 + x_2)^2}{(x_1 - x_2)^2}\epsilon_5.$$

The collection of terms with a bracket beneath them is special, as will be explained.

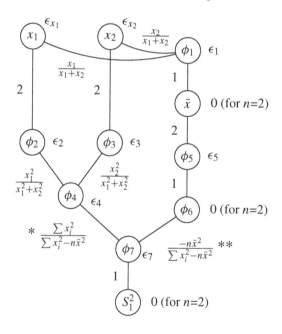

Figure 1.1 Bauer graph for S_1^2 with $n = 2$; Example 1.1.

Finally, before analyzing ΔS_2^2, note that there is a convenient graphical method to obtain the information embodied in (1.12). A "Bauer graph" [9] of algorithm 1 is given in Figure 1.1. A circle is drawn for each input variable and for each *elementary operation*, and lines are drawn to show how the data from one circle connects to the others. Associated with each circle is a rounding error, and associated with each line of the graph is the condition number of the elementary operation. The graph is "read" by tracing all paths to the final result, multiplying each rounding error by the product of condition numbers that follow it.

The second algorithm may be written: $n = 2$ and $\eta_1 = x_1 + x_2$, $\bar{x} = \eta_1/2$, $\eta_2 = x_1 - \bar{x}$, $\eta_3 = \eta_2^2$, $\eta_4 = x_2 - \bar{x}$, $\eta_5 = \eta_4^2$ $\eta_6 = \eta_4 + \eta_5$, $S_2^2 = \phi_6/(n-1)$. To analyze this algorithm we'll use its Bauer graph (Figure 1.2).

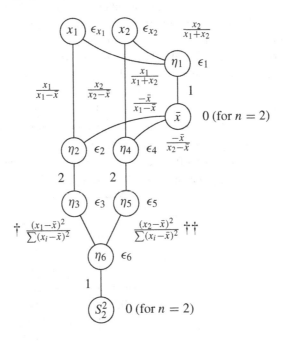

Figure 1.2 Bauer graph for S_2^2 with $n = 2$; Example 1.1.

Analyzing the graph gives the same result that would be obtained if each differential error expansion were written and combined:

$$\frac{\Delta S_2^2}{S_2^2} = \epsilon_6 + \underbrace{\frac{2x_1}{x_1 - x_2}\epsilon_{x_1} - \frac{2x_2}{x_1 - x_2}\epsilon_{x_2} + \epsilon_2} + \frac{1}{2}\epsilon_3 + \epsilon_4 + \frac{1}{2}\epsilon_5. \tag{1.14}$$

It is interesting to note that there is no dependence here at all on the relative error ϵ_1.

A collection of terms has been identified with a bracket in (1.13) and in (1.14), and these have the same bound. These bracketed errors are (1) the input errors ϵ_{x_1} and ϵ_{x_2}, propagated through to the result S^2, and (2) one roundoff error associated with the last step of the algorithm. This final error is included because, like (1.6), even an exact calculation will be subject to rounding. Any mathematically identical algorithm for calculating S^2 will contain these terms – they are completely unavoidable. They are called *inherent error*, to distinguish them from the algorithm-dependent errors that make up the remainder of (1.13) and (1.14).

We have, and will continue to assume in general, that $|\epsilon_i| \le \epsilon = 2^{-t}$ for all errors in the calculation. But, in fact, input error ϵ_x may be considerably worse than ϵ because of experimental errors. The inherent error may be used to analyze how experimental error affects the calculated solution.

Now, we compare the algorithm-dependent errors in (1.13) and (1.14). Algorithm 1 incurs large cancellation errors at the end of the algorithm: condition numbers (*) and (**) of Figure 1.1 are guaranteed ≥ 1 and ≤ -1, respectively, because actual subtraction occurs. No such large condition numbers appear in the algorithm-dependent part of the error for algorithm 2. In algorithm 2 the final steps are *numerically harmless*, and the amplification factors (†) and (††) of Figure 1.2 lie between 0 and 1. This results in a numerically stable algorithm because the roundoff errors grow like $|x_1 - x_2|^0$ while the inherent errors grow like $|x_1 - x_2|^{-1}$. Algorithm 2 is more numerically trustworthy than algorithm 1.

The lesson of this example is that when cancellation is unavoidable, do it as soon as possible. If it is postponed, the large condition numbers amplify numerical errors beyond those occurring in the inherent error.

To demonstrate the difference in these algorithms, the calculation is performed on the following two columns of numbers. Both columns have the same variance.

data set 1	data set 2
−0.329	$10^6 - 0.329$
0.582	$10^6 + 0.582$
−0.039	$10^6 - 0.039$
−0.156	$10^6 - 0.156$
−0.063	$10^6 - 0.063$
0.552	$10^6 + 0.552$
−0.927	$10^6 - 0.927$
−0.540	$10^6 - 0.540$
0.460	$10^6 + 0.460$
0.046	$10^6 + 0.046$

set		algorithm 1	algorithm 2
1	S^2	2.3655560000000003e−01	2.3655560000000003e−01
	ΔS^2	1.1733214354523381e−16	1.1733214354523381e−16
2	S^2	2.3676215277777779e−01	8.7316799001074320e−04
	ΔS^2	2.3655560001559353e−01	6.5919075558009081e−11

For the first set of data both algorithms compute the solution to approximately machine precision. With the second data set the differences are dramatic. The relative error of algorithm 1 is $\approx 10^7$ times worse than the relative error of algorithm 2.

Example 1.2 Consider two algorithms for the same function

$$y_1 = \frac{1 - \cos(x)}{x}$$

$$y_2 = \frac{\sin^2(x)}{x\,(1 + \cos(x))},$$

to be evaluated for $x \neq 0$ and $|x| \ll 1$.

Solution

Algorithm 1 may be written in terms of elementary operations as

$$\eta_1 = \cos(x)$$
$$\eta_2 = 1 - \eta_1 \qquad (1.15)$$
$$y_1 = \eta_2/x,$$

and algorithm 2 may be written in terms of elementary operations as

$$\xi_1 = \sin(x)$$
$$\xi_2 = \xi_1^2$$
$$\xi_3 = \cos(x)$$
$$\xi_4 = 1 + \xi_3 \qquad (1.16)$$
$$\xi_5 = x\xi_4$$
$$y_2 = \xi_2/\xi_5.$$

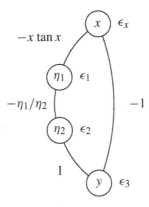

Figure 1.3 Bauer graph for algorithm 1, (1.15); Example 1.2.

The graph of algorithm 1 is given in Figure 1.3. The relative error associated with y_1 is found by summing the errors across the graph,

$$\frac{\Delta y_1}{y_1} = \underbrace{\left[\epsilon_3 + (-1)\epsilon_x + (1)\left(-\frac{\eta_1}{\eta_2}\right)(-x\tan x)\,\epsilon_x\right]}_{\text{inherent}} + \underbrace{(1)\epsilon_2 + (1)\left(-\frac{\eta_1}{\eta_2}\right)\epsilon_1}_{\text{algorithmic roundoff}},$$

then simplified by substituting values for η_i, and applying the limit $|x| \ll 1$:

$$\frac{\Delta y_1}{y_1} = \left[\epsilon_3 + \frac{x\sin x + \cos x - 1}{1 - \cos x}\epsilon_x\right] + \epsilon_2 - \frac{\cos x}{1 - \cos x}\epsilon_1$$

$$\frac{\Delta y_1}{y_1} \approx [\epsilon_3 - \epsilon_x] + \epsilon_2 - \frac{2}{x^2}\epsilon_1.$$

The inherent error is sometimes denoted with a superscript "0." The limit of this term is found by taking its absolute value, applying the triangle inequality, and simplifying with $|\epsilon_3| \leq \epsilon$ and $|\epsilon_x| \leq \epsilon$:

$$\left| \frac{\Delta y}{y} \right|^0 = |\epsilon_3 - \epsilon_x| \leq |\epsilon_3| + |\epsilon_x| = 2\epsilon.$$

Inherent error has nothing to do with the algorithm when the formulas being evaluated are mathematically identical: the same result applies to algorithm 2. Finally use the triangle inequality and the assumption $|\epsilon_i| < \epsilon$ to analyze the remaining error terms for algorithm 1:

$$\left| \frac{\Delta y_1}{y_1} \right| \leq \left| \frac{\Delta y}{y} \right|^0 + \epsilon + \frac{2}{x^2}\epsilon$$

$$\left| \frac{\Delta y_1}{y_1} \right| \leq \left(3 + \frac{2}{x^2} \right)\epsilon.$$

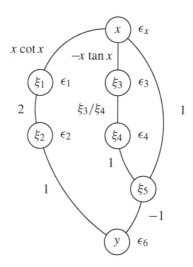

Figure 1.4 Bauer graph for algorithm 2, (1.16); Example 1.2.

To analyze algorithm 2: (1) sum errors across graph (Figure 1.4), (2) simplify in the appropriate limit, (3) take absolute value of both sides, (4) apply the triangle inequality, and (5) use $|\epsilon_i| \leq \epsilon$:

$$\frac{\Delta y_2}{y_2} = \left(\frac{\Delta y}{y} \right)^0 + \epsilon_2 + 2\epsilon_1 - \epsilon_5 - \epsilon_4 - \left(\frac{\xi_3}{\xi_4} \right)\epsilon_3$$

$$\frac{\Delta y_2}{y_2} \approx \left(\frac{\Delta y}{y} \right)^0 + \epsilon_2 + 2\epsilon_1 - \epsilon_5 - \epsilon_4 - \frac{1}{2}\epsilon_3$$

$$\left| \frac{\Delta y_2}{y_2} \right| \leq \left| \frac{\Delta y}{y} \right|^0 + |\epsilon_2| + 2|\epsilon_1| + |\epsilon_5| + |\epsilon_4| + \frac{1}{2}|\epsilon_3|$$

$$\left|\frac{\Delta y_2}{y_2}\right| \leq \frac{15}{2}\epsilon.$$

Algorithm 1 has a numerical error that grows without bound as $x \to 0$. Algorithm 2, which is mathematically identical, has an error bound that is constant and small – just a few times the inherent error. Algorithm 2 is more numerically trustworthy than algorithm 1 when $|x| \ll 1$.

1.4   Approximation error vs. algorithm error

Example 1.3   Using the Taylor series, one may approximate

$$y_e = e^x - 1$$

as

$$y_a = x + \frac{1}{2}x^2.$$

$R(x) = y_e - y_a$ represents the approximation error (see Appendix A.6)

$$R(x) = \frac{1}{6}x^3 e^\zeta \text{ for some } \zeta \in [0, x].$$

For what values of $x > 0$ is the total error bound (approximation plus numerical) involved in computing y_a less than or equal to the numerical error bound involved in computing y_e? In other words, when should we intentionally calculate the wrong thing to get the most reliable (numerically trustworthy) answer?

Solution
Compute y_e using the algorithm $\eta_1 = \exp(x)$, $y = \eta_1 - 1$ (see Figure 1.5). For this algorithm, the total error looks like:

$$\frac{\Delta y_e}{y_e} = \underbrace{\epsilon_2 + \frac{xe^x}{e^x - 1}\epsilon_x}_{\text{inherent}} + \frac{e^x}{e^x - 1}\epsilon_{\exp} \tag{1.17}$$

$$\left|\frac{\Delta y_e}{y_e}\right| \leq \left(1 + \left|\frac{xe^x}{e^x - 1}\right| + \left|\frac{e^x}{e^x - 1}\right|\right)\epsilon = \left(1 + (1 + |x|)\left|\frac{e^x}{e^x - 1}\right|\right)\epsilon.$$

Now let us compute y_a using the algorithm $\phi_1 = x^2$, $\phi_2 = \phi_1/2$, $y = \phi_2 + x$. For this algorithm (Figure 1.5), the total numerical error looks like:

$$\frac{\Delta y_a}{y_a} = \underbrace{\epsilon_3 + \frac{x}{x + \frac{1}{2}x^2}\epsilon_x + \frac{\frac{1}{2}x^2}{x + \frac{1}{2}x^2}(1)(2)\epsilon_x}_{\text{inherent}} + \frac{\frac{1}{2}x^2}{x + \frac{1}{2}x^2}\epsilon_2 + \frac{\frac{1}{2}x^2}{x + \frac{1}{2}x^2}(1)\epsilon_1.$$

Note that the inherent error for y_a is different from that for y_e – not because the algo-

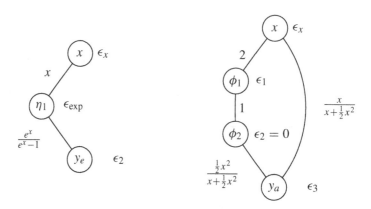

Figure 1.5 Bauer graphs for exact and approximate algorithms for $e^x - 1$; Example 1.3.

rithms differ, but because they are not mathematically the same functions. Note also that ϵ_2 in algorithm 2, the error attributed to division by 2, is zero. This is because division by 2 affects only the exponent, not the mantissa. Simplifying,

$$\frac{\Delta y_a}{y_a} = \epsilon_3 + \frac{x + x^2}{x + \frac{1}{2}x^2}\epsilon_x + \frac{\frac{1}{2}x^2}{x + \frac{1}{2}x^2}(1)\epsilon_1$$

$$= \epsilon_3 + \frac{2(1+x)}{(2+x)}\epsilon_x + \frac{x}{(2+x)}\epsilon_1$$

$$\left|\frac{\Delta y_a}{y_a}\right| \le \left(1 + \left|\frac{2(1+x)}{(2+x)}\right| + \left|\frac{x}{(2+x)}\right|\right)\epsilon.$$

We have now $|\tilde{y}_a - y_a|/|y_a|$ and $|\tilde{y}_e - y_e|/|y_e|$, where the tilde denotes the numerical evaluation, and y without a tilde is the mathematical result. These cannot be compared in any meaningful way, since they measure relative errors of mathematically different functions. What we need is $|\tilde{y}_a - y_e|/|y_e|$, which references the numerical result $\tilde{y}_a$ to the desired mathematical function y_e. We obtain this new measure of error through

$$\frac{\tilde{y}_a - y_e}{y_e} = \left(\frac{\tilde{y}_a - y_a}{y_a}\right)\left(\frac{y_a}{y_e}\right) + \left(\frac{y_a - y_e}{y_e}\right)$$

$$\left|\frac{\tilde{y}_a - y_e}{y_e}\right| \le \left|\frac{\Delta y_a}{y_a}\right|\left|\frac{y_a}{y_e}\right| + \left|\frac{R}{y_e}\right|,$$

where R is the Taylor series approximation error,

$$\left|\frac{\tilde{y}_a - y_e}{y_e}\right| \le \left(1 + \left|\frac{2(1+x)}{(2+x)}\right| + \left|\frac{x}{(2+x)}\right|\right)\left|\frac{x + \frac{1}{2}x^2}{e^x - 1}\right|\epsilon + \left|\frac{x^3 e^\zeta}{6(e^x - 1)}\right|, \quad (1.18)$$

and the greatest (most conservative) error bound comes from taking $\zeta = x$ when $x > 0$.

To determine the value of x for which the two functions have the same total error bound, we equate the bounds and solve for x. With $x > 0$,

$$\left(1 + \frac{(2 + 3x)}{(2 + x)}\right)\frac{x + \frac{1}{2}x^2}{e^x - 1}\epsilon + \frac{x^3 e^x}{6(e^x - 1)} = \left(1 + \frac{(1 + x)e^x}{e^x - 1}\right)\epsilon$$

$$x^3 = 6e^{-x}\left(e^x(2 + x) - 2x^2 - 2x - 1\right)\epsilon.$$

To leading order, x is proportional to $\epsilon^{1/3}$, which is about 4.8×10^{-6}. This suggests that we try approximating this solution by taking the limit $x \ll 1$ on the right-hand side:

$$x \approx (6\epsilon)^{1/3} \approx 8.7 \times 10^{-6}.$$

As $x \to 0$, the bound on $\Delta y_a / y_e$ (1.18) approaches 2ϵ, but the bound on $\Delta y_e / y_e$ grows without bound. Therefore, for values of x in the range $(0, 8.7 \times 10^{-6})$ we expect y_a to be a better solution, but for larger x the exact equation is expected to be more accurate. Figure 1.6 displays actual relative numerical errors for these algorithms. The quantitative estimate of the tradeoff point where the error bounds are equal is consistent with the data, which suggests that the assumptions (particularly (1.6)) were good for this example.

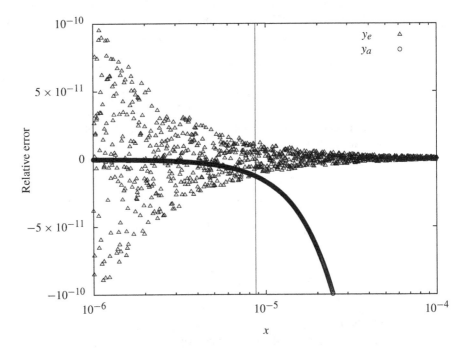

Figure 1.6 Actual measured relative error of two algorithms for calculating $\exp(x) - 1$; Example 1.3. The relative error for y_e is scattered, reflecting roundoff processes, whereas the relative error for y_a is systematic, reflecting approximation error.

1.5   An important example

The quadratic formula for the roots of $ax^2 + bx + c = 0$ is

$$x = \frac{-b \pm \sqrt{b^2 - 4ac}}{2a}.$$

You were probably taught this in high school. Mathematically, it is correct. Numerically, it is almost never the correct formula to use for both roots. This simple problem provides an "everyday" example of how roundoff error matters, and can sometimes be controlled by simply adjusting the algorithm.

Consider the following rearrangement,

$$x = -\operatorname{sign}(b)\frac{|b| \pm \sqrt{b^2 - 4ac}}{2a}.$$

Clearly when $+$ is chosen there is addition in the numerator, and that operation is safe because it has condition numbers in the range $(0, 1)$:

$$x_+ = -\operatorname{sign}(b)\frac{|b| + \sqrt{b^2 - 4ac}}{2a}.$$

When "$-$" is chosen there is subtraction for which the condition numbers have magnitude greater than 1. This subtraction can be avoided by a simple rearrangement:

$$x_- = -\operatorname{sign}(b)\frac{|b| - \sqrt{b^2 - 4ac}}{2a}\frac{|b| + \sqrt{b^2 - 4ac}}{|b| + \sqrt{b^2 - 4ac}}$$

$$= -\operatorname{sign}(b)\frac{2c}{|b| + \sqrt{b^2 - 4ac}}.$$

(1.19)

There remains a possible subtraction inside the square root, but there is nothing that can be done about that – it is inherent.

Example 1.4   Evaluate the roots of the quadratic equation $ax^2 + bx + c = 0$, with

$$a = 1$$
$$b = 6 + 3 \times 10^{-10}$$
$$c = 6 \times 3 \times 10^{-10}.$$

The exact roots are -6 and -3×10^{-10}.

Solution
Numerically, the quadratic formula gives

$$x_- = \frac{-b + \sqrt{b^2 - 4ac}}{2a} = -3.0000002482211130 \times 10^{-10}$$

$$x_+ = \frac{-b - \sqrt{b^2 - 4ac}}{2a} = -6.0000000000000000 \times 10^{+00},$$

with relative errors 8.2740×10^{-8}, or $7.45 \times 10^8 \epsilon$, and zero, using double precision. By putting root x_- in the numerically stable form (1.19) one obtains

$$x_- = -\frac{2c}{|b| + \sqrt{b^2 - 4ac}} = -3.0000000000000000 \times 10^{-10},$$

with relative error zero.

1.6   Backward error

The preceding sections have dealt with analyses that estimate the error of a calculation in a forward sense – following the errors from input to output through each step of the algorithm to answer the question "what is an error bound associated with the numerical (i.e., imperfect) evaluation of an algorithm?" Backward error analysis turns things around a bit to ask instead "the numerical solution is the exact answer to what different problem?" The distinction may best be appreciated through an example [244].

We consider the error associated with the evaluation of a degree n polynomial

$$p_n(x) = a_0 + a_1 x + a_2 x^2 + \cdots + a_n x^n$$

by *Horner's scheme* [111], also known as *nested multiplication*:

$$p_n(x) = a_0 + x(a_1 + x(a_2 + x(\cdots + x a_n))),$$

or

$$\phi_0 = a_n$$
$$\phi_1 = x\phi_0 + a_{n-1}$$
$$\vdots$$
$$\phi_i = x\phi_{i-1} + a_{n-i}$$
$$\vdots$$
$$\phi_n = x\phi_{n-1} + a_0.$$

The errors in x and the coefficients a_i are inherent, so we will focus on the algorithmic contributions. A forward error analysis of this algorithm might use

$$\tilde{\phi}_0 = a_n$$

$$\tilde{\phi}_1 = ((x\tilde{\phi}_0)(1+\epsilon_1) + a_{n-1})(1+\epsilon_2)$$

$$= xa_n(1+\epsilon_1)(1+\epsilon_2) + a_{n-1}(1+\epsilon_2)$$

$$\tilde{\phi}_2 = ((x\tilde{\phi}_1)(1+\epsilon_3) + a_{n-2})(1+\epsilon_4)$$

$$= x^2 a_n \prod_{k=1}^{4}(1+\epsilon_k) + xa_{n-1}\prod_{k=2}^{4}(1+\epsilon_k) + a_{n-2}(1+\epsilon_4)$$

$$\vdots$$

$$\tilde{\phi}_i = ((x\tilde{\phi}_{i-1})(1+\epsilon_{2i-1}) + a_{n-i})(1+\epsilon_{2i})$$

$$= x^i a_n \prod_{k=1}^{2i}(1+\epsilon_k) + x^{i-1}a_{n-1}\prod_{k=2}^{2i}(1+\epsilon_k)$$

$$+ x_{i-2}a_{n-2}\prod_{k=4}^{2i}(1+\epsilon_k) + \cdots + a_{n-i}(1+\epsilon_{2i})$$

$$\vdots$$

$$\tilde{\phi}_n = ((x\tilde{\phi}_{n-1})(1+\epsilon_{2n-1}) + a_0)(1+\epsilon_{2n})$$

$$= x^n a_n \prod_{k=1}^{2n}(1+\epsilon_k) + x^{n-1}a_{n-1}\prod_{k=2}^{2n}(1+\epsilon_k)$$

$$+ x^{n-2}a_{n-2}\prod_{k=4}^{2n}(1+\epsilon_k) + \cdots + a_0(1+\epsilon_{2n})$$

$$\tilde{\phi}_n = x^n a_n \prod_{k=1}^{2n}(1+\epsilon_k) + \sum_{j=1}^{n} x^{n-j}a_{n-j}\prod_{k=2j}^{2n}(1+\epsilon_k),$$

where the various terms ϵ_i correspond to relative errors associated with the elementary operations of multiplication and addition. $\tilde{\phi}_n$ is the exact value of the polynomial

$$p_n = b_0 + b_1 x + b_2 x^2 + \cdots + b_n x^n,$$

with

$$b_n = a_n \prod_{k=1}^{2n}(1+\epsilon_k)$$

$$b_{n-i} = a_{n-i}\prod_{k=2i}^{2n}(1+\epsilon_k) \quad i = 1, ..., n.$$

If terms of order ϵ^2 can be ignored, then

$$\frac{|b_n - a_n|}{|a_n|} \lesssim 2n\epsilon$$

$$\frac{|b_i - a_i|}{|a_i|} \lesssim [2i + 1]\epsilon \quad i = 0, ..., n - 1. \tag{1.20}$$

Since the numerical coefficients a should be expected to have some rounding error, i.e., $|\tilde{a} - a|/|a| \lesssim 1\epsilon$, the error implied by (1.20) is only $\mathcal{O}(n)$ times the inherent error. Backward error analysis leads to the conclusion that Horner's scheme is numerically stable.

This particular example of polynomial evaluation was first investigated by Wilkinson [244], who experimented with the polynomial

$$p_1(x) = (x - 1)(x - 2) \cdots (x - 20)$$

evaluated in the expanded form

$$p_2(x) = x^{20} - 210x^{19} + 20\,615x^{18} - 1\,256\,850x^{17} + \cdots + 2\,432\,902\,008\,176\,640\,000$$

by Horner's method. The evaluation of p_2 was dominated by error, but this error is *not* a consequence of Horner's method, which is numerically stable. It is a consequence of the expanded polynomial form which is very sensitive to perturbation of certain coefficients. This sensitivity will be addressed in Section 6.6.

Problems

1.1 Determine machine precision on your computer.

1.2 Find the condition numbers (amplification factors) for the following elementary operations:

   (i) $y = \cos(x)$.
  (ii) $y = \sin(x)$.
 (iii) $y = \tan(x)$.
 (iv) $y = \cot(x)$.
  (v) $\theta = \arccos(x)$.
 (vi) $\theta = \arcsin(x)$.
(vii) $\theta = \arctan(x)$ (corresponding to the C function atan(), which takes one argument and returns an angle in the range $(-\pi/2, +\pi/2]$).
(viii) $\theta = \arctan(y/x)$ (corresponding to the C function atan2(y, x) which takes two arguments and returns an angle in the range $(-\pi, \pi]$).
 (ix) $y = e^x$.
  (x) $y = x^\alpha$ (for machine-representable constant α).
 (xi) $z = x^y$.
(xii) $y = \ln(x)$.

1.3 Estimate the range of values over which $e^x - 1 \approx x$, for x of any sign, is a numerically trustworthy alternative.

1.4 [104, 122] For what x is

$$\left(\frac{x}{\ln \exp x}\right)(e^x - 1)$$

a numerically trustworthy alternative to $e^x - 1$?

1.5 [104, 115] For what $x > -1$ is

$$\left(\frac{x}{(1 + x) - 1}\right)\ln(1 + x)$$

a numerically trustworthy alternative to $\ln(1 + x)$?

1.6 For

$$y_1 = a - b,$$

$a, b > 0$, there is a cancellation error giving condition numbers of $a/(a - b)$ and $-b/(a - b)$ that are larger than 1 in magnitude. For example, with $a = 1$ and $b = 0.9$, the condition numbers are 10 and -9.

If we write

$$y_2 = \frac{a^2 - b^2}{a + b}$$

the subtraction $a^2 - b^2$ has condition numbers $a^2/(a^2 - b^2)$ and $-b^2/(a^2 - b^2)$ or $1/0.19 \approx 5.26$ and $-0.81/0.19 \approx -4.26$. The condition numbers are still larger than 1, because we're still doing subtraction, but they are lower by nearly a factor of 2 in this case.

Considering all numerical errors, when is y_2 more numerically trustworthy than y_1?

1.7 Consider the iteration sequence

$$p^{(1)} = 2\sqrt{2}$$

$$p^{(n+1)} = 2^{n+1}\sqrt{2 - \sqrt{4 - 2^{-2n}(p^{(n)})^2}}.$$

Rewrite this in a more numerically stable form. Calculate $p^{(30)}$ and $p^{(1024)}$ by both methods.

1.8 Find numerically stable alternatives to

$$y = \cos(x) - 1 \qquad \text{for } |x| \ll 1$$

other than the one used in Example 1.2.

1.9 Ignoring terms of order ϵ^2 in, e.g., the Horner analysis, is equivalent to the approximation

$$1 - n\epsilon \lesssim \prod_{i=1}^{n}(1 + \epsilon_i) \lesssim 1 + n\epsilon. \tag{1.21}$$

Wilkinson [242] avoids this assumption using the fact

$$1 - 1.06n\epsilon \le \prod_{i=1}^{n}(1 + \epsilon_i) \le 1 + 1.06n\epsilon \quad \text{if } n\epsilon < \frac{1}{10},$$

which implies that (1.21) is inaccurate by as much as 6% if $n \sim 1/10\epsilon$. Verify Wilkinson's formula.

1.10   (i) Characterize the approximation error associated with

$$f'(x) \approx \frac{f(x+h) - f(x-h)}{2h}.$$

    (ii) Characterize the numerical error associated with this formula, assuming that h is a power of 2.

    (iii) Considering both numerical and approximation errors, how should h be selected to obtain the most accurate estimate of the derivative?

    (iv) Verify your analysis using $f(x) = \sin(x)$ for $x = \pi/4$. Plot the error against h and compare your observation with your theory.

2   Direct solution of linear systems

Given a system of linear equations, e.g.,

$$a_{11}x_1 + a_{12}x_2 + \cdots + a_{1n}x_n = b_1$$
$$a_{21}x_1 + a_{22}x_2 + \cdots + a_{2n}x_n = b_2$$
$$\vdots$$
$$a_{m1}x_1 + a_{m2}x_2 + \cdots + a_{mn}x_n = b_m,$$

organized into matrix-vector form,

$$\begin{pmatrix} a_{11} & a_{12} & \cdots & a_{1n} \\ a_{21} & a_{22} & \cdots & a_{2n} \\ \vdots & \vdots & \vdots & \vdots \\ a_{m1} & a_{m2} & \cdots & a_{mn} \end{pmatrix} \begin{pmatrix} x_1 \\ x_2 \\ \vdots \\ x_n \end{pmatrix} = \begin{pmatrix} b_1 \\ b_2 \\ \vdots \\ b_m \end{pmatrix}$$

or

$$\mathbf{A}\mathbf{x} = \mathbf{b}, \tag{2.1}$$

we wish to find the solution $\mathbf{x}$.

For arbitrary $\mathbf{b}$, if $n = m$ there is exactly one solution if there are any solutions at all. For general square systems the most elementary method for solving this problem is Gaussian elimination.

If $n = m$ and also the matrix $\mathbf{A}$ is symmetric positive definite, Cholesky decomposition is an easy and stable technique.

When $n < m$ there are more equations than unknowns, and we do not expect an exact solution to (2.1). However, there is a best solution in a least squares sense, and Householder reduction is a great way to solve this problem. This will be taken up in Chapter 8 with other data fitting approaches. The Householder or QR method is also very important for the eigenvalue problem discussed later (Chapter 3).

It is rare that one would ever require the inverse of a matrix. The solution to (2.1) may be written mathematically as

$$\mathbf{x} = \mathbf{A}^{-1}\mathbf{b},$$

but $\mathbf{x}$ can be determined computationally with less effort and greater accuracy by direct methods.

2.1 Gaussian elimination

The basic idea behind Gaussian elimination is to process the equation $\mathbf{A}\mathbf{x} = \mathbf{b}$, by doing something to both the left- and right-hand sides of the equation, so that $\mathbf{A}$ becomes transformed into a right triangular matrix:

$$\mathbf{R} = $$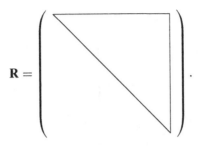

Here, $\mathbf{R}$ is zero below the diagonal, possibly nonzero above the diagonal, and ideally nonzero on the diagonal. If $\mathbf{R}$ has zeros on the diagonal, then $\mathbf{A}\mathbf{x} = \mathbf{b}$ cannot be solved for every $\mathbf{b}$: the matrix $\mathbf{A}$ is singular.

The significance of the triangular form is that it is very easy to solve. If $\mathbf{A}\mathbf{x} = \mathbf{b}$ is transformed into $\mathbf{R}\mathbf{x} = \mathbf{y}$, then

$$x_n = (y_n)/r_{nn}$$
$$x_{n-1} = (y_{n-1} - r_{n-1,n}x_n)/r_{n-1,n-1}$$
$$x_{n-2} = (y_{n-2} - r_{n-2,n}x_n - r_{n-2,n-1}x_{n-1})/r_{n-2,n-2}$$
$$\vdots \tag{2.2}$$
$$x_i = \left(y_i - \sum_{j=i+1}^{n} r_{i,j}x_j \right)/r_{ii}.$$

This process is called *back substitution*. Since the diagonal elements of $\mathbf{R}$ appear as denominators on the right-hand side of (2.2) you can see that a solution will not exist for arbitrary $\mathbf{b}$ if any diagonal elements are zero.

Gaussian elimination establishes the right triangular form through a sequence of three steps. Let $(\mathbf{A}, \mathbf{b})^{(d)}$ be matrix-vector pair at step d of the method. We begin with $(\mathbf{A}, \mathbf{b})^{(0)} = (\mathbf{A}, \mathbf{b})$. Then for $1 \leq d < n$,

- Find a row $i \geq d$ with $\mathbf{A}_{id}^{(d-1)} \neq 0$ (*pivot* selection).
- Permute $(\mathbf{A}, \mathbf{b})^{(d-1)}$ so that rows i and d are interchanged.
- Let $f_{kd} = \mathbf{A}_{kd}^{(d-1)}/\mathbf{A}_{dd}^{(d-1)}$ for each $d < k \leq n$. Subtract f_{kd} times row d from row k. If $\mathbf{A}$ is not singular, this number f_{kd} will exist since the first two steps of the method made sure that element dd of matrix $\mathbf{A}^{(d-1)}$ is not zero.

The result of these steps is the transformation of $(\mathbf{A}, \mathbf{b})^{(d-1)}$ into $(\mathbf{A}, \mathbf{b})^{(d)}$, with

$$(\mathbf{A}, \mathbf{b})^{(n-1)} = (\mathbf{R}, \mathbf{y}).$$

Gaussian elimination was developed by C. F. Gauss to solve least squares problems related to the motion of bodies about the sun using the normal equations [52].

Symbolically, step d of Gaussian elimination consists of multiplying the pair $(\mathbf{A}, \mathbf{b})$ from step $(d-1)$ by a permutation matrix $\mathbf{P}^{(d)}$, and a Frobenius matrix $\mathbf{G}^{(d)}$. If the permutation implements the second step by interchanging rows i and d, then it will have the form

$$\mathbf{P}^{(d)} = \mathbf{P}_{i \leftrightarrow d} = \begin{pmatrix} \ddots & & & & & & & \\ & 1 & & & & & & \\ & & 0 & 0 & \cdots & 0 & 1 & \\ & & 0 & 1 & & & 0 & \\ & & \vdots & & \ddots & & \vdots & \\ & & 0 & & & 1 & 0 & \\ & & 1 & 0 & \cdots & 0 & 0 & \\ & & & & & & & 1 & \\ & & & & & & & & \ddots \end{pmatrix}.$$

The Frobenius matrix takes the form

$$\mathbf{G}^{(d)} = \begin{pmatrix} 1 & & & & & \\ & \ddots & & & & \\ & & 1 & & & \\ & & & 1 & & \\ & & & -f_{d+1,d} & 1 & \\ & & & \vdots & & \ddots \\ & & & -f_{n,d} & & & 1 \end{pmatrix}, \tag{2.3}$$

and each step of the Gaussian elimination algorithm can be summarized as

$$(\mathbf{A}, \mathbf{b})^{(d)} = \mathbf{G}^{(d)} \mathbf{P}^{(d)} (\mathbf{A}, \mathbf{b})^{(d-1)}.$$

It is important to note that this symbolic representation of the algorithm is not what you actually implement in practice. For example, the action of the permutation operation is quickly accomplished by switching the rows of the pair $(\mathbf{A}, \mathbf{b})$. To create an $n \times n$ matrix $\mathbf{P}$, then multiply it with the pair $(\mathbf{A}, \mathbf{b})$ involves $\mathcal{O}(n^2)$ unnecessary storage and $\mathcal{O}(n^3)$ unnecessary multiplications.

Example 2.1   We wish to solve the linear system

$$0.03x_1 + 300x_2 + 8 \times 10^7 x_3 = 80\,000\,300.03$$
$$-0.995x_1 - 140\,000x_2 + 1.6 \times 10^7 x_3 = 15\,859\,999.005$$
$$0.01x_1 - 19\,000x_2 + 1.0 \times 10^6 x_3 = 981\,000.01.$$

Solution

As a matrix equation we have

$$
\underbrace{\begin{pmatrix} 0.3000 \times 10^{-1} & 0.3000 \times 10^{3} & 0.8000 \times 10^{8} \\ -0.9950 & -0.1400 \times 10^{6} & 0.1600 \times 10^{8} \\ 0.1000 \times 10^{-1} & -0.1900 \times 10^{5} & 0.1000 \times 10^{7} \end{pmatrix}}_{\mathbf{A}} \underbrace{\begin{pmatrix} x_1 \\ x_2 \\ x_3 \end{pmatrix}}_{\mathbf{x}} = \underbrace{\begin{pmatrix} 0.8000 \times 10^{8} \\ 0.1586 \times 10^{8} \\ 0.9810 \times 10^{6} \end{pmatrix}}_{\mathbf{b}}
$$

(2.4)

(here four decimal places are displayed, but the calculation is carried out to full machine precision). The first step is to find a row i so that A_{i1} is nonzero. With $i = 1$ we already satisfy this requirement, but it will be shown later that $i = 2$ is a better choice. With $i = 2$ our permutation is

$$
\mathbf{P}^{(1)} = \begin{pmatrix} 0 & 1 & 0 \\ 1 & 0 & 0 \\ 0 & 0 & 1 \end{pmatrix},
$$

and

$$
\mathbf{P}^{(1)} \left(\mathbf{A} \,|\, \mathbf{b} \right) = \begin{pmatrix} \underline{-0.9950} & -0.1400 \times 10^{6} & 0.1600 \times 10^{8} & | & 0.1586 \times 10^{8} \\ 0.3000 \times 10^{-1} & 0.3000 \times 10^{3} & 0.8000 \times 10^{8} & | & 0.8000 \times 10^{8} \\ 0.1000 \times 10^{-1} & -0.1900 \times 10^{5} & 0.1000 \times 10^{7} & | & 0.9810 \times 10^{6} \end{pmatrix}.
$$

(2.5)

The *pivot* -0.9950 is underlined in (2.4) and (2.5). Next, to put the first column in the right triangular form, we want to make the coefficients of x_1 be zero for the second and third rows. We will subtract $f_{21} = A_{21}/A_{11} = -0.3015 \times 10^{-1}$ times row 1 from row 2, and subtract $f_{31} = A_{31}/A_{11} = -0.1005 \times 10^{-1}$ times row 1 from row 3. For this first step we obtain

$$
\underbrace{\begin{pmatrix} 1 & 0 & 0 \\ -f_{21} & 1 & 0 \\ -f_{31} & 0 & 1 \end{pmatrix}}_{\mathbf{G}^{(1)}} \mathbf{P}^{(1)} \left(\mathbf{A} \,|\, \mathbf{b} \right)
$$

$$
= \begin{pmatrix} -0.9950 & -0.1400 \times 10^{6} & 0.1600 \times 10^{8} & | & 0.1586 \times 10^{8} \\ 0 & -0.3921 \times 10^{4} & 0.8048 \times 10^{8} & | & 0.8048 \times 10^{8} \\ 0 & \underline{-0.2041 \times 10^{5}} & 0.1161 \times 10^{7} & | & 0.1140 \times 10^{7} \end{pmatrix}.
$$

In the second step, we wish to find an index $i \geq 2$ so that A_{i2} is nonzero. The preferred choice is $i = 3$ giving

$$
\mathbf{P}^{(2)} = \begin{pmatrix} 1 & 0 & 0 \\ 0 & 0 & 1 \\ 0 & 1 & 0 \end{pmatrix},
$$

and

$$\mathbf{P}^{(2)}\mathbf{G}^{(1)}\mathbf{P}^{(1)}\,(\mathbf{A}|\,\mathbf{b})$$

$$= \left(\begin{array}{ccc|c} -0.9950 & -0.1400 \times 10^6 & 0.1600 \times 10^8 & 0.1586 \times 10^8 \\ 0 & -0.2041 \times 10^5 & 0.1161 \times 10^7 & 0.1140 \times 10^7 \\ 0 & -0.3921 \times 10^4 & 0.8048 \times 10^8 & 0.8048 \times 10^8 \end{array}\right).$$

Elimination proceeds with a Frobenius matrix based on $f_{32} = A_{32}/A_{22} = 0.1921$ and

$$\underbrace{\begin{pmatrix} 1 & 0 & 0 \\ 0 & 1 & 0 \\ 0 & -f_{32} & 1 \end{pmatrix}}_{\mathbf{G}^{(2)}} \mathbf{P}^{(2)}\mathbf{G}^{(1)}\mathbf{P}^{(1)}\,(\mathbf{A}|\,\mathbf{b})$$

$$= \left(\begin{array}{ccc|c} -0.9950 & -0.1400 \times 10^6 & 0.1600 \times 10^8 & 0.1586 \times 10^8 \\ 0 & -0.2041 \times 10^5 & 0.1161 \times 10^7 & 0.1140 \times 10^7 \\ 0 & 0 & 0.8026 \times 10^8 & 0.8026 \times 10^8 \end{array}\right).$$

The matrix is now triangular. The final $\mathbf{A}$ and $\mathbf{b}$ results can be expressed as

$$-0.9950x_1 - 0.1400 \times 10^6 x_2 + 0.1600 \times 10^8 x_3 = 0.1586 \times 10^8, \tag{2.6a}$$

$$-0.2041 \times 10^5 x_2 + 0.1161 \times 10^7 x_3 = 0.1140 \times 10^7, \tag{2.6b}$$

$$0.8026 \times 10^8 x_3 = 0.8026 \times 10^8. \tag{2.6c}$$

The back substitution process begins by looking at the last equation (2.6c), which gives $x_3 = 1.000$. Next, with x_3 determined use equation (2.6b) to determine x_2, and finally use (2.6a) to determine x_1. To full precision, the computed results are

$$x_1 = 9.9999999917631777\mathrm{e} - 01$$
$$x_2 = 1.0000000000000042\mathrm{e} + 00$$
$$x_3 = 1.0000000000000000\mathrm{e} + 00.$$

For this example, the exact result should be $x_1 = x_2 = x_3 = 1$.

At this point, we have a description of the Gaussian elimination method based on a sequence of matrix multiplications that transform a matrix $\mathbf{A}$ into a right triangular matrix $\mathbf{R}$, using permutation $\mathbf{P}$ and and elimination (Frobenius) matrices $\mathbf{G}$,

$$\mathbf{G}^{(n-1)}\mathbf{P}^{(n-1)} \cdots \mathbf{G}^{(1)}\mathbf{P}^{(1)}\mathbf{A} = \mathbf{R}. \tag{2.7}$$

The inverse of any permutation is its transpose, so

$$\mathbf{P}^{(j)}\mathbf{G}^{(k)} = \left(\mathbf{P}^{(j)}\mathbf{G}^{(k)}\mathbf{P}^{(j)\mathrm{T}}\right)\mathbf{P}^{(j)},$$

i.e., permutations and eliminations do not commute. However, if $j > k$ then the matrix $\mathbf{P}^{(j)}\mathbf{G}^{(k)}\mathbf{P}^{(j)\mathrm{T}}$ is a Frobenius matrix, with the subdiagonal column permuted by the action of $\mathbf{P}^{(j)}$. If one performs this commutation operation wherever possible, one transforms (2.7) into

$$\mathbf{G}^{(n-1)}\hat{\mathbf{G}}^{(n-2)} \cdots \hat{\mathbf{G}}^{(1)}\mathbf{P}^{(n-1)}\mathbf{P}^{(n-2)} \cdots \mathbf{P}^{(1)}\mathbf{A} = \mathbf{R},$$

where the caret (^) denotes that the subdiagonal $\mathbf{G}$ entries were affected by "moving" the permutations to the right. Next, multiplication of both sides by $(\mathbf{G}^{(n-1)})^{-1}$, then $(\hat{\mathbf{G}}^{(n-2)})$, etc., gives the form

$$\mathbf{P}^{(n-1)}\mathbf{P}^{(n-2)} \cdots \mathbf{P}^{(1)}\mathbf{A} = (\hat{\mathbf{G}}^{(1)})^{-1} \cdot (\hat{\mathbf{G}}^{(n-2)})^{-1}(\mathbf{G}^{(n-1)})^{-1}\mathbf{R}.$$

This transformation yields a very simple result using a few simple facts. First, the Frobenius matrices (2.3) have very simple inverses: one simply changes the sign of the subdiagonal entries:

$$(\mathbf{G}^{(d)})^{-1} = \begin{pmatrix} 1 & & & & & & \\ & \ddots & & & & & \\ & & 1 & & & & \\ & & & 1 & & & \\ & & & +f_{d+1,d} & 1 & & \\ & & & \vdots & & \ddots & \\ & & & +f_{n,d} & & & 1 \end{pmatrix}. \tag{2.8}$$

And second, for $j < k$,

$$(\mathbf{G}^{(j)})^{-1}(\mathbf{G}^{(k)})^{-1} = \begin{pmatrix} 1 & & & & & & & \\ & \ddots & & & & & & \\ & & 1 & & & & & \\ & & f_{j+1,j} & 1 & & & & \\ & & \vdots & & \ddots & & & \\ & & f_{k,j} & & & 1 & & \\ & & f_{k+1,j} & & & f_{k+1,k} & 1 & \\ & & \vdots & & & \vdots & & \ddots \\ & & f_{n,j} & & & f_{n,k} & & 1 \end{pmatrix}.$$

This means that the product $(\hat{\mathbf{G}}^{(1)})^{-1} \cdot (\hat{\mathbf{G}}^{(n-2)})^{-1}(\mathbf{G}^{(n-1)})^{-1}$ forms a left triangular matrix $\mathbf{L}$

$$\mathbf{L} = \begin{pmatrix} 1 & & & & & \\ \hat{f}_{21} & 1 & & & & \\ \hat{f}_{31} & \hat{f}_{32} & \ddots & & & \\ \vdots & \vdots & \ddots & 1 & & \\ \hat{f}_{n1} & \hat{f}_{n2} & \cdots & \hat{f}_{n,n-1} & 1 \end{pmatrix},$$

where again the caret indicates that the entries f_{ij} may have been permuted. The final simplification is that the product $\mathbf{P}^{(n-1)}\mathbf{P}^{(n-2)} \cdots \mathbf{P}^{(1)}$ is a composite permutation, which will be denoted $\mathbf{P}$ without superscript. All together, (2.7) can be represented as

$$\mathbf{PA} = \mathbf{LR}.$$

The zeros of $\mathbf{R}$ correspond to the subdiagonal elements of $\mathbf{L}$, and in practice one stores these together in the same matrix representation. In this composite representation, the matrix $\mathbf{A}$ may be converted directly into the combined $\mathbf{L}$ and $\mathbf{R}$, with additional storage required only for the composite permutation $\mathbf{P}$. Since the permutation simply describes a reordering of the rows, this information is represented more compactly as an integer list than as a matrix.

Example 2.2   Determine the LR decomposition of $\mathbf{A}$ from Example 2.1.

Solution
We begin with matrix $\mathbf{A}$, and a permutation list p:

$$\mathbf{A}^{(0)} = \begin{pmatrix} 0.3000 \times 10^{-1} & 0.3000 \times 10^3 & 0.8000 \times 10^8 \\ \underline{-0.9950} & -0.1400 \times 10^6 & 0.1600 \times 10^8 \\ 0.1000 \times 10^{-1} & -0.1900 \times 10^5 & 0.1000 \times 10^7 \end{pmatrix} \qquad p = \{1, 2, 3\}.$$

As before, the pivot in underlined. This indicates a permutation exchanging rows 1 and 2:

$$\mathbf{A}^{(0)} \rightarrow \begin{pmatrix} \underline{-0.9950} & -0.1400 \times 10^6 & 0.1600 \times 10^8 \\ 0.3000 \times 10^{-1} & 0.3000 \times 10^3 & 0.8000 \times 10^8 \\ 0.1000 \times 10^{-1} & -0.1900 \times 10^5 & 0.1000 \times 10^7 \end{pmatrix} \qquad p = \{2, 1, 3\}.$$

After this, we perform elimination with $f_{21} = -0.3015 \times 10^{-1}$ and $f_{31} = -0.1005 \times 10^{-1}$. From row 2 subtract f_{21} times row 1, and from row 3 subtract f_{31} times row 1. Store these f factors in the newly created zeros of the transformed matrix:

$$\mathbf{A}^{(1)} = \begin{pmatrix} -0.9950 & -0.1400 \times 10^6 & 0.1600 \times 10^8 \\ \hline -0.3015 \times 10^{-1} & -0.3921 \times 10^4 & 0.8048 \times 10^8 \\ -0.1005 \times 10^{-1} & -0.2041 \times 10^5 & 0.1161 \times 10^7 \end{pmatrix} \qquad p = \{2, 1, 3\}.$$

The partitioned column is the negative of the subdiagonal of the first Frobenius matrix $\mathbf{G}^{(1)}$ from the previous example. The new pivot suggests permutation,

$$\mathbf{A}^{(1)} \rightarrow \begin{pmatrix} -0.9950 & -0.1400 \times 10^6 & 0.1600 \times 10^8 \\ \hline -0.1005 \times 10^{-1} & -0.2041 \times 10^5 & 0.1161 \times 10^7 \\ -0.3015 \times 10^{-1} & -0.3921 \times 10^4 & 0.8048 \times 10^8 \end{pmatrix} \qquad p = \{2, 3, 1\}.$$

Permutation affects both partitions of the matrix. The next elimination is $f_{32} = 0.1921$, and we subtract from row 3 f_{32} times row 2, but only for that part lying to the right of the partition. Store the factor f_{32} in the newly created zero of $\mathbf{A}$:

$$\mathbf{A}^{(2)} = \begin{pmatrix} -0.9950 & -0.1400 \times 10^6 & 0.1600 \times 10^8 \\ -0.1005 \times 10^{-1} & -0.2041 \times 10^5 & 0.1161 \times 10^7 \\ -0.3015 \times 10^{-1} & 0.1921 & 0.8026 \times 10^8 \end{pmatrix} \qquad p = \{2, 3, 1\}.$$

This final result contains the subdiagonal component of $\mathbf{L}$ below the diagonal, and the remainder is the matrix $\mathbf{R}$:

$$\mathbf{L} = \begin{pmatrix} 1 & 0 & 0 \\ -0.1005 \times 10^{-1} & 1 & 0 \\ -0.3015 \times 10^{-1} & 0.1921 & 1 \end{pmatrix}$$

$$\mathbf{R} = \begin{pmatrix} -0.9950 & -0.1400 \times 10^6 & 0.1600 \times 10^8 \\ 0 & -0.2041 \times 10^5 & 0.1161 \times 10^7 \\ 0 & 0 & 0.8026 \times 10^8 \end{pmatrix}.$$

(2.9)

The permutation

$$\mathbf{P} = \begin{pmatrix} 0 & 1 & 0 \\ 0 & 0 & 1 \\ 1 & 0 & 0 \end{pmatrix}$$

is encoded in the list $\{2, 3, 1\}$.

With this so-called LR decomposition, one can solve any $\mathbf{Ax} = \mathbf{b}$ problem with two back substitution steps (i.e., $\mathbf{b}$ does not need to be processed along with $\mathbf{A}$, and can be changed after the decomposition is complete):

$$\mathbf{Ly} = \mathbf{Pb} \qquad \text{(solve for } \mathbf{y})$$
$$\mathbf{Rx} = \mathbf{y} \qquad \text{(solve for } \mathbf{x}).$$

Example 2.3 Given $\mathbf{A}$ with LR decomposition,

$$\underbrace{\begin{pmatrix} 0 & 1 & 0 \\ 0 & 0 & 1 \\ 1 & 0 & 0 \end{pmatrix}}_{\mathbf{P}} \underbrace{\begin{pmatrix} 1 & 4 & 2 \\ -2 & -8 & 3 \\ 0 & 1 & 1 \end{pmatrix}}_{\mathbf{A}} = \underbrace{\begin{pmatrix} 1 & 0 & 0 \\ 0 & 1 & 0 \\ -1/2 & 0 & 1 \end{pmatrix}}_{\mathbf{L}} \underbrace{\begin{pmatrix} -2 & -8 & 3 \\ 0 & 1 & 1 \\ 0 & 0 & 7/2 \end{pmatrix}}_{\mathbf{R}},$$

solve

$$\mathbf{Ax} = \mathbf{b} = \begin{pmatrix} 3 \\ 8 \\ 2 \end{pmatrix}.$$

Solution

First, solve $\mathbf{Ly} = \mathbf{Pb}$ for $\mathbf{y}$:

$$
\begin{pmatrix} 1 & 0 & 0 \\ 0 & 1 & 0 \\ -1/2 & 0 & 1 \end{pmatrix} \begin{pmatrix} y_1 \\ y_2 \\ y_3 \end{pmatrix} = \begin{pmatrix} 8 \\ 2 \\ 3 \end{pmatrix}.
$$

Back substitution yields $y_1 = 8$, $y_2 = 2$, and $y_3 = 7$.

Next, solve $\mathbf{Rx} = \mathbf{y}$ for $\mathbf{x}$:

$$
\begin{pmatrix} -2 & -8 & 3 \\ 0 & 1 & 1 \\ 0 & 0 & 7/2 \end{pmatrix} \begin{pmatrix} x_1 \\ x_2 \\ x_3 \end{pmatrix} = \begin{pmatrix} 8 \\ 2 \\ 7 \end{pmatrix}.
$$

Back substitution yields $x_3 = 2$, $x_2 = 0$, and $x_1 = -1$.

In addition to solving $\mathbf{Ax} = \mathbf{b}$, the LR decomposition is useful for calculating the determinant:

$$
\det(\mathbf{A}) = \det(\mathbf{P}^\mathsf{T})\det(\mathbf{L})\det(\mathbf{R}).
$$

The determinant of a matrix changes sign with each interchange of rows. Therefore, $\det(\mathbf{P}^\mathsf{T}) = \pm 1$ depending on whether an even $(+)$ or odd $(-)$ number of row interchanges occurred. The determinant of a triangular matrix is the product of its diagonal elements. Therefore, $\det(\mathbf{L}) = 1$ and $\det(\mathbf{R}) = R_{11} R_{22} \cdots R_{nn}$. All together,

$$
\det(\mathbf{A}) = \pm R_{11} R_{22} \cdots R_{nn}.
$$

2.2   Pivot selection

The previous discussion of Gaussian elimination did not fully specify the pivot selection criterion – in principle, any pivot could be chosen that does not result in a division by zero. Here we will examine some specific pivot selection strategies.

The strategy employed in the previous examples is called *partial pivoting* – at each step we chose the pivot from the column being reduced whose absolute value was greatest. The motivation for this choice concerns numerical error, about which more will be said later.

A sufficient strategy, which works perfectly well for rational integer manipulations (which are numerically exact), is *trivial pivoting*. This strategy selects any nonzero element as the pivot, regardless of its size relative to other candidate pivots.

With *complete pivoting* one uses both row and column permutations to choose the largest element in the $(n + 1 - d) \times (n + 1 - d)$ partition. The final decomposition is therefore $\mathbf{P}_L \mathbf{A} \mathbf{P}_R^\mathsf{T} = \mathbf{LR}$, where $\mathbf{P}_L$ is the left (row) permutation matrix, and $\mathbf{P}_R^\mathsf{T}$ is the right (column) permutation matrix.

Example 2.4 Perform Gaussian elimination with complete pivoting on the matrix $\mathbf{A}$ of Example 2.1.

Solution

As with Example 2.2, the matrix will be transformed in place, but this time with two lists for the two permutations.

$$
\mathbf{A}^{(0)} = \begin{pmatrix} 0.3000 \times 10^{-1} & 0.3000 \times 10^3 & 0.8000 \times 10^8 \\ -0.9950 & -0.1400 \times 10^6 & 0.1600 \times 10^8 \\ 0.1000 \times 10^{-1} & -0.1900 \times 10^5 & 0.1000 \times 10^7 \end{pmatrix} \quad \begin{array}{l} p_\ell = \{1, 2, 3\} \\ p_r = \{1, 2, 3\} \end{array}.
$$

The element of $\mathbf{A}^{(0)}$ largest in magnitude is $\mathbf{A}_{1,3}$, which suggests that no row permutation is required, but a column permutation is needed to exchange the first and third columns:

$$
\mathbf{A}^{(0)} \rightarrow \begin{pmatrix} 0.8000 \times 10^8 & 0.3000 \times 10^3 & 0.3000 \times 10^{-1} \\ 0.1600 \times 10^8 & -0.1400 \times 10^6 & -0.9950 \\ 0.1000 \times 10^7 & -0.1900 \times 10^5 & 0.1000 \times 10^{-1} \end{pmatrix} \quad \begin{array}{l} p_\ell = \{1, 2, 3\} \\ p_r = \{3, 2, 1\} \end{array}.
$$

Now, $f_{21} = 0.2000$ and $f_{31} = 0.0125$ so

$$
\mathbf{A}^{(1)} = \begin{pmatrix} 0.8000 \times 10^8 & 0.3000 \times 10^3 & 0.3000 \times 10^{-1} \\ 0.2000 & -0.1401 \times 10^6 & -0.1001 \times 10^1 \\ 0.1250 & -0.1900 \times 10^5 & 0.9625 \times 10^{-2} \end{pmatrix} \quad \begin{array}{l} p_\ell = \{1, 2, 3\} \\ p_r = \{3, 2, 1\} \end{array}.
$$

The number in the A_{22} position is the greatest in magnitude of all candidate pivots, so no further permutation is required. $f_{32} = 0.1357$ and

$$
\mathbf{A}^{(2)} = \begin{pmatrix} 0.8000 \times 10^8 & 0.3000 \times 10^3 & 0.3000 \times 10^{-1} \\ 0.2000 & -0.1401 \times 10^6 & -0.1001 \times 10^1 \\ 0.1250 & 0.1357 & 0.1454 \end{pmatrix} \quad \begin{array}{l} p_\ell = \{1, 2, 3\} \\ p_r = \{3, 2, 1\} \end{array}.
$$

We therefore have

$$\mathbf{P}_L = \begin{pmatrix} 1 & 0 & 0 \\ 0 & 1 & 0 \\ 0 & 0 & 1 \end{pmatrix}$$

$$\mathbf{L} = \begin{pmatrix} 1 & 0 & 0 \\ 0.2000 & 1 & 0 \\ 0.1250 & 0.1357 & 1 \end{pmatrix}$$

$$\mathbf{R} = \begin{pmatrix} 0.8000 \times 10^8 & 0.3000 \times 10^3 & 0.3000 \times 10^{-1} \\ 0 & -0.1401 \times 10^6 & -0.1001 \times 10^1 \\ 0 & 0 & 0.1454 \end{pmatrix}$$

(2.10)

$$\mathbf{P}_R = \begin{pmatrix} 0 & 0 & 1 \\ 0 & 1 & 0 \\ 1 & 0 & 0 \end{pmatrix}.$$

With the same vector $\mathbf{b}$ of Example 2.1, complete pivoting gives the solution

$$x_1 = 9.9999999980811705e - 01$$
$$x_2 = 1.0000000000000024e + 00$$
$$x_3 = 1.0000000000000000e + 00.$$

Variable x_1 has less than $1/4$ of the relative error incurred with partial pivoting, and x_2 has about $1/2$ the relative error.

The reason complete pivoting is generally superior to partial pivoting will be discussed in the next section. A contributing factor is the fact that both $\mathbf{L}$ and $\mathbf{R}$ are *diagonally dominant*, i.e., $|r_{ii}|$ is the greatest element of row i of $\mathbf{R}$, and $|\ell_{ii}|$ is the greatest element of row i of $\mathbf{L}$. Partial pivoting guarantees that $\mathbf{L}$ is diagonally dominant, but $\mathbf{R}$ may not be: compare (2.10) with (2.9).

In partial pivoting, there are $n - 1$ steps d, with $n + 1 - d$ candidate pivots in each step. A total of

$$\sum_{d=1}^{n-1} (n + 1 - d) = \frac{n(n+1)}{2} - 1 = \mathcal{O}(n^2)$$

candidate pivots are considered. With complete pivoting, the number of candidates for each step d is $(n + 1 - d)^2$, so a total of

$$\sum_{d=1}^{n-1} (n + 1 - d)^2 = \frac{n(n+1)(2n+1)}{6} - 1 = \mathcal{O}(n^3)$$

candidate pivots are considered. When n is very large, this can be become a significant expense. A pivoting strategy that makes $\mathbf{L}$ and $\mathbf{R}$ diagonally dominant with somewhat less cost is called the *rook's pivoting method*.

The rook's pivoting method [170] scans rows and columns in a systematic way to find the first element A_{rc} that is the greatest in magnitude of its row and of its column. The algorithm is this:

1. Let $j = d$, and find the greatest element in column j – say A_{rj}.
2. In row r, find the greatest element – say A_{rc}. If $c = j$, then A_{rj} is the greatest in its row and column: stop.
3. Otherwise, in column c, find the greatest element – say A_{ic}. If $i = r$ then A_{rc} is the pivot: stop.
4. Otherwise, let $r = i$ and $j = c$, and proceed to step 2.

The method is named for the chess piece, because the pivot search follows "L"-shaped trajectories like those of a rook. The worst case searching cost of rook's pivoting is the same as for complete pivoting, and the best case cost is two times the cost of partial pivoting (because a row and a column need to be searched). On average, the rook's strategy is only about three times the cost of partial pivoting [185].

To illustrate the impact of pivot selection on numerical errors, consider the equation [225]

$$\begin{pmatrix} 5 & 1000 \\ 1000 & 1000 \end{pmatrix} \begin{pmatrix} x_1 \\ x_2 \end{pmatrix} = \begin{pmatrix} 500 \\ 1000 \end{pmatrix}. \tag{2.11}$$

The exact solution is $x_1 = \frac{100}{199} \approx 0.50$ and $x_2 = \frac{99}{199} \approx 0.50$.

To emphasize the numerical errors, let's perform Gaussian elimination on this matrix rounding each operation to two decimal places. With partial pivoting, 1000 is the first pivot, so one first permutes the system,

$$\begin{pmatrix} 1000 & 1000 \\ 5 & 1000 \end{pmatrix} \begin{pmatrix} x_1 \\ x_2 \end{pmatrix} = \begin{pmatrix} 1000 \\ 500 \end{pmatrix},$$

and then subtracts 5/1000 times row 1 from row 2. Rounding to two decimal places, $\text{round}(1000-(5/1000)1000) = \text{round}(995) = 1000$; and $\text{round}(500-(5/1000)1000) = \text{round}(495) = 500$:

$$\begin{pmatrix} 1000 & 1000 \\ 0 & 1000 \end{pmatrix} \begin{pmatrix} x_1 \\ x_2 \end{pmatrix} = \begin{pmatrix} 1000 \\ 500 \end{pmatrix}.$$

Then, back substitution with rounding gives $x_2 = \text{round}(500/1000) = 0.50$ and $x_1 = \text{round}(\text{round}(1000 - \text{round}(1000 \times 0.50))/1000) = 0.50$. In this case, partial pivoting gave the correct solution.

Now, let's multiply the first row of (2.11) by 220 to obtain the equivalent system of equations:

$$\begin{pmatrix} 1100 & 220\,000 \\ 1000 & 1000 \end{pmatrix} \begin{pmatrix} x_1 \\ x_2 \end{pmatrix} = \begin{pmatrix} 110\,000 \\ 1000 \end{pmatrix}. \tag{2.12}$$

Since both sides of the equation represented by the first row were multiplied by the same factor, the exact solution is unchanged: $x_1 = x_2 = 0.50$ using two decimal places.

Using partial pivoting on this new equation, the pivot is 1100 so no permutation is required. The elimination step gives

$$\begin{pmatrix} 1100 & 220\,000 \\ 0 & -200\,000 \end{pmatrix} \begin{pmatrix} x_1 \\ x_2 \end{pmatrix} = \begin{pmatrix} 110\,000 \\ -99\,000 \end{pmatrix},$$

and back substitution yields $x_2 = \text{round}(-99\,000/-200\,000) = 0.50$; and $x_1 = \text{round}(\text{round}(110\,000-\text{round}(0.50 \times 220\,000))/1100) = 0.00$, which has 100% relative error.

This example shows that partial pivoting does not always select the pivot that will lead to the lowest roundoff error. Before we rescaled the matrix, partial pivoting did a fine job. After rescaling, partial pivoting led to a poor choice. One idea to address this problem of scale is to attempt to *equilibrate* the matrix, by imposing a uniform scaling before selecting a pivot.

The following discussion is very qualitative, and the algorithms for accomplishing this equilibration are rather heuristic. An algorithm based on simple row scaling follows. For each row i of the original matrix $\mathbf{A}$, we can construct a measure of scale,

$$S_i = \sum_j |A_{ij}|$$

(S_i is the discrete 1 norm (L^1) of row i. This idea of norms is developed in the following section, and an application of L^1 is described in Chapter 8). If we multiply $\mathbf{Ax} = \mathbf{b}$ by the diagonal matrix $\mathbf{D}$,

$$\mathbf{D} = \begin{pmatrix} S_1^{-1} & & \\ & \ddots & \\ & & S_n^{-1} \end{pmatrix},$$

then $\mathbf{A}' = \mathbf{DA}$ becomes *equilibrated* in the sense that $S_i' = 1$ – all rows have the same scale. For the 2×2 matrix we have been studying (2.11), $S_1 = 1005$ and $S_2 = 2000$, so

$$\mathbf{A}' = \begin{pmatrix} 1/201 & 200/201 \\ 1/2 & 1/2 \end{pmatrix}. \tag{2.13}$$

Based on this scaled matrix, partial pivoting would select the pivot from the second row – the choice that gave the better numerical result. Matrix $\mathbf{A}'$ (2.13) is also the equilibrated form of matrix (2.12), therefore this equilibration process compensated for the factor of 220 that led us to make a poor pivot selection.

Scaled partial pivoting refers to making the pivot selection based on the equilibrated matrix. To implement scaled partial pivoting, it is not necessary to actually scale the matrix. In fact, scaling should be avoided unless the numbers S_i are powers of two, since otherwise new numerical errors will be introduced. Instead, one selects the pivot A_{pd} by choosing the index p such that $|A_{pd}/S_p|$ is a maximum for all $p \in [d, \ldots, n]$. The row scaling numbers S_i are used only in this pivot selection process. Note that with implicit row scaling one may lose the diagonal dominance of $\mathbf{L}$.

Example 2.5 Use 2 decimal arithmetic to solve

$$\begin{pmatrix} 1 & 1 & 2 \times 10^9 \\ 2 & -1 & 10^9 \\ 1 & 2 & 0 \end{pmatrix} \mathbf{x} = \begin{pmatrix} 2 \times 10^9 + 2 \\ 10^9 + 1 \\ 3 \end{pmatrix}$$

with various pivoting strategies.

Solution
The exact result is $\mathbf{x} = (1, 1, 1)^T$. The two decimal numerical results are shown in Table 2.1.

Table 2.1 Results for Example 2.5.

P	L	R	x
trivial pivoting:			
$p_\ell = \{1, 2, 3\}$	$\begin{pmatrix} 1.0 & & \\ 2.0 & 1.0 & \\ 1.0 & -0.33 & 1.0 \end{pmatrix}$	$\begin{pmatrix} 1.0 & 1.0 & 2.0 \times 10^9 \\ & -3.0 & -3.0 \times 10^9 \\ & & -3.0 \times 10^9 \end{pmatrix}$	$\begin{pmatrix} 0.0 \\ 0.0 \\ 1.0 \end{pmatrix}$
partial pivoting:			
$p_\ell = \{2, 3, 1\}$	$\begin{pmatrix} 1.0 & & \\ 0.50 & 1.0 & \\ 0.50 & 0.60 & 1.0 \end{pmatrix}$	$\begin{pmatrix} 2.0 & -1.0 & 1.0 \times 10^9 \\ & 2.5 & -5.0 \times 10^8 \\ & & 1.8 \times 10^9 \end{pmatrix}$	$\begin{pmatrix} 0.0 \\ 0.0 \\ 1.0 \end{pmatrix}$
rook's pivoting and complete pivoting:			
$p_\ell = \{1, 3, 2\}$ $p_r = \{3, 2, 1\}$	$\begin{pmatrix} 1.0 & & \\ 0.0 & 1.0 & \\ 0.50 & -0.75 & 1.0 \end{pmatrix}$	$\begin{pmatrix} 2.0 \times 10^9 & 1.0 & 1.0 \\ & 2.0 & 1.0 \\ & & 2.3 \end{pmatrix}$	$\begin{pmatrix} 1.0 \\ 1.0 \\ 1.0 \end{pmatrix}$
implicit row scaling with partial pivoting:			
$p_\ell = \{3, 2, 1\}$	$\begin{pmatrix} 1.0 & & \\ 2.0 & 1.0 & \\ 1.0 & 0.20 & 1.0 \end{pmatrix}$	$\begin{pmatrix} 1.0 & 2.0 & 0.0 \\ & -5.0 & 1.0 \times 10^9 \\ & & 1.8 \times 10^9 \end{pmatrix}$	$\begin{pmatrix} 3.0 \\ 0.0 \\ 1.0 \end{pmatrix}$

This example, attributed to R. W. Hamming [69], illustrates the failure of row scaling. One can also see that rook's pivoting and complete pivoting give diagonally dominant triangular matrices, which contributes to their good performance.

There exist other pivoting strategies, and more sophisticated equilibration techniques. However, there are also deeper underlying problems of the Gaussian elimination technique that existing pivoting strategies do not fully compensate. The following section introduces some new tools to develop what is in the end a very simple error analysis for Gaussian elimination. It will be shown that the *condition number* of the matrix $\mathbf{A}^{(d)}$ is a key consideration, and the control of this condition number motivates the entirely different technique of *Householder reduction*.

2.3   Errors in Gaussian elimination

2.3.1   Norms

The goal of this discussion is to try to understand in a semi-quantitative way how the error analysis of Chapter 1 can be applied to the Gaussian elimination method without going into the detail of each multiplication and addition step.

Accordingly, the first thing we will do is find a single number to represent the n elements of an $\mathbb{R}^n$ vector $\mathbf{x}$ or $\mathbf{b}$, and a single number to represent the $n \times n$ elements of an $\mathbb{R}^n \times \mathbb{R}^n$ matrix $\mathbf{A}$. The numbers will be denoted $\|\mathbf{x}\|$ or $\|\mathbf{A}\|$ (for example). They are called *norms*, and they are characterized by the following properties:

1. Positivity: $\|\mathbf{x}\| \geq 0$ for all $\mathbf{x} \neq \mathbf{0}$.
2. Homogeneity: $\|c\mathbf{x}\| = |c|\,\|\mathbf{x}\|$ for all constants c.
3. Triangle inequality: $\|\mathbf{x} + \mathbf{y}\| \leq \|\mathbf{x}\| + \|\mathbf{y}\|$.

These same properties hold for matrix norms, even when the matrix is not square.

Matrix norms are *submultiplicative* if

$$\|\mathbf{AB}\| \leq \|\mathbf{A}\|\,\|\mathbf{B}\|.$$

If $\mathbf{B} = \mathbf{I}$, then we have $\|\mathbf{I}\| \geq 1$ for all submultiplicative matrix norms.

Matrix norms and vector norms are *consistent* if

$$\|\mathbf{Ax}\| \leq \|\mathbf{A}\|\,\|\mathbf{x}\|.$$

One way to construct consistent norms is to define the matrix norm in terms of the vector norms as follows. The *least upper bound norm* (or *subordinate matrix norm*, or *induced norm*, or *natural norm*) is

$$\text{lub}(\mathbf{A}) = \max_{\mathbf{x} \neq 0} \frac{\|\mathbf{Ax}\|}{\|\mathbf{x}\|}. \tag{2.14}$$

This particular matrix norm is the smallest consistent matrix norm for any choice of vector norm $\|\mathbf{x}\|$ (e.g., $\text{lub}(\mathbf{I}) = 1$, so $\|\mathbf{I}\| \geq \text{lub}(\mathbf{I}) = 1$ for any choice of vector norm). The least upper bound norm is also submultiplicative,

$$\text{lub}(\mathbf{AB}) = \max_{\mathbf{x} \neq 0} \frac{\|\mathbf{ABx}\|}{\|\mathbf{x}\|} = \max_{\mathbf{x} \neq 0} \frac{\|\mathbf{A(Bx)}\|}{\|\underbrace{(\mathbf{Bx})}_{\mathbf{y}}\|}\frac{\|\mathbf{Bx}\|}{\|\mathbf{x}\|}$$

$$\leq \max_{\mathbf{y} \neq 0} \frac{\|\mathbf{Ay}\|}{\|\mathbf{y}\|}\max_{\mathbf{x} \neq 0} \frac{\|\mathbf{Bx}\|}{\|\mathbf{x}\|} = \text{lub}(\mathbf{A})\text{lub}(\mathbf{B}).$$

Some particular norms are described below, and are distinguished by a subscript

$$\|\mathbf{x}\|_2 = \sqrt{\mathbf{x} \cdot \mathbf{x}}$$

$$\|\mathbf{A}\|_2 = \left(\sum_{i,j=1}^{n} |a_{ij}|^2 \right)^{1/2}$$

$$\text{lub}_2(\mathbf{A}) = \max_{\mathbf{x} \neq 0} \frac{\sqrt{\mathbf{x}^T \mathbf{A}^T \mathbf{A} \mathbf{x}}}{\sqrt{\mathbf{x}^T \mathbf{x}}}$$

($\|\mathbf{A}\|_2$ above is the *Frobenius norm*, a.k.a. the *Euclidean norm* or the *Schur norm*) and

$$\|\mathbf{x}\|_\infty = \max_i |x_i|$$

$$\|\mathbf{A}\|_\infty = \max_i \sum_{j=1}^{n} |a_{ij}| = \text{lub}_\infty(\mathbf{A}). \tag{2.15}$$

These matrix norms are submultiplicative and consistent with their corresponding norms.

A norm is *absolute* if, for all $n \times n$ diagonal matrices $\mathbf{D}$,

$$\|\mathbf{D}\| = \max_{1 \leq i \leq n} |d_i|.$$

The matrix norms $\| \cdot \|_2$, $\| \cdot \|_\infty$, $\text{lub}_2(\cdot)$ are absolute.

The norms $\| \cdot \|_1$, $\| \cdot \|_2$, and $\| \cdot \|_\infty$ are examples of L^p *Lebesgue norms*, which take the form

$$\|\mathbf{x}\|_p = \left(\sum_i |x_i|^p \right)^{1/p},$$

where p is arbitrary, although $p \geq 1$ is necessary for these norms to satisfy the triangle inequality. The common norms L^1, L^2, and L^∞ satisfy the following inequalities:

$$\|\mathbf{x}\|_\infty \leq \|\mathbf{x}\|_2 \leq \|\mathbf{x}\|_1 \tag{2.16}$$

$$\|\mathbf{x}\|_1 \leq \sqrt{n} \|\mathbf{x}\|_2 \leq n \|\mathbf{x}\|_\infty, \tag{2.17}$$

for $\mathbf{x} \in \mathbb{R}^n$.

2.3.2 Amplification factors for Gaussian elimination

With these definitions, let's consider the relative error in $\mathbf{x}$ due to a relative error in $\mathbf{b}$. For the exact solution we have $\mathbf{A}\mathbf{x} = \mathbf{b}$, and for the error $\Delta\mathbf{x}$ due to error $\Delta\mathbf{b}$ we have:

$$\mathbf{A}(\mathbf{x} + \Delta\mathbf{x}) = \mathbf{b} + \Delta\mathbf{b}$$

$$\Delta\mathbf{x} - \mathbf{A}^{-1}\Delta\mathbf{b} \tag{2.18}$$

$$\|\Delta\mathbf{x}\| = \|\mathbf{A}^{-1}\Delta\mathbf{b}\| \leq \|\mathbf{A}^{-1}\| \|\Delta\mathbf{b}\|,$$

using the consistency property. On the other hand,

$$\mathbf{b} = \mathbf{Ax}$$
$$\|\mathbf{b}\| = \|\mathbf{Ax}\| \le \|\mathbf{A}\| \|\mathbf{x}\|, \tag{2.19}$$

again using the consistency property. Combining (2.18) and (2.19) we get

$$\frac{\|\Delta\mathbf{x}\|}{\|\mathbf{x}\|} \le \|\mathbf{A}\| \|\mathbf{A}^{-1}\| \frac{\|\Delta\mathbf{b}\|}{\|\mathbf{b}\|}.$$

The product $\|\mathbf{A}\| \|\mathbf{A}^{-1}\|$ is called the *condition number*, cond($\mathbf{A}$), when matrix norms are defined as subordinate norms. With this definition,

$$\frac{\|\Delta\mathbf{x}\|}{\|\mathbf{x}\|} \le \text{cond}(\mathbf{A}) \frac{\|\Delta\mathbf{b}\|}{\|\mathbf{b}\|}. \tag{2.20}$$

Of course there are errors in $\mathbf{A}$ also. We will express the error in $\mathbf{A}$ as

$$\mathbf{A} + \Delta\mathbf{A} = \mathbf{A}(\mathbf{I} + \mathbf{F}).$$

If $\Delta\mathbf{x}$ is the error in $\mathbf{x}$ due only to errors in $\mathbf{A}$, then

$$\mathbf{A}(\mathbf{I} + \mathbf{F})(\mathbf{x} + \Delta\mathbf{x}) = \mathbf{b}$$
$$\mathbf{AFx} + \mathbf{A}(\mathbf{I} + \mathbf{F})\Delta\mathbf{x} = 0 \tag{2.21}$$
$$\Delta\mathbf{x} = -(\mathbf{I} + \mathbf{F})^{-1}\mathbf{Fx}.$$

If we assume submultiplicative and subordinate matrix norms then

$$1 = \|(\mathbf{I} + \mathbf{F})(\mathbf{I} + \mathbf{F})^{-1}\| = \|(\mathbf{I} + \mathbf{F})^{-1} + \mathbf{F}(\mathbf{I} + \mathbf{F})^{-1}\|$$
$$\ge \|(\mathbf{I} + \mathbf{F})^{-1}\| - \|\mathbf{F}(\mathbf{I} + \mathbf{F})^{-1}\|$$
$$\ge \|(\mathbf{I} + \mathbf{F})^{-1}\|(1 - \|\mathbf{F}\|)$$
$$\|(\mathbf{I} + \mathbf{F})^{-1}\| \le \frac{1}{1 - \|\mathbf{F}\|}. \tag{2.22}$$

The second step used (A.2) and assumes $\|\mathbf{F}\| < 1$. With (2.22), (2.21) becomes

$$\frac{\|\Delta\mathbf{x}\|}{\|\mathbf{x}\|} \le \frac{\|\mathbf{F}\|}{1 - \|\mathbf{F}\|}.$$

A different rearrangement expresses this result in terms of the condition number:

$$\mathbf{F} = \mathbf{A}^{-1}(\Delta\mathbf{A})$$
$$\|\mathbf{F}\| \le \|\mathbf{A}^{-1}\| \|\mathbf{A}\| \frac{\|\Delta\mathbf{A}\|}{\|\mathbf{A}\|},$$

then

$$\frac{\|\Delta\mathbf{x}\|}{\|\mathbf{x}\|} \le \frac{\text{cond}(\mathbf{A})\frac{\|\Delta\mathbf{A}\|}{\|\mathbf{A}\|}}{1 - \text{cond}(\mathbf{A})\frac{\|\Delta\mathbf{A}\|}{\|\mathbf{A}\|}}. \tag{2.23}$$

Like (2.20), we see that (to leading order) the amplification factor that relates a relative error in $\mathbf{A}$ to a relative error in the computed result $\mathbf{x}$ is the condition number cond($\mathbf{A}$).

We now have (2.20), which expresses the relative error in $\mathbf{x}$ due to errors in $\mathbf{b}$, and

(2.23), which expresses the relative error in $\mathbf{x}$ due to errors in $\mathbf{A}$. If the errors $\Delta\mathbf{A}$ are small enough, the denominator of (2.23) can be approximated as 1, and we have for both expressions that the relative error in $\mathbf{x}$ is proportional to the condition number cond($\mathbf{A}$).

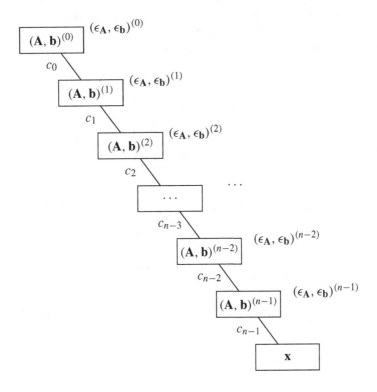

Figure 2.1 Bauer graph for Gaussian elimination.

The Gaussian elimination algorithm can be viewed as a succession of $(\mathbf{A}, \mathbf{b})^{(d)}$ pairs, each subject to some error $(\epsilon_\mathbf{A}, \epsilon_\mathbf{b})^{(d)}$ (scalars, when expressing errors by norms), and these errors propagate in the usual way. In Figure 2.1 the symbols c denote condition numbers in the sense of Chapter 1. The input error $(\epsilon_\mathbf{A}, \epsilon_\mathbf{b})^{(0)}$ propagates to the output with amplification $c_0 c_1 \cdots c_{n-1}$. However, by the analysis above, these input errors are amplified by cond($\mathbf{A}^{(0)}$). Therefore, the condition numbers cond($\mathbf{A}^{(d)}$) each represent a product of terms $c_d \cdots c_{n-1}$. The cumulative error is therefore approximately

$$\frac{\|\Delta\mathbf{x}\|}{\|\mathbf{x}\|} \leq \sum_{d=0}^{n-1} \text{cond}(\mathbf{A}^{(d)}) \left(|\epsilon_\mathbf{A}^{(d)}| + |\epsilon_\mathbf{b}^{(d)}| \right). \tag{2.24}$$

An important difference between this analysis and the analyses of Chapter 1 is that here the errors ϵ are not simply related to machine precision: each step of Gaussian elimination is not an elementary operation, and the errors can be large. These errors can be minimized with a good choice of pivot selection, but one has limited control over the process.

2.3.3 Rounding errors and backward error analysis

The preceding analysis explored the propagation of errors only – essentially just the inherent error of the procedure. To understand the numerical errors and the connection to pivoting strategies, one must apply differential error analysis methods of Chapter 1. Wilkinson [242] has shown that the matrix $\mathbf{E} = \tilde{\mathbf{L}}\tilde{\mathbf{R}} - \mathbf{P}_L \mathbf{A} \mathbf{P}_R^T$, which expresses the numerical errors committed in constructing $\mathbf{L}$ and $\mathbf{R}$ with prescribed permutations, has a bound

$$|\mathbf{E}| \le \frac{2g\epsilon}{1-\epsilon} \begin{pmatrix} 0 & 0 & 0 & \cdots & 0 & 0 \\ 1 & 1 & 1 & \cdots & 1 & 1 \\ 1 & 2 & 2 & \cdots & 2 & 2 \\ 1 & 2 & 3 & \cdots & 3 & 3 \\ \vdots & \vdots & \vdots & & \vdots & \vdots \\ 1 & 2 & 3 & \cdots & n-1 & n-1 \end{pmatrix}, \tag{2.25}$$

where $\mathbf{A}$ is $n \times n$, and g is the largest element of $\mathbf{A}^{(d)}$ of all steps $d = 0, ..., n-1$:

$$g = \max_{i,j,d} |a_{ij}^{(d)}|.$$

The *growth factor* g is one the the main connections between the pivoting strategy and the error of the method. Some bounds on g are

$$\frac{g}{\max_{ij} |a_{ij}^{(0)}|} \le 2^{(n-1)} \qquad\qquad\qquad \text{partial pivoting} \tag{2.26}$$

$$\frac{g}{\max_{ij} |a_{ij}^{(0)}|} \le \sqrt{n[2^1 3^{1/2} 4^{1/3} \cdots n^{1/(n-1)}]} \qquad \text{complete pivoting}$$

$$\sim Cn^{1/2} n^{1/4 \log n},$$

but these are **very** pessimistic for matrices occurring in practice. For partial pivoting, the bound (2.26) is achieved for the matrices of the form

$$\begin{pmatrix} 1 & 0 & 0 & 0 & 1 \\ -1 & 1 & 0 & 0 & 1 \\ -1 & -1 & 1 & 0 & 1 \\ -1 & -1 & -1 & 1 & 1 \\ -1 & -1 & -1 & -1 & 1 \end{pmatrix},$$

but for complete pivoting $g \lesssim n \max_{ij} |a_{ij}^{(0)}|$ (for large n) is a practical upper bound (e.g., [106]).

Since we determine $\mathbf{x}$ via the numerically determined $\tilde{\mathbf{L}}\tilde{\mathbf{R}}$ decomposition, $\mathbf{E}$ of (2.25) represents the error $\Delta\mathbf{A}$ in (2.23). Given a choice of pivot strategy, one could estimate g, hence $\|\Delta\mathbf{A}\| = \|\mathbf{E}\|$, and given $\text{cond}(\mathbf{A})$, one then obtains a bound on $\|\Delta\mathbf{x}\|$. A simpler approach is to use backward error analysis.

The equations we actually solve in computing $\mathbf{x}$ are $\tilde{\mathbf{L}}\tilde{\mathbf{R}}(\mathbf{P}_R\tilde{\mathbf{x}}) = (\mathbf{P}_L\mathbf{b})$, or

$$(\mathbf{P}_L\mathbf{A}\mathbf{P}_R^T + \mathbf{E})(\mathbf{P}_R\tilde{\mathbf{x}}) = \mathbf{b},$$

so $\tilde{\mathbf{x}}$ is the exact answer to the wrong problem, and $\mathbf{E}$ shows how different the problems might be. Recall that in backward analysis we are concerned with the error in the problem solved, not the error in the solution.

$\mathbf{E}$ conveniently expresses through backward error analysis the error associated with decomposing $\mathbf{A}$ into permutations, $\tilde{\mathbf{L}}$, and $\tilde{\mathbf{R}}$. There are also errors associated with the two back substitution steps. Wilkinson also derives these. For $\mathbf{Ly} = (\mathbf{P}_L \mathbf{b})$, $\tilde{\mathbf{y}}$ solves exactly

$$(\tilde{\mathbf{L}} + \Delta\tilde{\mathbf{L}})\tilde{\mathbf{y}} = \mathbf{Pb}, \tag{2.27}$$

with

$$|\Delta\mathbf{L}| \le \epsilon \begin{pmatrix} |\ell_{11}| & & & & & \\ 2|\ell_{21}| & 2|\ell_{22}| & & & & \\ 3|\ell_{31}| & 2|\ell_{32}| & 2|\ell_{33}| & & & \\ 4|\ell_{41}| & 3|\ell_{42}| & 2|\ell_{43}| & 2|\ell_{44}| & & \\ \vdots & \vdots & \vdots & \vdots & \ddots & \\ n|\ell_{n1}| & (n-1)|\ell_{n2}| & (n-2)|\ell_{n3}| & (n-3)|\ell_{n4}| & \cdots & 2|\ell_{nn}| \end{pmatrix}.$$

And, for $\mathbf{R}(\mathbf{P}_R\tilde{\mathbf{x}}) = \mathbf{y}$ one has

$$(\tilde{\mathbf{R}} + \Delta\mathbf{R})\mathbf{P}_R\tilde{\mathbf{x}} = \mathbf{y}, \tag{2.28}$$

with

$$|\Delta\mathbf{R}| \le \epsilon \begin{pmatrix} 2|r_{11}| & \cdots & (n-2)|r_{1,n-2}| & (n-1)|r_{1,n-1}| & n|r_{1n}| \\ & \ddots & \vdots & \vdots & \vdots \\ & & 2|r_{n-2,n-2}| & 2|r_{n-2,n-1}| & 3|r_{n-2,n}| \\ & & & 2|r_{n-1,n-1}| & 2|r_{n-1,n}| \\ & & & & |r_{n,n}| \end{pmatrix}.$$

The combined errors in constructing $\mathbf{L}$ and $\mathbf{R}$ (2.25), in performing the $\mathbf{L}$ back substitution (2.27), and in performing the $\mathbf{R}$ back substitution (2.28), are

$$(\tilde{\mathbf{R}} + \Delta\mathbf{R})\mathbf{P}_R\mathbf{x} = \mathbf{y}$$

$$(\tilde{\mathbf{L}} + \Delta\mathbf{L})(\tilde{\mathbf{R}} + \Delta\mathbf{R})\mathbf{P}_R\mathbf{x} = (\tilde{\mathbf{L}} + \Delta\mathbf{L})\mathbf{y} = \mathbf{P}_L\mathbf{b}$$

$$(\tilde{\mathbf{L}}\tilde{\mathbf{R}} + \tilde{\mathbf{L}}\Delta\mathbf{R} + \Delta\mathbf{L}\tilde{\mathbf{R}} + \Delta\mathbf{L}\Delta\mathbf{R})\mathbf{P}_R\mathbf{x} = \mathbf{P}_L\mathbf{b}$$

$$(\mathbf{P}_L\mathbf{A}\mathbf{P}_R^{\mathrm{T}} + \mathbf{E} + \tilde{\mathbf{L}}\Delta\mathbf{R} + \Delta\mathbf{L}\tilde{\mathbf{R}} + \Delta\mathbf{L}\Delta\mathbf{R})\mathbf{P}_R\mathbf{x} = \mathbf{P}_L\mathbf{b},$$

and from this one could construct an estimate of the norm of the *residual* $\mathbf{r} = \mathbf{b} - \mathbf{Ax}$

$$\mathbf{r} = \mathbf{P}_L^{\mathrm{T}}(\mathbf{E} + \tilde{\mathbf{L}}\Delta\mathbf{R} + \Delta\mathbf{L}\tilde{\mathbf{R}} + \Delta\mathbf{L}\Delta\mathbf{R})\mathbf{P}_R\mathbf{x}$$

$$\|\mathbf{r}\| \le \left(\|\mathbf{E}\| + \|\tilde{\mathbf{L}}\|\|\Delta\mathbf{R}\| + \|\Delta\mathbf{L}\|\|\tilde{\mathbf{R}}\| + \|\Delta\mathbf{L}\|\|\Delta\mathbf{R}\| \right) \|\mathbf{x}\|, \tag{2.29}$$

with

$$\|\mathbf{E}\|_\infty \le g\epsilon n(n+1) \sim g\epsilon n^2$$

$$\|\mathbf{L}\|_\infty \le n$$

$$\|\Delta\mathbf{L}\|_\infty \le \epsilon\left(1 + \frac{n(n+1)}{2}\right) \sim \frac{\epsilon n^2}{2}$$

$$\|\mathbf{R}\|_\infty \le gn$$

$$\|\Delta\mathbf{R}\|_\infty \le g\epsilon\left(1 + \frac{n(n+1)}{2}\right) \sim \frac{g\epsilon n^2}{2}$$

$$\|\mathbf{r}\|_\infty \le g\epsilon n^2(n+1)\|\mathbf{x}\|_\infty, \tag{2.30}$$

neglecting terms of order ϵ^2. The leading error ($\propto n^3$) comes from the back substitution steps.

The main value of this final result is its emphasis on system size, $\|\mathbf{r}\|_\infty \propto n^3$ (so $\|\mathbf{r}\|_2 \propto n^{5/2}$ by (2.17)), and its sensitivity to g – which in turn is strongly sensitive to the pivoting strategy.

2.3.4   Diagonal dominance

In contrast with the thorough analysis of Wilkinson, the concept of diagonal dominance is best appreciated though a simple graphical argument [184].

The back substitution equation

$$r_j x_j + r_i x_i = y \tag{2.31}$$

for variable x_j is the intersection of the line given by (2.31) with the line $x_i = $ constant. If $|r_i| \gg |r_j|$ then the solution is very sensitive to errors in input x_i (see Figure 2.2). This undesirable circumstance corresponds to the off-diagonal component of $\mathbf{R}$ being much greater than the diagonal component – a lack of diagonal dominance.

On the other hand, if x_i is the variable being solved for, then the circumstance $|r_i| \gg |r_j|$ causes input errors in x_j to be attenuated. This desirable circumstance corresponds to the case of diagonal dominance.

2.3.5   Summary

Three views of the error associated with Gaussian elimination have been presented. The third view was geometrical, emphasizing the value of diagonal dominance for controlling errors. The errors associated with this geometrical problem are fully accounted for in the analysis of Wilkinson, e.g., (2.27) and (2.28), and (2.29), but that analysis focused attention of the growth factor g rather than the source of these errors in relation to $\mathbf{L}$ and $\mathbf{R}$. The L^∞ norm used in (2.30) is a particularly blunt instrument that is completely insensitive to the distribution of numbers in a row. Diagonal dominance in $\mathbf{L}$ is enforced by partial pivoting, rook's pivoting, and complete pivoting. Diagonal dominance in $\mathbf{R}$

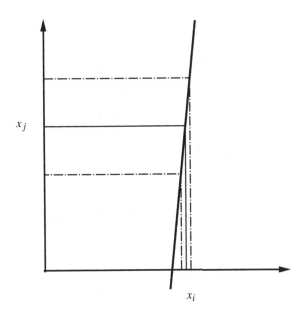

Figure 2.2 A geometric interpretation of why **L**, **R** should be diagonally dominant.

is enforced by rook's pivoting and complete pivoting. For these reasons, the rook's pivoting method or complete pivoting method are numerically superior to other pivoting alternatives.

The first view, Section 2.3.2, emphasizes growth of the condition number in the propagation of errors (2.24). The second, Section 2.3.3, emphasizes the growth factor g. These are different perspectives on the same problem, so to the extent that g is well-controlled by a good pivoting strategy one might conclude that the condition number does not grow dramatically during the elimination process. Some growth cannot be ruled out (see Problem 2.2) and one strategy to controlling the error in solving $\mathbf{Ax} = \mathbf{b}$ is to seek an approach that holds cond($\mathbf{A}$) fixed – a motivation for the Householder reduction algorithm.

Scaling can diminish the condition number cond($\mathbf{A}^{(0)}$). Bauer [8] shows that the condition number is minimized when row and column scaling are combined to make both $\mathbf{A}$ and $\mathbf{A}^{-1}$ equilibrated in rows and in columns. An algorithm to achieve this scaling is not known. Experience suggests that the principal effect of scaling (and the only effect of implicit scaling) is on the choice of pivots [69, 104]. The most direct way to control error with Gaussian elimination is to select a favorable pivoting strategy (e.g., rook's or complete).

2.4 Householder reduction

Householder reduction [112] transforms matrix $\mathbf{A}$ into the product of a unitary matrix $\mathbf{Q}$ and a right triangular matrix $\mathbf{R}$. A matrix is unitary if $\mathbf{QQ}^H = \mathbf{Q}^H\mathbf{Q} = \mathbf{I}$, or if real, then $QQ^T = Q^TQ = I$. So, given Q and R,

$$Ax = QRx = b$$
$$Rx = Q^Tb,$$

which is easily solved by back substitution.

If $Q_1, ..., Q_n$ are each unitary, then the product $Q_1Q_2 \cdots Q_{n-2}Q_{n-1}$ is also unitary. We will seek a sequence of unitary Q_i that will successively eliminate the subdiagonal elements of the matrix A.

The benefit of using unitary matrices concerns the condition number. Using $Q^TQ = I$,

$$\text{lub}_2(QA) = \max_{x \neq 0} \frac{\sqrt{x^TA^TQ^TQAx}}{\sqrt{x^Tx}} = \max_{x \neq 0} \frac{\sqrt{x^TA^TAx}}{\sqrt{x^Tx}} = \text{lub}_2(A),$$

and

$$\text{lub}_2((QA)^{-1}) = \max_{x \neq 0} \frac{\sqrt{x^TQA^{-T}A^{-1}Q^Tx}}{\sqrt{x^Tx}}$$

$$= \max_{x \neq 0} \frac{\sqrt{(Q^Tx)^TA^{-T}A^{-1}(Q^Tx)}}{\sqrt{(Q^Tx)^T(Q^Tx)}}$$

$$= \max_{x' \neq 0} \frac{\sqrt{x'^TA^{-T}A^{-1}x'}}{\sqrt{x'^Tx'}} = \text{lub}_2(A^{-1}).$$

In combination,

$$\text{cond}_2(QA) = \text{lub}_2(QA)\text{lub}_2((QA)^{-1}) = \text{cond}_2(A).$$

Thus, the triangularization of A by unitary matrices Q keeps the condition number constant.

Each of the component matrices Q_i that make up $Q = Q_1 \cdots Q_{n-1}$ will have a block partitioned form

$$Q_i = \left(\begin{array}{c|c} I_{(i-1) \times (i-1)} & \\ \hline & \tilde{Q}_{i\,(n+1-i) \times (n+1-i)} \end{array} \right) \tag{2.32}$$

$$\tilde{Q}_i = I - 2w_iw_i^T, \tag{2.33}$$

with $w_i \in \mathbb{R}^{n+1-i}$. Note that $Q_i = Q_i^T$. To be unitary, there is a constraint on the vectors w:

$$Q_iQ_i^T = \left(\begin{array}{c|c} I & \\ \hline & \tilde{Q}_i\tilde{Q}_i^T \end{array} \right)$$

$$\tilde{Q}_i\tilde{Q}_i^T = I - 4w_iw_i^T + 4w_iw_i^Tw_iw_i^T$$

$$= I \quad \text{if } w_i^Tw_i = 1,$$

so the goal is to find $\mathbf{w}_i$ such that $\mathbf{w}_i \cdot \mathbf{w}_i = 1$, and so that the product of $\mathbf{Q}_i$s accomplishes the triangularization of $\mathbf{A}$.

If $\mathbf{a}_i$ is vector made up of the diagonal and subdiagonal elements of the ith column of $\mathbf{A}$, then $\mathbf{Q}_i$ will triangularize $\mathbf{A}$ if

$$\tilde{\mathbf{Q}}_i \mathbf{a}_i = k_i \mathbf{e}_1, \tag{2.34}$$

where $\mathbf{e}_1 = (1, \ 0, \ ...)^T$ is the unit basis vector in $\mathbb{R}^{n+1-i}$.

To determine k_i, multiply (2.34) against its transpose:

$$(\tilde{\mathbf{Q}}_i \mathbf{a}_i)^T (\tilde{\mathbf{Q}}_i \mathbf{a}_i) = \mathbf{a}_i^T \tilde{\mathbf{Q}}_i^T \tilde{\mathbf{Q}}_i \mathbf{a}_i = \|\mathbf{a}_i\|_2^2 = k_i^2$$

$$k_i = \pm \|\mathbf{a}_i\|_2.$$

Then, to find $\mathbf{w}_i$,

$$\tilde{\mathbf{Q}}_i \mathbf{a}_i = k_i \mathbf{e}_1$$

$$\mathbf{a}_i - 2\mathbf{w}_i (\mathbf{w}_i \cdot \mathbf{a}_i) = k_i \mathbf{e}_1,$$

then take the inner product of $2\mathbf{w}_i (\mathbf{w}_i \cdot \mathbf{a}_i) = \mathbf{a}_i - k_i \mathbf{e}_1$ with itself, using $\mathbf{w}_i^T \mathbf{w}_i = 1$, to obtain

$$\tag{2.35}$$

$$4(\mathbf{w}_i \cdot \mathbf{a}_i)^2 = \|\mathbf{a}_i - k_i \mathbf{e}_1\|^2,$$

which determines $(\mathbf{w}_i \cdot \mathbf{a}_i)$, giving

$$\mathbf{w}_i = \frac{\mathbf{a}_i - k_i \mathbf{e}_1}{\|\mathbf{a}_i - k_i \mathbf{e}_1\|_2}. \tag{2.36}$$

There are two possible values for k_i: one is more numerically stable than the other as regards the evaluation of the denominator of (2.36):

$$\|\mathbf{a}_i - k_i \mathbf{e}_1\|_2 = \sqrt{(a_1 - k_i)^2 + a_2^2 + \cdots + a_n^2},$$

where a_1 is the first element of $\mathbf{a}_i$, etc. The sign of k_i should be chosen in order to make $a_1 - k_i$ be addition rather than subtraction:

$$k_i = -\,\text{sign}(a_1) \|\mathbf{a}_i\|_2. \tag{2.37}$$

Geometrically, a Householder matrix is an example of a reflection. Imagine throwing an elastic ball at the floor with a velocity $\mathbf{v}$. The component of the velocity directed at the ground is $(\mathbf{v} \cdot \mathbf{n})\mathbf{n}$ where $\mathbf{n}$ is a unit vector perpendicular to the ground. If the ball bounces in a perfectly elastic manner, then after bouncing the ball's velocity in direction $\mathbf{n}$ will be reversed: it will now be $-(\mathbf{v} \cdot \mathbf{n})\mathbf{n}$. The overall velocity after bouncing is therefore $\mathbf{v} - 2(\mathbf{v} \cdot \mathbf{n})\mathbf{n}$, or $(\mathbf{I} - 2\mathbf{n}\mathbf{n}^T)\mathbf{v}$ (cf. (2.33)).

In the QR method, the vector $\mathbf{w}$ is chosen to make the vector $\mathbf{a}$ be oriented in direction $\pm \mathbf{e}_1$. There are two choices for $\mathbf{w}$, depending on the sign; call these $\mathbf{w}'$ and $\mathbf{w}''$. In order that the normalization factor $\|\mathbf{a} - k\mathbf{e}_1\|$ embody addition rather than subtraction, the reflection is chosen to accomplish an obtuse "bounce" rather than the acute bounce.

Figure 2.3 shows these possibilities in two dimensions. $\mathbf{a}'$ is the result $\mathbf{Qa}$ of the acute bounce with normal vector $\mathbf{n} = \mathbf{w}'$, and $a_x - a'_x$ is subtraction subject to cancellation error. If $\mathbf{n} = \mathbf{w}''$ is chosen, the result is $\mathbf{Qa} = \mathbf{a}''$, and $a_x - a''_x$ is addition. The vectors $\mathbf{w}'$ and $\mathbf{w}''$ are orthogonal.

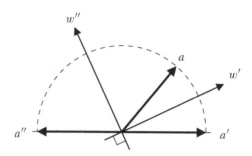

Figure 2.3 Geometric interpretation of the Householder matrix as a reflection.

The Householder algorithm works like this:

- Begin with $\mathbf{A}_1 = \mathbf{A}$ and $\mathbf{Q} = \mathbf{I}$.
- For $i = 1$ to n
  - $\mathbf{a}_i$ is the vector of length $n+1-i$ made up of the diagonal and subdiagonal elements of matrix $\mathbf{A}_i$, column i;
  - calculate k_i from (2.37) and $\mathbf{w}_i$ from (2.36);
  - $\mathbf{Q}_i$ is constructed according to (2.32) with (2.33);
  - $\mathbf{A}_{i+1} = \mathbf{Q}_i\mathbf{A}_i$ and $\mathbf{Q} := \mathbf{Q}\mathbf{Q}_i$.
- On completion, $\mathbf{R} = \mathbf{A}_n$, and $\mathbf{Q}$ is a unitary matrix such that $\mathbf{A} = \mathbf{QR}$.

There are two special qualities that recommend the QR method over competitors. First, since $\mathbf{Q}$ is unitary, we have $\mathrm{cond}(\mathbf{Q}) = 1$: the condition number associated with this transformation is as small as possible. Therefore, numerical errors are not amplified. Second, although $\mathbf{Q}$ is always square, $\mathbf{A}$ need not be. This will make the Householder method useful for the problem of least squares data fitting (Chapter 8).

As a practical matter, note that the matrices $\mathbf{Q}$ do not need to be constructed – one only needs to know the vector $\mathbf{w}$ that defines the matrix $\mathbf{Q}$. For example, the product $\mathbf{Qx}$ may be decomposed as follows. In block factored form,

$$\mathbf{Qx} = \mathbf{y}$$

$$\begin{pmatrix} \mathbf{I} & 0 \\ 0 & \mathbf{I} - 2\mathbf{ww}^{\mathrm{T}} \end{pmatrix} \begin{pmatrix} \mathbf{x}_1 \\ \mathbf{x}_2 \end{pmatrix} = \begin{pmatrix} \mathbf{y}_1 \\ \mathbf{y}_2 \end{pmatrix}$$

$$\mathbf{y}_1 = \mathbf{x}_1$$

$$d = \mathbf{w} \cdot \mathbf{x}_2 \qquad\qquad (2.38a)$$

$$\mathbf{y}_2 = \mathbf{x}_2 - 2d\mathbf{w}. \qquad\qquad (2.38b)$$

Example 2.6   Solve by Householder reduction

$$
\begin{pmatrix}
2 & -1 & 1 \\
\underline{-1} & 10^{-10} & 10^{-10} \\
\underline{1} & 10^{-10} & 10^{-10}
\end{pmatrix}
\begin{pmatrix}
x_1 \\
x_2 \\
x_3
\end{pmatrix}
=
\begin{pmatrix}
2(1 + 10^{-10}) \\
-10^{-10} \\
10^{-10}
\end{pmatrix}.
$$

Solution

The first step is to convert the first column of A (underlined),

$$
a_1 = \begin{pmatrix} 2 \\ -1 \\ 1 \end{pmatrix},
$$

into $k_1 e_1$. From (2.37) one calculates $k_1 = -2.4495$, then with (2.36) one calculates

$$
w_1 = \begin{pmatrix} 0.95302 \\ -0.21419 \\ 0.21419 \end{pmatrix}.
$$

Multiplying $Ax = b$ from the left by $Q = I - 2ww^T$ (column-wise, using (2.38a) and (2.38b), without actually making Q) one gets

$$
\begin{pmatrix}
-2.4495 & 0.81650 & -0.81650 \\
0 & -0.40825 & 0.40825 \\
0 & \underline{0.40825} & -0.40825
\end{pmatrix} x =
\begin{pmatrix}
-1.6330 \\
0.81650 \\
-0.81650
\end{pmatrix}. \tag{2.39}
$$

For the second step, we select the second column of (2.39) for processing. We are concerned only with the diagonal and below. So,

$$
a_2 = \begin{pmatrix} -0.40825 \\ 0.40825 \end{pmatrix},
$$

which gives $k_2 = 0.57735$ and

$$
w_2 = \begin{pmatrix} -0.92388 \\ 0.38268 \end{pmatrix}.
$$

Multiplying through by Q (block partitioned like (2.32)), equation (2.39) becomes

$$
\begin{pmatrix}
-2.4495 & 0.81650 & -0.81650 \\
0 & 0.57735 & -0.57735 \\
0 & 0 & 2.8284 \times 10^{-10}
\end{pmatrix} x =
\begin{pmatrix}
-1.6330 \\
-1.1547 \\
2.8284 \times 10^{-10}
\end{pmatrix}.
$$

At this point, the problem may be solved by back substitution:

$$
x = \begin{pmatrix} 1.0000 \times 10^{-10} \\ -1.0000 \\ 1.0000 \end{pmatrix}.
$$

In an early application of Householder reduction to least squares problems, Golub [90] noted that the numerical accuracy of the method can be improved somewhat if at each stage of the reduction columns are permuted in such a way as to make $(\mathbf{A}_k)_{kk}$, the emerging diagonal of $\mathbf{R}$, be as large as possible. This makes $\mathbf{R}$ diagonally dominant in the same way that complete pivoting does for Gaussian elimination, and thereby reduces the errors associated with back substitution. Through an analogy to Cholesky decomposition, Section 2.5, he showed that this is accomplished by choosing for column k the column q that maximizes

$$\sum_{i=k,n} \mathbf{A}_{iq}^2.$$

Subsequent analysis by Powell and Reid [186] showed that row permutations also aid numerical stability. As an example [44, 186], when $|\lambda| \gg 1$ the first column of matrix

$$\mathbf{A}_1 = \begin{pmatrix} 0 & 2 & 1 \\ \lambda & \lambda & 0 \\ \lambda & 0 & \lambda \\ 0 & 1 & 1 \end{pmatrix}$$

satisfies Golub's criterion. Proceeding with the reduction in exact math then rounding,

$$\mathbf{A}_2 = \begin{pmatrix} -\lambda\sqrt{2} & -\lambda/\sqrt{2} & -\lambda/\sqrt{2} \\ 0 & \frac{1}{2}\lambda - \sqrt{2} & -\frac{1}{2}\lambda - 1/\sqrt{2} \\ 0 & -\frac{1}{2}\lambda - \sqrt{2} & \frac{1}{2}\lambda - 1/\sqrt{2} \\ 0 & 1 & 1 \end{pmatrix} \xrightarrow{\text{round}} \begin{pmatrix} -\lambda\sqrt{2} & -\lambda/\sqrt{2} & -\lambda/\sqrt{2} \\ 0 & \frac{1}{2}\lambda & -\frac{1}{2}\lambda \\ 0 & -\frac{1}{2}\lambda & \frac{1}{2}\lambda \\ 0 & 1 & 1 \end{pmatrix},$$

where the rounding step assumes $\lambda > 2\sqrt{2}/\epsilon$. This rounded matrix is the same as would have been obtained from matrix

$$\tilde{\mathbf{A}}_1 = \begin{pmatrix} 0 & 0 & 0 \\ \lambda & \lambda & 0 \\ \lambda & 0 & \lambda \\ 0 & 1 & 1 \end{pmatrix},$$

so the effect of rounding in this case is equivalent to the loss of the entire first row of the initial matrix. The remedy proposed by Powell and Reid [186] is to select at each stage k, the row i that maximizes $|\mathbf{A}_k|_{ik}$; essentially a partial pivot step as in Gaussian elimination. With this reordering, the first step is

$$\begin{pmatrix} \lambda & \lambda & 0 \\ 0 & 2 & 1 \\ \lambda & 0 & \lambda \\ 0 & 1 & 1 \end{pmatrix} \rightarrow \begin{pmatrix} -\lambda\sqrt{2} & -\lambda/\sqrt{2} & -\lambda/\sqrt{2} \\ 0 & 2 & 1 \\ 0 & -\lambda/\sqrt{2} & \lambda/\sqrt{2} \\ 0 & 1 & 1 \end{pmatrix}.$$

No column ordering is indicated, but a row exchange is indicated, giving

$$
\begin{pmatrix}
-\lambda\sqrt{2} & -\lambda/\sqrt{2} & -\lambda/\sqrt{2} \\
0 & -\lambda/\sqrt{2} & \lambda/\sqrt{2} \\
0 & 2 & 1 \\
0 & 1 & 1
\end{pmatrix}
\rightarrow
\begin{pmatrix}
-\lambda\sqrt{2} & -\lambda/\sqrt{2} & -\lambda/\sqrt{2} \\
0 & \lambda/\sqrt{2} & -\lambda/\sqrt{2} \\
0 & 0 & 3 \\
0 & 0 & 2
\end{pmatrix}
$$

$$
\rightarrow
\begin{pmatrix}
-\lambda\sqrt{2} & -\lambda/\sqrt{2} & -\lambda/\sqrt{2} \\
0 & \lambda/\sqrt{2} & -\lambda/\sqrt{2} \\
0 & 0 & -\sqrt{13} \\
0 & 0 & 0
\end{pmatrix}.
$$

This result is equivalent to performing the decomposition with exact math, then rounding the final result. Thus, row pivoting is essential to minimize the error associated with this decomposition.

Björck [17] hypothesized that one could sort the rows of A_1 prior to reduction, in decreasing order of L^∞ magnitude. Cox and Higham [44] showed that the backward error bound for Powell–Reid pivoting is the same as the backward error bound for Björck pre-ordering.

Wilkinson [243] derived a backward error estimate for triangularization with Householder matrices. A slightly different columnwise analysis is provided by Higham [104] and presented here. If $E = \hat{Q}\hat{R} - A$, then for $A \in \mathbb{R}^{m \times n}$,

$$
\|Ee_j\|_2 \leq \frac{cnm\epsilon}{1 - cnm\epsilon} \|Ae_j\|_2, \quad j = 1, ..., n, \tag{2.40}
$$

where c is a small constant independent of A and its dimensions. And, for the solution of $Ax = b$ ($A \in \mathbb{R}^{n \times n}$), the numerical solution $\hat{x}$ satisfies

$$
(A + E)\hat{x} = b + \Delta b,
$$

with E given by (2.40) with $m = n$, and

$$
\|\Delta b\|_2 \leq \frac{cn^2\epsilon}{1 - cn^2\epsilon} \|b\|_2.
$$

2.5   Cholesky decomposition

If A is symmetric positive definite, then a left triangular matrix L exists such that

$$
A = LL^H.
$$

Here the Hermitian transpose is indicated to allow for the possibility that A is complex. This decomposition was discovered by A. L. Cholesky [13] to facilitate solving least squares problems in surveying by the normal equations.

$$\mathbf{L} = \begin{pmatrix} \ \end{pmatrix} \qquad \mathbf{L}^{\mathrm{H}} = \begin{pmatrix} \ \end{pmatrix}.$$

Here $\mathbf{L}$ has the same shape as the matrix $\mathbf{L}$ in Gaussian elimination, but now we allow $\mathbf{L}$ to have any value on the diagonal.

If $\mathbf{A} = \mathbf{L}\mathbf{L}^{\mathrm{H}}$ then

$$a_{ij} = \sum_{k=1}^{n} \ell_{ik}\ell_{jk}^* = \sum_{k=1}^{\min(i,j)} \ell_{ik}\ell_{jk}^*,$$

and expanded out:

$$
\begin{aligned}
a_{11} &= \ell_{11}\ell_{11}^* \\
a_{12} &= \ell_{11}\ell_{21}^* \\
&\ \ \vdots \\
a_{1n} &= \ell_{11}\ell_{n1}^* \\
a_{22} &= \ell_{21}\ell_{21}^* + \ell_{22}\ell_{22}^* \\
&\ \ \vdots \\
a_{2n} &= \ell_{21}\ell_{n1}^* + \ell_{22}\ell_{n2}^* \\
a_{33} &= \ell_{31}\ell_{31}^* + \ell_{32}\ell_{32}^* + \ell_{33}\ell_{33}^* \\
&\ \ \vdots \\
a_{3n} &= \ell_{31}\ell_{n1}^* + \ell_{32}\ell_{n2}^* + \ell_{33}\ell_{n3}^* \\
&\ \ \vdots
\end{aligned}
\tag{2.41}
$$

In each of the equations (2.41) there is only one new unknown element of $\mathbf{L}$, so $\mathbf{L}$ can be determined by analyzing these equations line by line.

Assuming now that $\mathbf{A}$, hence $\mathbf{L}$, is real:

$$
\begin{aligned}
\ell_{11} &= \sqrt{a_{11}} \\
\ell_{21} &= a_{12}/\ell_{11} \\
&\ \ \vdots \\
\ell_{n1} &= a_{1n}/\ell_{11} \\
\ell_{22} &= \sqrt{a_{22} - \ell_{21}^2}
\end{aligned}
$$

$$\vdots$$

$$\ell_{n2} = (a_{2n} - \ell_{21}\ell_{n1})/\ell_{22}$$

$$\ell_{33} = \sqrt{a_{33} - \ell_{31}^2 - \ell_{32}^2}$$

$$\vdots$$

$$\ell_{n3} = (a_{3n} - \ell_{31}\ell_{n1} - \ell_{32}\ell_{n2})/\ell_{33}$$

$$\vdots$$

In general, for $i = 1$ to n,

- solve for ℓ_{ii}:

$$\ell_{ii} = \sqrt{a_{ii} - \sum_{k=1}^{i-1} \ell_{ik}^2};$$

- for $j = i$ to n, solve for ℓ_{ji}:

$$\ell_{ji} = \left(a_{ij} - \sum_{k=1}^{i-1} \ell_{jk}\ell_{ik} \right) /\ell_{ii}.$$

Example 2.7 Use Cholesky decomposition to solve $\mathbf{Ax} = \mathbf{b}$ with

$$\mathbf{A} = \begin{pmatrix} 4 & -1 & 0 & 0 \\ -1 & 4 & -1 & 0 \\ 0 & -1 & 4 & -1 \\ 0 & 0 & -1 & 4 \end{pmatrix} \qquad \mathbf{b} = \begin{pmatrix} 1 \\ 2 \\ 2 \\ 1 \end{pmatrix}.$$

Solution
First, find the matrix $\mathbf{L}$:

$$\ell_{11} = \sqrt{a_{11}} = 2$$

$$\ell_{21} = a_{12}/\ell_{11} = -\frac{1}{2}$$

$$\ell_{31} = a_{13}/\ell_{11} = 0$$

$$\ell_{41} = a_{14}/\ell_{11} = 0$$

$$\ell_{22} = \sqrt{a_{22} - \ell_{21}^2} = \frac{1}{2}\sqrt{15}$$

$$\ell_{32} = (a_{23} - \ell_{21}\ell_{31})/\ell_{22} = -2\sqrt{\frac{1}{15}}$$

$$\ell_{42} = (a_{24} - \ell_{21}\ell_{41})/\ell_{22} = 0$$

$$\ell_{33} = \sqrt{a_{33} - \ell_{31}^2 - \ell_{32}^2} = 2\sqrt{\frac{14}{15}}$$

$$\ell_{43} = (a_{34} - \ell_{31}\ell_{41} - \ell_{32}\ell_{42})/\ell_{33} = -\frac{1}{2}\sqrt{\frac{15}{14}}$$

$$\ell_{44} = \sqrt{a_{44} - \ell_{41}^2 - \ell_{42}^2 - \ell_{43}^2} = \frac{1}{2}\sqrt{\frac{209}{14}},$$

or

$$\mathbf{L} = \begin{pmatrix} 2.0000 & 0 & 0 & 0 \\ -0.500\,00 & 1.9365 & 0 & 0 \\ 0 & -0.516\,40 & 1.9322 & 0 \\ 0 & 0 & -0.517\,55 & 1.9382 \end{pmatrix}.$$

The problem $\mathbf{LL}^H\mathbf{x} = \mathbf{b}$ is now decomposed into the two back substitution steps

$$\mathbf{Ly} = \mathbf{b}$$
$$\mathbf{L}^H\mathbf{x} = \mathbf{y},$$

which give

$$\mathbf{y} = \begin{pmatrix} 0.500\,00 \\ 1.1619 \\ 1.3456 \\ 0.878\,13 \end{pmatrix}$$

$$\mathbf{x} = \begin{pmatrix} 0.454\,55 \\ 0.818\,18 \\ 0.818\,18 \\ 0.454\,55 \end{pmatrix}.$$

Wilkinson [242] derived the error associated with the Cholesky decomposition. With $\mathbf{E} = \hat{\mathbf{L}}\hat{\mathbf{L}}^T - \mathbf{A}$,

$$|\mathbf{E}| \leq \frac{\epsilon}{2} \begin{pmatrix} 2\ell_{11} & \ell_{11} & \ell_{11} & \ell_{11} \\ \ell_{11} & 2\ell_{22} & \ell_{22} & \ell_{22} \\ \ell_{11} & \ell_{22} & 2\ell_{33} & \ell_{33} \\ \ell_{11} & \ell_{22} & \ell_{33} & 2\ell_{44} \end{pmatrix}.$$

For solution of $\mathbf{LL}^T\mathbf{x} = \mathbf{b}$, the two back substitution steps have the errors already given in (2.27) and (2.28).

2.6 The residual correction method

The idea behind the *residual correction method*, a.k.a. the *method of iterative refinement*, is to detect the errors committed in solving $\mathbf{Ax} = \mathbf{b}$, then correct the solution $\mathbf{x}$ to compensate for these errors. Specifically,

$$\mathbf{Ax} \approx \mathbf{b} \qquad\qquad \text{imperfect solution,} \qquad (2.42a)$$

$$\mathbf{r} = \mathbf{b} - \mathbf{Ax} \qquad\qquad \text{measure of error,} \qquad (2.42b)$$

$$\mathbf{A}\delta\mathbf{x} = \mathbf{r} \qquad\qquad \text{correction,} \qquad (2.42c)$$

$$\mathbf{x} := \mathbf{x} + \delta\mathbf{x} \qquad\qquad \text{compensate for error.} \qquad (2.42d)$$

Here, (2.42a) represents the numerical determination of $\mathbf{x}$ by, for example, Gaussian elimination or Householder reduction. Because of numerical errors, the *residual* $\mathbf{r}$ (2.42b) may not be zero. We use the residual as a new right-hand side in (2.42c) to find a vector of corrections $\delta\mathbf{x}$. If an LR or QR decomposition of the matrix is made in step (2.42a), then that information can be used in step (2.42c) to expedite solution. The corrected solution (2.42d) is expected to be improved:

$$\mathbf{A}(\mathbf{x} + \delta\mathbf{x}) = \underbrace{\mathbf{Ax}}_{\mathbf{b}-\mathbf{r}} + \underbrace{\mathbf{A}\delta\mathbf{x}}_{\mathbf{r}} = \mathbf{b}.$$

Iterative refinement was first described by Wilkinson [242] in the context of Gaussian elimination, and Jankowski and Woźniakowski [120] generalized the approach to other solution techniques. They show that when the product of relative solution error and condition number is less than unity,

$$\frac{\|\Delta\mathbf{x}\|}{\|\mathbf{x}\|}\mathrm{cond}(\mathbf{A}) < 1,$$

iterative refinement is stable. Since the error in the initial solution will be of order $\mathrm{cond}(\mathbf{A})\epsilon$, the stability requirement can also be stated

$$\mathrm{cond}(\mathbf{A})^2 < \mathcal{O}(\epsilon^{-1}).$$

Example 2.8   Use iterative refinement to improve the solution obtained in Example 2.5 with partial pivoting.

Solution
The solution

$$\mathbf{x} = \begin{pmatrix} 0.0 \\ 0.0 \\ 1.0 \end{pmatrix}$$

gives a residual vector

$$\mathbf{r} = \begin{pmatrix} 2 \times 10^9 + 2 \\ 10^9 + 1 \\ 3 \end{pmatrix} - \begin{pmatrix} 1 & 1 & 2 \times 10^9 \\ 2 & -1 & 10^9 \\ 1 & 2 & 0 \end{pmatrix}\begin{pmatrix} 0.0 \\ 0.0 \\ 1.0 \end{pmatrix} = \begin{pmatrix} 0.0 \\ 0.0 \\ 3.0 \end{pmatrix},$$

again rounding all calculations to two decimal places. Now determine the correction

$$\begin{pmatrix} 1.0 & & \\ 0.50 & 1.0 & \\ 0.50 & 0.60 & 1.0 \end{pmatrix} \left(\begin{pmatrix} 2.0 & -1.0 & 1.0 \times 10^9 \\ & 2.5 & -5.0 \times 10^8 \\ & & 1.8 \times 10^9 \end{pmatrix} \begin{pmatrix} \delta x_1 \\ \delta x_2 \\ \delta x_3 \end{pmatrix} \right) = \begin{pmatrix} 0 & 1 & 0 \\ 0 & 0 & 1 \\ 1 & 0 & 0 \end{pmatrix} \begin{pmatrix} 0.0 \\ 0.0 \\ 3.0 \end{pmatrix}$$

$$\begin{pmatrix} \delta x_1 \\ \delta x_2 \\ \delta x_3 \end{pmatrix} = \begin{pmatrix} 1.0 \\ 1.0 \\ -1.0 \times 10^{-9} \end{pmatrix},$$

using two decimal arithmetic. The corrected solution $\mathbf{x} + \delta\mathbf{x} = (1.0, \ 1.0, \ 1.0)^\mathsf{T}$ is perfect.

The solution obtained in Example 2.5 with implicit scaled pivoting is not improved by iterative refinement because with 2 decimal arithmetic the residual vector is zero, although the difference $\mathbf{x} - \mathbf{x}_{\text{exact}}$ is large. In the present circumstance this is only because of the severe rounding, but in general caution must be used when using $\mathbf{r}$ as a measure of solution quality.

The next chapter introduces some ideas that help understand why a small residual does not always signify a small error (see Problem 3.4). It also helps set the stage for a variety of purely iterative approaches for solving $\mathbf{Ax} = \mathbf{b}$ without any elimination or matrix decomposition. Those methods will be taken up in Chapter 4

Problems

2.1 Show
(i) $\text{cond}_2(\mathbf{A}) \le n \ \text{cond}_\infty(\mathbf{A})$, and
(ii) $\text{lub}_\infty(\mathbf{A}) = \max_i \sum_k |A_{ik}|$.

2.2 Use Problem 2.1 and representation (2.1) to show

$$\text{cond}_\infty(\mathbf{A}^{(d)}) \le 4 \ \text{cond}_\infty(\mathbf{A}^{(d-1)})$$
$$\text{cond}_2(\mathbf{A}^{(d)}) \le 4n \ \text{cond}_2(\mathbf{A}^{(d-1)})$$

using partial pivoting.

2.3 The Vandermonde matrix $\mathbf{V}(\mathbf{y})$, $\mathbf{y} \in \mathbb{R}^n$, is

$$\mathbf{V}(\mathbf{y}) = \begin{pmatrix} 1 & y_1 & y_1^2 & \cdots & y_1^{n-1} \\ 1 & y_2 & y_2^2 & \cdots & y_2^{n-1} \\ 1 & y_3 & y_3^2 & \cdots & y_3^{n-1} \\ \vdots & \vdots & \vdots & \ddots & \vdots \\ 1 & y_n & y_n^2 & \cdots & y_n^{n-1} \end{pmatrix}. \tag{2.43}$$

For $n = 5, 10, 15, 20$ let $\mathbf{y}$ be given by

$$y_i = \frac{n-i}{n-1}, \quad i = 1, \ldots, n,$$

and let

$$b_i = \sum_{j=0}^{n-1} y_i^j,$$

i.e., such that $\mathbf{V}\mathbf{x} = \mathbf{b}$ has exact solution $\mathbf{x} = \mathbf{1}$. Solve these problems with Gaussian elimination and Householder reduction with a variety of pivoting methods.

2.4   Show that pivoting is not required for Gaussian elimination if

$$|a_{ii}| > \sum_{j \neq i} |a_{ij}|.$$

2.5   [69] Let $\mathbf{y}$ and $\mathbf{b}$ be any two vectors in $\mathbb{R}^n$, and define

$$\mathbf{e} = \mathbf{y} - \mathbf{A}^{-1}\mathbf{b} = -\mathbf{A}^{-1}\mathbf{r}$$
$$\mathbf{r} = \mathbf{b} - \mathbf{A}\mathbf{y} = -\mathbf{A}\mathbf{e}$$

and define

$$\rho_x = \frac{\|\mathbf{e}\|}{\|\mathbf{A}^{-1}\mathbf{b}\|}$$
$$\rho_r = \frac{\|\mathbf{r}\|}{\|\mathbf{b}\|}.$$

Show that

$$\frac{1}{\text{cond}(\mathbf{A})} \leq \frac{\rho_x}{\rho_r} \leq \text{cond}(\mathbf{A}),$$

which may be used to bound the condition number of $\mathbf{A}$.

3 Eigenvalues and eigenvectors

The eigenvalues of matrix $\mathbf{A}$ are the numbers λ that satisfy

$$\mathbf{A}\mathbf{x} = \lambda\mathbf{x}, \tag{3.1}$$

for some $\mathbf{x} \neq 0$, and the vectors $\mathbf{x}$ are the eigenvectors corresponding to eigenvalue λ.

The are many computational approaches for determining eigenvalues and eigenvectors. In this chapter the emphasis will be on methods that involve manipulations of the matrix $\mathbf{A}$. Another set of approaches is based on seeking roots of the characteristic polynomial

$$\phi(\lambda) = \det(\mathbf{A} - \lambda\mathbf{I}).$$

One such method due to Hyman will be described.

When talking about eigenvalues λ and eigenvectors $\mathbf{x}$ it is not uncommon to begin with the assumption that $n \times n$ matrix $\mathbf{A}$ has n eigenvalues and n eigenvectors, each obeying (3.1). If there are n eigenvectors, we can organize them column-wise to make a matrix $\mathbf{X}$, and make $\Lambda = \mathrm{diag}(\lambda_1, \ldots, \lambda_n)$ to give

$$\mathbf{A}\mathbf{X} = \mathbf{X}\Lambda$$

$$\mathbf{A} = \mathbf{X}\Lambda\mathbf{X}^{-1}.$$

Such matrices are called *diagonalizable*.

Diagonal matrices are diagonalizable, permutation matrices are diagonalizable, and "*normal*" matrices are diagonalizable. A matrix $\mathbf{A}$ is normal if $\mathbf{A}^H\mathbf{A} = \mathbf{A}\mathbf{A}^H$. Unitary, Hermitian, and skew-Hermitian matrices are normal.

The *spectral radius* of matrix $\mathbf{A}$, written $\rho(\mathbf{A})$, is the greatest magnitude of the eigenvalues:

$$\rho(\mathbf{A}) = \max_{i=1}^{n} |\lambda_i|.$$

It is bounded by the subordinate matrix norm for any vector norm: $\rho(\mathbf{A}) \leq \mathrm{lub}(\mathbf{A})$.

It is easy to show that not all $n \times n$ matrices have n eigenvectors. The simplest example

$$\mathbf{C} = \begin{pmatrix} \lambda & 1 & & & \\ & \lambda & 1 & & \\ & & \ddots & \ddots & \\ & & & \lambda & 1 \\ & & & & \lambda \end{pmatrix} \tag{3.2}$$

is called a *Jordan block*. Since $n \times n$ matrix $\mathbf{C}$ is triangular it should be obvious that the number λ is the eigenvalue of multiplicity n.

If $\mathbf{C}$ has an eigenvector $\mathbf{x}$ it will satisfy $(\mathbf{C} - \lambda \mathbf{I})\mathbf{x} = 0$, or

$$
\begin{pmatrix} 0 & 1 & & & \\ & 0 & 1 & & \\ & & \ddots & \ddots & \\ & & & 0 & 1 \\ & & & & 0 \end{pmatrix} \begin{pmatrix} x_1 \\ x_2 \\ \vdots \\ x_{n-1} \\ x_n \end{pmatrix} = \begin{pmatrix} x_2 \\ x_3 \\ \vdots \\ x_n \\ 0 \end{pmatrix} = \begin{pmatrix} 0 \\ 0 \\ \vdots \\ 0 \\ 0 \end{pmatrix}, \tag{3.3}
$$

and the solution is

$$
\mathbf{x} = \begin{pmatrix} x_1 \\ 0 \\ \vdots \\ 0 \\ 0 \end{pmatrix}.
$$

We can pick any $x_1 \neq 0$ to get an eigenvector, but only one, so $\mathbf{C}$ is missing $n - 1$ eigenvectors. The vectors $\mathbf{e}_2, \ldots, \mathbf{e}_n$ are called the *"principal vectors."* The set {eigenvectors, principal vectors} spans the space $\mathbb{C}^n$.

An arbitrary square matrix $\mathbf{A}$ is similar to a matrix in the *Jordan normal form*,

$$
\mathbf{SAS}^{-1} = \begin{pmatrix} \mathbf{C}_1 & & & \\ & \mathbf{C}_2 & & \\ & & \ddots & \\ & & & \mathbf{C}_m \end{pmatrix}.
$$

Here the submatrices $\mathbf{C}_i$ are blocks, each of the form (3.3) with $\lambda = \lambda_i$. The number of blocks m is the number of eigenvectors possessed by $\mathbf{A}$. If all $\mathbf{C}_i$ are 1×1, then there are n eigenvectors and $\mathbf{A}$ is diagonalizable. An $n \times n$ matrix with fewer than n eigenvectors is called *defective*.

From this brief introduction it should be clear that the general problem of finding all eigenvalues and all eigenvectors of an arbitrary matrix is a difficult one. We begin with an approximate technique, Gerschgorin's estimate, by which a crude estimate of eigenvalues can be had with little effort. Next, power methods are described. Complexities due to complex eigenvalues, repeated eigenvalues, and eigenvector deficiencies are explored. The QR method is then discussed with less rigor. Its resemblance to power methods is emphasized, and enough details are given to create a respectable implementation for symmetric matrices with distinct eigenvalues. The closely related singular value decomposition algorithm, SVD, is also briefly discussed.

3.1   Gerschgorin's estimate

Gerschgorin's theorem provides a very quick estimate of the distribution of eigenvalues in the complex plane, a.k.a. the spectrum of the eigenvalues.

Let λ be an eigenvalue of square matrix $\mathbf{A}$, and let $\mathbf{x}$ be the corresponding eigenvector. Then

$$\mathbf{A}\mathbf{x} = \lambda\mathbf{x}$$

or, in terms of elements of $\mathbf{A}$ and $\mathbf{x}$,

$$a_{i1}x_1 + \cdots + a_{i,i-1}x_{i-1} + a_{ii}x_i + a_{i,i+1}x_{i+1} + \cdots + a_{in}x_n = \lambda x_i. \tag{3.4}$$

Let's pick i to be the index of the largest element of vector $\mathbf{x}$:

$$|x_i| \geq |x_j| \quad \forall\, j, \tag{3.5}$$

("$\forall$" means "for all") and assume that $x_i \neq 0$ (if $x_i = 0$ then (3.5) means that $\mathbf{x} = 0$, which is the trivial solution to the eigenvector problem). Then, (3.4) may be rearranged to give

$$(\lambda - a_{ii})x_i = a_{i1}x_1 + \cdots + a_{i,i-1}x_{i-1} + a_{i,i+1}x_{i+1} + \cdots + a_{in}x_n$$

$$(\lambda - a_{ii}) = a_{i1}\frac{x_1}{x_i} + \cdots + a_{i,i-1}\frac{x_{i-1}}{x_i} + a_{i,i+1}\frac{x_{i+1}}{x_i} + \cdots + a_{in}\frac{x_n}{x_i}$$

$$|\lambda - a_{ii}| \leq |a_{i1}|\left|\frac{x_1}{x_i}\right| + \cdots + |a_{i,i-1}|\left|\frac{x_{i-1}}{x_i}\right| + |a_{i,i+1}|\left|\frac{x_{i+1}}{x_i}\right| + \cdots + |a_{in}|\left|\frac{x_n}{x_i}\right|$$

using the triangle inequality in the last step. Now with condition (3.5) each of the factors $|x_j/x_i|$ must be less than 1. Therefore

$$|\lambda - a_{ii}| \leq \sum_{j \neq i} |a_{ij}|. \tag{3.6}$$

Geometrically, (3.6) says that for some index i there is an eigenvalue λ that lies inside a circle in the complex plane centered at a_{ii}, and with radius $\sum_{j \neq i} |a_{ij}|$.

Let disk D_i be the set of complex numbers μ that satisfy the equation $|\mu - a_{ii}| \leq \sum_{j \neq i} |a_{ij}|$:

$$D_i = \left\{ \mu \in \mathbb{C} \mid\, ; |\mu - a_{ii}| \leq \sum_{j \neq i} |a_{ij}| \right\}. \tag{3.7}$$

We do not know that any particular disk D_i contains an eigenvalue, because we do not know that the particular index i satisfies (3.5) for any eigenvector. However, we do know that every eigenvector has *some* index i that satisfies (3.5), and so if we take the union of all disks $\bigcup_{i=1}^{n} D_i$ we are sure to have identified the whereabouts of all the eigenvalues. This is Gerschgorin's theorem [84]: every eigenvalue of the matrix $\mathbf{A}$ lies inside the union of disks D_i (3.7).

Now consider the matrix

$$\mathbf{B}(t) = \mathbf{D} + t(\mathbf{A} - \mathbf{D})$$

$$\mathbf{D} = \mathrm{diag}(a_{11}, a_{22}, \ldots, a_{nn}),$$

and let's assume that the diagonal elements are distinct. $\mathbf{B}$ is a smoothly varying function of parameter t. When $t = 1$, $\mathbf{B} = \mathbf{A}$. Since $\mathbf{B}(t)$ is a smoothly varying function of t, we expect the eigenvalues of $\mathbf{B}(t)$ to also be smoothly varying functions of t.

When $t = 0$, **B** is the diagonal matrix **D** consisting of the diagonal elements of **A**. The eigenvalues of an arbitrary diagonal matrix are the diagonal elements of the matrix d_{ii} (and the associated eigenvectors are the unit basis vectors e_i). Therefore, the eigenvalues of **B**(0) will be a_{11}, a_{22}, ..., a_{nn}. In terms of Gerschgorin's disks, the disks of **B**(0) are just points in the complex plane. Associated with each such point is one eigenvalue.

As t increases from 0, the disk radii will grow, and disks that did not formerly overlap may begin to overlap. Until an overlap occurs, we know each disk will contain one eigenvalue.

If, when $t = 1$ (hence **B** = **A**) there is a disjoint set of m disks (the union of these m disks does not overlap the union of the remaining $n - m$ disks), then that set will contain at least m eigenvalues.

Example 3.1   Estimate the eigenvalues of the matrix

$$\begin{pmatrix} 5 & 1 & 1 \\ 0 & 1 & 1 \\ 0 & -1/2 & 0 \end{pmatrix}. \tag{3.8}$$

Solution

In Figure 3.1 the Gerschgorin disks for $B(t)$ are drawn for $t = 0, 0.5, 1$. For all t the disks are centered at 0, 1, 5 – the diagonal elements of the matrix. When $t = 0$, the disks have zero radius and there is one eigenvalue per disk. The eigenvalues are indicated by the bold dots. When $t = 0.5$, the disk radii have grown and the eigenvalues have changed. The disks do not overlap, so we still have one eigenvalue per disk. When $t = 1$ the disk centered at 0 and the disk centered at 1 overlap – we can no longer claim that there is one eigenvalue per disk. Indeed, the two eigenvalues associated with these disks reside only in the disk centered at 1. The disk centered at 5 remains isolated, and it still contains a single eigenvalue.

The Gerschgorin estimate states:

$$D_1 = \left\{ \mu \in \mathbb{C} \,\middle|\, |\lambda| \leq 1/2 \right\}$$

$$D_2 = \left\{ \mu \in \mathbb{C} \,\middle|\, |\lambda - 1| \leq 1 \right\}$$

$$D_3 = \left\{ \mu \in \mathbb{C} \,\middle|\, |\lambda - 5| \leq 2 \right\}$$

$$\lambda_1, \lambda_2 \in D_1 \cup D_2$$

$$\lambda_3 \in D_3.$$

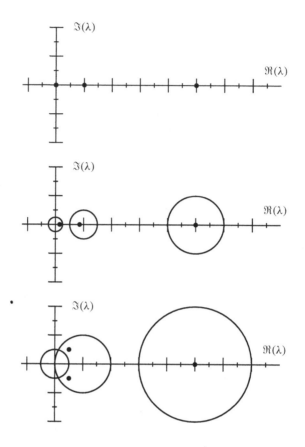

Figure 3.1 Gerschgorin disks for matrix (3.8) for $t = 0, 0.5, 1$; Example 3.1.

The similarity transformation $\mathbf{SAS}^{-1}$ changes the matrix but does not change its eigenvalues. By choosing a simple transformation such as

$$
\begin{pmatrix} d_1 & & & \\ & d_2 & & \\ & & \ddots & \\ & & & d_n \end{pmatrix}
\begin{pmatrix} a_{11} & a_{12} & \cdots & a_{1n} \\ a_{21} & a_{22} & \cdots & a_{2n} \\ \vdots & \vdots & \vdots & \vdots \\ a_{n1} & a_{n2} & \cdots & a_{nn} \end{pmatrix}
\begin{pmatrix} d_1^{-1} & & & \\ & d_2^{-1} & & \\ & & \ddots & \\ & & & d_n^{-1} \end{pmatrix}
$$

$$
= \begin{pmatrix} a_{11}\frac{d_1}{d_1} & a_{12}\frac{d_1}{d_2} & \cdots & a_{1n}\frac{d_1}{d_n} \\ a_{21}\frac{d_2}{d_1} & a_{22}\frac{d_2}{d_2} & \cdots & a_{2n}\frac{d_2}{d_n} \\ \vdots & \vdots & \vdots & \vdots \\ a_{n1}\frac{d_n}{d_1} & a_{n2}\frac{d_n}{d_2} & \cdots & a_{nn}\frac{d_n}{d_n} \end{pmatrix}
$$

it is possible to manipulate the radii of the Gerschgorin disks in order to make some smaller and others larger. By careful choice of the coefficients d_i one can improve the Gerschgorin estimates.

Example 3.2   Use scaling to improve the Gerschgorin estimates for matrix (3.8).

Solution
With

$$\mathbf{D} = \begin{pmatrix} d_1 & 0 & 0 \\ 0 & d_2 & 0 \\ 0 & 0 & d_3 \end{pmatrix}$$

matrix $\mathbf{DAD}^{-1}$ becomes

$$\begin{pmatrix} 5 & \frac{d_1}{d_2} & \frac{d_1}{d_3} \\ 0 & 1 & \frac{d_2}{d_3} \\ 0 & -\frac{d_3}{2d_2} & 0 \end{pmatrix}.$$

The disk centered at 5 can be made arbitrarily small by allowing $d_1 \rightarrow 0$. The disks centered at 0 and 1 cannot be adjusted in this way – making one small makes the other large. If we let the radii become equal, then $d_2 = \sqrt{2}$ and $d_3 = 2$, resulting in

$$\begin{pmatrix} 5 & 0 & 0 \\ 0 & 1 & \frac{\sqrt{2}}{2} \\ 0 & -\frac{\sqrt{2}}{2} & 0 \end{pmatrix}$$

with the disks shown in Figure 3.2. The disk areas are considerably sharpened relative to the unscaled results of Figure 3.1.

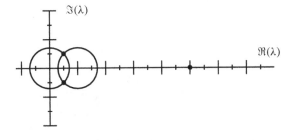

Figure 3.2 Scaled Gerschgorin disks for matrix (3.8); Example 3.2.

3.2   The power method

Some simple methods are available to estimate the eigenvalue of greatest magnitude and the eigenvalue of least magnitude. These are simple iterative techniques, and with some care the corresponding eigenvectors can be determined.

To begin, let us assume that $n \times n$ matrix $\mathbf{A}$ is diagonalizable: $\mathbf{A}$ has n distinct eigenvectors.

Let $\mathbf{z}^{(0)}$ be a random vector that is expressible in terms of eigenvectors $\mathbf{x}_i$ of matrix $\mathbf{A}$:

$$\mathbf{z}^{(0)} = \alpha_1 \mathbf{x}_1 + \alpha_2 \mathbf{x}_2 + \cdots + \alpha_n \mathbf{x}_n.$$

Since $\mathbf{z}$ is chosen randomly, we assume that none of the coefficients α is zero.

Next, compute the product $\mathbf{A}\mathbf{z}^{(0)}$. Since $\mathbf{A}\mathbf{x}_i = \lambda_i \mathbf{x}_i$ by definition of the eigenvalue λ_i and eigenvector $\mathbf{x}_i$,

$$\mathbf{z}^{(1)} = \mathbf{A}\mathbf{z}^{(0)} = \alpha_1 \lambda_1 \mathbf{x}_1 + \alpha_2 \lambda_2 \mathbf{x}_2 + \cdots + \alpha_n \lambda_n \mathbf{x}_n.$$

If we repeat this procedure iteratively, we obtain

$$\mathbf{z}^{(k)} = \mathbf{A}^k \mathbf{z}^{(0)} = \alpha_1 \lambda_1^k \mathbf{x}_1 + \alpha_2 \lambda_2^k \mathbf{x}_2 + \cdots + \alpha_n \lambda_n^k \mathbf{x}_n.$$

If $|\lambda_1| > |\lambda_2| > \cdots > |\lambda_n|$ then when k is big enough the term $\alpha_1 \lambda_1^k$ will be the greatest, and the other terms will be negligible by comparison. We therefore obtain

$$\lim_{k \to \infty} \mathbf{z}^{(k)} \propto \mathbf{x}_1$$

$$\lim_{k \to \infty} \frac{\mathbf{z}^{(k)\mathsf{T}} \mathbf{A} \mathbf{z}^{(k)}}{\mathbf{z}^{(k)\mathsf{T}} \mathbf{z}^{(k)}} = \lambda_1.$$

The term $\mathbf{z}^\mathsf{T} \mathbf{A} \mathbf{z} / \mathbf{z}^\mathsf{T} \mathbf{z}$ is called the *Rayleigh quotient*.

Of course, if $|\lambda_1| > 1$ then we also run into the problem that $|\mathbf{z}^{(k)}|$ will grow without bound. And, if $|\lambda_1| < 1$, then $\|\mathbf{z}^{(k)}\|$ will tend to zero. We can fix this problem by devising an algorithm that controls the size of $\mathbf{z}^{(k)}$. For convenience, we will renormalize $\mathbf{z}^{(n)}$ to have unit length at each step of the algorithm.

$$\mathbf{z}^{(0)} = \text{random}$$

$$\mathbf{z}^{(0)} := \frac{\mathbf{z}^{(0)}}{\sqrt{\mathbf{z}^{(0)} \cdot \mathbf{z}^{(0)}}}$$

$$\mathbf{z}^{(i+1)} = \mathbf{A}\mathbf{z}^{(i)}$$

$$\mathbf{z}^{(i+1)} := \frac{\mathbf{z}^{(i+1)}}{\sqrt{\mathbf{z}^{(i+1)} \cdot \mathbf{z}^{(i+1)}}} \qquad (3.9)$$

$$\lambda_1 = \lim_{n \to \infty} \mathbf{z}^{(n)\mathsf{T}} \mathbf{A} \mathbf{z}^{(n)}.$$

Note for future reference that the iteration (3.9) can be described by

$$\mathbf{z}^{(i+1)} = \text{normalize}\left(\mathbf{A}\mathbf{z}^{(i)}\right). \qquad (3.10)$$

This procedure is called the *power method*, or *simple vector iteration*.

The eigenvalues of $\mathbf{A}^{-1}$ are the inverse of the eigenvalues of $\mathbf{A}$. We could therefore obtain the smallest eigenvalue of $\mathbf{A}$ through an iteration $\mathbf{z}^{(n)} = \mathbf{A}^{-1}\mathbf{z}^{(n-1)}$, with appropriate renormalization, etc.

$$\mathbf{z}^{(0)} = \text{random}$$

$$\mathbf{z}^{(0)} := \frac{\mathbf{z}^{(0)}}{\sqrt{\mathbf{z}^{(0)} \cdot \mathbf{z}^{(0)}}}$$

$$\mathbf{z}^{(i+1)} = \mathbf{A}^{-1}\mathbf{z}^{(i)}$$

$$\mathbf{z}^{(i+1)} := \frac{\mathbf{z}^{(i+1)}}{\sqrt{\mathbf{z}^{(i+1)} \cdot \mathbf{z}^{(i+1)}}}$$

$$\lambda_n^{-1} = \lim_{n \to \infty} \mathbf{z}^{(n)\mathrm{T}}\mathbf{A}^{-1}\mathbf{z}^{(n)},$$

or

$$\mathbf{z}^{(i+1)} = \text{normalize}\left(\mathbf{A}^{-1}\mathbf{z}^{(i)}\right), \tag{3.11}$$

in analogy to (3.10). This inverse iteration procedure assumes that $\mathbf{A}$ is invertible so $\lambda_n \neq 0$, while the forward iteration method did not make this assumption.

The inverse iteration method [238] can be adapted to singular matrices, and to finding arbitrary eigenvalues/eigenvectors, by the following observation: if λ is an eigenvalue of $\mathbf{A}$, then $\lambda - \mu$ is an eigenvalue of $(\mathbf{A} - \mu\mathbf{I})$. This is referred to as *eigenvalue shifting*.

Now, let $\mu \approx \lambda_i$, a particular eigenvalue, but with μ not equal to any eigenvalue of $\mathbf{A}$. With these assumptions, $(\mathbf{A} - \mu\mathbf{I})$ will be invertible, and $\lambda_i - \mu$ can be found by inverse iteration. We will also determine the corresponding eigenvector $\mathbf{x}_i$.

This procedure requires an estimate of the eigenvalues. If the characteristic equation of $\mathbf{A}$ is known, root finding can provide the requisite estimates. The Gerschgorin analysis also provides approximations.

Example 3.3 The $n \times n$ matrix

$$\begin{pmatrix} 2 & -1 & & & & \\ -1 & 2 & \ddots & & & \\ & -1 & \ddots & -1 & & \\ & & \ddots & 2 & -1 \\ & & & -1 & 2 \end{pmatrix}$$

has eigenvalues $\lambda_i = 2(1 - \cos\frac{i}{n+1}\pi)$, for $i = 1, n$. Find the eigenvector corresponding to λ_4 with $n = 10$.

Solution

Define the shift $\mu = \lambda_4 - \epsilon$, where ϵ is much smaller than the separation between eigenvalues, but large enough that the shifted matrix will be nonsingular. We will try

$\epsilon = 10^{-4}$. This should make λ_4' (the fourth eigenvalue of $(\mathbf{A} - \mu\mathbf{I})^{-1}$) be the largest in magnitude, and finite.

Begin the inverse vector iteration process with "random" normalized vector $\mathbf{z} = (1/\sqrt{10}, ..., 1/\sqrt{10})$. The inverse iteration is performed by first constructing the QR Householder decomposition of $(\mathbf{A} - \mu\mathbf{I})$, then

$$(\mathbf{A} - \mu\mathbf{I})\mathbf{z}^{(k+1)} = \mathbf{z}^{(k)}$$

is implemented by solving

$$\mathbf{R}\mathbf{z}^{(k+1)} = \mathbf{Q}^T\mathbf{z}^{(k)}$$

by back substitution – the QR decomposition is only done once.

After five iterations, the observed Rayleigh quotient $\mathbf{z}^T\mathbf{A}^{-1}\mathbf{z}$ is $10\,000$ ($1/\epsilon$ – the expected value) and $\mathbf{z}^{(5)}$ converges to eigenvector $\mathbf{x}_4$:

$$\mathbf{z}^{(5)} = \begin{pmatrix} 0.387\,87 \\ 0.322\,25 \\ -0.120\,13 \\ -0.422\,06 \\ -0.230\,53 \\ 0.230\,53 \\ 0.422\,06 \\ 0.120\,13 \\ -0.322\,25 \\ -0.387\,87 \end{pmatrix}.$$

To understand the limitations of vector iteration it is useful to consider some special cases. First, what if $\lambda_1 = \lambda_2$, with $|\lambda_1| > |\lambda_3| \geq |\lambda_4| \geq \cdots$? In this case, we will recover exactly one eigenvector, even if there are two available. If a second exists, one way to find it is through orthogonalization. Let $\mathbf{x}_1$ be an eigenvector associated with λ_1, λ_2. Then,

$$\tilde{\mathbf{z}} = \mathbf{z} - \frac{\mathbf{x}_1 \cdot \mathbf{z}}{\mathbf{x}_1 \cdot \mathbf{x}_1}\mathbf{x}_1 \tag{3.12}$$

will be orthogonal to $\mathbf{x}_1$:

$$\tilde{\mathbf{z}} \cdot \mathbf{x}_1 = \mathbf{z} \cdot \mathbf{x}_1 - \frac{\mathbf{x}_1 \cdot \mathbf{z}}{\mathbf{x}_1 \cdot \mathbf{x}_1}\mathbf{x}_1 \cdot \mathbf{x}_1 = 0.$$

This procedure it called Gram–Schmidt orthogonalization after [95, 210]. It provides an opportunity to find $\mathbf{x}_2$, a second eigenvector associated with λ_1, λ_2:

$$\mathbf{z}^{(0)} = \text{random}$$

$$\mathbf{z}^{(0)} := \mathbf{z}^{(0)} - \frac{\mathbf{x}_1 \cdot \mathbf{z}^{(0)}}{\mathbf{x}_1 \cdot \mathbf{x}_1} \mathbf{x}_1$$

$$\mathbf{z}^{(0)} := \frac{\mathbf{z}^{(0)}}{\sqrt{\mathbf{z}^{(0)} \cdot \mathbf{z}^{(0)}}}$$

$$\mathbf{z}^{(i+1)} = \mathbf{A}\mathbf{z}^{(i)}$$

$$\mathbf{z}^{(i+1)} := \mathbf{z}^{(i+1)} - \frac{\mathbf{x}_1 \cdot \mathbf{z}^{(i+1)}}{\mathbf{x}_1 \cdot \mathbf{x}_1} \mathbf{x}_1 \tag{3.13}$$

$$\mathbf{z}^{(i+1)} := \frac{\mathbf{z}^{(i+1)}}{\sqrt{\mathbf{z}^{(i+1)} \cdot \mathbf{z}^{(i+1)}}}$$

$$\lambda_2 = \lim_{n \to \infty} \mathbf{z}^{(n)\mathrm{T}} \mathbf{A}\mathbf{z}^{(n)}$$

$$\mathbf{x}_2 = \lim_{n \to \infty} \mathbf{z}^{(n)}.$$

Since the iteration combines orthogonalization and normalization, we can write (3.13) as

$$\mathbf{z}^{(i+1)} = \text{orthonormalize} \left(\mathbf{A}\mathbf{z}^{(i)} \right). \tag{3.14}$$

In addition to its application to the degenerate ($\lambda_1 = \lambda_2$) case, orthogonalization in combination with simple vector iteration can be used to find the next largest eigenvalue/eigenvector pair by removing $\mathbf{x}_1$ from the $\mathbf{z}$ sequence. Applied recursively this allows for the determination of all eigenvalues and eigenvectors in the well-separated real case.

Example 3.4 Find the eigenvalues and eigenvectors of

$$\begin{pmatrix} 2 & -1 & 1 \\ -1 & 10^{-10} & 10^{-10} \\ 1 & 10^{-10} & 10^{-10} \end{pmatrix}. \tag{3.15}$$

Solution

We will use vector iteration with orthonormalization. As a "random" starting vector, guess $\mathbf{z}_1^{(0)} = \mathbf{e}_1$, $\mathbf{z}_2^{(0)} = \mathbf{e}_2$, and $\mathbf{z}_3^{(0)} = \mathbf{e}_3$. The results of this iteration are presented in Table 3.1.

If $\mathbf{z}_1 \cdot \mathbf{z}_2 = 0$, and $\mathbf{z}_3$ is orthogonal to both $\mathbf{z}_1$ and $\mathbf{z}_2$, then $\mathbf{z}_3$ should be uniquely determined from $\mathbf{z}_1$ and $\mathbf{z}_2$ and no iterations should be required. This mathematical circumstance is not strictly observed numerically, but it is very nearly so in this case.

Table 3.1: Eigenvector and eigenvalue computation by the power method; Example 3.4.

k	$\mathbf{z}_1$			$\mathbf{z}_1^{(k-1)\mathrm{T}}\mathbf{A}\mathbf{z}_1^{(k-1)}$
1	+8.1650e-01	-4.0825e-01	+4.0825e-01	+2.0000e+00
2	+9.0453e-01	-3.0151e-01	+3.0151e-01	+2.6667e+00
3	+8.8345e-01	-3.3129e-01	+3.3129e-01	+2.7273e+00
$\vdots$				
10	+8.8807e-01	-3.2506e-01	+3.2506e-01	+2.7321e+00
$\vdots$				

k	$\mathbf{z}_2$			$\mathbf{z}_2^{(k-1)\mathrm{T}}\mathbf{A}\mathbf{z}_2^{(k-1)}$
1	-4.5970e-01	-6.2796e-01	+6.2796e-01	+1.0000e-10
2	+4.5970e-01	+6.2796e-01	-6.2796e-01	-7.3205e-01
3	-4.5970e-01	-6.2796e-01	+6.2796e-01	-7.3205e-01
$\vdots$				

k	$\mathbf{z}_3$			$\mathbf{z}_3^{(k-1)\mathrm{T}}\mathbf{A}\mathbf{z}_3^{(k-1)}$
1	+3.9252e-07	+7.0711e-01	+7.0711e-01	+1.0000e-10
2	+0.0000e+00	+7.0711e-01	+7.0711e-01	+2.0031e-10
3	-2.4881e-17	+7.0711e-01	+7.0711e-01	+2.0000e-10
$\vdots$				

A more difficult situation occurs if $\lambda_1 = -\lambda_2$, with $|\lambda_1| > |\lambda_3| \geq |\lambda_4| \geq \cdots$. In this case,

$$\mathbf{z}^{(1)} \propto \alpha_1\mathbf{x}_1 - \alpha_2\mathbf{x}_2 + \cdots + \frac{\lambda_i}{\lambda_1}\alpha_i\mathbf{x}_i + \cdots$$

$$\mathbf{z}^{(2)} \propto \alpha_1\mathbf{x}_1 + \alpha_2\mathbf{x}_2 + \cdots + \frac{\lambda_i^2}{\lambda_1^2}\alpha_i\mathbf{x}_i + \cdots$$

$$\vdots$$

$$\mathbf{z}^{(2m)} \propto \alpha_1\mathbf{x}_1 + \alpha_2\mathbf{x}_2$$
$$\mathbf{z}^{(2m+1)} \propto \alpha_1\mathbf{x}_1 - \alpha_2\mathbf{x}_2,$$

and so

$$\mathbf{x}_1 \propto \mathbf{z}^{(2m)} + \mathbf{z}^{(2m+1)}$$
$$\mathbf{x}_2 \propto \mathbf{z}^{(2m)} - \mathbf{z}^{(2m+1)}.$$

Finally, the most common problem is that $\mathbf{A}$ is real, but $|\lambda_1| = |\lambda_2|$, $|\lambda_1| > |\lambda_3| \geq$

$|\lambda_4| \geq \cdots$, with both λ_1 and λ_2 being complex conjugate pairs, i.e., $\lambda_1^* = \lambda_2$. Let us write $\lambda_1 = Me^{i\theta}$ so then $\lambda_2 = Me^{-i\theta}$. In this case, $\mathbf{x}_1^* = \mathbf{x}_2$. If $\mathbf{z}$ is real, then $\alpha_1^* = \alpha_2$.

After conducting enough iterations that only $\mathbf{x}_1$ and $\mathbf{x}_2$ contribute to $\mathbf{z}^{(m)}$ we will have a situation like

$$\mathbf{z}^{(m)} = c\mathbf{x}_1 + c^*\mathbf{x}_1^* = 2\Re(c\mathbf{x}_1).$$

At this point, let $c\mathbf{x}_1 = \rho e^{i\boldsymbol{\beta}}$. In this notation, ρ is a vector consisting of the magnitude of the components of the vector $c\mathbf{x}_1$, and $\boldsymbol{\beta}$ is a phase angle. Formally, $\boldsymbol{\beta}$ is a diagonal matrix of phase angles, one per component of $\mathbf{x}_1$. With this notation,

$$\mathbf{z}^{(m)} = 2\rho \cos(\boldsymbol{\beta}) \tag{3.16}$$

where the cosine of the diagonal matrix $\boldsymbol{\beta}$ is a diagonal matrix of the cosines. Here we are lumping c and $\mathbf{x}_1$, but this is OK since if $\mathbf{x}_1$ is an eigenvector associated with λ_1, then also $c\mathbf{x}_1$ is an eigenvector associated with λ_1 even if c is complex.

The next iteration will give, after normalization,

$$\mathbf{z}^{(m+1)} = 2\rho \cos(\boldsymbol{\beta} + \theta). \tag{3.17}$$

And the next iteration will give, after normalization,

$$\mathbf{z}^{(m+2)} = 2\rho \cos(\boldsymbol{\beta} + 2\theta). \tag{3.18}$$

Equations (3.16), (3.17), and (3.18) constitute $3n$ equations in $2n + 1$ variables. Therefore, if a suitable method can be determined one could in principle obtain the n coefficients ρ, the n phase angles $\boldsymbol{\beta}$, and the 1 phase θ of the eigenvalue.

With some elementary trigonometric manipulations, it is easy to show that

$$2\cos(\theta)\mathbf{z}^{(m+1)} = \mathbf{z}^{(m)} + \mathbf{z}^{(m+2)}, \tag{3.19}$$

from which the angle θ may be deduced.

Next, write

$$\mathbf{p} = \frac{1}{2}\mathbf{z}^{(m)} = \rho \cos(\boldsymbol{\beta}) \propto \Re(c\mathbf{x}_1), \tag{3.20}$$

$$\mathbf{q} = \frac{1}{2}\left[\cot(\theta)\mathbf{z}^{(m)} - \csc(\theta)\mathbf{z}^{(m+1)}\right] = \rho \sin(\boldsymbol{\beta}) \propto \Im(c\mathbf{x}_1). \tag{3.21}$$

One could deduce ρ and $\boldsymbol{\beta}$ by taking one component of the vector at a time,

$$\rho_i = \sqrt{p_i^2 + q_i^2}$$

$$\beta_i = \arctan\frac{q_i}{p_i}.$$

However, the results (3.20), (3.21) are sufficient.

Example 3.5 As a trivial example, consider the matrix

$$\mathbf{A} = \begin{pmatrix} 1/2 & \sqrt{3}/2 \\ -\sqrt{3}/2 & 1/2 \end{pmatrix},$$

and a "random" normalized initial vector

$$\mathbf{z}^{(0)} = \begin{pmatrix} 1/\sqrt{5} \\ -2/\sqrt{5} \end{pmatrix}.$$

The first two iterations calculate

$$\mathbf{z}^{(1)} = \begin{pmatrix} \frac{1-2\sqrt{3}}{2\sqrt{5}} \\ \frac{-2-\sqrt{3}}{2\sqrt{5}} \end{pmatrix}$$

$$\mathbf{z}^{(2)} = \begin{pmatrix} \frac{-1-2\sqrt{3}}{2\sqrt{5}} \\ \frac{2-\sqrt{3}}{2\sqrt{5}} \end{pmatrix}.$$

Equation (3.19) gives

$$2\cos(\theta) \begin{pmatrix} \frac{1-2\sqrt{3}}{2\sqrt{5}} \\ \frac{-2-\sqrt{3}}{2\sqrt{5}} \end{pmatrix} = \begin{pmatrix} \frac{1-2\sqrt{3}}{2\sqrt{5}} \\ \frac{-2-\sqrt{3}}{2\sqrt{5}} \end{pmatrix},$$

so $\theta = \pi/3$. Now $\mathbf{p}$ and $\mathbf{q}$ are

$$\mathbf{p} = \begin{pmatrix} \frac{1}{2\sqrt{5}} \\ \frac{-1}{\sqrt{5}} \end{pmatrix}$$

$$\mathbf{q} = \begin{pmatrix} \frac{1}{\sqrt{5}} \\ \frac{1}{2\sqrt{5}} \end{pmatrix}.$$

This implies eigenvectors

$$\mathbf{x}_1 \propto \mathbf{p} + i\mathbf{q} \propto \begin{pmatrix} 1 + 2i \\ -2 + i \end{pmatrix}$$

$$\mathbf{x}_2 = \mathbf{x}_1^* \propto \mathbf{p} - i\mathbf{q} \propto \begin{pmatrix} 1 - 2i \\ -2 - i \end{pmatrix}.$$

The magnitude of the eigenvalue λ_1 comes from the Rayleigh quotient.

$$M^2 = \|\mathbf{A}\mathbf{z}\|_2 = \mathbf{z}^\mathrm{T}\mathbf{A}^\mathrm{T}\mathbf{A}\mathbf{z} = 1,$$

so

$$\lambda_1 = Me^{i\theta} = \frac{1}{2} + \frac{\sqrt{3}i}{2}$$

$$\lambda_2 = \lambda_1^* = \frac{1}{2} - \frac{\sqrt{3}i}{2}.$$

If $\mathbf{A}$ is not diagonalizable, then it is similar to a block Jordan form, with one or more blocks larger than 1×1. For simplicity, consider a matrix similar to a single Jordan block:

$$\mathbf{SAS}^{-1} = \mathbf{C}$$

$$\mathbf{C} = \begin{pmatrix} \lambda & 1 & & & \\ & \lambda & 1 & & \\ & & \ddots & \ddots & \\ & & & \lambda & 1 \\ & & & & \lambda \end{pmatrix},$$

with a single eigenvector $\mathbf{e}_1$.

If we conducted vector iteration on this matrix, the process would be equivalent to the following (for further simplicity, we will omit the renormalization step):

$$\mathbf{A}\mathbf{z}^k = \mathbf{z}^{(k+1)}$$

$$\mathbf{S}^{-1}\mathbf{A}\mathbf{S}\mathbf{S}^{-1}\mathbf{z}^k = \mathbf{S}^{-1}\mathbf{z}^{(k+1)},$$

or

$$\mathbf{C}\mathbf{y}^{(k)} = \mathbf{y}^{(k+1)}$$

$$\mathbf{C}^k\mathbf{y}^{(0)} = \mathbf{y}^{(k)},$$

with $\mathbf{y} = \mathbf{S}^{-1}\mathbf{z}$. The kth power of the matrix $\mathbf{C}$ is

$$\mathbf{C}^k = \begin{pmatrix} \lambda^k & \binom{k}{1}\lambda^{k-1} & \cdots & \binom{k}{n-2}\lambda^{k+2-n} & \binom{k}{n-1}\lambda^{k+1-n} \\ & \lambda^k & \cdots & \binom{k}{n-3}\lambda^{k+3-n} & \binom{k}{n-2}\lambda^{k+2-n} \\ & & \ddots & \vdots & \vdots \\ & & & \lambda^k & \binom{k}{1}\lambda^{k-1} \\ & & & & \lambda^k \end{pmatrix} \quad k \geq 0, \quad (3.22)$$

with

$$\binom{k}{j} = \begin{cases} \frac{k!}{(k-j)!j!} & \text{if } k \geq 0, \quad 0 \leq j \leq k \\ 0 & \text{otherwise,} \end{cases}$$

and if the iteration converges, $\mathbf{y}^{(\infty)} \propto \mathbf{e}_1$.

The kth power of $\mathbf{C}$ contains powers of λ^k in all rows, and therefore the components of $\mathbf{e}_2$ in $\mathbf{y}$ (for example) will not diminish in proportion to some power of λ. This is completely different from the behavior observed when $|\lambda_1| > |\lambda_2| > \cdots |\lambda_n|$.

If $\mathbf{y}^{(0)} = \mathbf{e}_2$ then after k iterations of the vector iteration method, the normalized result will be

$$\mathbf{y}^{(k)} = \frac{k\mathbf{e}_1 + \lambda\mathbf{e}_2}{\sqrt{k^2 + \lambda^2}} \approx \left(\mathbf{e}_1 + \frac{\lambda}{k}\mathbf{e}_2\right)\left(1 - \frac{1}{2}\frac{\lambda}{k} + \frac{3}{8}\frac{\lambda^2}{k^2} \cdots\right), \quad (3.23)$$

which approaches the eigenvector $\mathbf{e}_1$ only algebraically (i.e., in proportion to k, versus geometrically: in proportion to c^k for some constant c). The rate of approach is very slow unless shifting is used with a shift parameter $\mu \approx \lambda$.

Do Jordan blocks fare any better with inverse vector iteration? The inverse of Jordan block $\mathbf{C}$ is

$$
\mathbf{C}^{-1} =
\begin{pmatrix}
\lambda^{-1} & -\lambda^{-2} & \cdots & (-1)^n \lambda^{1-n} & (-1)^{n+1} \lambda^{-n} \\
 & \lambda^{-1} & \cdots & (-1)^{n-1} \lambda^{2-n} & (-1)^n \lambda^{1-n} \\
 & & \ddots & \vdots & \vdots \\
 & & & \lambda^{-1} & -\lambda^{-2} \\
 & & & & \lambda^{-1}
\end{pmatrix},
$$

and the kth power of this block looks like

$$
\mathbf{C}^{-k} =
\begin{pmatrix}
\lambda^{-k} & -k\lambda^{-(k+1)} & \cdots & (-1)^n \binom{k+n-3}{n-2} \lambda^{-(k+n-2)} & (-1)^{n+1} \binom{k+n-2}{n-1} \lambda^{-(k+n-1)} \\
 & \lambda^{-k} & \cdots & (-1)^{n-1} \binom{k+n-4}{n-3} \lambda^{-(k+n-3)} & (-1)^n \binom{k+n-3}{n-2} \lambda^{-(k+n-2)} \\
 & & \ddots & \vdots & \vdots \\
 & & & \lambda^{-k} & -k\lambda^{-(k+1)} \\
 & & & & \lambda^{-k}
\end{pmatrix}.
$$

Again, assume $\mathbf{y}^{(0)} = \mathbf{e}_2$, then by inverse vector iteration the result of k iterations will be

$$
\mathbf{y}^{(k)} = \frac{-k\mathbf{e}_1 + \lambda\mathbf{e}_2}{\sqrt{k^2 + \lambda^2}},
$$

which has the same poor convergence properties as (3.23).

What is occurring with these Jordan blocks is that, from an iteration perspective, the amplification factors are like λ^k, $k\lambda^k$, ..., up to $k^{n-1}\lambda^k$ for a size n block. When this defect does not occur, the amplification factors are simply proportional to λ^k.

The point of this discussion is that simple vector iteration can be applied with good success when one knows that the eigenvalues are real and nondegenerate. For more complicated situations, one can salvage the technique, but the approach taken will depend on the structure of the problem: you more or less need to know the answer in advance of choosing the method.

For well-separated systems, i.e., $|\lambda_1| > |\lambda_2| > \cdots > |\lambda_n|$, vector iteration will take $\mathcal{O}(n^2)$ operations to compute a single iteration for a single eigenvector-eigenvalue pair, or $\mathcal{O}(n^3)$ operations per iteration to determine all n pairs. One might hope to do better, which is a motivation for the QR algorithm.

3.3   The QR algorithm

The QR method for eigenvalue and eigenvector determination is a very powerful technique that seeks to diagonalize a matrix through a succession of similarity transformations. This section presents a very simple application of the QR approach. Without modification it will find the eigenvalues of any square matrix whose eigenvalues are distinct and real ($|\lambda_1| > |\lambda_2| > \cdots > |\lambda_n|$). The method was developed by J. G. F. Francis [72, 73], building on the broadly similar LR method of H. Rutishauser [206]. C. Jacobi developed a strikingly similar method over 100 years earlier in the course of studying the secular perturbation of planetary motion [119].

We have seen that the Householder reduction method transforms a matrix $\mathbf{A}$ into a unitary matrix $\mathbf{Q}$ and a right triangular matrix $\mathbf{R}$:

$$\mathbf{A} = \mathbf{QR}.$$

Recall that $\mathbf{Q}$ being unitary means $\mathbf{QQ}^H = \mathbf{Q}^H\mathbf{Q} = \mathbf{I}$.

The matrix $\mathbf{A}^{(1)}$,

$$\mathbf{A}^{(1)} = \mathbf{RQ},$$

is similar to the matrix $\mathbf{A}$, and therefore it has the same eigenvalues as $\mathbf{A}$:

$$\mathbf{A}^{(1)} = \mathbf{RQ} = \mathbf{IRQ} = \mathbf{Q}^H \underbrace{\mathbf{QR}}_{\mathbf{A}} \mathbf{Q}.$$

The QR method for eigenvalue evaluation consists of the iteration sequence

$$\mathbf{A}^{(0)} = \mathbf{A} = \mathbf{Q}^{(0)}\mathbf{R}^{(0)}$$
$$\mathbf{A}^{(i)} = \mathbf{R}^{(i-1)}\mathbf{Q}^{(i-1)} = \mathbf{Q}^{(i)}\mathbf{R}^{(i)},$$

which generates a sequence of matrices $\mathbf{A}^{(i)}$ that are each similar to the original matrix $\mathbf{A}$:

$$\mathbf{A}^{(i)} = \mathbf{Q}^{(i-1)H}\mathbf{A}^{(i-1)}\mathbf{Q}^{(i-1)}.$$

To understand what this algorithm accomplishes, notice that the kth power of the matrix $\mathbf{A}$ is related to the QR sequence generated:

$$\mathbf{A}^1 = \mathbf{Q}^{(0)}\mathbf{R}^{(0)}$$
$$\mathbf{A}^2 = \mathbf{Q}^{(0)}\mathbf{R}^{(0)}\mathbf{Q}^{(0)}\mathbf{R}^{(0)} = \mathbf{Q}^{(0)}\mathbf{A}^{(1)}\mathbf{R}^{(0)} = \mathbf{Q}^{(0)}\mathbf{Q}^{(1)}\mathbf{R}^{(1)}\mathbf{R}^{(0)}$$
$$\mathbf{A}^3 = \mathbf{Q}^{(0)}\mathbf{R}^{(0)}\mathbf{Q}^{(0)}\mathbf{Q}^{(1)}\mathbf{R}^{(1)}\mathbf{R}^{(0)} = \mathbf{Q}^{(0)}\mathbf{A}^{(1)}\mathbf{Q}^{(1)}\mathbf{R}^{(1)}\mathbf{R}^{(0)}$$
$$= \mathbf{Q}^{(0)}\mathbf{Q}^{(1)}\mathbf{R}^{(1)}\mathbf{Q}^{(1)}\mathbf{R}^{(1)}\mathbf{R}^{(0)} = \mathbf{Q}^{(0)}\mathbf{Q}^{(1)}\mathbf{A}^{(2)}\mathbf{R}^{(1)}\mathbf{R}^{(0)}$$
$$= \mathbf{Q}^{(0)}\mathbf{Q}^{(1)}\mathbf{Q}^{(2)}\mathbf{R}^{(2)}\mathbf{R}^{(1)}\mathbf{R}^{(0)}$$
$$\vdots$$
$$\mathbf{A}^k = \mathbf{Q}^{(0)} \cdots \mathbf{Q}^{(k-1)}\mathbf{R}^{(k-1)} \cdots \mathbf{R}^{(0)} = \mathbf{P}^{(k)}\mathbf{U}^{(k)},$$

with

$$\mathbf{P}^{(k)} = \mathbf{Q}^{(0)} \cdots \mathbf{Q}^{(k-1)}$$
$$\mathbf{U}^{(k)} = \mathbf{R}^{(k-1)} \cdots \mathbf{R}^{(0)}.$$

With this notation, the sequence $\mathbf{P}^{(k)}$ obeys the recursion

$$\mathbf{AP}^{(k)} = \mathbf{P}^{(k+1)}\mathbf{R}^{(k)}. \tag{3.24}$$

The derivation of (3.24) follows:

$$\begin{aligned}
\mathbf{AP}^{(k)} &= \mathbf{AQ}^{(0)} \cdots \mathbf{Q}^{(k-1)} \\
&= \mathbf{Q}^{(0)} \mathbf{A}^{(1)} \mathbf{Q}^{(1)} \cdots \mathbf{Q}^{(k-1)} \\
&= \mathbf{Q}^{(0)} \mathbf{Q}^{(1)} \cdots \mathbf{A}^{(k-1)} \mathbf{Q}^{(k-1)} \\
&= \mathbf{Q}^{(0)} \mathbf{Q}^{(1)} \cdots \mathbf{Q}^{(k-1)} \mathbf{A}^{(k)} \\
&= \mathbf{Q}^{(0)} \mathbf{Q}^{(1)} \cdots \mathbf{Q}^{(k-1)} \mathbf{Q}^{(k)} \mathbf{R}^{(k)} \\
&= \mathbf{P}^{(k+1)} \mathbf{R}^{(k)}.
\end{aligned}$$

The value of this is that it shows $\mathbf{P}^{(k)}$ to be the result of some kind of vector iteration. The usual vector iteration is $\mathbf{Az}^{(k)} = \mathbf{z}^{(k+1)}$, but here we're dealing with a collection of vectors, and we're also doing a rearrangement $\mathbf{P}^{(k+1)} \mathbf{R}^{(k)}$. This rearrangement is an orthogonalization. The vectors making up $\mathbf{P}^{(k)}$ are orthogonal, and the vectors $\mathbf{AP}^{(k)}$ are not orthogonal in general (though they would be if they were eigenvectors of $\mathbf{A}$). So, we're essentially saying

$$\mathbf{P}^{(k+1)} = \text{orthonormalize}\left(\mathbf{AP}^{(k)}\right).$$

Now, in comparison with (3.10) and (3.14), a correspondence between the QR algorithm and vector iteration should be clear.

This orthogonalization has a nice property that can be seen by block partitioning equation (3.24)

$$\mathbf{A}\left(\; \mathbf{P}_1 \;\middle|\; \mathbf{P}_2 \;\right)^{(k)} = \left(\; \mathbf{P}_1 \;\middle|\; \mathbf{P}_2 \;\right)^{(k+1)} \left(\begin{array}{c|c} \mathbf{R}_1 & \mathbf{R}_3 \\ \hline 0 & \mathbf{R}_2 \end{array}\right)^{(k)}$$

where $\mathbf{P}_1$ contains r columns, and block $\mathbf{R}_1$ is $r \times r$. This means that, block-wise,

$$\mathbf{AP}_1^{(k)} = \mathbf{P}_1^{(k+1)} \mathbf{R}_1^{(k)}.$$

Define the *column space* of $\mathbf{P}_1$ to be the set of all vectors that can be made by linear combinations of the columns of $\mathbf{P}_1$: the set $\{\mathbf{P}_1\mathbf{z} \mid \mathbf{z} \in \mathbb{C}^r\}$, where r is the number of columns of the partition $\mathbf{P}_1$ (i.e., "the set of all vectors $\mathbf{P}_1\mathbf{z}$ made from all $\mathbf{z}$ that are r-dimensional vectors of complex numbers").

The column space spanned by $\mathbf{AP}_1^{(k)}$ lies in the column space spanned by $\mathbf{P}_1^{(k+1)}$:

$$\text{column space}\left(\mathbf{AP}_1^{(k)}\right) = \text{column space}\left(\mathbf{P}_1^{(k+1)}\mathbf{R}_1^{(k)}\right) \subseteq \text{column space}\left(\mathbf{P}_1^{(k+1)}\right)$$

($A \subseteq B$ means "A is contained in B," i.e., every element of A is an element of B). If $\mathbf{A}$ is invertible, then the rank of $\mathbf{AP}_1^{(k)}$ will be r, and therefore the rank of $\mathbf{P}_1^{(k+1)}\mathbf{R}_1^{(k)}$ must be r, which implies that $\mathbf{R}_1^{(k)}$ is rank r and therefore invertible. In this case, the above inequality becomes

$$\text{column space}\left(\mathbf{AP}_1^{(k)}\right) = \text{column space}\left(\mathbf{P}_1^{(k+1)}\right). \tag{3.25}$$

The significance of this result is that one can interpret QR as being a *subspace iteration* method, by analogy to simple vector iteration.

To take the analogy further, in the limit $r = 1$, the partition $\mathbf{P}_1$ is a vector – the vector $\mathbf{z}$ of our vector iteration scheme. The block $\mathbf{R}_1$ becomes a constant – the constant we introduced to renormalize our vectors so they maintain unit length.

If we assume $|\lambda_1| > |\lambda_2| > \cdots > |\lambda_n|$, then the analogy between vector iteration and QR with $r = 1$ suggests that the first column of $\mathbf{P}$ will evolve into the eigenvector $\mathbf{x}_1$, and the diagonal element $\mathbf{R}_1$ will become λ_1.

We might expect the second column of $\mathbf{P}$ will become $\mathbf{x}_2$, and $(\mathbf{R}_1)_{22}$ will be λ_2 and $(\mathbf{R}_1)_{12}$ should be zero. Considering the other columns of $\mathbf{P}$ and the associated behavior of $\mathbf{R}$, when $|\lambda_1| > |\lambda_2| > \cdots > |\lambda_n|$ we should expect $\mathbf{R}$ to become the diagonal matrix of eigenvalues, and columns of $\mathbf{P}$ will be the eigenvectors.

It is also possible to contemplate the QR method from the perspective of inverse vector iteration. Beginning with (3.24),

$$\mathbf{A}\mathbf{P}^{(k)} = \mathbf{P}^{(k+1)}\mathbf{R}^{(k)}$$

$$\mathbf{P}^{(k)\mathrm{T}}\mathbf{A}^{\mathrm{T}} = \mathbf{R}^{(k)\mathrm{T}}\mathbf{P}^{(k+1)\mathrm{T}}$$

$$\mathbf{A}^{-\mathrm{T}}\mathbf{P}^{(k)} = \mathbf{P}^{(k+1)}(\mathbf{R}^{(k)})^{-\mathrm{T}}$$

$$\mathbf{A}^{-\mathrm{T}}\left(\;\mathbf{P}_1 \mid \mathbf{P}_2\;\right)^{(k)} = \left(\;\mathbf{P}_1 \mid \mathbf{P}_2\;\right)^{(k+1)}\left(\begin{array}{c|c}\hat{\mathbf{R}}_1 & 0 \\ \hline \hat{\mathbf{R}}_3 & \hat{\mathbf{R}}_2\end{array}\right)^{(k)},$$

where $\mathbf{P}_2$ contains r columns, and block $\hat{\mathbf{R}}_2$ is $r \times r$ (the inverse of right triangular $\mathbf{R}^{(k)}$ is also right triangular; its transpose is therefore left triangular). This means that, block-wise,

$$\mathbf{A}^{-\mathrm{T}}\mathbf{P}_2^{(k)} = \mathbf{P}_2^{(k+1)}\hat{\mathbf{R}}_2^{(k)}.$$

In terms of column spaces,

$$\text{column space}\left(\mathbf{A}^{-\mathrm{T}}\mathbf{P}_2^{(k)}\right) = \text{column space}\left(\mathbf{P}_2^{(k+1)}\hat{\mathbf{R}}_2\right) \subseteq \text{column space}\left(\mathbf{P}_2^{(k+1)}\right),$$

and with $\mathbf{A}$ invertible,

$$\text{column space}\left(\mathbf{A}^{-\mathrm{T}}\mathbf{P}_2^{(k)}\right) = \text{column space}\left(\mathbf{P}_2^{(k+1)}\right) \tag{3.26}$$

so QR is simultaneously a subspace iteration (3.25), and an inverse subspace iteration (3.26). We can also write

$$\mathbf{P}^{(k+1)} = \text{orthonormalize}\left(\mathbf{A}^{-\mathrm{T}}\mathbf{P}^{(k)}\right),$$

in analogy to (3.11) with orthogonalization.

As described so far, the QR algorithm requires $\mathcal{O}(n^3)$ operations for a single iteration, which offers no benefit relative to the power method.

The most important improvement will consist of the following two changes: (1) begin the process with a similarity transformation that makes $\mathbf{A}$ Hessenberg, or tridiagonal if $\mathbf{A}$ is Hermitian, and (2) choose a unitary transformation matrix $\mathbf{Q}$ that retains the Hessenberg (or tridiagonal) shape. To facilitate presentation the subsequent discussion will focus exclusively on the Hermitian case.

The creation of the tridiagonal form is easily accomplished using a sequence of

Householder matrices that are designed to eliminate elements below the first subdiagonal.

If we construct a Householder matrix designed to eliminate the last $n - 2$ elements of the first column,

$$\mathbf{Q}_1 = \left(\begin{array}{c|c} 1 & 0 \\ \hline 0 & \mathbf{I} - 2\mathbf{w}_1\mathbf{w}_1^H \end{array} \right),$$

then because $\mathbf{A}$ is Hermitian,

$$\mathbf{A}_1 = \mathbf{Q}_1\mathbf{A}\mathbf{Q}_1^H = \begin{pmatrix} * & * & 0 & \cdots & 0 \\ * & * & * & \cdots & * \\ 0 & * & * & \cdots & * \\ \vdots & \vdots & \vdots & \vdots & \vdots \\ 0 & * & * & \cdots & * \end{pmatrix}. \tag{3.27}$$

The resulting matrix is similar to $\mathbf{A}$, and since $\mathbf{A}$ is Hermitian $\mathbf{A}_1$ will be Hermitian too. Now, construct a Householder matrix to zero the last $n - 3$ elements of the second column,

$$\mathbf{Q}_2 = \left(\begin{array}{cc|c} 1 & 0 & \\ 0 & 1 & \mathbf{0} \\ \hline & \mathbf{0} & \mathbf{I} - 2\mathbf{w}_2\mathbf{w}_2^H \end{array} \right).$$

This matrix will transform $\mathbf{A}_1$ into $\mathbf{A}_2$:

$$\mathbf{A}_2 = \mathbf{Q}_2\mathbf{A}_1\mathbf{Q}_2^H = \begin{pmatrix} * & * & 0 & 0 & \cdots & 0 \\ * & * & * & 0 & \cdots & 0 \\ 0 & * & * & * & \cdots & * \\ 0 & 0 & * & * & \cdots & * \\ \vdots & \vdots & \vdots & \vdots & \vdots & \vdots \\ 0 & 0 & * & * & \cdots & * \end{pmatrix}, \tag{3.28}$$

which is also similar to $\mathbf{A}$. After $n - 2$ such transformations Hermitian $\mathbf{A}$ is transformed into a Hermitian tridiagonal matrix

$$\mathbf{A}_{n-2} = \begin{pmatrix} * & * & & \\ * & \ddots & \ddots & \\ & \ddots & * & * \\ & & * & * \end{pmatrix}.$$

For the tridiagonal-preserving $\mathbf{Q}$, one uses a sequence of appropriate Givens reflection matrices [85] (first used in this context by Jacobi [119])

$$
G_{i,i+1} =
\begin{pmatrix}
I & & \\
 & c & s & \\
 & s & -c & \\
 & & & I
\end{pmatrix},
\tag{3.29}
$$

where c and s appear in rows and columns i and $i + 1$. If these numbers are chosen so that $c^2 + s^2 = 1$, then this matrix is unitary. We pick the values of c and s so that

$$
\begin{pmatrix}
I & & \\
 & c & s & \\
 & s & -c & \\
 & & & I
\end{pmatrix}
\begin{pmatrix}
* \\
y_i \\
y_{i+1} \\
*
\end{pmatrix}
=
\begin{pmatrix}
* \\
k \\
0 \\
*
\end{pmatrix}.
$$

A numerically stable algorithm for the calculation of c, s, and k is

$$
\mu = \max(|y_i|, |y_{i+1}|)
$$

$$
k = \text{sign}(y_i)\mu\sqrt{(y_i/\mu)^2 + (y_{i+1}/\mu)^2},
\tag{3.30a}
$$

$$
c = y_i/k,
\tag{3.30b}
$$

$$
s = y_{i+1}/k.
\tag{3.30c}
$$

The choice of sign in (3.30a) is based on the following idea: repeated calculation of

$$
z_1 = cy_1 + sy_2
$$
$$
z_2 = sy_1 - cy_2
$$

(for fixed c, s) takes four multiplications per vector $\mathbf{y}$, but

$$
z_1 = cy_1 + sy_2
$$
$$
z_2 = v(y_1 + z_1) - y_2,
$$

with

$$
v = \frac{s}{1 + c},
\tag{3.31}
$$

takes three multiplications per $\mathbf{y}$. The sign in (3.30a) is chosen to make $c > 0$ so that (3.31) is free of cancellation errors.

Matrices $G_{i,i+1}$ are designed to eliminate a particular element $y_{i+1} = a_{i+1,j}$ by mixing elements $y_i = a_{i,j}$ and $y_{i+1} = a_{i+1,j}$. So, when a sequence of G matrices is written, it is understood from the subscripts which rows are affected, and from the context which column is used to determine c and s. With this in mind,

$$\underbrace{\mathbf{G}_{n-2,n-1}\cdots\mathbf{G}_{23}\mathbf{G}_{12}}_{\mathbf{Q}^{(i)\mathrm{H}}}\underbrace{\begin{pmatrix} * & * & & & & \\ * & \ddots & \ddots & & & \\ & \ddots & \ddots & \ddots & & \\ & & \ddots & \ddots & \ddots & \\ & & & \ddots & \ddots & * \\ & & & & * & * \end{pmatrix}}_{\mathbf{A}^{(i)}} = \underbrace{\begin{pmatrix} * & * & * & & & \\ & \ddots & \ddots & \ddots & & \\ & & \ddots & \ddots & \ddots & \\ & & & \ddots & \ddots & * \\ & & & & \ddots & * \\ & & & & & * \end{pmatrix}}_{\mathbf{R}^{(i)}},\quad (3.32)$$

and, we will show

$$\underbrace{\begin{pmatrix} * & * & * & & & \\ & \ddots & \ddots & \ddots & & \\ & & \ddots & \ddots & \ddots & * \\ & & & \ddots & \ddots & * \\ & & & & & * \end{pmatrix}}_{\mathbf{R}^{(i)}}\underbrace{\mathbf{G}_{12}^{\mathrm{H}}\mathbf{G}_{23}^{\mathrm{H}}\cdots\mathbf{G}_{n-2,n-1}^{\mathrm{H}}}_{\mathbf{Q}^{(i)}} = \underbrace{\begin{pmatrix} * & * & & & & \\ * & \ddots & \ddots & & & \\ & \ddots & \ddots & \ddots & & \\ & & \ddots & \ddots & \ddots & * \\ & & & & * & * \end{pmatrix}}_{\mathbf{A}^{(i+1)}},\quad (3.33)$$

preserving the tridiagonal form.

In (3.33), it should be clear that one subdiagonal is created by the action of the $\mathbf{G}$ matrices, which makes $\mathbf{A}^{(k+1)}$ Hessenberg. The reason we know the tridiagonal form is preserved is that $\mathbf{A}^{(k)}$ is not just tridiagonal, but also Hermitian so

$$\mathbf{A}^{(i+1)} = \mathbf{R}^{(i)}\mathbf{Q}^{(i)} = \underbrace{\mathbf{Q}^{(i)\mathrm{H}}\mathbf{A}^{(i)}\mathbf{Q}^{(i)}}_{\text{Hermitian}}.$$

A matrix that is both Hessenberg and Hermitian must be tridiagonal.

By maintaining $\mathbf{A}$ in the tridiagonal form, and by using Givens matrices to make up $\mathbf{Q}$, there are $\mathcal{O}(n)$ operations per iteration, which is remarkably efficient. If the steps (3.32) and (3.33) are performed in the order

$$\cdots\left(\mathbf{G}_{23}\left(\mathbf{G}_{12}\mathbf{A}^{(i)}\mathbf{G}_{12}^{\mathrm{H}}\right)\mathbf{G}_{23}^{\mathrm{H}}\right)\cdots,$$

then the initially tridiagonal matrix undergoes the shape change

$$(3.34)$$

which motivates some to describe this algorithm as "chasing the bulge."

Example 3.6 Compute the eigenvalues and eigenvectors of the matrix

$$\begin{pmatrix} 2 & -1 & 1 \\ \underline{-1} & 10^{-10} & 10^{-10} \\ \underline{1} & 10^{-10} & 10^{-10} \end{pmatrix}.$$

Solution
Start with a matrix **P**

$$\mathbf{P} = \begin{pmatrix} 1 & 0 & 0 \\ 0 & 1 & 0 \\ 0 & 0 & 1 \end{pmatrix}$$

that will transform into the matrix of eigenvectors.

The first step is to use Householder matrices to convert the matrix **A** into the tridiagonal form. The first Householder step is designed to convert the underlined elements into $k_1 \mathbf{e}_1$ (cf. Example 2.6). With

$$\mathbf{a}_1 = \begin{pmatrix} -1 \\ 1 \end{pmatrix}$$

one computes $k_1 = 1.4142$ and

$$\mathbf{w}_1 = \begin{pmatrix} -0.923\,88 \\ 0.382\,68 \end{pmatrix}.$$

Multiplication of **A** from the left by the Householder matrix gives

$$\begin{pmatrix} 2.0000 & -1.0000 & 1.0000 \\ 1.4142 & 3.8774 \times 10^{-26} & 3.8774 \times 10^{-26} \\ 0 & 1.4142 \times 10^{-10} & 1.4142 \times 10^{-10} \end{pmatrix},$$

and multiplication of **A** from the right by the same Householder matrix (it is its own transpose) gives

$$\begin{pmatrix} 2.0000 & 1.4142 & 0.0000 \\ 1.4142 & 1.1479 \times 10^{-41} & 6.6192 \times 10^{-26} \\ 0.0000 & 6.6192 \times 10^{-26} & 2.0000 \times 10^{-10} \end{pmatrix}. \tag{3.35}$$

(This result is symmetrized, $\mathbf{A} := (\mathbf{A} + \mathbf{A}^T)/2$. Numerically, perfect symmetry is unlikely to be observed because the sequence of steps leading to A_{ij} is different than the sequence of steps leading to A_{ji} – the algorithms are different, resulting in different numerical error.)

Multiplication of **P** from the right by the Householder matrix gives

$$\mathbf{P}^{(0)} = \begin{pmatrix} 1 & 0 & 0 \\ 0 & -0.707\,11 & 0.707\,11 \\ 0 & 0.707\,11 & 0.707\,11 \end{pmatrix}.$$

$\mathbf{PAP}^{\mathrm{T}}$ is the original matrix $\mathbf{A}$ of (3.15). A similarity transformation has occurred, leaving $\mathbf{A}$ in the tridiagonal form. This completes the tridiagonalization that is preliminary to the QR iterations.

Now unitary matrices $\mathbf{Q}$ are made up of Givens reflections. The first reflection is designed to operate on the two underlined elements of (3.35), transforming them into $(*, 0)^{\mathrm{T}}$. With $c = 0.816\,50$, and $s = 0.577\,35$, $\mathbf{G}_{12}\mathbf{AG}_{12}^{\mathrm{T}}$ becomes

$$
\begin{pmatrix}
2.6667 & 0.471\,40 & 3.8216 \times 10^{-26} \\
\underline{0.471\,40} & -0.666\,67 & -5.4045 \times 10^{-26} \\
\underline{3.8216 \times 10^{-26}} & -5.4045 \times 10^{-26} & 2.0000 \times 10^{-10}
\end{pmatrix}, \tag{3.36}
$$

which is no longer tridiagonal. A second Givens matrix is designed to operate on the underlined elements of (3.36), to restore the tridiagonal form. With $c = 1.0000$, $s = 8.1068 \times 10^{-26}$, $\mathbf{G}_{23}\mathbf{G}_{12}\mathbf{AG}_{12}^{\mathrm{T}}\mathbf{G}_{23}$ becomes

$$
\mathbf{A}^{(0)} = \begin{pmatrix}
2.6667 & 0.471\,40 & 0 \\
0.471\,40 & -0.666\,67 & -1.6214 \times 10^{-35} \\
0 & -1.6214 \times 10^{-35} & 2.0000 \times 10^{-10}
\end{pmatrix}.
$$

$\mathbf{P}$, multiplied from the right by $\mathbf{G}_{12}^{\mathrm{T}}\mathbf{G}_{23}^{\mathrm{T}}$ becomes

$$
\mathbf{P}^{(1)} = \begin{pmatrix}
0.816\,50 & 0.577\,35 & 4.6804 \times 10^{-26} \\
-0.408\,25 & 0.577\,35 & -0.707\,11 \\
0.408\,25 & -0.577\,35 & -0.707\,11
\end{pmatrix}.
$$

This completes one iteration of the QR method.

After five iterations, the diagonal elements of $\mathbf{A}$ give the eigenvalues discovered in Example 3.4:

$$
\mathbf{A}^{(5)} = \begin{pmatrix}
2.7321 & 6.6364 \times 10^{-4} & 0 \\
6.6364 \times 10^{-4} & -0.732\,05 & 2.4445 \times 10^{-83} \\
0 & 2.4445 \times 10^{-83} & 2.0000 \times 10^{-10}
\end{pmatrix}. \tag{3.37}
$$

However, the algorithm has not converged, as judged by the magnitude of the off-diagonal elements. After about 30 iterations,

$$
\mathbf{P}^{(30)} = \begin{pmatrix}
0.888\,07 & 0.459\,70 & 4.6804 \times 10^{-26} \\
-0.325\,06 & 0.627\,96 & -0.707\,11 \\
0.325\,06 & -0.627\,96 & -0.707\,11
\end{pmatrix}
$$

$$
\mathbf{A}^{(30)} = \begin{pmatrix}
2.7321 & 3.3361 \times 10^{-18} & 0 \\
3.3361 \times 10^{-18} & -0.732\,05 & 3.3889 \times 10^{-308} \\
0 & 3.3889 \times 10^{-308} & 2.0000 \times 10^{-10}
\end{pmatrix}.
$$

Here the magnitude of off-diagonal elements is $\mathcal{O}(2^{-t})$ times the magnitude of the neighboring diagonals, which signifies good numerical convergence.

Additional improvements, and a firmer basis for deciding when to stop iterating, come from consideration of a *reducible* Hessenberg matrix, e.g.,

$$
\begin{pmatrix}
* & * & * & * & * \\
* & * & * & * & * \\
0 & * & * & * & * \\
\hline
0 & 0 & \underline{0} & * & * \\
0 & 0 & 0 & * & *
\end{pmatrix}.
\tag{3.38}
$$

The key feature is that one of the subdiagonal elements (underlined) is zero. Now follow the evolution of this matrix through one iteration of the QR algorithm, where the unitary matrices are chosen to be Givens matrices. Left- and right-multiplication by G_{12} gives the matrices,

$$
\rightarrow
\begin{pmatrix}
* & * & * & * & * \\
0 & * & * & * & * \\
0 & * & * & * & * \\
\hline
0 & 0 & \underline{0} & * & * \\
0 & 0 & 0 & * & *
\end{pmatrix}
\rightarrow
\begin{pmatrix}
* & * & * & * & * \\
* & * & * & * & * \\
* & * & * & * & * \\
\hline
0 & 0 & \underline{0} & * & * \\
0 & 0 & 0 & * & *
\end{pmatrix}.
$$

Then left- and right-multiplication by G_{23} gives

$$
\rightarrow
\begin{pmatrix}
* & * & * & * & * \\
* & * & * & * & * \\
0 & * & * & * & * \\
\hline
0 & 0 & \underline{0} & * & * \\
0 & 0 & 0 & * & *
\end{pmatrix}
\rightarrow
\begin{pmatrix}
* & * & * & * & * \\
* & * & * & * & * \\
0 & * & * & * & * \\
\hline
0 & 0 & \underline{0} & * & * \\
0 & 0 & 0 & * & *
\end{pmatrix}.
$$

Because of the zero in position 4, 3, right-multiplication by G_{23} did not create a 4, 2 entry, so $G_{34} = I$ in the next step because no elimination is required. Right-multiplication by I does not spoil the Hessenberg form, so no further elimination will be required.

Because the matrix was reducible, the upper left and lower right blocks were completely decoupled. In this particular example, the upper left block would evolve to the diagonal matrix of eigenvalues, but the lower right block would remain unchanged. However, the two eigenvalues that will emerge from the 2×2 lower right block are independent of the upper left block (see (A.18)), and they can be determined immediately by solution of the quadratic characteristic equation,

$$
\det \begin{pmatrix}
A_{n-1,n-1} - \lambda & A_{n-1,n} \\
A_{n,n-1} & A_{n,n} - \lambda
\end{pmatrix} = 0.
$$

Once these two eigenvalues are determined, the last two rows of **A**, and the last two columns, can be removed so all subsequent iterations happen in the smaller remaining 3×3 space.

This example shows that reducibility can be used to reduce the size of the system being considered as eigenvalues are discovered in the lower right position, which speeds up the calculation. It also shows that reducibility must be detected, or the sequence of operations beginning at the upper left will not affect the lower right at all, and the method could fail to converge.

The practical questions that must be answered are now (i) "How small must a subdiagonal entry be for the Hessenberg matrix to be considered reducible?," and (ii) "What steps can be taken to encourage the reducible form?"

The preceding example also raises the question "when is it OK to stop iterating?" Consider two matrices, $\mathbf{A}$ and $\mathbf{B}$ (diagonalizable), and assume that the eigenvalues of λ of $\mathbf{A}$ are not exact eigenvalues of $\mathbf{B}$. Then

$$(\mathbf{A} - \mathbf{B})\mathbf{x} = (\lambda\mathbf{I} - \mathbf{B})\mathbf{x}, \tag{3.39a}$$

$$(\lambda\mathbf{I} - \mathbf{B})^{-1}(\mathbf{A} - \mathbf{B})\mathbf{x} = \mathbf{x}, \tag{3.39b}$$

$$\text{lub}\left[(\lambda\mathbf{I} - \mathbf{B})^{-1}(\mathbf{A} - \mathbf{B})\right] \geq 1. \tag{3.39c}$$

Equation (3.39a) is exact. Because λ is (by assumption) not an eigenvalue of $\mathbf{B}$, $(\lambda\mathbf{I} - \mathbf{B})$ is invertible, from which (3.39b) follows. This equation says that $\mathbf{x}$ is an eigenvector of eigenvalue 1 of the matrix $[(\lambda\mathbf{I} - \mathbf{B})^{-1}(\mathbf{A} - \mathbf{B})]$, so the spectral radius ρ of this matrix is ≥ 1. Equation (3.39c) follows by the definition of the least upper bound norm (2.14).

Diagonalizable $\mathbf{B}$ can be written $\mathbf{S}\Lambda(\mathbf{B})\mathbf{S}^{-1}$, so

$$\text{lub}(\lambda\mathbf{I} - \mathbf{B})^{-1} \leq \text{lub}(\lambda\mathbf{I} - \Lambda(\mathbf{B}))^{-1}\text{cond}(\mathbf{S})$$

(recall that least upper bound norms are submultiplicative). Then, (3.39c) becomes

$$1 \leq \text{lub}(\mathbf{A} - \mathbf{B})\text{lub}(\lambda\mathbf{I} - \Lambda(\mathbf{B}))^{-1}\text{cond}(\mathbf{S}),$$

and, for any *absolute norm*, for any eigenvalue $\lambda(\mathbf{A})$ of matrix $\mathbf{A}$, there is an eigenvalue $\lambda(\mathbf{B})$ of matrix $\mathbf{B}$ such that

$$|\lambda(\mathbf{A}) - \lambda(\mathbf{B})| \leq \text{lub}(\mathbf{A} - \mathbf{B})\text{cond}(\mathbf{S}). \tag{3.40}$$

If $\mathbf{B}$ is normal, then $\text{cond}(\mathbf{S}) = 1$.

The point of (3.40) is this. Let $\mathbf{B}$ be $\mathbf{A}$ with small elements rounded to zero, then the exact eigenvalues of $\mathbf{B}$ are close to the eigenvalues of $\mathbf{A}$.

In Example 3.6 iteration was stopped when $\mathbf{A}$ was nearly diagonal. In this case, we think of $\mathbf{B}$ as being the diagonal part of $\mathbf{A}$, and with $\mathbf{B}$ diagonal we have $\text{cond}(\mathbf{S}) = 1$. The error associated with taking the diagonal elements of (3.37) as eigenvalues is $\mathcal{O}(6 \times 10^{-4})$.

However, this error bound does not apply equally for all eigenvalues. The error associated with treating (3.37) as being reducible is $\mathcal{O}(10^{-83})$ – entirely negligible. According to the discussion of reducibility, this means that the eigenvalue 2×10^{-10} is essentially exact, so the error $\mathcal{O}(6 \times 10^{-4})$ applies only to the eigenvalues of $\mathcal{O}(1)$ from the upper left 2×2 block.

Since QR is analogous to inverse vector iteration, the nth eigenvalue will emerge very quickly if shifting is used. Shifting in the present circumstance is equivalent to

$$\mathbf{A}^{(k)} - \mu^{(k)}\mathbf{I} \rightarrow \mathbf{Q}^{(k)}\mathbf{R}^{(k)}$$

$$\mathbf{R}^{(k)}\mathbf{Q}^{(k)} + \mu^{(k)}\mathbf{I} \rightarrow \mathbf{A}^{(k+1)}, \tag{3.41}$$

but the implementation for Hessenberg matrices is much simpler than these formulas imply. After the Hessenberg form is established, the first operation of each iteration is the construction of a Givens reflection $\mathbf{G}_{12}$ designed to operate on the two matrix elements $A_{1,1}$ and $A_{2,1}$. All subsequent operations are governed by the outcome of this first reflection (see, e.g., (3.34)). Therefore, constructing $\mathbf{G}_{12}$ as if it were to operate on the two elements $A_{1,1} - \mu$ and $A_{2,1}$ sets into play the operations implied by sequence (3.41).

Let us assume that the eigenvalues are real and simple. If the shift $\mu^{(k)}$ is chosen to expedite convergence of the smallest (in magnitude) eigenvalue, then $\mathbf{A}$ will quickly evolve into a reduced form, with the lower right block being 1×1 and consisting of the smallest eigenvalue. Once this circumstance is detected, the matrix $\mathbf{A}$ can be reduced in dimension so subsequent operations will proceed more rapidly.

This strategy suggests the simple choice $\mu^{(k)} = A_{n,n}$. However, $A_{n,n}$ is not a good estimate of the eigenvalue unless $A_{n-1,n}$ is small. A better estimate is to compute the eigenvalues μ of

$$\begin{pmatrix} A_{n-1,n-1} & A_{n-1,n} \\ A_{n-1,n} & A_{n,n} \end{pmatrix},$$

and choose the value closest to $A_{n,n}$.

Jacobi's method [119] for symmetric positive definite systems also employs so-called Givens matrices, but sequentially as follows:

- Begin with $\mathbf{S} = \mathbf{I}$.
- Find the largest (in magnitude) off-diagonal element of $\mathbf{A}_{ij}$.
- If the largest off-diagonal is zero the matrix has been diagonalized. Columns $\mathbf{S}$ are eigenvectors, and diagonal elements of $\mathbf{A}$ are eigenvalues.
- Otherwise, find a Givens matrix $\mathbf{G}_{ij}$ to eliminate element $\mathbf{A}_{ij}$ by the following similarity transformation:

$$\mathbf{A} := \mathbf{G}\mathbf{A}\mathbf{G}^{\mathrm{T}}.$$

Perform this transformation, and update $\mathbf{S}$ with

$$\mathbf{S} := \mathbf{S}\mathbf{G}^{\mathrm{T}}.$$

Under some circumstances this simple method is more accurate than QR [57].

The backward error associated the symmetric QR eigenvalue problem is [243]

$$\|\tilde{\mathbf{A}} - \tilde{\mathbf{P}}^{-1}\mathbf{A}\tilde{\mathbf{P}}\|_2 \le Kn\epsilon\|\mathbf{A}\|_2, \tag{3.42}$$

for some constant K, where $\|\cdot\|_2$ is the Frobenius norm, $\mathbf{A} \in \mathbb{R}^{n \times n}$, $\tilde{\mathbf{A}}$ is the numerical eigenvalue result, and $\tilde{\mathbf{P}}$ is the numerical matrix of eigenvectors.

3.4   Singular value decomposition

The singular value decomposition (SVD) of a matrix $\mathbf{A}$ takes the form

$$\mathbf{A} = \mathbf{U}\boldsymbol{\Sigma}\mathbf{V}^{\mathrm{T}},$$

where $\mathbf{U}$ and $\mathbf{V}$ are unitary, and $\mathbf{\Sigma}$ is diagonal. Singular value decomposition applies to any matrix $\mathbf{A}$, even matrices that are not square. If $\mathbf{A}$ is $m \times n$, then $\mathbf{U}$ is $m \times m$, $\mathbf{\Sigma}$ is $m \times n$, and $\mathbf{V}^\mathrm{T}$ is $n \times n$. The discovery of this form of matrix reduction occurred independently by Beltrami [12] and Jordan [121], and the first algorithm was developed by Golub and Kahan [91]. A historical review was researched by Stewart [223]. The algorithm presented below is due to Golub and Reinsch [92].

The diagonal entries of $\mathbf{\Sigma}$ are called the singular values. They are the positive square roots of the eigenvalues of matrix $\mathbf{A}^\mathrm{T}\mathbf{A}$:

$$\mathbf{A}^\mathrm{T}\mathbf{A} = (\mathbf{U}\mathbf{\Sigma}\mathbf{V}^\mathrm{T})^\mathrm{T}(\mathbf{U}\mathbf{\Sigma}\mathbf{V}^\mathrm{T}) = \mathbf{V}\mathbf{\Sigma}^\mathrm{T}\mathbf{\Sigma}\mathbf{V}^\mathrm{T}.$$

This shows that $\mathbf{\Sigma}^\mathrm{T}\mathbf{\Sigma}$ is a diagonal matrix similar to $\mathbf{A}^\mathrm{T}\mathbf{A}$, and therefore the diagonal entries σ_i^2 of matrix $\mathbf{\Sigma}^\mathrm{T}\mathbf{\Sigma}$ are eigenvalues of $\mathbf{A}^\mathrm{T}\mathbf{A}$.

If one wanted only the singular values of $\mathbf{A}$, one could build the matrix $\mathbf{A}^\mathrm{T}\mathbf{A}$ and find its eigenvalues with vector iteration or the QR method. Don't do this. This is not a good idea because the condition number of $\mathbf{A}^\mathrm{T}\mathbf{A}$ is the square of the condition number of $\mathbf{A}$, so numerical errors are unnecessarily amplified.

Instead, the approach we will take is to mimic the steps of QR for a Hermitian matrix (because $\mathbf{A}^\mathrm{T}\mathbf{A}$ is Hermitian), without actually ever making $\mathbf{A}^\mathrm{T}\mathbf{A}$.

In QR, our transformation of $\mathbf{A}^{(i)}$ to $\mathbf{A}^{(i+1)}$ is a similarity transformation, $\mathbf{A}^{(i+1)} = (\mathbf{Q}^{(i)})^\mathrm{H}\mathbf{A}^{(i)}\mathbf{Q}^{(i)}$. For SVD, we can write

$$\mathbf{A}^{(i+1)} = \mathbf{P}^{(i)}\mathbf{A}^{(i)}\mathbf{Q}^{(i)}, \tag{3.43}$$

where $\mathbf{P}$ and $\mathbf{Q}$ are suitably chosen unitary matrices. They cannot be the same in general since $\mathbf{A}$ need not be square, and they need not be the same even if $\mathbf{A}$ is square because we're really seeking to transform $\mathbf{A}^\mathrm{T}\mathbf{A}$. With (3.43), the Hermitian matrix $\mathbf{A}^\mathrm{T}\mathbf{A}$ (which we never actually make) is transformed as

$$(\mathbf{A}^{(i+1)})^\mathrm{T}\mathbf{A}^{(i+1)} = (\mathbf{Q}^{(i)})^\mathrm{T}(\mathbf{A}^{(i)})^\mathrm{T}(\mathbf{P}^{(i)})^\mathrm{T}\mathbf{P}^{(i)}\mathbf{A}^{(i)}\mathbf{Q}^{(i)} = (\mathbf{Q}^{(i)})^\mathrm{T}(\mathbf{A}^{(i)\mathrm{T}}\mathbf{A}^{(i)})\mathbf{Q}^{(i)} \tag{3.44}$$

because $\mathbf{P}$ is unitary.

The first step of QR for a Hermitian matrix is its reduction to the tridiagonal form, using Householder matrices. The first step of SVD will be to reduce $\mathbf{A}$ to a bidiagonal matrix $\mathbf{J}$, such that $\mathbf{J}^\mathrm{T}\mathbf{J}$ is the same as the tridiagonal matrix that would have been built by QR. For QR, this consisted of special Householder matrices $\mathbf{Q}_i$, so

$$\mathbf{Q}_{n-1} \cdots \mathbf{Q}_2\mathbf{A}\mathbf{Q}_2^\mathrm{H} \cdots \mathbf{Q}_{n-1}^\mathrm{H}$$

resulted in the tridiagonal form (e.g., (3.27), (3.28)). As (3.44) suggests, this first step of SVD will use different Householder matrices on the left ($\mathbf{P}$) and on the right ($\mathbf{Q}$), and the matrices used on the right are exactly the same as would be used if we were doing QR on $\mathbf{A}^\mathrm{T}\mathbf{A}$.

The QR iteration for tridiagonal symmetric matrices consists of select Givens reflections that preserve the tridiagonal form. In SVD we will again choose select givens matrices, now preserving the bidiagonal form.

This brief discussion asserts that the QR algorithm will transform $\mathbf{J}$ into a matrix such that $\mathbf{J}^T\mathbf{J}$ is diagonal. However, this does not mean that $\mathbf{J}$ is diagonal. Consider for instance

$$\mathbf{J} = \begin{pmatrix} 0 & 1 \\ 0 & 0 \end{pmatrix}$$

$$\mathbf{J}^T\mathbf{J} = \begin{pmatrix} 0 & 0 \\ 0 & 1 \end{pmatrix}.$$

So, the SVD algorithm requires one additional step to assure that $\mathbf{J}$ reduces to the positive diagonal matrix $\mathbf{\Sigma}$.

Consider the general case

$$\mathbf{J} = \begin{pmatrix} \ddots & & \ddots & \\ & d_i & s_i & \\ & & \ddots & \ddots \end{pmatrix}$$

$$\mathbf{J}^T\mathbf{J} = \begin{pmatrix} d_0^2 & d_0 s_0 & & & & \\ d_0 s_0 & \ddots & & \ddots & & \\ & \ddots & d_i^2 + s_{i-1}^2 & d_i s_i & & \\ & & d_i s_i & \ddots & & \ddots \\ & & & \ddots & d_{n-1}^2 + s_{n-2}^2 & d_{n-1} s_{n-1} \\ & & & & d_{n-1} s_{n-1} & d_n^2 + s_{n-1}^2 \end{pmatrix}.$$

This pattern shows that for $\mathbf{J}^T\mathbf{J}$ to be diagonal, with $\mathbf{J}$ not diagonal, it must be that there is a zero diagonal entry, $d_i = 0$, with corresponding superdiagonal entry nonzero, $s_i \neq 0$. This can be rectified by applying suitable Givens matrices from the left, with a corresponding modification to $\mathbf{U}$.

Example 3.7   Find the singular value decomposition of

$$\mathbf{A} = \begin{pmatrix} 1 & 0 & 5 \\ -10 & 2 & -10 \\ 4 & 1 & 0 \end{pmatrix}.$$

Solution

First, use Householder matrices to transform $\mathbf{A}$ into a bidiagonal form. Matrix $\mathbf{J}$ will transform into the singular value matrix $\mathbf{\Sigma}$.

$$\mathbf{A} = \mathbf{U}\mathbf{J}\mathbf{V}^T$$

$$\mathbf{U} = \begin{pmatrix} -0.092\,450 & -0.715\,39 & -0.692\,58 \\ 0.924\,50 & 0.196\,65 & -0.326\,54 \\ -0.369\,80 & 0.670\,48 & -0.643\,20 \end{pmatrix}$$

$$J = \begin{pmatrix} -10.8171 & -9.8193 & 0 \\ 0 & -5.6405 & -0.216\,56 \\ 0 & 0 & 1.3112 \end{pmatrix}$$

$$V^T = \begin{pmatrix} 1.0000 & 0.0000 & 0.0000 \\ 0.0000 & -0.150\,64 & 0.988\,59 \\ 0.0000 & -0.988\,59 & -0.150\,64 \end{pmatrix}.$$

The iteration sequence begins in analogy to QR. First, a shift can be determined from the lower 2×2 block of the matrix $J^T J$:

$$J^T J = \begin{pmatrix} * & * & 0 \\ * & 128.234 & -1.2215 \\ 0 & -1.2215 & 1.766\,14 \end{pmatrix}.$$

The eigenvalues of the lower block are 128.25 and $1.754\,34$: the latter is closest to $(J^T J)_{33}$ and is therefore the better shift.

First, a Givens rotation is constructed to diagonalize the implicitly shifted matrix $(J^T J - \mu I)$, i.e.,

$$\begin{pmatrix} c & s \\ s & -c \end{pmatrix} \begin{pmatrix} (J^T J)_{11} - \mu \\ (J^T J)_{21} \end{pmatrix} = \begin{pmatrix} * \\ 0 \end{pmatrix},$$

with parameters c, s selected by the algorithm (3.30). This Givens matrix operates on J from the right, and V^T from the left, i.e.,

$$A = U(J G_{12})(G_{12} V^T),$$

giving

$$J := \begin{pmatrix} -14.608 & -0.109\,92 & 0 \\ \underline{-3.8225} & 4.1477 & 0.216\,56 \\ 0 & 0 & 1.3112 \end{pmatrix}$$

$$V^T := \begin{pmatrix} 0.735\,34 & -0.102\,09 & 0.669\,97 \\ 0.677\,70 & 0.110\,77 & -0.726\,95 \\ 0.000\,00 & 0.988\,59 & -0.150\,64 \end{pmatrix}.$$

This operation created a nonzero element (underlined) in the 2, 1 position of J, spoiling its bidiagonal form. We reestablish the bidiagonal form by creating a new Givens operation G'_{12} designed to operate on J from the left, i.e.,

$$A = (U G'_{12})(G'_{12} J) V^T,$$

giving

$$U := \begin{pmatrix} -0.270\,54 & 0.668\,69 & 0.692\,58 \\ 0.944\,17 & 0.043\,78 & 0.326\,54 \\ -0.188\,03 & -0.742\,25 & 0.643\,20 \end{pmatrix}$$

$$J := \begin{pmatrix} -15.100 & 0.943\,61 & \underline{0.054\,822} \\ 0 & -4.0404 & -0.209\,51 \\ 0 & 0 & 1.3112 \end{pmatrix}.$$

A new nonzero element (underlined) was created that spoils the bidiagonal form. This is fixed by operating on $\mathbf{J}$ with a Givens rotation $\mathbf{G}_{23}$ from the right, i.e.,

$$\mathbf{A} = \mathbf{U}(\mathbf{J}\mathbf{G}_{23})(\mathbf{G}_{23}\mathbf{V}^{\mathsf{T}}),$$

giving

$$\mathbf{J} := \begin{pmatrix} -15.100 & 0.94521 & 0 \\ 0 & -4.0457 & -2.5185 \\ 0 & \underline{0.076\,052} & -1.3090 \end{pmatrix}$$

$$\mathbf{V}^{\mathsf{T}} := \begin{pmatrix} 0.735\,34 & -0.102\,09 & 0.669\,97 \\ 0.676\,56 & 0.167\,92 & -0.716\,99 \\ 0.039\,307 & -0.980\,50 & -0.192\,55 \end{pmatrix}.$$

Again the symmetry is spoiled, and fixed by operation from the left of $\mathbf{J}$ with a Givens rotation $\mathbf{G}'_{23}$, i.e.,

$$\mathbf{A} = (\mathbf{U}\mathbf{G}'_{23})(\mathbf{G}'_{23}\mathbf{J})\mathbf{V}^{\mathsf{T}},$$

giving

$$\mathbf{U} := \begin{pmatrix} -0.269\,19 & 0.657\,52 & 0.703\,71 \\ 0.944\,26 & 0.036\,399 & 0.327\,19 \\ -0.189\,52 & -0.741\,87 & -0.643\,20 \end{pmatrix}$$

$$\mathbf{J} := \begin{pmatrix} -15.100 & 0.94521 & 0 \\ 0 & -4.0465 & -5.7793 \times 10^{-4} \\ 0 & 0 & 1.3093 \end{pmatrix}.$$

This completes one iteration. After three more iterations

$$\mathbf{U} = \begin{pmatrix} -0.282\,28 & -0.650\,62 & -0.704\,99 \\ 0.943\,34 & -0.054\,633 & -0.327\,30 \\ -0.174\,43 & 0.757\,44 & -0.629\,17 \end{pmatrix}$$

$$\mathbf{J} = \begin{pmatrix} 15.132 & 0 & 0 \\ 0 & 4.0379 & 0 \\ 0 & 0 & 1.3093 \end{pmatrix} = \Sigma$$

$$\mathbf{V}^{\mathsf{T}} = \begin{pmatrix} -0.688\,17 & 0.113\,15 & -0.716\,68 \\ 0.724\,49 & 0.160\,52 & -0.670\,33 \\ 0.039\,191 & -0.980\,53 & -0.192\,44 \end{pmatrix}.$$

In this final result, $\mathbf{J} = \Sigma$ has been multiplied by a diagonal matrix $\mathrm{diag}(\pm 1, \pm 1, \cdots)$ to respect the convention that the singular values Σ_{ii} be positive. The matrix $\mathbf{V}^{\mathsf{T}}$ is pre-multiplied by this same diagonal phase factor.

The backward error associated with SVD [93] is similar to that of the symmetric QR method, (3.42),

$$\|\Sigma - \tilde{\mathbf{U}}^{\mathsf{T}}\tilde{\mathbf{A}}\tilde{\mathbf{V}}\|_2 \leq K\sqrt{nm}\epsilon\|\mathbf{A}\|_2,$$

for some constant K, where $\mathbf{A} \in \mathbb{R}^{m \times n}$. Unfortunately, this error analysis does not give a favorable absolute bound on the singular values themselves, particularly the smallest ones. If $\mathbf{A}$ is very poorly conditioned, $\mathrm{cond}_2(\mathbf{A})$ may be very different from $\tilde{\sigma}_1/\tilde{\sigma}_n$ because of the numerical inaccuracy of $\tilde{\sigma}_n$ (see Example 3.8 below). Demmel *et al.* [56] have proposed ways of computing SVD with high relative accuracy for a number of poorly conditioned matrices.

The most significant application of SVD is to the solution of linear equations that are poorly conditioned or even singular. When $\mathbf{A}$ is singular, some of the singular values will be zero, and the decomposed matrix equation will have the form

$$\mathbf{A}\mathbf{x} = \mathbf{U} \left(\begin{array}{c|c} \boldsymbol{\Sigma}_1 & \\ \hline & \mathbf{0} \end{array} \right) \left(\begin{array}{c} \mathbf{V}_1^\mathsf{T} \\ \hline \mathbf{V}_2^\mathsf{T} \end{array} \right) \mathbf{x} = \mathbf{b},$$

and

$$\mathbf{x} = \mathbf{V}_1 \boldsymbol{\Sigma}_1^{-1} \mathbf{U}^\mathsf{T} \mathbf{b} \tag{3.45}$$

is the solution obtained from the upper partition. This solution does not necessarily solve $\mathbf{A}\mathbf{x} = \mathbf{b}$:

$$\mathbf{A}\mathbf{x} = \mathbf{U} \left(\begin{array}{c|c} \boldsymbol{\Sigma}_1 & \\ \hline & \mathbf{0} \end{array} \right) \left(\begin{array}{c} \mathbf{V}_1^\mathsf{T} \\ \hline \mathbf{V}_2^\mathsf{T} \end{array} \right) \mathbf{V}_1 \boldsymbol{\Sigma}_1^{-1} \mathbf{U}^\mathsf{T} \mathbf{b}$$

$$= \left(\begin{array}{c|c} \mathbf{U}_1 & \mathbf{U}_2 \end{array} \right) \left(\begin{array}{c|c} \mathbf{I} & \\ \hline & \mathbf{0} \end{array} \right) \left(\begin{array}{c} \mathbf{U}_1^\mathsf{T} \\ \hline \mathbf{U}_2^\mathsf{T} \end{array} \right) \mathbf{b}$$

$$= \mathbf{U}_1 \mathbf{U}_1^\mathsf{T} \mathbf{b} = (\mathbf{I} - \mathbf{U}_2 \mathbf{U}_2^\mathsf{T}) \mathbf{b}$$

$$\neq \mathbf{b} \qquad \text{if } \mathbf{U}_2^\mathsf{T} \mathbf{b} \neq 0.$$

And, for any vector $\mathbf{c}$,

$$\mathbf{A}\mathbf{x} = \mathbf{A}(\mathbf{x} + \mathbf{V}_2 \mathbf{c}). \tag{3.46}$$

From this we see that the component of $\mathbf{b}$ that lies in the span of $\mathbf{U}_2$ cannot influence $\mathbf{x}$. By varying $\mathbf{c}$, we obtain an infinite number of solutions $\mathbf{y}$ to $\mathbf{A}\mathbf{y} = \mathbf{U}_1 \mathbf{U}_1^\mathsf{T} \mathbf{b}$, but $\mathbf{y} = \mathbf{x}$, given by (3.45), is the smallest solution in L^2.

When a matrix is very poorly conditioned, the solution error is dominated by the smallest singular values [175]. For this reason, it is sometimes advantageous to simply omit very small singular values [91] by replacing them with zero.

Unfortunately, there is not a simple way to choose which singular values are significant, and which can be ignored. If one omitted all singular values less than some fraction r of the greatest singular value, σ_1, then one would be modifying $\mathbf{A}$ to be some related matrix $\tilde{\mathbf{A}}$ which, while singular, would have the numerical error associated with condition number $\mathrm{cond}_2(\tilde{\mathbf{A}}) < r^{-1}$. This is clearly advantageous from the numerical error perspective, but the problem solved $\tilde{\mathbf{A}}\mathbf{x} = \mathbf{b}$ is not the original problem $\mathbf{A}\mathbf{x} = \mathbf{b}$: an approximation error is committed. And, since $\mathbf{A}$ is singular, or nearly so, the approximation error is unbounded. The choice of singular value truncation criterion is therefore not really a mathematical one: it must be justified in the context of the application.

Nonetheless, there is a good computational strategy to assess benefits and costs of truncating the singular value spectrum. Lawson and Hanson [143] advocate computing $\mathbf{x}_k$, the solution using the greatest k singular values only, and plotting the residual $\|\mathbf{r}_k\|_2$ against the norm of the solution $\|\mathbf{x}_k\|_2$. Often, what is seen is a cusp in the plot where increasing k increases $\|\mathbf{x}_k\|_2$ without diminishing $\|\mathbf{r}_k\|_2$. This signals the case where the solution is growing in the null space of the matrix (the case $\mathbf{c} \neq 0$ of (3.46)).

Example 3.8 The $n \times n$ Hilbert Matrix $\mathbf{H}_n$,

$$
\mathbf{H}_n = \begin{pmatrix}
1 & \frac{1}{2} & \frac{1}{3} & \cdots & \frac{1}{n} \\
\frac{1}{2} & \frac{1}{3} & \frac{1}{4} & \cdots & \frac{1}{n+1} \\
\frac{1}{3} & \frac{1}{4} & \frac{1}{5} & \cdots & \frac{1}{n+2} \\
\vdots & & & & \vdots \\
\frac{1}{n} & \frac{1}{n+1} & \frac{1}{n+2} & \cdots & \frac{1}{2n-1}
\end{pmatrix},
$$

is a notoriously poorly conditioned symmetric positive definite matrix. For large n [241],

$$
\text{cond}(\mathbf{H}_n) \sim \frac{(1+\sqrt{2})^{4n}}{\sqrt{n}},
$$

and $\text{cond}(\mathbf{H}_{20}) = 2.45 \times 10^{28}$, significantly larger than $1/\epsilon$. For $\mathbf{H}_{20}$, a relative rounding error of 1ϵ in $\mathbf{b}$ will result in a relative solution error bound of $\mathcal{O}(10^{12})$ in $\mathbf{x}$. Use truncated SVD to solve

$$
\mathbf{H}_{20}\mathbf{x} = \mathbf{b}
$$

$$
\mathbf{b}_i = \sum_{j=1}^{20} \frac{1}{i+j-1},
$$

i.e., such that $\mathbf{x} = \mathbf{1}$ is the exact solution.

Solution
The SVD is computed, and the solution computed including the k greatest singular values, $k = 1, ..., 20$, is shown in Figure 3.3. When $k = 10$, there is a distinct cusp in $\|\mathbf{r}_2\|$. As k increases beyond that point, the residual does not improve, thought the solution vector increases.

For this particular problem we know the solution, $\mathbf{x} = \mathbf{1}$, so we can also look at the norm $\|\mathbf{x} - \mathbf{1}\|_2$, Figure 3.4. In this figure it is clear that $k = 10$ achieves the optimum solution error, validating Lawson and Hanson's truncation criterion for this extremely poorly conditioned problem.

The numerical condition number σ_1/σ_n is 3×10^{18}. This can be trusted to conclude that $\mathbf{H}_{20}$ is poorly conditioned, but the condition number itself is in error by a factor of 10^{10} owing to numerical error in σ_n.

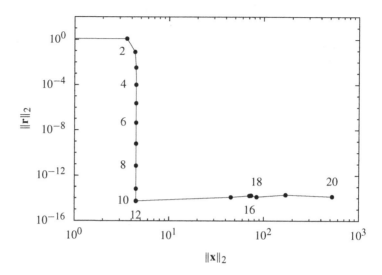

Figure 3.3 Residual versus solution norm for truncated SVD; Example 3.8.

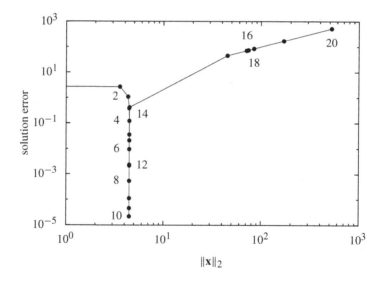

Figure 3.4 Solution error versus solution norm for truncated SVD; Example 3.8.

Additional applications of SVD are described in Chapter 8.

3.5 Hyman's method

Hyman's method for the determination of eigenvalues is based on root finding methods for polynomials, applied to a function with the same roots as the characteristic equation. It is attributed to a conference presentation by M. A. Hyman of the Naval Ordnance Laboratory [116]. The stability of Hyman's method is shown by Wilkinson [242], and its advantages are described in [76]. It is particularly advantageous relative to QR for defective matrices.

We assume that the matrix $\mathbf{A}$ is upper Hessenberg. If it is not, it can be made upper Hessenberg with a similarity transformation using Householder matrices, as for the QR eigenvalue method.

The characteristic equation, $\det(\mathbf{A} - \lambda\mathbf{I}) = 0$, is unaffected by adding multiples of one column to another. Thus,

$$\mathbf{A} - \lambda\mathbf{I} = \begin{pmatrix} a_{11} - \lambda & a_{12} & \cdots & a_{1,n-1} & a_{1n} \\ a_{21} & a_{22} - \lambda & \cdots & a_{2,n-1} & a_{2n} \\ & a_{32} & \cdots & a_{3,n-1} & a_{3n} \\ & & \ddots & \vdots & \vdots \\ & & & a_{n,n-1} & a_{nn} - \lambda \end{pmatrix}$$

and

$$\mathbf{B} = \begin{pmatrix} a_{11} - \lambda & a_{12} & \cdots & a_{1,n-1} & b \\ a_{21} & a_{22} - \lambda & \cdots & a_{2,n-1} & 0 \\ & a_{32} & \cdots & a_{3,n-1} & 0 \\ & & \ddots & \vdots & \vdots \\ & & & a_{n,n-1} & 0 \end{pmatrix}$$

have the same determinant if $\mathbf{B}$ is created from $(\mathbf{A} - \lambda\mathbf{I})$ by adding multiples of columns 1 through $n - 1$ of $(\mathbf{A} - \lambda\mathbf{I})$ to column n of $(\mathbf{A} - \lambda\mathbf{I})$, i.e., if

$$(\mathbf{A} - \lambda\mathbf{I}) \begin{pmatrix} x_1 \\ x_2 \\ \vdots \\ x_{n-1} \\ 1 \end{pmatrix} = \begin{pmatrix} b \\ 0 \\ \vdots \\ 0 \\ 0 \end{pmatrix}. \tag{3.47}$$

Note that b will depend not only on the elements of $\mathbf{A}$, but also on the parameter λ.

The determinant of $\mathbf{B}$ is particularly easy to determine by expanding in cofactors of the nth column:

$$\det(\mathbf{A} - \lambda\mathbf{I}) = \det\mathbf{B} = (-1)^n a_{21} a_{32} \cdots a_{n,n-1} b(\lambda).$$

So, $b(\lambda)$ is an auxiliary function whose roots are the same as the characteristic equation of $\mathbf{A}$.

An algorithm for the determination of the factors x_i comes from analyzing (3.47) one row at a time beginning with the last:

$$x_n = 1$$
$$x_{n-1} = -(a_{nn} - \lambda)x_n/a_{n,n-1}$$
$$x_{n-2} = -((a_{n-1,n-1} - \lambda)x_{n-1} + a_{n-1,n}x_n)/a_{n-1,n-2}$$
$$\vdots$$
$$x_i = -\left((a_{i+1,i+1} - \lambda)x_{i+1} + \sum_{j=i+2}^{n} a_{i+1,j}x_j\right)/a_{i+1,i}$$
$$\vdots$$

for $i = n - 3, ..., 1$. Then, b is given by

$$b(\lambda) = (a_{11} - \lambda)x_1 + a_{12}x_2 + \cdots + a_{1,n-1}x_{n-1} + a_{1n}.$$

The eigenvalues λ of matrix $\mathbf{A}$ are the roots of the polynomial function $b(\lambda)$, i.e., those numbers λ such that $b(\lambda) = 0$. Numerical root finding methods will be described in Chapter 6. If it is known that the eigenvalues are simple and real, the method of zero suppression is well-suited for determining all roots of $b(\lambda)$. That method requires also $b'(\lambda)$, which is easily found by differentiating (3.47):

$$(\mathbf{A} - \lambda\mathbf{I})\begin{pmatrix} x'_1 \\ x'_2 \\ \vdots \\ x'_{n-1} \\ 0 \end{pmatrix} - \begin{pmatrix} x_1 \\ x_2 \\ \vdots \\ x_{n-1} \\ 1 \end{pmatrix} = \begin{pmatrix} b'(\lambda) \\ 0 \\ \vdots \\ 0 \\ 0 \end{pmatrix}.$$

The factors x' are then given by

$$x'_n = 0$$
$$x'_{n-1} = 1/a_{n,n-1}$$
$$x'_{n-2} = \left(1 - (a_{n-1,n-1} - \lambda)x'_{n-1}\right)/a_{n-1,n-2}$$
$$\vdots$$
$$x'_i = \left(1 - (a_{i+1,i+1} - \lambda)x'_{i+1} - \sum_{j=i+2}^{n-1} (a_{i+1,j}x'_j)\right)/a_{i+1,i}$$
$$\vdots$$

and

$$b'(\lambda) = (a_{11} - \lambda)x'_1 + \sum_{j=2}^{n} a_{1j}x'_j - x_1.$$

Problems

3.1   Show the relation between the 2×2 Givens reflection matrix and a Householder Q matrix designed to operate on the same $\mathbb{R}^2$ vector.

3.2   Show that the **A** defined by (3.41) are similar to each other.

3.3   This problem concerns finding the smallest resonant frequency of a rod of length L which is clamped at both ends, like a guitar string. The motion of the rod is governed by the wave equation,

$$\frac{\partial^2 u}{\partial t^2} = c(x)^2 \frac{\partial^2 u}{\partial x^2}.$$

Here, $c(x)$ is the sound speed which can vary with coordinate. If we assume periodic motion $u(x, t) = U(x)\cos(\omega t)$, the wave equation becomes the Helmholtz equation

$$c(x)^2 \frac{\partial^2 U}{\partial x^2} + \omega^2 U = 0.$$

One way to express this as a numerical problem is to (i) introduce the discretization $x_i = ih$, $h = L/(N+1)$, $i = 0, ..., N+1$, for some integer N; (ii) approximate the second derivative with respect to x using finite differences

$$\left.\frac{\partial^2 U}{\partial x^2}\right|_{x_i} \approx \frac{U_{i+1} - 2U_i + U_{i-1}}{h^2};$$

and (iii) apply the boundary conditions $U_0 = U_{N+1} = 0$ (where $U_k = U(x_k) = U(kh)$). With these steps, the Helmholtz equation becomes a matrix equation

$$\begin{pmatrix} 2\frac{c_1^2}{h^2} - \omega^2 & -\frac{c_1^2}{h^2} & & & \\ -\frac{c_2^2}{h^2} & 2\frac{c_2^2}{h^2} - \omega^2 & -\frac{c_2^2}{h^2} & & \\ & \ddots & \ddots & \ddots & \\ & & -\frac{c_{N-1}^2}{h^2} & 2\frac{c_{N-1}^2}{h^2} - \omega^2 & -\frac{c_{N-1}^2}{h^2} \\ & & & -\frac{c_N^2}{h^2} & 2\frac{c_N^2}{h^2} - \omega^2 \end{pmatrix} \begin{pmatrix} U_1 \\ U_2 \\ \vdots \\ U_{N-1} \\ U_N \end{pmatrix} = 0,$$

which is an eigenvalue problem where the eigenvalues are the special values of ω^2 where resonance occurs.

Find the smallest resonant frequency for a rod of length $L = 1$ with sound speed $c(x) = 3 + (x/L)^2$.

3.4   Let $\bar{x}$ be the exact solution to $\mathbf{A}\mathbf{x} = \mathbf{b}$. Determine the relation between the solution error vector

$$\mathbf{e} = \mathbf{x} - \bar{\mathbf{x}}$$

and the residual

$$\mathbf{r} = \mathbf{b} - \mathbf{A}\mathbf{x}$$

in the L^2 norm. Hint: expand **e** in columns of **V** – the unitary matrix of the SVD decomposition.

3.5 Use Hyman's method to construct $b(\lambda)$ for

$$
\mathbf{A} = \begin{pmatrix}
0 & 0 & 0 & 0 & 0 & -a_0 \\
1 & 0 & 0 & 0 & 0 & -a_1 \\
0 & 1 & 0 & 0 & 0 & -a_2 \\
0 & 0 & 1 & 0 & 0 & -a_3 \\
0 & 0 & 0 & 1 & 0 & -a_4 \\
0 & 0 & 0 & 0 & 1 & -a_5
\end{pmatrix},
$$

which is already in the Hessenberg form (cf. (A.22)).

3.6 Construct the $n = 20$ Vandermonde matrix and the corresponding vector $\mathbf{b}$ as instructed in Problem 2.3. Solve $\mathbf{Vx} = \mathbf{b}$ for $\mathbf{x}$ using truncated SVD. Also estimate the condition number of this matrix.

3.7 [58] Let $\mathbf{A}$ be a matrix of rank k, and let $\mathbf{B}$ be a matrix of rank $s < k$. Show that the matrix $\mathbf{B}$ closest to $\mathbf{A}$ in the least squares sense is the matrix

$$
\mathbf{B} = \mathbf{U} \operatorname{diag}(\sigma_1, \sigma_2, ..., \sigma_s, 0, ..., 0), \mathbf{V}^\mathrm{T}
$$

where $\mathbf{U\Sigma V}^\mathrm{T}$ is the singular value decomposition of $\mathbf{A}$. Here, closeness in the least squares sense means that

$$
\|\mathbf{A} - \mathbf{B}\|_2 = \sqrt{\operatorname{tr}\left((\mathbf{A} - \mathbf{B})^\mathrm{T}(\mathbf{A} - \mathbf{B})\right)}
$$

is a minimum. $\| \cdot \|_2$ is the Frobenius norm.

4 Iterative approaches for linear systems

There are two fundamentally different types of iterative algorithms for the solution of $\mathbf{Ax} = \mathbf{b}$ with $\mathbf{A}$ $n \times n$: those that converge to the exact answer in $\mathcal{O}(n)$ steps (assuming exact math), and those that approach the solution geometrically. The conjugate gradient method [103] typifies the former category. The latter category are known as relaxation methods: Jacobi and Gauss–Seidel are two simple relaxation strategies that will be described.

All of these iterative approaches have the benefit that the matrix $\mathbf{A}$ is not modified as the solution is developed. This is especially important when $\mathbf{A}$ is very large, and when $\mathbf{A}$ is very sparse. This is the most significant distinction between iterative approaches and elimination type approaches described in Chapter 2.

4.1 Conjugate gradient

We assume initially that $\mathbf{A}$ is symmetric positive definite and consider a strategy to solve $\mathbf{Ax} = \mathbf{b}$ based on the error formula

$$f(\mathbf{x}) = (\mathbf{x} - \bar{\mathbf{x}})\mathbf{A}(\mathbf{x} - \bar{\mathbf{x}}),$$

where $\bar{\mathbf{x}}$ is the exact solution. $f > 0$ when $\mathbf{x} \neq \bar{\mathbf{x}}$ because $\mathbf{A}$ is symmetric positive definite, so $\mathbf{x} = \bar{\mathbf{x}}$ is the unique root of the error function. The gradient of $f(\mathbf{x})$ gives the direction in which f increases most rapidly:

$$\nabla f = 2\mathbf{A}(\mathbf{x} - \bar{\mathbf{x}}) = -2\mathbf{r},$$

where

$$\mathbf{r} = \mathbf{b} - \mathbf{Ax}$$

is the *residual*.

This implies that for any vector $\mathbf{x}^{(0)}$ one could find a better vector $\mathbf{x}^{(1)}$ using a search direction $\mathbf{p}_0 = \mathbf{r}_0$:

$$\mathbf{x}^{(1)} = \mathbf{x}^{(0)} + \alpha_0 \mathbf{p}_0, \tag{4.1}$$

where α_0 is a positive parameter. In fact, α_0 can be determined exactly by finding the value for which

$$\frac{df(\mathbf{x}^{(0)} + \alpha_0 \mathbf{p}_0)}{d\alpha_0} = 0.$$

Expanding f

$$f = (\mathbf{x}^{(0)} - \bar{x})^{\mathrm{T}} \mathbf{A} (\mathbf{x}^{(0)} - \bar{x}) + 2\alpha_0 \mathbf{p}_0^{\mathrm{T}} \mathbf{A} (\mathbf{x}^{(0)} - \bar{x}) + \alpha_0^2 \mathbf{p}_0^{\mathrm{T}} \mathbf{A} \mathbf{p}_0,$$

then differentiating,

$$\frac{df}{d\alpha_0} = 2\mathbf{p}_0^{\mathrm{T}} \mathbf{A} (\mathbf{x}^{(0)} - \bar{x}) + 2\alpha_0 \mathbf{p}_0^{\mathrm{T}} \mathbf{A} \mathbf{p}_0,$$

gives the optimum value

$$\alpha_0 = \frac{-\mathbf{p}_0^{\mathrm{T}} \mathbf{A} (\mathbf{x}^{(0)} - \bar{x})}{\mathbf{p}_0^{\mathrm{T}} \mathbf{A} \mathbf{p}_0} = \frac{\mathbf{p}_0^{\mathrm{T}} \mathbf{r}_0}{\mathbf{p}_0^{\mathrm{T}} \mathbf{A} \mathbf{p}_0}. \tag{4.2}$$

The new residual $\mathbf{r}_1$ is

$$\mathbf{r}_1 = \mathbf{b} - \mathbf{A}\mathbf{x}^{(1)} = \mathbf{b} - \mathbf{A}\mathbf{x}^{(0)} - \alpha_0 \mathbf{A} \mathbf{p}_0$$

$$\mathbf{r}_1 = \mathbf{r}_0 - \alpha_0 \mathbf{A} \mathbf{p}_0. \tag{4.3}$$

The inner product of (4.3) with $\mathbf{p}_0$, simplified with the definition of α_0, implies

$$\mathbf{p}_0^{\mathrm{T}} \mathbf{r}_1 = \mathbf{r}_0^{\mathrm{T}} \mathbf{r}_1 = 0. \tag{4.4}$$

Given the improved value of $\mathbf{x}$ and its residual, one then seeks an improved solution. Let's leave unspecified, for the moment, exactly how to calculate the new search direction $\mathbf{p}_1$. Instead, consider the proposed update

$$\mathbf{x}^{(2)} = \mathbf{x}^{(1)} + \alpha_1 \mathbf{p}_1 + \alpha_0 \mathbf{p}_0. \tag{4.5}$$

Here α_0 is the weight of $\mathbf{p}_0$, to be determined, not to be confused with α_0 in (4.1). Instead, α_0, α_1 are chosen to minimize $f(\mathbf{x}^{(2)})$:

$$\frac{\partial f(\mathbf{x}^{(1)} + \alpha_1 \mathbf{p}_1 + \alpha_0 \mathbf{p}_0)}{\partial \alpha_0} = \frac{\partial f(\mathbf{x}^{(1)} + \alpha_1 \mathbf{p}_1 + \alpha_0 \mathbf{p}_0)}{\partial \alpha_1} = 0.$$

Following the previous methodology, these equations can be written

$$\frac{\partial f}{\partial \alpha_0} = 0 = 2\alpha_1 \mathbf{p}_0^{\mathrm{T}} \mathbf{A} \mathbf{p}_1 + 2\alpha_0 \mathbf{p}_0^{\mathrm{T}} \mathbf{A} \mathbf{p}_0 - 2\mathbf{p}_0^{\mathrm{T}} \mathbf{r}_1, \tag{4.6a}$$

$$\frac{\partial f}{\partial \alpha_1} = 0 = 2\alpha_0 \mathbf{p}_1^{\mathrm{T}} \mathbf{A} \mathbf{p}_0 + 2\alpha_1 \mathbf{p}_1^{\mathrm{T}} \mathbf{A} \mathbf{p}_1 - 2\mathbf{p}_1^{\mathrm{T}} \mathbf{r}_1. \tag{4.6b}$$

If $\mathbf{p}_1$ were $\mathbf{A}$-orthogonal (or, *conjugate* to) $\mathbf{p}_1$, i.e., if

$$\mathbf{p}_0^{\mathrm{T}} \mathbf{A} \mathbf{p}_1 = 0,$$

then (4.6b) would give

$$\alpha_1 = \frac{\mathbf{p}_1^{\mathrm{T}} \mathbf{r}_1}{\mathbf{p}_1^{\mathrm{T}} \mathbf{A} \mathbf{p}_1},$$

and (4.6a) would give $\alpha_0 = 0$ (the first term is zero by conjugacy, and the third term is zero by (4.4)). This suggests that by making $\mathbf{p}_1$ be conjugate to $\mathbf{p}_0$, the one-dimensional search

$$\mathbf{x}^{(2)} = \mathbf{x}^{(1)} + \alpha_1 \mathbf{p}_1$$

would generate the same solution as the two-dimensional model (4.5).

The vector $\mathbf{p}_1$ closest to $\mathbf{r}_1$ but conjugate to $\mathbf{p}_0$ is given by the projection:

$$\mathbf{p}_1 = \mathbf{r}_1 - \gamma \mathbf{p}_0,$$

with γ given by the condition of conjugacy:

$$\mathbf{p}_0^T \mathbf{A} \mathbf{p}_1 = 0 = \mathbf{p}_0^T \mathbf{A} \mathbf{r}_1 - \gamma \mathbf{p}_0^T \mathbf{A} \mathbf{p}_0$$

$$\gamma = \frac{\mathbf{p}_0^T \mathbf{A} \mathbf{r}_1}{\mathbf{p}_0^T \mathbf{A} \mathbf{p}_0},$$

or (cf. (3.12))

$$\mathbf{p}_1 = \mathbf{r}_1 - \frac{\mathbf{p}_0^T \mathbf{A} \mathbf{r}_1}{\mathbf{p}_0^T \mathbf{A} \mathbf{p}_0} \mathbf{p}_0. \tag{4.7}$$

The inner product of (4.7) with $\mathbf{r}_1$, simplified using (4.4), gives

$$\mathbf{r}_1^T \mathbf{p}_1 = \mathbf{r}_1^T \mathbf{r}_1.$$

Given the development to this point, it is instructive to consider the algorithm (this is *not yet* the conjugate gradient method – do not use this)

$$\mathbf{r}_k = \mathbf{b} - \mathbf{A} \mathbf{x}^{(k)} \tag{4.8a}$$

and terminate if $\mathbf{r}_k = 0$. Otherwise,

$$\mathbf{p}_k = \begin{cases} \mathbf{r}_k - \underbrace{\dfrac{\mathbf{p}_{k-1}^T \mathbf{A} \mathbf{r}_k}{\mathbf{p}_{k-1}^T \mathbf{A} \mathbf{p}_{k-1}}}_{\gamma_{k-1}} \mathbf{p}_{k-1} & k > 0 \\[4mm] \mathbf{r}_0 & k = 0, \end{cases} \tag{4.8b}$$

$$\mathbf{x}^{(k+1)} = \mathbf{x}^{(k)} + \underbrace{\frac{\mathbf{p}_k^T \mathbf{r}_k}{\mathbf{p}_k^T \mathbf{A} \mathbf{p}_k}}_{\alpha_k} \mathbf{p}_k. \tag{4.8c}$$

This algorithm follows exactly the method inferred for $k = 0, 1$, and suggests its use for all steps.

We have established the properties

$$\mathbf{p}_j^T \mathbf{r}_\ell = 0 \qquad\qquad 0 \le j < \ell, \tag{4.9a}$$

$$\mathbf{r}_\ell^T \mathbf{r}_\ell = \mathbf{r}_\ell^T \mathbf{p}_\ell, \tag{4.9b}$$

$$\mathbf{r}_j^T \mathbf{r}_\ell = 0 \qquad\qquad 0 \le j < \ell, \tag{4.9c}$$

$$\mathbf{p}_j^T \mathbf{A} \mathbf{p}_\ell = 0 \qquad\qquad 0 \le j < \ell, \tag{4.9d}$$

for $\ell \le k = 1$. The question now arises, "will these properties be obeyed for all subsequent $\ell \le k + 1, k + 2, \ldots$ using algorithm (4.8)?" We will show that these properties do hold using induction.

(a) Following the steps of (4.3), residual vector $\mathbf{r}_k$ can be written

$$\mathbf{r}_{k+1} = \mathbf{r}_k - \alpha_k \mathbf{A} \mathbf{p}_k, \tag{4.10}$$

and the inner product of this equation with $\mathbf{p}_j$ is

$$\mathbf{p}_j^{\mathsf{T}}\mathbf{r}_{k+1} = \mathbf{p}_j\mathbf{r}_k - \alpha_k\mathbf{p}_j^{\mathsf{T}}\mathbf{A}\mathbf{p}_k. \tag{4.11}$$

This is zero when $j = k$ because of the definition of α_k:

$$\mathbf{p}_k^{\mathsf{T}}\mathbf{r}_{k+1} = 0. \tag{4.12}$$

When $j < k$, the first term on the right-hand side of (4.11) is zero by application of (4.9a), and the second is zero by application of (4.9d). Therefore, if (4.9) is true for $\ell \leq k$, then (4.9a) is true for $\ell \leq k + 1$.

(b) Equation (4.8b) can be written

$$\mathbf{p}_{k+1} = \mathbf{r}_{k+1} - \gamma_k\mathbf{p}_k,$$

so

$$\mathbf{r}_{k+1}^{\mathsf{T}}\mathbf{p}_{k+1} = \mathbf{r}_{k+1}^{\mathsf{T}}\mathbf{r}_{k+1} - \gamma_k\mathbf{r}_{k+1}^{\mathsf{T}}\mathbf{p}_k,$$

where the last term on the right-hand side is zero using (4.12). Therefore, (4.9b) is true for $\ell \leq k + 1$.

(c) The inner product of $\mathbf{r}_{k+1}$ with a rearrangement of (4.8b) gives

$$\mathbf{r}_{k+1}^{\mathsf{T}}\mathbf{r}_{j\leq k} = \mathbf{r}_{k+1}^{\mathsf{T}}(\mathbf{p}_{j\leq k} + \gamma_{j-1}\mathbf{p}_{j-1<k}),$$

and this is zero by application of (4.9a). Therefore, (4.9c) is true for all $\ell \leq k + 1$.

(d) $\mathbf{p}_{k+1}^{\mathsf{T}}\mathbf{A}\mathbf{p}_k = 0$ by design: this is the purpose of equation (4.8b). And,

$$\mathbf{p}_{k+1}^{\mathsf{T}}\mathbf{A}\mathbf{p}_{j<k} = \mathbf{p}_{j<k}^{\mathsf{T}}\mathbf{A}(\mathbf{r}_{k+1} - \gamma_k\mathbf{p}_k) = \mathbf{p}_{j<k}^{\mathsf{T}}\mathbf{A}\mathbf{r}_{k+1}$$

using (4.9d) (with $\ell \leq k$). Substitution of (4.10) then gives

$$\alpha_{k+1}\mathbf{p}_{k+1}^{\mathsf{T}}\mathbf{A}\mathbf{p}_{j<k} = \alpha_{k+1}\mathbf{r}_{k+1}^{\mathsf{T}}\mathbf{A}\mathbf{p}_{j<k}$$
$$= \mathbf{r}_{k+1}^{\mathsf{T}}(\mathbf{r}_j - \mathbf{r}_{j+1\leq k}) = 0$$

using (4.9c). Therefore, (4.9d) is true for all $\ell \leq k + 1$ if α_{k+1} is not zero. We know this to be true because $\mathbf{p}_{k+1}^{\mathsf{T}}\mathbf{r}_{k+1} = \mathbf{r}_{k+1}^{\mathsf{T}}\mathbf{r}_{k+1} \geq 0$, and is zero only for $\mathbf{r}_{k+1} = 0$ which is prevented by the termination criterion.

Taken together, these steps establish the validity of (4.9) for all $\ell \leq k$, for any k, given the algorithm (4.8).

The conjugacy condition, $\mathbf{p}_j^{\mathsf{T}}\mathbf{A}\mathbf{p}_k = 0$ for $j \neq k$, means that

$$\mathbf{p}_k \not\subset \text{column space}(\mathbf{p}_0, ..., \mathbf{p}_{k-1}) \tag{4.13}$$

unless $\mathbf{r}_k = 0$. That is, each new search direction does not lie entirely in the space of prior search directions. The dimensionality of the searched space increases with each iteration. Of course, the dimensionality of this space cannot exceed n for $\mathbf{A}$ $n \times n$. Therefore, the algorithm must terminate (with exact math) in n steps.

Algorithm (4.8) will therefore converge to the exact solution. The conjugate gradient algorithm is a rearrangement of (4.8) that reduces the number of operations and the storage requirements.

1. Choose a starting value $\mathbf{x}^{(0)}$, and begin the computation with

$$\mathbf{p}_0 = \mathbf{r}_0 = \mathbf{b} - \mathbf{A}\mathbf{x}^{(0)}. \tag{4.14a}$$

Then, for $k = 1, 2, 3, \ldots$ until the termination criterion is met:

2. if $\mathbf{p}_k = 0$ then stop. Otherwise
3. perform the following steps:

$$\alpha_{k-1} = \frac{\mathbf{r}_{k-1}^T \mathbf{r}_{k-1}}{\mathbf{p}_{k-1}^T (\mathbf{A}\mathbf{p}_{k-1})}, \tag{4.14b}$$

$$\mathbf{x}^{(k)} = \mathbf{x}^{(k-1)} + \alpha_{k-1}\mathbf{p}_{k-1}, \tag{4.14c}$$

$$\mathbf{r}_k = \mathbf{r}_{k-1} - \alpha_{k-1}(\mathbf{A}\mathbf{p}_{k-1}), \tag{4.14d}$$

$$\gamma_{k-1} = -\frac{\mathbf{r}_k^T \mathbf{r}_k}{\mathbf{r}_{k-1}^T \mathbf{r}_{k-1}}, \tag{4.14e}$$

$$\mathbf{p}_k = \mathbf{r}_k - \gamma_{k-1}\mathbf{p}_{k-1}. \tag{4.14f}$$

Note that there is only one matrix multiplication $(\mathbf{A}\mathbf{p}_{k-1})$ per iteration.

The equivalence of algorithm (4.14) and (4.8) depends on the correspondence of the equations for variables α and γ. The α equations are equal because of (4.9b). The equivalence of the γ formulas comes from

$$\gamma_j = \frac{\mathbf{p}_j^T \mathbf{A}\mathbf{r}_{j+1}}{\mathbf{p}_j^T \mathbf{A}\mathbf{p}_j} = \frac{\alpha_j \mathbf{r}_{j+1}^T \mathbf{A}\mathbf{p}_j}{\alpha_j \mathbf{p}_j^T \mathbf{A}\mathbf{p}_j} = \frac{\mathbf{r}_{j+1}^T (\mathbf{r}_j - \mathbf{r}_{j+1})}{\mathbf{p}_j^T (\mathbf{r}_j - \mathbf{r}_{j+1})} = -\frac{\mathbf{r}_{j+1}^T \mathbf{r}_{j+1}}{\mathbf{r}_j^T \mathbf{r}_j}.$$

A variety of important modern methods are derived from this algorithm, including ones for arbitrary (not symmetric positive definite) matrices.

Example 4.1   Solve

$$\begin{pmatrix} 75 & 43 \\ 43 & 26 \end{pmatrix} \mathbf{x} = \begin{pmatrix} -15 \\ 8 \end{pmatrix} \tag{4.15}$$

by the conjugate gradient method.

Solution
With initial guess $\mathbf{x}^{(0)} = (0, 0)$, the initial residual and search direction are

$$\mathbf{p}_0 = \mathbf{r}_0 = \mathbf{b} - \mathbf{A}\mathbf{x}^{(0)} = \begin{pmatrix} -15 \\ 8 \end{pmatrix}.$$

The optimum solution in this search direction is calculated to be

$$\alpha_0 = 3.5162 \times 10^{-2}$$

$$\mathbf{x}^{(1)} = \mathbf{x}^{(0)} + \alpha_0 \mathbf{p}_0 = \begin{pmatrix} -5.2744 \times 10^{-1} \\ 2.8130 \times 10^{-1} \end{pmatrix},$$

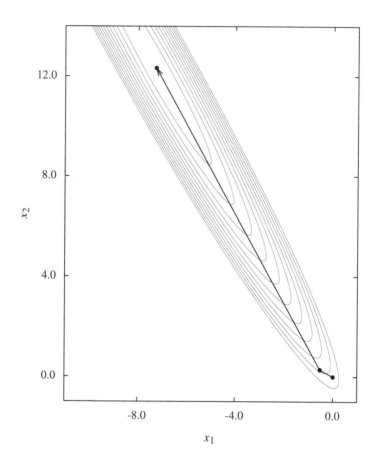

Figure 4.1 Conjugate gradient solution of Example 4.1.

and the new residual and search direction are computed to be

$$\mathbf{r}_1 = \mathbf{r}_0 - \alpha_0(\mathbf{A}\mathbf{p}_0) = \begin{pmatrix} 12.462 \\ 23.366 \end{pmatrix}$$

$$\gamma_0 = -2.4265$$

$$\mathbf{p}_1 = \mathbf{r}_1 - \gamma_0\mathbf{p}_0 = \begin{pmatrix} -23.936 \\ 42.778 \end{pmatrix}.$$

This completes one iteration. The second iteration begins with

$$\alpha_1 = 0.28158$$

$$\mathbf{x}^{(2)} = \mathbf{x}^{(1)} + \alpha_1\mathbf{p}_1 = \begin{pmatrix} -7.2673 \\ 12.327 \end{pmatrix}.$$

The residual is

$$\mathbf{r}_2 = \mathbf{r}_1 - \alpha_1(\mathbf{Ap}_1) = \begin{pmatrix} 6.5725 \times 10^{-14} \\ 3.5527 \times 10^{-14} \end{pmatrix}.$$

In exact math this residual $\mathbf{r}_n$ would be zero.

The solution is displayed in Figure 4.1, with some contours of the error function $f(x)$.

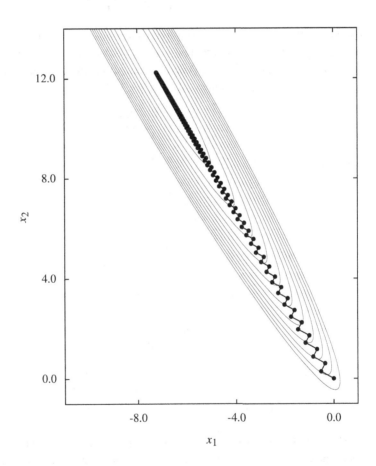

Figure 4.2 Steepest descents solution of (4.15).

Because of numerical errors, the solution may not be adequate after n iterations. One could proceed with the algorithm as given for $n + 1$, $n + 2$, etc., iterations. However, Hestenes and Stiefel [103] recommend that one should start anew beginning with the last estimate obtained.

The method of *steepest descents* is essentially the first step of the conjugate gradient method, repeated iteratively. Each search direction is the local residual, and no conjugacy condition is enforced. For the same problem evaluated in Example 4.1, the first 200 iterations of the steepest descent solution are shown graphically in Figure 4.2. Because conjugacy is not enforced, each iteration undoes some of the progress made by previous iterations. The solution does proceed toward the optimum, but requires significantly more than n iterations.

When $\mathbf{A}$ is not symmetric positive definite, the conjugate gradient algorithm presented above does not apply. However, with slight modification,

$$\mathbf{A}^{\mathrm{T}}\mathbf{A}\mathbf{x} = \mathbf{A}^{\mathrm{T}}\mathbf{b},$$

one can write a linear system with symmetric positive definite matrix $\mathbf{A}^{\mathrm{T}}\mathbf{A}$ that can be treated by the conjugate gradient method. The specific algorithm is not (4.14) with $\mathbf{A}^{\mathrm{T}}\mathbf{A}, \mathbf{A}^{\mathrm{T}}\mathbf{b}$ substituted for $\mathbf{A}, \mathbf{b}$. This would amplify numerical errors since the condition number $\mathrm{cond}(\mathbf{A}^{\mathrm{T}}\mathbf{A})$ is the square of $\mathrm{cond}(\mathbf{A})$. Instead, use

$$\mathbf{r}_0 = \mathbf{b} - \mathbf{A}\mathbf{x}^{(0)}$$
$$\mathbf{s}_0 = \mathbf{p}_0 = \mathbf{A}^{\mathrm{T}}\mathbf{r}_0$$
$$\mathbf{q}_0 = \mathbf{A}\mathbf{p}_0$$
$$\alpha_{k-1} = \frac{\mathbf{s}_{k-1}^{\mathrm{T}}\mathbf{s}_{k-1}}{\mathbf{q}_{k-1}^{\mathrm{T}}\mathbf{q}_{k-1}}$$
$$\mathbf{x}^{(k)} = \mathbf{x}^{(k-1)} + \alpha_{k-1}\mathbf{p}_{k-1} \qquad (4.16)$$
$$\mathbf{r}_k = \mathbf{r}_{k-1} - \alpha_{k-1}\mathbf{q}_{k-1}$$
$$\mathbf{s}_k = \mathbf{A}^{\mathrm{T}}\mathbf{r}_k$$
$$\gamma_{k-1} = -\frac{\mathbf{s}_k\mathbf{s}_k}{\mathbf{s}_{k-1}\mathbf{s}_{k-1}}$$
$$\mathbf{p}_k = \mathbf{s}_k - \gamma_{k-1}\mathbf{p}_{k-1}$$
$$\mathbf{q}_k = \mathbf{A}\mathbf{p}_k.$$

This formulation does not require that $\mathbf{A}$ be square – only that $\mathbf{A}^{\mathrm{T}}\mathbf{A}$ be invertible.

The idea of applying the conjugate gradient method to a modified system is very important and makes this method, and related methods derived from it, amongst the most important techniques in common use.

Although in exact math the method converges to the exact solution in no more than n steps, understanding the rate at which this solution is approached leads to an important enhancement of the method. The first step of the method is steepest descents, and gave a residual vector $\mathbf{r}_0$. In the second step the method provided the optimal solution in the vector space $\mathrm{span}[\mathbf{r}_0, \mathbf{A}\mathbf{r}_0]$. The inductive proof, in particular (4.13), implies that in step k the solution is optimal in the vector space

$$K_k(\mathbf{r}_0, \mathbf{A}) = \mathrm{span}[\mathbf{r}_0, \mathbf{A}\mathbf{r}_0, ..., \mathbf{A}^{k-1}\mathbf{r}_0],$$

the kth *Krylov space* generated by matrix $\mathbf{A}$ and initial vector $\mathbf{r}_0$. In fact, solution $\mathbf{x}^k$

minimizes $f(\mathbf{x})$ over all vectors in this space. To make this precise, if $\mathbf{e}^{(k)} = \mathbf{x}^{(k)} - \bar{\mathbf{x}}$ is the error at step k, then the function $\mathbf{f}$ can be interpreted as the square of a particular vector norm, the $\mathbf{A}$-norm,

$$f(\mathbf{x}) = \mathbf{e}^T \mathbf{A} \mathbf{e} \equiv \|\mathbf{e}\|_{\mathbf{A}}^2. \tag{4.17}$$

The solution at step k is a particular linear combination of the initial solution and vectors of the Krylov space

$$\mathbf{x}^{(k)} = \mathbf{x}^{(0)} + a_0 \mathbf{r}_0 + a_1 \mathbf{A} \mathbf{r}_0 + \cdots + a_{k-1} \mathbf{A}^{k-1} \mathbf{r}_0.$$

Now subtract from this the exact solution, and note that $\mathbf{r}_0 = -\mathbf{A}\mathbf{e}^{(0)}$,

$$\mathbf{e}^{(k)} = \mathbf{e}^{(0)} - a_0 \mathbf{A}\mathbf{e}^{(0)} - a_1 \mathbf{A}^2 \mathbf{e}^{(0)} - \cdots - a_{k-1} \mathbf{A}^k \mathbf{e}^{(0)}$$
$$= P_k(\mathbf{A})\mathbf{e}^{(0)} \qquad P_k(0) = 1.$$

The coefficients in this expansion, hence the coefficients of the polynomial P_k, are chosen to minimize the $\mathbf{A}$-norm of the error. Therefore,

$$\|\mathbf{e}^{(k)}\|_{\mathbf{A}}^2 = \min_{P_k | P(0) = 1} \|P_k(\mathbf{A})\mathbf{e}^{(0)}\|_{\mathbf{A}}^2.$$

Next, expand $\mathbf{e}^{(0)}$ in the orthonormal eigenvectors $\mathbf{y}$ of the matrix $\mathbf{A}$ (these exist with $\mathbf{A}$ symmetric positive definite):

$$\mathbf{e}^{(0)} = \sum_i c_i \mathbf{y}_i,$$

in which case

$$P_k(\mathbf{A})\mathbf{e}^{(0)} = \sum_i c_i P_k(\lambda_i)\mathbf{y}_i,$$

and exploiting the orthonormality property,

$$\|P_k(\mathbf{A})\mathbf{e}^{(0)}\|_{\mathbf{A}}^2 = \sum_i c_i^2 \lambda_i P_k(\lambda_i)^2$$
$$\leq \max_j P_k(\lambda_j)^2 \sum_i c_i^2 \lambda_i = \max_j P_k(\lambda_j)^2 \|\mathbf{e}^{(0)}\|_{\mathbf{A}}^2,$$

and finally

$$\frac{\|\mathbf{e}^{(k)}\|_{\mathbf{A}}^2}{\|\mathbf{e}^{(0)}\|_{\mathbf{A}}^2} = \min_{P_k | P_k(0) = 1} \max_j P_k(\lambda_j)^2 \leq \min_{P_k | P_k(0) = 1} \max_{\lambda \in [\lambda_{min}, \lambda_{max}]} P_k(\lambda)^2. \tag{4.18}$$

The polynomial that minimizes this function turns out to be describable by Chebyshev polynomials ($T_0 = 1$, $T_1 = x$, $T_{k+1} = 2xT_k - T_{k-1}$) which have the trigonometric representation

$$T_k(x) = \cos(k \arccos x). \tag{4.19}$$

P_k is given by mapping the range $[\lambda_{min}, \lambda_{max}]$ to $[-1, 1]$ and normalizing,

$$P_k(\lambda) = \frac{T_k\left(\frac{2\lambda - (\lambda_{max} + \lambda_{min})}{\lambda_{max} - \lambda_{min}}\right)}{T_k\left(\frac{0 - (\lambda_{max} + \lambda_{min})}{\lambda_{max} - \lambda_{min}}\right)}. \tag{4.20}$$

First, clearly $P_k(0) = 1$. Second, function $P_k(\lambda)$ has the smallest maximum in the range $[\lambda_{\min}, \lambda_{\max}]$. To show this, suppose there were a different polynomial Q_k of degree k, with $Q_k(0) = 1$ and with the property

$$\max_{\lambda} Q_k(\lambda)^2 < \max_{\lambda} P_k(\lambda)^2 = T_k \left(\frac{\lambda_{\max} + \lambda_{\min}}{\lambda_{\min} - \lambda_{\max}} \right)^{-2}.$$

This follows from (4.20) by noting that the numerator in (4.20) lies in $[-1, 1]$ for all $\lambda \in [\lambda_{\min}, \lambda_{\max}]$. In fact, $|P_k|$ takes its extreme value k times over the λ interval. If $|Q_k| \le |P_k|$ on $[\lambda_{\min}, \lambda_{\max}]$ then $P_k - Q_k$ must have at least k zeros in this range. Outside this range, $P_k(0) - Q_k(0) = 0$ also. Therefore, the degree k polynomial $P_k - Q_k$ has at least $k + 1$ zeros: $P_k - Q_k$ must be identically zero by the fundamental theorem of algebra (Figure 4.3).

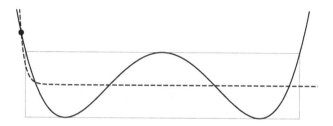

Figure 4.3 The Chebychev polynomial optimizes (4.18). The box is the region in which the minimum polynomial is to be determined, with horizontal extent $[\lambda_{\min}, \lambda_{\max}]$. The solid curve shows the polynomial $P_4(\lambda)$ (4.20) and the filled circle is $P_4(0)$. The dashed curve is a function $Q(\lambda)$ which satisfies $Q(0) = P_4(0)$, and $|Q| < |P_4|$ in the box. $Q - P_4 = 0$ at 5 points, so Q cannot be a polynomial of degree four.

Now returning to (4.18), the error bound may be written

$$\frac{\|e^{(k)}\|_A}{\|e^{(0)}\|_A} \le \left| T_k \left(\frac{\lambda_{\max} + \lambda_{\min}}{\lambda_{\min} - \lambda_{\max}} \right)^{-1} \right| = \left| T_k \left(\frac{C + 1}{C - 1} \right)^{-1} \right|, \tag{4.21}$$

where $C = \lambda_{\max}/\lambda_{\min}$ is the condition number for symmetric positive definite matrix **A**.

A more useful estimate comes from the analysis of (4.19) by Euler's formula:

$$T_k(x) = \frac{e^{ik\theta} + e^{-ik\theta}}{2},$$

where

$$x = \frac{e^{i\theta} + e^{-i\theta}}{2},$$

and $i = \sqrt{-1}$. Writing argument x from (4.21) as

$$\frac{C + 1}{C - 1} = \frac{1}{2} \left[\frac{\sqrt{C} + 1}{\sqrt{C} - 1} + \frac{\sqrt{C} - 1}{\sqrt{C} + 1} \right],$$

i.e., $e^{i\theta} = (\sqrt{C} + 1)/(\sqrt{C} - 1)$, one determines

$$\frac{\|\mathbf{e}^{(k)}\|_{\mathbf{A}}}{\|\mathbf{e}^{(0)}\|_{\mathbf{A}}} \leq 2 \left[\left(\frac{\sqrt{C} + 1}{\sqrt{C} - 1} \right)^k + \left(\frac{\sqrt{C} - 1}{\sqrt{C} + 1} \right)^k \right]^{-1} \leq 2 \left(\frac{\sqrt{C} + 1}{\sqrt{C} - 1} \right)^k,$$

since $C \geq 1$.

A powerful consequence of this result is that if one could *precondition* the matrix $\mathbf{A}$ in such a way as to reduce its condition number, this would have a strong effect on accelerating convergence.

For example, suppose we have an approximate inverse $\mathbf{B}$, $\mathbf{BA} \approx \mathbf{I}$, then $\mathbf{A}' = \mathbf{B}^{1/2}\mathbf{AB}^{1/2}$ is symmetric positive definite and may have a smaller condition number than $\mathbf{A}$. The original matrix problem $\mathbf{Ax} = \mathbf{b}$ is transformed into $\mathbf{A}'\mathbf{x}' = \mathbf{b}'$, with $\mathbf{x}' = \mathbf{B}^{-1/2}\mathbf{x}$ and $\mathbf{b}' = \mathbf{B}^{1/2}\mathbf{b}$. Write the conjugate gradient algorithm for these new variables, then rearrange to simplify. The result is:

- Begin with $\mathbf{x}^{(0)}$ and compute

$$\mathbf{r}_0 = \mathbf{b} - \mathbf{Ax}^{(0)}$$

$$\mathbf{p}_0 = \mathbf{q}_0 = \mathbf{Br}_0.$$

Then, for $k = 1, 2, 3, \ldots$ until the termination criterion is met:
- If $p_k = 0$ then stop. Otherwise,
- perform the following steps:

$$\alpha_{k-1} = \frac{\mathbf{r}_{k-1}^{\mathrm{T}}\mathbf{q}_{k-1}}{\mathbf{p}_{k-1}^{\mathrm{T}}(\mathbf{Ap}_{k-1})}$$

$$\mathbf{x}^{(k)} = \mathbf{x}^{(k-1)} + \alpha_{k-1}\mathbf{p}_{k-1}$$

$$\mathbf{r}_k = \mathbf{r}_{k-1} - \alpha_{k-1}(\mathbf{Ap}_{k-1})$$

$$\mathbf{q}_k = \mathbf{Br}_k$$

$$\beta_k = \frac{\mathbf{r}_k^{\mathrm{T}}\mathbf{q}_k}{\mathbf{r}_{k-1}^{\mathrm{T}}\mathbf{q}_{k-1}}$$

$$\mathbf{p}_k = \mathbf{q}_k + \beta\mathbf{p}_{k-1}.$$

There is one matrix multiplication per iteration, as before, and one operation associated with preconditioning, $\mathbf{q} = \mathbf{Br}$, which is either a multiplication if $\mathbf{B}$ is known, or a matrix solution if instead $\mathbf{B}^{-1}$ is known. For this method to be efficient, it is important that one have an inexpensive approximation $\mathbf{B}$. One example of a very simple approximation is the *Jacobi preconditioner*, $\mathbf{B} = \mathrm{diag}(A_{11}^{-1}, A_{22}^{-1}, \ldots, A_{nn}^{-1})$.

4.2 Relaxation methods

Relaxation methods are iterative techniques that, with exact math, require an infinite number of iterations to generate an exact solution. We want to solve

$$\mathbf{Ax} = \mathbf{b}, \tag{4.22}$$

when $\mathbf{A}$ is too large to treat by usual techniques. Instead, we will write

$$\mathbf{Bx} = (\mathbf{B} - \mathbf{A})\mathbf{x} + \mathbf{b}, \tag{4.23}$$

where $\mathbf{B}$ has a very simple structure – diagonal or triangular. Equations (4.22) and (4.23) are mathematically identical. Equation (4.23) has a simpler left-hand side, but this comes at the expense of having a right-hand side that depends on the solution. If $\mathbf{B}$ is chosen properly, (4.23) can be implemented in an iterative fashion. Beginning with $\mathbf{x}^{(0)} = 0$,

$$\mathbf{Bx}^{(k+1)} = (\mathbf{B} - \mathbf{A})\mathbf{x}^{(k)} + \mathbf{b}. \tag{4.24}$$

Equation (4.24) is the form we will use to actually solve our iterative equations. For the purpose of understanding the iterative behavior, it is useful to write this equation in the form

$$\mathbf{x}^{(k+1)} = (\mathbf{I} - \mathbf{B}^{-1}\mathbf{A})\mathbf{x}^{(k)} + \mathbf{B}^{-1}\mathbf{b}. \tag{4.25}$$

Again, I stress that (4.25) is not the form that you would use to implement these methods. This is a mathematical convenience to study the behavior. Let $\tilde{\mathbf{x}}$ be the exact solution to (4.22). If (4.24) converges, i.e., if $\lim_{k\to\infty} \mathbf{x}^{(k)} = \tilde{\mathbf{x}}$, then

$$\tilde{\mathbf{x}} = (\mathbf{I} - \mathbf{B}^{-1}\mathbf{A})\tilde{\mathbf{x}} + \mathbf{B}^{-1}\mathbf{b}. \tag{4.26}$$

Now write

$$\mathbf{e}^{(k)} = \mathbf{x}^{(k)} - \tilde{\mathbf{x}}$$
$$\mathbf{e}^{(k+1)} = (\mathbf{I} - \mathbf{B}^{-1}\mathbf{A})\mathbf{e}^{(k)}$$
$$\mathbf{e}^{(k)} = (\mathbf{I} - \mathbf{B}^{-1}\mathbf{A})^k \mathbf{e}^{(0)}$$

by subtracting (4.26) from (4.25).

For the iterative method (4.24) to converge, the eigenvalues λ of $(\mathbf{I} - \mathbf{B}^{-1}\mathbf{A})$ must satisfy $-1 < \lambda < 1$. If this constraint is not satisfied, say $|\lambda_i| > 1$ for some eigenvalue i, then if $\mathbf{e}$ has a component proportional to the corresponding eigenvector $\mathbf{x}_i$, that component will grow with each iteration (see Section 3.2).

Qualitatively, the more $\mathbf{B}$ resembles $\mathbf{A}$, the closer $(\mathbf{I} - \mathbf{B}^{-1}\mathbf{A})$ will be to zero (see discussion beginning on page 80). This recommends matrices $\mathbf{B}$ that are simply related to $\mathbf{A}$ while being easy to solve. In the simplest approximation $\mathbf{B}$ is made of the diagonal elements of $\mathbf{A}$. This is called Jacobi iteration. The next approximation, Gauss–Seidel, takes $\mathbf{B}$ to be the lower triangular elements of $\mathbf{A}$.

To facilitate discussion, we will decompose the arbitrary $n \times n$ matrix $\mathbf{A}$ as follows:

$$\mathbf{A} = \mathbf{D} - \mathbf{E} - \mathbf{F}$$

$$\mathbf{D} = \begin{pmatrix} a_{11} & & & \\ & a_{22} & & \\ & & \ddots & \\ & & & a_{nn} \end{pmatrix}$$

$$\mathbf{E} = -\begin{pmatrix} 0 & & & \\ a_{21} & \ddots & & \\ \vdots & \ddots & \ddots & \\ a_{n1} & \cdots & a_{n,n-1} & 0 \end{pmatrix}$$

$$\mathbf{F} = -\begin{pmatrix} 0 & a_{12} & \cdots & a_{1n} \\ & \ddots & \ddots & \vdots \\ & & \ddots & a_{n-1,n} \\ & & & 0 \end{pmatrix},$$

and further define

$$\mathbf{L} = \mathbf{D}^{-1}\mathbf{E} = -\begin{pmatrix} 0 & & & \\ \frac{a_{21}}{a_{22}} & \ddots & & \\ \vdots & \ddots & \ddots & \\ \frac{a_{n1}}{a_{nn}} & \cdots & \frac{a_{n,n-1}}{a_{nn}} & 0 \end{pmatrix}$$

$$\mathbf{U} = \mathbf{D}^{-1}\mathbf{F} = -\begin{pmatrix} 0 & \frac{a_{12}}{a_{11}} & \cdots & \frac{a_{1n}}{a_{11}} \\ & \ddots & \ddots & \vdots \\ & & \ddots & \frac{a_{n-1,n}}{a_{n-1,n-1}} \\ & & & 0 \end{pmatrix}.$$

4.3   Jacobi

The Jacobi method chooses $\mathbf{B}$ to be the diagonal elements of the matrix $\mathbf{A}$:

$$\mathbf{B} = \mathbf{D} = \text{diag}(a_{11}, a_{22}, ..., a_{nn}). \tag{4.27}$$

C. G. J. Jacobi [118] developed this method to solve least squares problems in astronomy by the normal equations.

Example 4.2   Use Jacobi iteration to solve $\mathbf{Ax} = \mathbf{b}$ with

$$
\mathbf{A} = \begin{pmatrix} 4 & -1 & 0 & 0 \\ -1 & 4 & -1 & 0 \\ 0 & -1 & 4 & -1 \\ 0 & 0 & -1 & 4 \end{pmatrix}
$$

$$
\mathbf{b} = \begin{pmatrix} 1 \\ 2 \\ 2 \\ 1 \end{pmatrix}.
$$

Solution

In the form (4.24) one has

$$
\begin{pmatrix} 4 & 0 & 0 & 0 \\ 0 & 4 & 0 & 0 \\ 0 & 0 & 4 & 0 \\ 0 & 0 & 0 & 4 \end{pmatrix} \begin{pmatrix} x_1 \\ x_2 \\ x_3 \\ x_4 \end{pmatrix}^{(k+1)} = \begin{pmatrix} 0 & 1 & 0 & 0 \\ 1 & 0 & 1 & 0 \\ 0 & 1 & 0 & 1 \\ 0 & 0 & 1 & 0 \end{pmatrix} \begin{pmatrix} x_1 \\ x_2 \\ x_3 \\ x_4 \end{pmatrix}^{(k)} + \begin{pmatrix} 1 \\ 2 \\ 2 \\ 1 \end{pmatrix},
$$

so the solution would proceed as follows:

$$
\begin{aligned}
x_1^{(k+1)} &= \frac{1}{4}\left(x_2^{(k)} + 1\right) \\
x_2^{(k+1)} &= \frac{1}{4}\left(x_1^{(k)} + x_3^{(k)} + 2\right) \\
x_3^{(k+1)} &= \frac{1}{4}\left(x_2^{(k)} + x_4^{(k)} + 2\right) \\
x_4^{(k+1)} &= \frac{1}{4}\left(x_3^{(k)} + 1\right).
\end{aligned}
\tag{4.28}
$$

Table 4.1 Jacobi iteration results from Example 4.2.

k	x_1	x_2	x_3	x_4
0	0.000 000	0.000 000	0.000 000	0.000 000
1	0.250 000	0.500 000	0.500 000	0.250 000
2	0.375 000	0.687 500	0.687 500	0.375 000
3	0.421 875	0.765 625	0.765 625	0.421 875
4	0.441 406	0.796 875	0.796 875	0.441 406
5	0.449 219	0.809 570	0.809 570	0.449 219
6	0.452 393	0.814 697	0.814 697	0.452 393
7	0.453 674	0.816 772	0.816 772	0.453 674
8	0.454 193	0.817 612	0.817 612	0.454 193
9	0.454 403	0.817 951	0.817 951	0.454 403
10	0.454 488	0.818 089	0.818 089	0.454 488
$\vdots$				
42	0.454 545	0.818 182	0.818 182	0.454 545

Note that to implement this iteration, one needs to maintain two copies of the vector $\mathbf{x}$: $x_j^{(k)}$ may be used after $x_j^{(k+1)}$ has been calculated, so the calculation cannot be done in place. To begin the cycle, choose $\mathbf{x}^{(0)} = \mathbf{0}$, then the first several iterations give the results in Table 4.1. The iteration sequence converges in double precision after 42 iterations.

Note because the matrix $\mathbf{A}$ had a simple structure, the iteration sequence (4.28) had a very compact form that did not reference all $n \times n$ elements of $\mathbf{A}$. If n were very large, say millions, then this sparseness would be a real advantage.

To understand the convergence of Jacobi iteration, we write

$$\mathbf{J} = (\mathbf{I} - \mathbf{D}^{-1}\mathbf{A}) = \mathbf{D}^{-1}\mathbf{E} + \mathbf{D}^{-1}\mathbf{F} = \mathbf{L} + \mathbf{U}.$$

$\mathbf{J}$ has zeros along the diagonal, and off-diagonal entries are simply related to the elements of $\mathbf{A}$: $\mathbf{J}_{i,j \neq i} = -a_{ij}/a_{ii}$.

The Jacobi method converges if all eigenvalues of $\mathbf{J}$ satisfy $|\lambda_i| < 1$, or in terms of the spectral radius, $\rho(\mathbf{J}) < 1$. A simple test to determine if $\rho(\mathbf{J}) < 1$ is the strong row sum criterion. Since $\mathbf{J}$ has zeros on the diagonal the Gerschgorin disks are all centered at the origin. For $\rho(\mathbf{J}) < 1$ we need to show that the disk radii are less than 1:

$$\sum_{j \neq i} \left| \frac{a_{ij}}{a_{ii}} \right| < 1,$$

or *strong row sum criterion* [234] if

$$|a_{ii}| > \sum_{j \neq i} |a_{ij}|, \quad i = 1, \ldots, n, \tag{4.29}$$

then matrix $\mathbf{A}$ will converge by the Jacobi method.

If $\mathbf{A}$ satisfies the strong row sum criterion, then the Jacobi iteration method will converge. Unfortunately, some very important and common matrices fail this test. Failure to satisfy the strong row sum criterion does not mean that $\rho(\mathbf{J}) \geq 1$. It simply means that we do not know that $\rho(\mathbf{J}) < 1$. The weak row sum criterion is another test that applies to many important matrices.

Weak row sum criterion [82] if

$$|a_{ii}| \geq \sum_{j \neq i} |a_{ij}|, \quad i = 1, \ldots, n, \tag{4.30}$$

with inequality holding for at least one index i, and if there is no permutation matrix $\mathbf{P}$ such that $\mathbf{PAP}^\mathrm{T}$ has the block structure

$$\left(\begin{array}{c|c} \mathbf{Q} & \mathbf{R} \\ \hline \mathbf{0} & \mathbf{S} \end{array} \right), \tag{4.31}$$

with $\mathbf{Q}$ and $\mathbf{S}$ square, then $\rho(\mathbf{J}) < 1$.

To understand this test, assume $\rho(\mathbf{J}) = 1$, and assume that (4.30) is true with inequality holding for at least one index i.

Because $\rho(\mathbf{J}) = 1$, there is an eigenvector $\mathbf{x}$ such that $\mathbf{J}\mathbf{x} = \gamma\mathbf{x}$ with $|\gamma| = 1$, or equivalently $\mathbf{D}\mathbf{J}\mathbf{x} = \gamma\mathbf{D}\mathbf{x}$, which takes the form

$$-\gamma a_{11}x_1 + a_{12}x_2 + \cdots + a_{1n}x_n = 0$$
$$a_{21}x_1 - \gamma a_{22}x_2 + \cdots + a_{2n}x_n = 0$$
$$\vdots \qquad\qquad\qquad (4.32)$$
$$a_{n1}x_1 + a_{n2}x_2 + \cdots - \gamma a_{nn}x_n = 0.$$

Let ℓ be a row for which inequality holds in (4.30), i.e.,

$$|a_{\ell\ell}| > \sum_{j\neq\ell} |a_{\ell j}|. \qquad (4.33)$$

But, from the ℓth equation of (4.32),

$$|a_{\ell\ell}||x_\ell| \leq \sum_{j\neq\ell} |a_{\ell j}||x_j| \qquad (4.34)$$

using the triangle inequality. To reconcile (4.33) and (4.34),

$$\overbrace{\underbrace{\sum_{j\neq\ell} |a_{\ell j}||x_\ell| < |a_{\ell\ell}||x_\ell|}_{(4.33)} \leq \sum_{j\neq\ell} |a_{\ell j}||x_j|}^{(4.34)},$$

it cannot be true that all numbers $|x_i|$ are equal (unless they are all zero, but this trivial solution is incompatible with the assumption that $\mathbf{x}$ is an eigenvector). If r is an index such that $|x_r| \geq |x_i|$ for all i, then there must be at least one variable index k such that $|x_r| > |x_k|$.

If r is the index corresponding to the greatest element $|x_r|$, then from (4.32)

$$|a_{rr}| \leq \sum_{j\neq r} |a_{rj}|\frac{|x_j|}{|x_r|},$$

and from (4.30)

$$|a_{rr}| \geq \sum_{j\neq r} |a_{rj}|.$$

To reconcile these inequalities,

$$\sum_{j\neq r} |a_{rj}| \leq \sum_{j\neq r} |a_{rj}|\frac{|x_j|}{|x_r|},$$

we need $a_{rj} = 0$ whenever $|x_j| < |x_r|$ ($|x_j| \not> |x_r|$ by definition of r). If there are p indices j such that $|x_j| = |x_r|$, then there are $(n-p)$ indices j such that $|x_j| < |x_r|$.

We require a block of $p \times q$ zeros to reconcile (4.30) with $\rho(\mathbf{J}) = 1$. However, this contradicts the assumption of irreducibility. That is, there is an ordering of the variables such that $\mathbf{Q}$ is $p \times p$, $\mathbf{S}$ is $(n-p) \times (n-p)$ and the zero block of (4.31) has size $(n-p) \times p$.

4.4　Irreducibility

The solution of many physics and engineering problems involves matrices that are large and sparse, and that do not satisfy the strong row sum criterion because equality holds in tests (4.29), versus the inequality that is required for that theory. Because such matrices are large and sparse, iterative solutions are more desirable than direct solution techniques. The test for applicability of simple iterative techniques hinges entirely on whether or not a permutation $\mathbf{P}$ exists that transforms $\mathbf{A}$ into the form (4.31). If such a permutation exists, $\mathbf{A}$ is called *reducible*. If no such permutation exists, $\mathbf{A}$ is *irreducible*. If $\mathbf{A}$ is irreducible, and fails the strong row sum criterion only because of some equalities, then $\mathbf{A}$ will satisfy the weak row sum criterion and therefore be amenable to solution by, e.g., Jacobi iteration.

There is a simple test of irreducibility that uses a distinctly different approach than the other analysis methods in this book. This new technique is a type of *graph theory*. It may be attributed to D. König [134], but the historical study of H. Schneider [211] shows that the history is a bit murky. A contender for the origin of this technique "depending on one's criteria" [211, p. 148] is R. S. Varga's very accessible book [231].

To understand the technique, we first consider what the permutation $\mathbf{PAP}^{\mathrm{T}}$ accomplishes. Since the same permutation matrix $\mathbf{P}$ is applied from the left and from the right, the action of these permutations is to relabel the indices A_{ij} in a consistent manner: the relabeling that applies to the first index i, also applies to the second index j. All diagonal entries of $\mathbf{A}$, A_{ii}, get transformed into different diagonal entries $A_{i'i'}$, and off-diagonal entries $A_{i,j\neq i}$ remain off-diagonal entries, $A_{i',j'\neq i'}$.

With this in mind, we seek a characterization of the reduced matrix

$$\left(\begin{array}{c|c} \mathbf{Q} & \mathbf{R} \\ \hline 0 & \mathbf{S} \end{array} \right) \tag{4.35}$$

that will not be affected by permutations (cf. (3.38)).

Let us suppose that block $\mathbf{Q}$ has dimension $p \times p$, and that block $\mathbf{S}$ has dimension $(n - p) \times (n - p)$. $\mathbf{Q}$ has p diagonal elements, and $\mathbf{S}$ has $(n - p)$ diagonal elements.

Let us construct a *directed graph* as follows:

1. Draw one *node* for each diagonal entry, and label the node corresponding to $\mathbf{A}_{ii}$ by the index i.

2. Draw an arrow from node i to node j if element A_{ij} is nonzero.

The graph of matrix (4.35) will contain n nodes, and there may be directed paths between any of the nodes $\alpha \in [1, ..., p]$ and any other node $\beta \in [1, ..., p]$, $\beta \neq \alpha$, depending on whether $Q_{\alpha\beta}$ is zero or not. Likewise, nodes corresponding to block $\mathbf{S}$ may be connected depending on the structure of $\mathbf{S}$.

However, there will be no directed graphs from nodes of $\mathbf{S}$ to nodes of $\mathbf{Q}$ because the corresponding elements of $\mathbf{PAP}^{\mathrm{T}}$ in (4.35) are all zero.

For example,

$$
\mathbf{PAP}^{\mathrm{T}} =
\left(
\begin{array}{ccc|cccc}
q_{11} & q_{12} & q_{13} & r_{11} & r_{12} & r_{13} & r_{14} \\
q_{21} & q_{22} & q_{23} & r_{21} & r_{22} & r_{23} & r_{24} \\
q_{31} & q_{32} & q_{33} & r_{31} & r_{32} & r_{33} & r_{34} \\
0 & 0 & 0 & s_{11} & s_{12} & s_{13} & s_{14} \\
0 & 0 & 0 & s_{21} & s_{22} & s_{23} & s_{24} \\
0 & 0 & 0 & s_{31} & s_{32} & s_{33} & s_{34} \\
0 & 0 & 0 & s_{41} & s_{42} & s_{43} & s_{44}
\end{array}
\right),
\tag{4.36}
$$

where q_{ij}, r_{ij}, and s_{ij} are all nonzero will have the graph of Figure 4.4. There is no directed path (a path following the direction indicated by the arrows) from any node $4, 5, 6, 7$ to any node $1, 2, 3$. This is a characteristic of a reducible matrix.

Since a permutation $\mathbf{PAP}^{\mathrm{T}}$ only relabels the nodes, the topology of this graph will be unaffected by the choice of permutation: any permutation of matrix (4.36) will have a directed graph that is topologically identical to the graph of Figure 4.4.

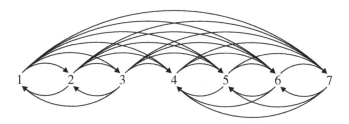

Figure 4.4 Directed graph of matrix (4.36).

Example 4.3 The matrix

$$
\mathbf{A} =
\begin{pmatrix}
-2 & 1 & & & & & & \\
1 & -2 & 1 & & & & & \\
& 1 & -2 & 1 & & & & \\
& & 1 & -2 & 1 & & & \\
& & & 1 & -2 & 1 & & \\
& & & & 1 & -2 & 1 & \\
& & & & & 1 & -2 & 1 \\
& & & & & & 1 & -2
\end{pmatrix}
\tag{4.37}
$$

arises in the discretization of the operator $\frac{d^2}{dx^2}$ using finite differences. This one-dimensional operator occurs in, for example, the diffusion or heat equations. Can one use Jacobi iteration to solve $\mathbf{Ax} = \mathbf{b}$?

Solution
First, we note that this matrix fails the strong row sum criterion because $|1| + |1| \not< |-2|$.

However, this matrix would satisfy the weak row sum criterion if it were irreducible. This is because we have several equalities $|1| + |1| = |-2|$, and inequality holds for the first and last rows: $|1| < |-2|$.

To assess irreducibility, we draw the graph of this matrix. The graph is shown in Figure 4.5. This graph is clearly irreducible because one can pass from·any node to any other node by following the arcs in the direction of the arrows. Therefore, this matrix can be solved by the Jacobi method.

Figure 4.5 Directed graph of matrix (4.37); Example 4.3.

4.5   Gauss–Seidel

The Gauss–Seidel method follows the template (4.23) with the choice that $\mathbf{B}$ consists of the lower triangular parts of $\mathbf{A}$:

$$\mathbf{B} = \mathbf{D} - \mathbf{E}.$$

C. F. Gauss developed this method in ca. 1823 to facilitate solving least squares problems in surveying [79, Letter #163, Dec. 26, 1823] by the normal equations. The English translation of this letter by Forsythe [68] contains the quote

> You will hardly ever again eliminate directly, at least not when you have more than 2 unknowns. The indirect procedure can be done while half asleep, or while thinking about other things.

The method is also attributed to L. Seidel [214].

Example 4.4   Solve the problem of Example 4.2 using the Gauss–Seidel method.

Solution
In the form (4.24) we have

$$
\begin{pmatrix} 4 & 0 & 0 & 0 \\ -1 & 4 & 0 & 0 \\ 0 & -1 & 4 & 0 \\ 0 & 0 & -1 & 4 \end{pmatrix}
\begin{pmatrix} x_1 \\ x_2 \\ x_3 \\ x_4 \end{pmatrix}^{(k+1)}
=
\begin{pmatrix} 0 & 1 & 0 & 0 \\ 0 & 0 & 1 & 0 \\ 0 & 0 & 0 & 1 \\ 0 & 0 & 0 & 0 \end{pmatrix}
\begin{pmatrix} x_1 \\ x_2 \\ x_3 \\ x_4 \end{pmatrix}^{(k)}
-
\begin{pmatrix} 1 \\ 2 \\ 2 \\ 1 \end{pmatrix},
$$

so the solution would proceed as follows:

$$x_1^{(k+1)} = \frac{1}{4}\left(x_2^{(k)} + 1\right)$$

$$x_2^{(k+1)} = \frac{1}{4}\left(x_1^{(k+1)} + x_3^{(k)} + 2\right)$$

$$x_3^{(k+1)} = \frac{1}{4}\left(x_2^{(k+1)} + x_4^{(k)} + 2\right)$$

$$x_4^{(k+1)} = \frac{1}{4}\left(x_3^{(k+1)} + 1\right).$$

These equations are almost identical to the ones solved by Jacobi (4.28). The only difference is that some terms on the right-hand side are taken from iteration $(k+1)$ instead of iteration (k). Those terms affected correspond to elements of $\mathbf{x}$ that have already been calculated in the current iteration. This means that in Gauss–Seidel one needs to maintain only one vector $\mathbf{x}$ – one can overwrite the (k) values with $(k + 1)$ values as the iteration proceeds.

Some results from this method are given in Table 4.2. The sequence converges completely in 22 iterations, about twice as fast as Jacobi.

Table 4.2 Gauss–Seidel iteration results from Example 4.4.

k	x_1	x_2	x_3	x_4
0	0.000 000	0.000 000	0.000 000	0.000 000
1	0.250 000	0.562 500	0.640 625	0.410 156
2	0.390 625	0.757 812	0.791 992	0.447 998
3	0.439 453	0.807 861	0.813 965	0.453 491
4	0.451 965	0.816 483	0.817 493	0.454 373
5	0.454 121	0.817 904	0.818 069	0.454 517
6	0.454 476	0.818 136	0.818 163	0.454 541
7	0.454 534	0.818 174	0.818 179	0.454 545
8	0.454 544	0.818 181	0.818 181	0.454 545
9	0.454 545	0.818 182	0.818 182	0.454 545
10	0.454 545	0.818 182	0.818 182	0.454 545
⋮				
22	0.454 545	0.818 182	0.818 182	0.454 545

To analyze this method we consider

$$\mathbf{H} = (\mathbf{I} - \mathbf{B}^{-1}\mathbf{A}) = (\mathbf{I} - \mathbf{L})^{-1}\mathbf{U}.$$

Because of the special shape of $\mathbf{L}$, it is true that

$$\mathbf{L}^n = |\mathbf{L}|^n = 0$$
$$(\mathbf{I} - \mathbf{L})^{-1} = \mathbf{I} + \mathbf{L} + \cdots + \mathbf{L}^{n-1}$$
$$(\mathbf{I} - |\mathbf{L}|)^{-1} = \mathbf{I} + |\mathbf{L}| + \cdots + |\mathbf{L}^{n-1}| \qquad (4.38)$$
$$\|(\mathbf{I} - \mathbf{L})^{-1}\| \le \|(\mathbf{I} - |\mathbf{L}|)^{-1}\|$$
$$\|\mathbf{I}\| \le \|(\mathbf{I} - |\mathbf{L}|)^{-1}\|.$$

We will use this to estimate $\rho(\mathbf{H})$ in the case that $\mathbf{A}$ satisfies the strong row sum criterion:

$$\mathbf{U} = \mathbf{J} - \mathbf{L}$$
$$|\mathbf{U}| = |\mathbf{J}| - |\mathbf{L}|$$
$$|\mathbf{H}| \le (\mathbf{I} - |\mathbf{L}|)^{-1}(|\mathbf{J}| - |\mathbf{L}|)$$
$$|\mathbf{H}| \le (\mathbf{I} - |\mathbf{L}|)^{-1}[(|\mathbf{J}| - \mathbf{I}) + (\mathbf{I} - |\mathbf{L}|)]$$
$$|\mathbf{H}| - \mathbf{I} \le (\mathbf{I} - |\mathbf{L}|)^{-1}(|\mathbf{J}| - \mathbf{I}).$$

Next introduce $\mathbf{e}^{\mathrm{T}} = (1, \ldots, 1)$. If $\mathbf{A}$ satisfies the strong row sum criterion, then $|\mathbf{J}|\mathbf{e} \le \|\mathbf{J}\|_{\infty}\mathbf{e} < \mathbf{I}\mathbf{e}$. Also, $(\mathbf{I} - |\mathbf{L}|)^{-1}$ is greater than $\mathbf{I}$ (4.38). Continuing,

$$|\mathbf{H}|\mathbf{e} - \mathbf{e} \le (\mathbf{I} - |\mathbf{L}|)^{-1} \underbrace{(\|\mathbf{J}\|_{\infty} - \mathbf{I})}_{\text{negative}} \mathbf{e} \le (\|\mathbf{J}\|_{\infty} - 1)\mathbf{e}$$

$$|\mathbf{H}|\mathbf{e} \le \|\mathbf{J}\|_{\infty}\mathbf{e}.$$

All elements of $\|\mathbf{J}\|_{\infty}\mathbf{e}$ are equal. The greatest element of $|\mathbf{H}|\mathbf{e}$ is $\|\mathbf{H}\|_{\infty}$. From the row containing that element,

$$\rho(\mathbf{H}) \le \|\mathbf{H}\|_{\infty} \le \|\mathbf{J}\|_{\infty} < 1.$$

So if a matrix satisfies the strong row sum criterion then it will converge with both Jacobi and Gauss–Seidel. Since $\rho(\mathbf{J}) \le \|\mathbf{J}\|_{\infty}$ this analysis does not compare the relative convergence of these methods. To obtain such an estimate, further restrictions on $\mathbf{A}$ are required. The remainder of this section summarizes the results of P. Stein and R. L. Rosenberg [222].

We will say that $\mathbf{J}$ is positive if each of its elements $J_{ik} \ge 0$. When $\mathbf{J}$ is positive, some stronger convergence results can be had. This would seem to be an overly restrictive condition, but most matrices arising from physical problems have this positivity character. To analyze this case, we need two facts.

First, if $\mathbf{A}$ is positive, and $a_{ij} \ge |b_{ij}|$, then $\rho(\mathbf{A}) \ge \rho(\mathbf{B})$. Let $\beta = \rho(\mathbf{B})$ be the maximal eigenvalue of $\mathbf{B}$, with corresponding eigenvector $\mathbf{x}$:

$$\beta\mathbf{x} = \mathbf{B}\mathbf{x}$$
$$|\beta||\mathbf{x}| \le |\mathbf{B}||\mathbf{x}| \le \mathbf{A}|\mathbf{x}|,$$

therefore $\rho(\mathbf{A}) \ge |\beta|$.

Second, if $\mathbf{A}$ is positive, then its maximal eigenvalue and corresponding eigenvector are positive. Let $\mathbf{x}$ be the eigenvector whose eigenvalue α equals $\rho(\mathbf{A})$,

$$\alpha \mathbf{x} = \mathbf{A}\mathbf{x}.$$

Assume $\mathbf{x}$ has mixed signs, then

$$\alpha \mathbf{x} = \underbrace{\mathbf{A}\, \text{diag}\, (\text{sign}(x_1),\ ...,\ \text{sign}(x_n))}_{\mathbf{B}}\, |\mathbf{x}|$$

$$|\alpha||\mathbf{x}| \leq \mathbf{A}|\mathbf{x}|,$$

so there exists a positive eigenvector, with positive eigenvalue $\geq \alpha$. As this contradicts that the maximum eigenvector had mixed signs it completes the desired proof.

Let $\mathbf{x}$ be the positive maximal eigenvector of positive $\mathbf{J}$, with corresponding eigenvalue $\lambda = \rho(\mathbf{J})$:

$$(\mathbf{L} + \mathbf{U})\mathbf{x} = \lambda \mathbf{x}$$

$$\mathbf{U}\mathbf{x} = \lambda(\mathbf{I} - \lambda^{-1}\mathbf{L})\mathbf{x}$$

$$(\mathbf{I} - \lambda^{-1}\mathbf{L})^{-1}\mathbf{U}\mathbf{x} = \lambda \mathbf{x}.$$

If $\lambda > 1$, then

$$\lambda \mathbf{x} = (\mathbf{I} - \lambda^{-1}\mathbf{L})^{-1}\mathbf{U}\mathbf{x} < (\mathbf{I} - \mathbf{L})^{-1}\mathbf{U}\mathbf{x} = \mathbf{H}\mathbf{x}, , , \tag{4.39}$$

so $\mu = \rho(\mathbf{H}) > \lambda$. If Jacobi fails to converge for positive $\mathbf{J}$, then Gauss–Seidel will fail more egregiously. If we carry out the previous analysis assuming $\lambda \geq 1$, we conclude $\mu \geq \lambda$.

Now let $\mathbf{y}$ be the maximal positive eigenvector of positive $\mathbf{H}$, with positive eigenvalue $\mu = \rho(\mathbf{H})$:

$$\begin{aligned} \mathbf{H}\mathbf{y} &= \mu \mathbf{y} \\ (\mathbf{I} - \mathbf{L})^{-1}\mathbf{U}\mathbf{y} &= \mu \mathbf{y} \\ \mathbf{U}\mathbf{y} &= \mu(\mathbf{I} - \mathbf{L})\mathbf{y} \\ (\mu\mathbf{L} + \mathbf{U})\mathbf{y} &= \mu \mathbf{y}. \end{aligned} \tag{4.40}$$

If $\mu > 1$, then

$$\mu \mathbf{y} = (\mu\mathbf{L} + \mathbf{U})\mathbf{y} < \mu(\mathbf{L} + \mathbf{U})\mathbf{y} = \mu \mathbf{J}\mathbf{y} \leq \mu\lambda \mathbf{y}, \tag{4.41}$$

so when $\mu > 1$, $\lambda > 1$ also (if $\mu \geq 1$, then $\lambda \geq 1$). Together, (4.39) and (4.41) say that for positive $\mathbf{J}$, failure of Jacobi implies failure of Gauss–Seidel, and vice versa.

But, if $\mu < 1$, then (4.40) gives

$$\mu \mathbf{y} = (\mu\mathbf{L} + \mathbf{U})\mathbf{y} < (\mathbf{L} + \mathbf{U})\mathbf{y} = \mathbf{J}\mathbf{y},$$

so when $\mu < 1$, we have $\lambda > \mu$. But, it must also be true that $\lambda < 1$, or by (4.39) μ would be > 1. So the case must be $\mu < \lambda < 1$.

In the trivial case $\mathbf{J} = 0$, we get $\lambda = \mu = 0$. Summarizing the above findings, for positive $\mathbf{J}$ one of the following four circumstances must be true:

$$\mu = \lambda = 0$$
$$\mu < \lambda < 1$$
$$\lambda = \mu = 1$$

or

$$\mu > \lambda > 1.$$

The key result is that when $\mathbf{J}$ is positive, and if $\rho(\mathbf{J}) < 1$ by the strong or weak row sum criterion, then use Gauss–Seidel because it will converge more rapidly and use memory more effectively.

4.6 Multigrid

A technique to accelerate convergence of relaxation methods is the *multigrid method*. The technique was first developed by Fedorenko [61, 62], and later Brandt [24, 25]. The basic idea is very simple. To make a physical analogy, think of the solution vector $\mathbf{x}$ as being a discretization of a solution on some interval, e.g., $x_1 = x(h)$, $x_2 = x(2h)$, ..., $x_{n-1} = x((n-1)h)$. Then, with a suitable choice of relaxation method one can attenuate the short wavelength error in $\mathbf{x}$ rapidly, leaving residual error only at long wavelengths. Long wavelength error can be represented on a coarser discretization (a smaller dimensional vector space), and solved with a residual correction method, Section 2.6. To implement the multigrid method one needs three tools. The *smoother* is the relaxation method that rapidly solves short wavelength error. The operation that maps the residual to a coarser domain is called *restriction* (or simply *coarsening*), and the operation that maps the correction to the original fine domain is called *prolongation* (or *interpolation*).

The simplest plan implements iterative refinement on two grids,

$$\begin{aligned}
\mathbf{x}_h : \quad & \mathbf{A}_h \mathbf{x}_h \approx \mathbf{b}_h \\
\mathbf{r}_h &= \mathbf{b}_h - \mathbf{A}_h \mathbf{x}_h \\
\mathbf{b}_{2h} &= I_h^{2h} \mathbf{r}_h \\
\mathbf{x}_{2h} : \quad & \mathbf{A}_{2h} \mathbf{x}_{2h} = \mathbf{b}_{2h} \\
\mathbf{x}_h &:= \mathbf{x}_h + I_{2h}^h \mathbf{x}_{2h}
\end{aligned} \qquad (4.42)$$

(this is a general outline, not yet the algorithm). Here I_h^{2h} is the restriction operator, and I_{2h}^h is the prolongation operator. It is understood that $\mathbf{A}_h \mathbf{x}_h = \mathbf{b}_h$ is not solved exactly, but with a relaxation method.

Some insight into the relation between I_h^{2h} and I_{2h}^h, and into the nature of $\mathbf{A}_{2h}$ in relation to $\mathbf{A}_h$ comes by considering a quadratic function

$$F(\mathbf{u}) = \frac{1}{2}\mathbf{u}^{\mathrm{T}}\mathbf{A}_h\mathbf{u} - \mathbf{u}^{\mathrm{T}}\mathbf{b}_h,$$

and assuming that $\mathbf{A}_h$ is symmetric positive definite (cf. Section 4.1). Then, the $\mathbf{u}$ that minimizes F solves $\mathbf{A}_h\mathbf{u} = \mathbf{b}_h$. For $\mathbf{u}$, write

$$\mathbf{u} = \mathbf{x}_h + I_{2h}^h\mathbf{x}_{2h},$$

and expand $F(\mathbf{u})$ with the residual from (4.42)

$$F(\mathbf{u}) = F(\mathbf{x}_h) + \frac{1}{2}(\mathbf{x}_{2h}^{\mathrm{T}}{I_{2h}^h}^{\mathrm{T}}\mathbf{A}_h I_{2h}^h\mathbf{x}_{2h}) - \mathbf{x}_{2h}^{\mathrm{T}}{I_{2h}^h}^{\mathrm{T}}\mathbf{r}_h.$$

The minimum of $F(\mathbf{u})$ with respect to $\mathbf{x}_{2h}^{\mathrm{T}}$ occurs when

$${I_{2h}^h}^{\mathrm{T}}\mathbf{A}_h I_{2h}^h\mathbf{x}_{2h} = {I_{2h}^h}^{\mathrm{T}}\mathbf{r}_h,$$

but $\mathbf{x}_{2h}$ is computed in (4.42) from

$$\mathbf{A}_{2h}\mathbf{x}_{2h} = I_h^{2h}\mathbf{r}_h,$$

which suggests we should choose $\mathbf{A}_{2h}$ and I_h^{2h} to obey

$$\begin{aligned} I_h^{2h} &= c_h {I_{2h}^h}^{\mathrm{T}} \\ \mathbf{A}_{2h} &= I_h^{2h}\mathbf{A}_h I_{2h}^h, \end{aligned} \tag{4.43}$$

where c_h is a constant that may depend on h. These are called the *variational conditions*. Here, the development used matrices, but by the Galerkin variational method we would have used operators. In operator parlance the transpose in (4.43) is called the *adjoint*, and multigrid jargon tends to use that nomenclature also; I_h^{2h} should be proportional to the adjoint of I_{2h}^h.

4.6.1   The smoother

Consider $n-1 \times n-1$ matrix (4.37) with eigenvalues $\lambda_i = 2(\cos\frac{i\pi}{n} - 1) = -4\sin^2\frac{i\pi}{2n}$, and corresponding eigenvectors

$$\mathbf{y}_i = \left(\sin\frac{i\pi}{n}, \ \sin\frac{2i\pi}{n}, \ ..., \ \sin\frac{(n-1)i\pi}{n}\right)^{\mathrm{T}}. \tag{4.44}$$

A plot of eigenvectors (Figure 4.6) shows that $\mathbf{y}_i$ is $i/2$ periods of the sine function of wavelength $2n/i$.

For this matrix, $\mathbf{J} = \mathbf{I} + \mathbf{A}/2$, the eigenvalues of $\mathbf{J}$ are

$$\mu_i = 1 + \lambda_i/2 = \cos\frac{i}{n}\pi,$$

and the eigenvectors of $\mathbf{J}$ are the eigenvectors of $\mathbf{A}$. The fastest modes to converge under Jacobi relaxation will be those modes $i \approx n/2$ with wavelength ≈ 2. Jacobi, unmodified, does not preferentially suppress the short wavelength component of the solution.

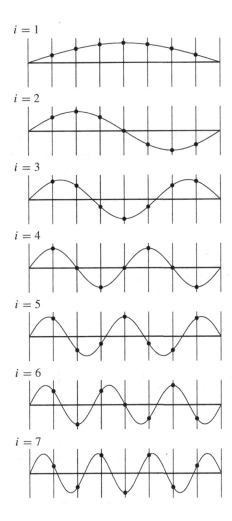

Figure 4.6 Eigenvectors of the 7×7 discrete Laplace matrix (4.37).

To preferentially damp the short wavelength modes, consider the generalized Jacobi method,

$$\mathbf{B}(\omega) = \frac{1}{\omega}\mathbf{D}$$

(compare with (4.27)). For matrix (4.37), this gives

$$\mathbf{J}(\omega) = \mathbf{I} + \frac{\omega}{2}\mathbf{A}$$

$$\mu_i(\omega) = 1 + \omega - \omega\cos\frac{i\pi}{n} = 1 - 2\omega\sin^2\frac{i}{2n}\pi.$$

For the associated function

$$f(x, \omega) = 1 - 2\omega\sin^2\frac{x\pi}{2},$$

the minimum value of

$$s = \max_{1/2 \le x \le 1} |f(x, \omega)|,$$

occurs when

$$f(1/2, \omega) = -f(1, \omega).$$

This equality is satisfied with $\omega = 2/3$, and $s = |f(1/2, 2/3)| = 1/3$. That is, with $\omega = 2/3$, the *damped Jacobi* method will preferentially damp the large i, short wavelength, components of the solution error. The maximum value of $|\mu_i(\omega)|$ in this range is $1/3$.

The early work on multigrid by Fedorenko used Gauss–Seidel with no modification as a smoother.

4.6.2   Restriction and prolongation

Fundamentally, the prolongation and restriction operators perform interpolation, so with a model in mind that connects grid indices with geometry, e.g., Figure 4.7, some models are easy to construct.

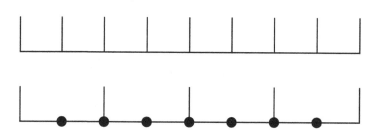

Figure 4.7 A fine and coarse grid function representation, for node-centered Poisson.

Consider prolongation first, and interpolate coarse grid data to obtain fine grid data. One possibility,

$$I_{2h}^{h} = \begin{pmatrix} \frac{3}{2} & -\frac{1}{2} \\ 1 & \\ \frac{1}{2} & \frac{1}{2} \\ & 1 \\ & \frac{1}{2} & \frac{1}{2} \\ & & 1 \\ & -\frac{1}{2} & \frac{3}{2} \end{pmatrix},$$

is based on second-order interpolation. The negative values occur where one is extrapolating the coarse data set to obtain the first and last fine data, which lie outside the geometric support of the coarse data. This is not a good idea: do not use this. With nonuniform signs, there is the possibility of instability (this will be explored in more detail in Chapter 9, page 244).

The coarse grid $\mathbf{A}\mathbf{x} = \mathbf{b}$ problem is solving for a correction, and we can reasonably assume that no correction is required outside the geometric domain: the corrections on the left and right boundaries of Figure 4.7 should have numerical value zero. We could use those zeros to perform genuine interpolation, rather than extrapolation, at the endpoints. The resulting operator is

$$
I_{2h}^{h} = \begin{pmatrix} \frac{1}{2} \\ 1 \\ \frac{1}{2} & \frac{1}{2} \\ & 1 \\ & \frac{1}{2} & \frac{1}{2} \\ & & 1 \\ & & \frac{1}{2} \end{pmatrix},
\tag{4.45}
$$

and this is stable.

Using the variational conditions, we would choose

$$
I_{h}^{2h} = c_h \begin{pmatrix} \frac{1}{2} & 1 & \frac{1}{2} \\ & & \frac{1}{2} & 1 & \frac{1}{2} \\ & & & & \frac{1}{2} & 1 & \frac{1}{2} \end{pmatrix}.
\tag{4.46}
$$

For the constant, think of this restriction operator as performing interpolation. The sum of the interpolation weights should be one, so $c_h = 1/2$.

With these choices, the matrix $\mathbf{A}_{2h}$ is given by the variational conditions:

$$
\mathbf{A}_{2h} = I_{h}^{2h} \mathbf{A}_h I_{2h}^{h} = \frac{1}{4} \begin{pmatrix} -2 & 1 & \\ 1 & -2 & 1 \\ & 1 & -2 \end{pmatrix}.
$$

We have the very nice result that $\mathbf{A}_{2h}$ follows the same pattern as $\mathbf{A}_h$, and in particular it has the same sparseness.

Early work on multigrid did not rely on the variational conditions. Instead, practitioners were motivated to use high-order accurate interpolation schemes (e.g., [25]). So, the operators described above are not essential to the success of the method. However, the variational conditions are theoretically compelling, lead to stable and robust methods, and require that only one interpolation problem be solved.

McCormick [155] notes that when (4.43) holds, the column space of I_{2h}^{h} is orthogonal to the null space of I_{h}^{2h}: all error transferred from the coarse grid to the fine grid can be passed back to the coarse grid. If the converse occurred, an instability could develop where long wavelength error generated on the coarse grid could accumulate in the fine grid solution. Thus, property (4.43) is stabilizing.

4.6.3 The two-grid method

The two-grid method formalizes the outline (4.42). One step of the two-grid method consists of:

1. Perform ν relaxation steps with damped Jacobi, $\omega = 2/3$.

2. Compute the residual, and solve $\mathbf{A}_{2h}\mathbf{x}_{2h} = I_h^{2h}\mathbf{r}_h$ for correction $\mathbf{x}_{2h}$. An exact solution is assumed.

3. Correct the solution: $\mathbf{x}_h := \mathbf{x}_h + I_{2h}^h\mathbf{x}_{2h}$.

To analyze the performance of this algorithm, let $\bar{\mathbf{x}}_h$ be the exact solution, and $\mathbf{e}_h$ the error. After ν steps of damped Jacobi, one has

$$\mathbf{e}_h = J(\omega)^\nu \mathbf{e}_h^0,$$

and the residual is

$$\mathbf{r}_h = -\mathbf{A}_h \mathbf{e}_h.$$

Therefore, the correction on the coarse grid is

$$\mathbf{x}_{2h} = -\mathbf{A}_{2h}^{-1} I_h^{2h} \mathbf{A}_h \mathbf{e}_h = -\mathbf{A}_{2h}^{-1} I_h^{2h} \mathbf{A}_h J(\omega)^\nu \mathbf{e}_h^0,$$

and for the result $\mathbf{x}_h + I_{2h}^h\mathbf{x}_{2h}$ one has the error

$$\mathbf{e}_h := \left(\mathbf{I} - I_{2h}^h \mathbf{A}_{2h}^{-1} I_h^{2h} \mathbf{A}_h \right) \mathbf{e}_h \tag{4.47}$$

$$\mathbf{e}_h = \left(\mathbf{I} - I_{2h}^h \mathbf{A}_{2h}^{-1} I_h^{2h} \mathbf{A}_h \right) J(\omega)^\nu \mathbf{e}_h^0.$$

For $\mathbf{A}$ given by (4.37), eigenvector $\mathbf{y}_i$ is given by (4.44), and for I_h^{2h} given by (4.46),

$$I_h^{2h}(\mathbf{y}_h)_i = \frac{1}{4} \begin{pmatrix} \sin\frac{\pi i}{n} + 2\sin\frac{2\pi i}{n} + \sin\frac{3\pi i}{n} \\ \sin\frac{3\pi i}{n} + 2\sin\frac{4\pi i}{n} + \sin\frac{5\pi i}{n} \\ \vdots \\ \sin\frac{(n-3)\pi i}{n} + 2\sin\frac{(n-2)\pi i}{n} + \sin\frac{(n-1)\pi i}{n} \end{pmatrix}$$

$$= \frac{(1 + \cos\frac{\pi i}{n})}{2} \begin{pmatrix} \sin\frac{2\pi i}{n} \\ \sin\frac{4\pi i}{n} \\ \vdots \\ \sin\frac{(n-2)\pi i}{n} \end{pmatrix} = \cos^2\frac{\pi i}{2n}(\mathbf{y}_{2h})_i.$$

However, $(\mathbf{y}_h)_{n-i}$ also maps to $(\mathbf{y}_{2h})_i$, and with simple trigonometric manipulations,

$$I_h^{2h}(\mathbf{y}_h)_{n-i} = -\sin^2\frac{\pi i}{2n}(\mathbf{y}_{2h})_i.$$

With I_{2h}^h given by (4.45),

$$I_{2h}^h(\mathbf{y}_{2h})_i = \frac{1}{2}\begin{pmatrix} \sin\frac{2\pi i}{n} \\ 2\sin\frac{2\pi i}{n} \\ \sin\frac{2\pi i}{n} + \sin\frac{4\pi i}{n} \\ 2\sin\frac{4\pi i}{n} \\ \vdots \\ \sin\frac{(n-4)\pi i}{n} + \sin\frac{(n-2)\pi i}{n} \\ 2\sin\frac{(n-2)\pi i}{n} \\ \sin\frac{(n-2)\pi i}{n} \end{pmatrix}$$

$$= \cos^2\frac{\pi i}{2n}\begin{pmatrix} \sin\frac{\pi i}{n} \\ \sin\frac{2\pi i}{n} \\ \sin\frac{3\pi i}{n} \\ \sin\frac{4\pi i}{n} \\ \vdots \\ \sin\frac{(n-3)\pi i}{n} \\ \sin\frac{(n-2)\pi i}{n} \\ \sin\frac{(n-1)\pi i}{n} \end{pmatrix} - \sin^2\frac{\pi i}{2n}\begin{pmatrix} \sin\frac{\pi(n-i)}{n} \\ \sin\frac{2\pi(n-i)}{n} \\ \sin\frac{3\pi(n-i)}{n} \\ \sin\frac{4\pi(n-i)}{n} \\ \vdots \\ \sin\frac{(n-3)\pi(n-i)}{n} \\ \sin\frac{(n-2)\pi(n-i)}{n} \\ \sin\frac{(n-1)\pi(n-i)}{n} \end{pmatrix}$$

$$= \cos^2\frac{\pi i}{2n}(\mathbf{y}_h)_i - \sin^2\frac{\pi i}{2n}(\mathbf{y}_h)_{n-i}.$$

Then, expanding $\mathbf{e}_h$ in eigenvectors of $\mathbf{A}_h$, the above results allow one to determine how each component of the error is attenuated. A matrix $\mathbf{M}$ is constructed to represent the effect of the terms in parentheses in (4.47), block partitioned to act on $\mathbf{y}_i$ and $\mathbf{y}_{n-1}$. The matrix $\mathbf{A}_h$ is represented by the diagonal matrix of eigenvalues at resolution h, and the matrix $\mathbf{A}_{2h}^{-1}$ is represented by the inverse eigenvalue:

$$\mathbf{M} = \begin{pmatrix} 1 & 0 \\ 0 & 1 \end{pmatrix}$$

$$- \begin{pmatrix} \cos^2\frac{\pi i}{2n} \\ -\sin^2\frac{\pi i}{2n} \end{pmatrix}\left(-\frac{1}{\sin^2\frac{i\pi}{n}}\right)\left(\cos^2\frac{\pi i}{2n} \quad -\sin^2\frac{\pi i}{2n}\right)\begin{pmatrix} -4\sin^2\frac{i\pi}{2n} & 0 \\ 0 & -4\sin^2\frac{(n-i)\pi}{2n} \end{pmatrix}$$

$$= \begin{pmatrix} \sin^2\frac{i\pi}{2n} & \cos^2\frac{i\pi}{2n} \\ \sin^2\frac{i\pi}{2n} & \cos^2\frac{i\pi}{2n} \end{pmatrix},$$

and

$$|\mathbf{M}| \leq \begin{pmatrix} \frac{1}{2} & 1 \\ \frac{1}{2} & 1 \end{pmatrix}$$

componentwise. With $|\mu_i| < 1/3$ for $i > n/2$, and $|\mu_i| < 1$ otherwise, including the effect of the damped Jacobi operator in (4.47) gives

$$|\mathbf{M}J^\nu| \le \begin{pmatrix} \frac{1}{2} & \frac{1}{3^\nu} \\ \frac{1}{2} & \frac{1}{3^\nu} \end{pmatrix} \quad \text{componentwise.}$$

Therefore,

$$\|\mathbf{e}_h\|_\infty \le \left(\frac{1}{2} + \frac{1}{3^\nu} \right) \|\mathbf{e}_h^0\|_\infty$$

(with the L^∞ norm, (2.15)) for one pass of the two-grid method.

This iterative method has first-order convergence independent of h. In contrast, Gauss–Seidel alone, or Jacobi alone, have spectral radii that are very sensitive to h, and very close to unity when n is large. In this respect, the two-grid convergence is amazing. However, to achieve this result one has to solve exactly a problem of only modestly smaller size, the $\mathbf{A}_{2h}$ problem. To improve this cost, one could defer long wavelength error to successively coarser grids, which leads to the multigrid V-cycle.

4.6.4   The V-cycle

The V-cycle is the generalization of the two-grid method to multiple grids. Let $\ell = 0, \ldots, L$ be the level of resolution. On the finest level, $\ell = 0$, one has $n = 2^K$ with $\mathbf{x}_0 \in \mathbb{R}^{2^K-1}$. On level ℓ, $n = 2^{K-\ell}$, and $\mathbf{x}_\ell \in \mathbb{R}^{2^{K-\ell}-1}$, with $K \ge L$. The algorithm follows.

1. Begin with $\mathbf{r}_0 = \mathbf{b}$.
2. For $\ell = 0, \ldots, L-1$,
    (i) Compute $\mathbf{x}_\ell$ with ν smoothing iterations:

    $$\mathbf{A}_\ell \mathbf{x}_\ell = \mathbf{r}_\ell.$$

    (ii) Compute the residual on grid $\ell + 1$:

    $$\mathbf{r}_{\ell+1} = I_\ell^{\ell+1}(\mathbf{r}_\ell - \mathbf{A}_\ell \mathbf{x}_\ell).$$

3. At the coarsest level, L, solve the remaining problem exactly:

    $$\mathbf{A}_L \mathbf{x}_L = \mathbf{r}_L.$$

    This is called the *bottom solver*.
4. For $\ell = L-1, \ldots, 0$,
    (i) Correct the level ℓ solution

    $$\mathbf{x}_\ell := \mathbf{x}_\ell + I_{\ell+1}^\ell \mathbf{x}_{\ell+1}.$$

    (ii) This correction can produce some high-frequency error on the grid, due to in-accuracy of the prolongation operator. Remove this new high-frequency error with ν iterations of the smoother:

    $$\mathbf{A}_\ell \mathbf{x}_\ell = \mathbf{r}_\ell.$$

If $K = L$, $\mathbf{A}_L$ is 1×1 and the bottom solver is trivial.

Example 4.5   Solve $\mathbf{Ax} = \mathbf{1}$ for $\mathbf{x} \in \mathbb{R}^{511}$, with the matrix (4.37).

Solution

The solution is computed with multigrid using the full V-cycle (to coarsest resolution $\mathbf{x} \in \mathbb{R}^1$) using both damped Jacobi and Gauss–Seidel as smoothers, with parameter $\nu = 3$. For comparison, the solution is computed also with Jacobi ($\omega = 1$) and Gauss–Seidel on a single grid. In all cases, the iterations begin with $\mathbf{x} = \mathbf{0}$ and terminate when the residual is 10^{-4} in L^2.

The results displayed in the following table speak for themselves and clearly show the performance benefits of the multigrid strategy.

method	iterations
V-cycle, damped Jacobi $J(2/3)$	5
V-cycle, Gauss–Seidel	4
Jacobi, $J(1)$	649 379
Gauss–Seidel	324 691

The sample code that generated these results is given in Appendix B.8.

An excellent resource for practical multigrid techniques, including the full approximation scheme (FAS multigrid) for nonlinear problems, is the tutorial by Briggs, Henson, and McCormick [26].

Problems

4.1   Prove (4.38).

4.2   [232] For the $n \times n$ matrix

$$\mathbf{A} = \begin{pmatrix} 1 & & & & a \\ & 1 & \ddots & & \\ & & \ddots & \ddots & \\ b & & & 1 & 1 \end{pmatrix},$$

show that
(i)   Jacobi converges, and Gauss–Seidel diverges, if

$$a = \tfrac{1}{2}(-1)^{n+1} \qquad\qquad b = (-1)^n;$$

and
(ii)   Jacobi diverges, and Gauss–Seidel converges, if

$$a \geq 1 \qquad\qquad b = (-1)^{n+1},$$

when $n \geq 3$.

Use Rouché's theorem [202]: if $f(z)$ and $g(z)$ are analytic in and on a closed contour C, and $|f(z)| > |g(z)|$ everywhere on C, then $f(z)$ and $f(z) + g(z)$ have the same number of zeros inside C.

4.3   Write the discretization of the two-dimensional Laplace problem

$$\frac{\partial^2 u}{\partial x^2} + \frac{\partial u^2}{\partial y^2} = 0$$

$$u(x, 0) = x^2$$

$$u(x, 1) = x^2 - 1$$

$$u(0, y) = -y^2$$

$$u(1, y) = 1 - y^2$$

on a node-centered grid such as the one shown below.

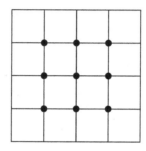

Use variable ordering following the pattern

$$\mathbf{u}^{\mathsf{T}} = (u_{00}, u_{01}, u_{02}, u_{10}, u_{11}, u_{12}, u_{20}, u_{21}, u_{22}),$$

where there are $N \times N$ cells of length $h = 1/N$, and $u_{ij} = u((i+1)h, (j+1)h)$. Show that this discretization is amenable to solution by Jacobi and Gauss–Seidel relaxation.

4.4   The matrix obtained in Problem 4.3 is either positive definite or negative definite, depending on the sign conventions you used. For any convention, Problem 4.3 can be solved using conjugate gradients. Find the numerical solution using conjugate gradients when $N = 64$, i.e., when there are $63 \times 63 = 3969$ variables and $\mathbf{A}$ is 3969×3969. With $\mathbf{x} = 0$ as an initial guess, stop when $\|\mathbf{p}\|_2 < 10^{-3}$.

4.5   Solve Problem 4.3 using multigrid, stopping when $\|\mathbf{r}\|_2 < 10^{-3}$.

4.6   Solve Problem 4.3 using one multigrid V-cycle as a preconditioner for conjugate gradients.

5 Interpolation

The goal of interpolation is to answer the following question. Given *support points*

$$(x_0, y_0), \ (x_1, y_1), \ ..., \ (x_n, y_n), \tag{5.1}$$

what is $y(x)$ for an arbitrary value of x? Strictly speaking, this is interpolation only if the arbitrary x lies in the interval spanning the *support abscissas* $x_0, \ ..., \ x_n$; i.e., $x \in I(x_0, ..., x_n)$. If x lies outside this interval, the problem is called extrapolation. When we have more data than model parameters, the problem is data fitting, Chapter 8.

We will assume that the abscissas x_i are distinct. With that assumption, one could find a polynomial $p(x)$ of degree n such that $p(x_i) = y_i$ for each of the given support points. Then, $y = p(x)$ solves the interpolation problem.

There is a well-known approach to this problem called the Lagrange interpolation formula:

$$
\begin{aligned}
p(x) &= \sum_{i=0}^{n} y_i L_i(x) \\
L_i(x) &= \frac{(x - x_0) \cdots (x - x_{i-1})(x - x_{i+1}) \cdots (x - x_n)}{(x_i - x_0) \cdots (x_i - x_{i-1})(x_i - x_{i+1}) \cdots (x_i - x_n)} \\
&= \prod_{j=0, j \neq i}^{n} \frac{(x - x_j)}{(x_i - x_j)}.
\end{aligned}
\tag{5.2}
$$

This clearly accomplishes the stated goal because (i) it is a polynomial of degree n – each L_i contains n products $(x - x_j)$; (ii) it reproduces the given data – $L_i(x_i) = 1$, and $L_i(x_j) = 0$ when $j \neq i$. The method is attributed to J. L. Lagrange who published it in 1795 [140], although it seems to have been discovered earlier by E. Waring [236]. A nice history of interpolation is given by E. Meijering [160].

Although the Lagrange interpolation formula accomplishes the stated goal, it is cumbersome to apply and not numerically stable. It is mostly useful as a theoretical construction – to numerically solve a given problem it is almost never a good choice. Do not use it.

The polynomial of degree n solving the interpolation problem is unique. If it were not unique, then $P_n(x)$ and $Q_n(x)$ are two different polynomials of degree n and $P_n(x_0) = Q_n(x_0) = y_0, \ ..., \ P_n(x_n) - Q_n(x_n) = y_n$ since they both solve the given problem. This implies that the degree n polynomial $P_n(x) - Q_n(x)$ has $n + 1$ zeros. According to the fundamental theorem of algebra, this is impossible unless $P_n(x) - Q_n(x)$ is zero identically, which contradicts the supposition that they are different.

The error associated with polynomial interpolation is easily expressed. If $p(x)$ is the unique degree n polynomial that reproduces data (5.1), then

$$p(x) + K(x - x_0)(x - x_1) \cdots (x - x_n)$$

is a degree $n + 1$ polynomial that also reproduces data (5.1). With the extra degree of freedom, we can also fit one additional point, say $(\bar{x}, \bar{y})$, which means that the function $F(x)$

$$F(x) = p(x) + K(x - x_0) \cdots (x - x_n) - y(x)$$

has $n + 2$ zeros at $x_0, \ldots, x_n, \bar{x}$. By Rolle's theorem, $F'(x)$ has $n + 1$ zeros in the interval $I(x_0, \ldots, x_n, \bar{x})$, and $F''(x)$ has n zeros, etc. Finally, $F^{(n+1)}$ has one zero in $I(x_0, \ldots, x_n, \bar{x})$. Call this special point ξ, then

$$\frac{d^{n+1}}{dx^{n+1}} \left[p(x) + K(x - x_0) \cdots (x - x_n) - y(x) \right]_{x=\xi} = 0.$$

But, $p(x)$ is only a degree n polynomial, so $p^{(n+1)}(x) = 0$. And, the product $(x - x_0) \cdots (x - x_n) = x^{n+1} +$ lower order terms, so $d^{n+1}/dx^{n+1}(x - x_0) \cdots (x - x_n) = (n + 1)!$. Therefore,

$$y^{(n+1)}(\xi) = (n + 1)!K$$

and

$$p(x) - y(x) = -\frac{y^{(n+1)}(\xi)}{(n + 1)!}(x - x_0) \cdots (x - x_n), \tag{5.3}$$

for some $\xi \in I(x_0, \ldots, x_n, x)$. This is the approximation error associated with polynomial interpolation. There are also numerical errors, which will be different for different methods of constructing $p(x)$.

A different estimate of the error comes from the integral form of the Taylor remainder. This analysis uses the "Peano kernel" [178], which works as follows. Let $\mathcal{L}$ be an operator, acting on a function through discrete and integral terms. In the present circumstance, we will define $\mathcal{L}$ to be the difference between a function and its polynomial representation

$$\mathcal{L}f(x) = f(x) - P_n(x) = f(x) - \sum_{i=0}^{n} f(x_i)L_i(x), \tag{5.4}$$

here expressed as a degree n Lagrange interpolation polynomial.

Now, use the Taylor series of f with integral remainder term. The number of Taylor terms is chosen to match the degree of the polynomial. Let "a" represent the smallest abscissa at which one might be interested in interpolating:

$$f(x) = f(a) + \cdots + \frac{(x - a)^n}{n!} f^{(n)} + \int_a^x \frac{(x - t)^n}{n!} f^{(n+1)}(t)dt.$$

Since the first terms on the right-hand side are a degree n polynomial, they can be represented exactly by the polynomial approximation under consideration. Therefore,

$$\mathcal{L}f = \mathcal{L}\int_a^x \frac{(x-t)^n}{n!}f^{(n+1)}(t)dt. \tag{5.5}$$

This term can be rearranged as follows. First, we replace the upper limit of integration with some constant b chosen so that for any x of interest $a \le x \le b$. In so doing, we need to restrict the integrand so that it becomes zero when $x \le t \le b$. We accomplish this by writing

$$(x-t)_+^n = \begin{cases} (x-t)^n & \text{if } x \ge t \\ 0 & \text{if } x < t. \end{cases} \tag{5.6}$$

Now, (5.5) may be written

$$\mathcal{L}f = \mathcal{L}\int_a^b \frac{(x-t)_+^n}{n!}f^{(n+1)}(t)dt.$$

Next, we note that $\mathcal{L}$ operates on functions of x, so interchanging operation $\mathcal{L}$ with integration,

$$\mathcal{L}f = \int_a^b K_n(t)f^{(n+1)}(t)dt \tag{5.7}$$

$$K_n(t) = \mathcal{L}\frac{(x-t)_+^n}{n!}. \tag{5.8}$$

So, the idea is to construct the function $K_n(t)$ (5.8), which embodies the particulars of our interpolation scheme, and then use $K_n(t)$ to determine the error from (5.7). $K_n(t)$ is called the Peano kernel.

Example 5.1   Use the Peano kernel to find the error associated with linear interpolation $n = 1$.

Solution
The Peano kernel (5.8) is

$$K_1(t) = \mathcal{L}(x-t)_+$$

$$= (x-t)_+ - \frac{(x-x_1)}{(x_0-x_1)}(x_0-t)_+ - \frac{(x-x_0)}{(x_1-x_0)}(x_1-t)_+,$$

where we have substituted $(x-t)_+$ into the definition (5.4). It can be shown with some difficulty that the function $K_1(t)$ is not positive when $x \le x_0$, and not negative when $x \ge x_0$ (this is demonstrated graphically in Figure 5.1). Since the sign does not depend on t, the mean value theorem may be applied:

$$\mathcal{L}f = f''(\xi)\int_a^b K_t(t)dt = \frac{(x-x_0)(x-x_1)}{2}f''(\xi)$$

and

$$P_1(x) - f(x) = -\frac{f''(\xi)}{2}(x - x_0)(x - x_1),$$

for some $\xi \in [a, b]$. This result agrees with the error bound given by (5.3).

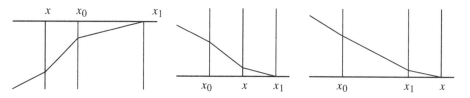

Figure 5.1 Peano interpolation kernel $K_1(t)$, Example 5.1.

5.1   Modified Lagrange interpolation and the barycentric form

While use of the well-known formula (5.2) is inadvisable, simple rearrangements yield useful forms.

With the definitions

$$L(x) = \prod_{i=0}^{n}(x - x_i)$$

$$w_j = \frac{1}{\prod_{k=0,\, k \neq j}(x_j - x_k)},$$

one can write

$$L_i(x) = L(x)\frac{w_i}{x - x_i}$$

$$P(x) = L(x)\sum_{i=0}^{n}\frac{w_i}{x - x_i}y_i. \tag{5.9}$$

This rearrangement is called the *modified Lagrange formula*. The main advantage of this form concerns efficiency, but it also has good stability properties. The calculation of $L_i(x)$ (5.2) involves $\mathcal{O}(2n)$ operations (counting only multiplication and division), so $P(x)$ requires $\mathcal{O}(2n^2)$ calculations for each evaluation point x. The revised form (5.9) takes $\mathcal{O}(n^2)$ calculations to set up, by evaluating the numbers w_j. Then, for each evaluation point x there are only $\mathcal{O}(n)$ calculations to make $L(x)$ and another $\mathcal{O}(n)$ calculations to evaluate the polynomial. Compared to the form (5.2) this modified form is $\mathcal{O}(n) \times$ faster.

Still more efficiency can be had by considering the interpolation problem consisting of interpolating unity:

$$1 = L(x)\sum_{i=0}^{n}\frac{w_i}{x - x_i}1,$$

which permits the evaluation of $L(x)$ with fewer operations. The setup phase of the calculation still takes $\mathcal{O}(n^2)$ steps to make the w_j. The evaluation phase uses $\mathcal{O}(n)$ calculations to make the ratios $w_i/(x - x_i)$, so $L(x)$ takes only $\mathcal{O}(1)$ additional operation:

$$L(x) = \frac{1}{\sum_{i=0}^{n} \frac{w_i}{x - x_i}}.$$

This final *barycentric form* is written

$$P(x) = \frac{\sum_{i=0}^{n} \frac{w_i}{x - x_i} y_i}{\sum_{i=0}^{n} \frac{w_i}{x - x_i}}.$$

All Lagrange type formulae have an advantage over other polynomial interpolation methods in that the computed quantities (in the barycentric case, w_j, and in the original case L_i) do not depend on the support ordinates y_i. The work involved in setting up the calculation can therefore be recycled when considering related problems using the same support abscissas.

Higham [105] examined the numerical stability of the modified Lagrange formula and the barycentric form, and showed the modified Lagrange formula to have good stability properties. The barycentric form is not unconditionally stable, but circumstances under which instability occur are considered rare. Favorable properties [15] and numerical stability [105] recommend the modified Lagrange formula as the method of choice.

5.2   Neville's algorithm

The idea behind Neville's algorithm [171] is this: define the $n + 1$ degree zero polynomials

$$P_0(x) = y_0$$

$$\vdots$$

$$P_n(x) = y_n,$$

to each interpolate one point.

The n degree one polynomials

$$P_{01}(x) = \frac{(x - x_0)P_1(x) - (x - x_1)P_0(x)}{x_1 - x_0}$$

$$\vdots \qquad\qquad\qquad\qquad\qquad (5.10)$$

$$P_{n-1,n}(x) = \frac{(x - x_{n-1})P_n(x) - (x - x_n)P_{n-1}(x)}{x_n - x_{n-1}}$$

each linearly interpolate between the points indicated by their subscripts.

The $n - 1$ degree two polynomials

$$P_{012}(x) = \frac{(x - x_0)P_{12}(x) - (x - x_2)P_{01}(x)}{x_2 - x_0}$$

$$\vdots \tag{5.11}$$

$$P_{n-2,n-1,n}(x) = \frac{(x - x_{n-2})P_{n-1,n}(x) - (x - x_n)P_{n-2,n-1}(x)}{x_n - x_{n-2}}$$

each quadratically interpolate between the three points indicated by their subscripts. Carrying on similarly, one would finally obtain the single polynomial of degree n,

$$P_{0,\ldots,n}(x) = \frac{(x - x_0)P_{1,\ldots,n}(x) - (x - x_n)P_{0,\ldots,n-1}(x)}{x_n - x_0},$$

that interpolates all $n + 1$ points.

Numerically, one does not keep $P_{\ldots}(x)$ as a function, but rather as a number constructed for a given choice of x.

Example 5.2 Estimate the value $y(1)$ given the following data:

i	x_i	y_i
0	0	0
1	2	2.864 66
2	4	4.981 68
3	6	6.997 52

Solution

The completed Neville tableau for this problem is as follows:

x_0	P_0					
0	0					
		P_{01}				
		1.4323				
x_1	P_1		P_{012}			
2	2.864 66		1.5258			
		P_{12}		P_{0123}		
		1.8062		1.5662		
x_2	P_2		P_{123}			
4	4.981 68		1.7682			
		P_{23}				
		1.9579				
x_3	P_3					
6	6.997 52					

The four data points define a degree three polynomial, whose value at $x = 1$ is 1.5662. Note that each number in the tableau is an estimate of $y(1)$ using different support data.

The calculations underlying this tableau are:

$$P_{01} = \frac{(x - x_0)P_1 - (x - x_1)P_0}{x_1 - x_0} = \frac{(1 - 0)2.8647 - (1 - 2)0}{2 - 0} = 1.4323$$

$$P_{12} = \frac{(x - x_1)P_2 - (x - x_2)P_1}{x_2 - x_1} = \frac{(1 - 2)4.9817 - (1 - 4)2.8647}{4 - 2} = 1.8062$$

$$P_{23} = \frac{(x - x_2)P_3 - (x - x_3)P_2}{x_3 - x_2} = \frac{(1 - 4)6.9975 - (1 - 6)4.9817}{6 - 4} = 1.9579$$

$$P_{012} = \frac{(x - x_0)P_{12} - (x - x_2)P_{01}}{x_2 - x_0} = \frac{(1 - 0)1.8062 - (1 - 4)1.4323}{4 - 0} = 1.5258$$

$$P_{123} = \frac{(x - x_1)P_{23} - (x - x_3)P_{12}}{x_3 - x_1} = \frac{(1 - 2)1.9579 - (1 - 6)1.8062}{6 - 2} = 1.7682$$

$$P_{0123} = \frac{(x - x_0)P_{123} - (x - x_3)P_{012}}{x_3 - x_0} = \frac{(1 - 0)1.7682 - (1 - 6)1.5258}{6 - 0} = 1.5662.$$

5.3 Newton

Since the interpolation polynomials are unique, there is no problem with assuming the particular form

$$p_n(x) = a_0 + a_1(x - x_0) + a_2(x - x_0)(x - x_1) + \cdots + a_n(x - x_0)\cdots(x - x_{n-1}).$$

Now, define a_0 so that $p_0(x) = a_0$ "interpolates" the point (x_0, y_0). That is, choose $a_0 = y_0$.

Next, choose a_1 so that $p_1(x) = a_0 + a_1(x - x_0)$ interpolates the two points (x_0, y_0) and (x_1, y_1). Rearranging $p_1(x_1) = y_1$ gives

$$a_1 = \frac{y_1 - y_0}{x_1 - x_0}.$$

This is a divided difference that will be related to the derivative y' in the interval $I(x_0, x_1)$.

If a_2 is chosen so that $p_2(x) = a_0 + a_1(x - x_0) + a_2(x - x_0)(x - x_1)$ interpolates the first three points, then from $p_2(x_2) = y_2$ one obtains a result that may be written

$$a_2 = \frac{\frac{y_2 - y_1}{x_2 - x_1} - \frac{y_1 - y_0}{x_1 - x_0}}{x_2 - x_0},$$

where a_2 is a divided difference of divided differences, and will be related to $y''/2$ in the interval $I(x_0, x_1, x_2)$.

This recursive calculation is organized in a tableau as follows:

$$
\begin{array}{lll}
x_0 & y[x_0] & \\
 & & y[x_0, x_1] \\
x_1 & y[x_1] & & y[x_0, x_1, x_2] \\
 & & y[x_1, x_2] & & y[x_0, x_1, x_2, x_3]. \\
x_2 & y[x_2] & & y[x_1, x_2, x_3] \\
 & & y[x_2, x_3] & \\
x_3 & y[x_3] & &
\end{array}
$$

The first column consists of the abscissas x_i of the support data, and the second column consists of the corresponding support ordinates $y_i = y[x_i]$. The third column contains first *divided differences*, constructed from the adjacent data sets:

$$
y[x_i, x_{i+1}] = \frac{y[x_{i+1}] - y[x_i]}{x_{i+1} - x_i}.
$$

The fourth column consists of second divided differences, constructed from the adjacent first divided differences, and with a denominator given by the most extreme indices, e.g.,

$$
y[x_{i-1}, x_i, x_{i+1}] = \frac{y[x_i, x_{i+1}] - y[x_{i-1}, x_i]}{x_{i+1} - x_{i-1}},
$$

and so on. The interpolation polynomial is now constructed from the entries at the top of each column, as follows:

$$
\begin{aligned}
p(x) = {} & y[x_0] + y[x_0, x_1](x - x_0) + y[x_0, x_1, x_2](x - x_0)(x - x_1) \\
& + y[x_0, x_1, x_2, x_3](x - x_0)(x - x_1)(x - x_2),
\end{aligned}
$$

or

$$
\begin{aligned}
p(x) = {} & y[x_0] + (x - x_0)\Big(y[x_0, x_1] + (x - x_1)\Big(y[x_0, x_1, x_2] \\
& + (x - x_2)y[x_0, x_1, x_2, x_3]\Big)\Big),
\end{aligned}
\tag{5.12}
$$

in the Horner form.

The divided differences $y[]$ have a special meaning, derived by consideration of the error formula (5.3). Polynomial

$$
p_0(x) = y[x_i]
$$

is a degree zero polynomial that exactly matches the support data (x_i, y_i). If we add a term $K(x - x_i)$ then we can choose K to match exactly one other point. If x_{i+1} is the other point we want to match, then $K = y'(\xi)$ for some $\xi \in I[x_i, x_{i+1}]$, and

$$
p_1(x) = y[x_i] + y'(\xi)(x - x_i) \quad \text{for some } \xi \in I[x_i, x_{i+1}].
\tag{5.13}
$$

On the other hand, from the Newton tableau we would have written

$$
p_1(x) = y[x_i] + y[x_i, x_{i+1}](x - x_i)
\tag{5.14}
$$

as the polynomial that interpolates (x_i, y_i) and (x_{i+1}, y_{i+1}). Comparing (5.13) and (5.14) we conclude that

$$
y[x_i, x_{i+1}] = y'(\xi) \quad \text{for some } \xi \in I[x_i, x_{i+1}].
$$

By this same analysis, we find

$$y[\underbrace{\qquad \cdots \qquad}_{\substack{n+1 \text{ support} \\ \text{abscissas}}}] = \frac{y^{(n)}(\xi)}{n!} \quad \text{for some } \xi \in I[\underbrace{\qquad \cdots \qquad}_{\substack{\text{the same } n+1 \\ \text{support abscissas}}}].$$

The finite difference brackets $y[]$ are symmetric with respect to interchange of the arguments, i.e., $y[x_0, x_1, x_2] = y[x_1, x_0, x_2]$, etc.

For a given evaluation point x, the polynomial (5.12) in the Horner form is going to have different numerical errors depending on the ordering or permutation of the support points. The last step of the evaluation of the Horner form could be written

$$y(x) = y[x_0] + (x - x_0)\alpha_0,$$

where $\alpha_0 = y[x_0, x_1] + (x - x_1)\alpha_1 + \cdots$. To assess the importance of ordering, let us consider all divided differences $y[]$ to be free of error. Then for this trivial step we have an error analysis like

$$\tilde{y}(x) = y[x_0](1 + \epsilon_3) + (x - x_0)\alpha_0(1 + \epsilon_1)(1 + \epsilon_2)(1 + \epsilon_3),$$

where ϵ_1 is the relative error associated with the subtraction $(x - x_0)$, ϵ_2 is the relative error associated with the multiplication $(x - x_0)\alpha_0$, and ϵ_3 is the relative error associated the final addition. The key points are (i) that the left-hand side $y(x)$ is independent of x_0; and (ii) the first term on the right-hand side $y_0 = y[x_0]$ has less relative error associated with it than does the second term $(x - x_0)\alpha_0$, which can also be written $y(x) - y_0$. The most numerically stable choice of x_0 is the choice that maximizes the contribution of the first term, and minimizes the contribution of the second term. As $x_0 \to x$, the magnitude of the second term $|y(x) - y_0|$ will tend to zero (assuming the function to be continuous and differentiable). Therefore, choosing x_0 to minimize $|x - x_i|$ over support abscissas x_i is the best choice, numerically.

Note that α_0 can be written

$$\alpha_0 = \frac{y[x] - y[x_0]}{x - x_0} = y[x_0, x].$$

This permits the second-to-last step of the Horner evaluation to be written

$$y[x_0, x] = y[x_0, x_1] + (x - x_1)\alpha_1.$$

Again, the left-hand side is independent of x_1, and the first term on the right-hand side carries less relative numerical error than does the second. As $x_1 \to x$, the magnitude of the second term $|y[x_0, x] - y[x_0, x_1]|$ tends to zero (if $y(x) \in C^2$) so the contribution of the second term, which carries the greatest relative error, is minimized. Choosing x_1 to minimize $|x - x_i|$ over the support abscissas $x_i \neq x_0$ is the best choice, numerically.

This analysis suggests that the most numerically trustworthy way to evaluate a Newton polynomial occurs when the support points are first permuted so that $|x - x_0| \leq |x - x_1| \leq \cdots \leq |x - x_n|$. This in turn suggests the unattractive circumstance that a given tableau should not be used for all x. However, with the initial ordering $x_0 \leq x_1 \leq$

$x_2 \leq \ldots \leq x_n$ the necessary terms can be obtained from the resulting tableau [225]. This will be shown in Example 5.4.

First, to illustrate the relevance of this implicit permutation algorithm, consider the interpolation of $f = 1/(x^2 + 1)$ on the interval $[-1, 1]$ using 8 regularly spaced abscissas. For each of the 8! support point permutations, the Newton tableau is constructed with full machine precision, then evaluated at $x = 0.91$ without implicit permutation. Only four decimal places are used to evaluate the tableau with the Horner scheme. This exaggerates the evaluation error so it can be readily distinguished from the errors associated with creation of the finite difference tableau. The distribution of evaluation errors for this problem is displayed in Figure 5.2. The minimum error *is* observed for the permutation associated with the ordering $|x - x_0| \leq |x - x_1| \leq \cdots$ justified above, and errors as much as $20\times$ greater are seen with other permutations.

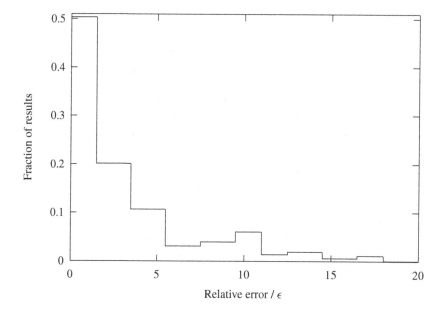

Figure 5.2 Histogram of observed Horner evaluation errors.

Example 5.3 Solve the problem of Example 5.2 by first constructing the interpolation polynomial by Newton's method.

Solution

The Newton tableau is

$$
\begin{array}{llll}
x_0 & y[x_0] \\
0 & 0 \\
& & y[x_0,x_1] \\
& & 1.4323 \\
x_1 & y[x_1] & & y[x_0,x_1,x_2] \\
2 & 2.864\,66 & & -0.093\,456 \\
& & y[x_1,x_2] & & y[x_0,x_1,x_2,x_3] \\
& & 1.0585 & & 0.013\,468 \\
x_2 & y[x_2] & & y[x_1,x_2,x_3] \\
4 & 4.981\,68 & & -0.012\,648 \\
& & y[x_2,x_3] \\
& & 1.0079 \\
x_3 & y[x_3] \\
6 & 6.997\,52
\end{array}
$$

The underlying calculations are:

$$y[x_0, x_1] = \frac{y[x_1] - y[x_0]}{x_1 - x_0} = \frac{2.8647 - 0}{2 - 0} = 1.4323$$

$$y[x_1, x_2] = \frac{y[x_2] - y[x_1]}{x_2 - x_1} = \frac{4.9817 - 2.8647}{4 - 2} = 1.0585$$

$$y[x_2, x_3] = \frac{y[x_3] - y[x_2]}{x_3 - x_2} = \frac{6.9975 - 4.9817}{6 - 4} = 1.0079$$

$$y[x_0, x_1, x_2] = \frac{y[x_1, x_2] - y[x_0, x_1]}{x_2 - x_0} = \frac{1.0585 - 1.4323}{4 - 0} = -0.093\,456$$

$$y[x_1, x_2, x_3] = \frac{y[x_2, x_3] - y[x_1, x_2]}{x_3 - x_1} = \frac{1.0079 - 1.0585}{6 - 2} = -0.012\,648$$

$$y[x_0, x_1, x_2, x_3] = \frac{y[x_1, x_2, x_3] - y[x_0, x_1, x_2]}{x_3 - x_0}$$

$$= \frac{-0.012\,648 + 0.093\,456}{6 - 0} = 0.013\,468.$$

From the completed tableau the interpolation polynomial may be constructed:

$$P(x) = 0 + (x - 0)\,(1.4323 + (x - 2)\,(-0.093\,456 + (x - 4)\,(0.013\,468)))$$

and $P(1) = 1.5662$, which is identical to the Neville solution.

Example 5.4 For the support data of Example 5.2, find the most numerically stable representation of Newton's interpolation polynomial for $x \approx 3.5$.

Solution

The previous example constructed the tableau with properly ordered data $0 \le 2 \le 4 \le 6$. For $x = 3.5$, the preferred order is $|x - 4| \le |x - 2| \le |x - 6| \le |x - 0|$, so the interpolation polynomial would be constructed

$$y[4] + (x - 4)\big[y[2, 4] + (x - 2)\big[y[2, 4, 6] + (x - 6)y[0, 2, 3, 6]\big]\big].$$

The indicated difference terms $y[]$ were given in the tableau already created.

$$
\begin{array}{lllll}
x_0 & y[x_0] & & & \\
0 & 0 & & & \\
& & y[x_0,x_1] & & \\
& & 1.4323 & & \\
x_1 & y[x_1] & & y[x_0,x_1,x_2] & \\
2 & 2.864\,66 & & -0.093\,456 & \\
& & y[x_1,x_2] & & y[x_0,x_1,x_2,x_3] \\
& & 1.0585 & & \underline{0.0134\,68} \\
x_2 & y[x_2] & & y[x_1,x_2,x_3] & \\
4 & 4.981\,68 & & -0.012\,648 & \\
& & y[x_2,x_3] & & \\
& & 1.0079 & & \\
x_3 & y[x_3] & & & \\
6 & 6.997\,52 & & &
\end{array}
$$

That information is copied above, with the relevant terms underlined.

Sir Isaac Newton's work on interpolation, ca. 1675, is summarized and translated (from Latin) by D. C. Fraser in [75].

5.4   Hermite

C. Hermite first considered the interpolation problem in which support data contains not only ordinates $y(x_i)$, but also derivatives $y'(x_i)$ [101]. The Hermite interpolation tableau is basically a Newton interpolation tableau, where one or more piece of derivative information is included in the support data.

Since $y[x_0, x_1]$ is a finite difference approximation to y' in the interval $I(x_0, x_1)$, it follows that

$$y'(x_0) = y[x_0, x_0]$$

by taking the limit $x_1 \rightarrow x_0$. So, if the support data consisted of (x_0, y_0, y_0'), (x_1, y_1), ..., (x_n, y_n), then this derivative information could be incorporated by writing a tableau with point (x_0, y_0) included twice, and with $y'(0)$ inserted into the location where the finite difference $y[x_0, x_0]$ would otherwise appear; e.g.,

$$
\begin{array}{llll}
x_0 & y_0 & & \\
& y_0' & & \\
x_0 & y_0 & y[x_0, x_0, x_1] & \\
& y[x_0, x_1] & & y[x_0, x_0, x_1, x_2] \\
x_1 & y_1 & y[x_0, x_1, x_2] & \\
& y[x_1, x_2] & & \\
x_2 & y_2 & &
\end{array}
$$

The remaining finite differences $y[]$ are computed in the usual manner, e.g.,

$$y[x_0, x_0, x_1] = \frac{y[x_0, x_1] - y_0'}{x_1 - x_0},$$

giving a polynomial

$$p(x) = y_0 + (x - x_0)\left(y_0' + (x - x_0)\left(y[x_0, x_0, x_1] + (x - x_1)\left(y[x_0, x_0, x_1, x_2]\right)\right)\right),$$

with error

$$p(x) - y(x) = -\frac{y^{(4)}(\xi)}{4!}(x - x_0)^2(x - x_1)(x - x_2).$$

Second derivatives can be included if the first derivatives are also included, by substituting $y''(x_i)/2!$ for $y[x_i, x_i, x_i]$, etc.

Example 5.5   Estimate the value $y(1)$ given the following data:

i	x_i	y_i	y_i'
0	0	0	2
1	2	2.864 66	
2	4	4.981 68	
3	6	6.997 52	

Solution

The completed tableau is

x_0	$y[x_0]$				
0	0				
		$y[x_0, x_0]$			
		2			
x_0	$y[x_0]$		$y[x_0, x_0, x_1]$		
0	0		$-0.283\,83$		
		$y[x_0, x_1]$		$y[x_0, x_0, x_1, x_2]$	
		1.4323		$0.047\,595$	
x_1	$y[x_1]$		$y[x_0, x_1, x_2]$		$y[x_0, x_0, x_1, x_2, x_3]$
2	2.864 66		$-0.093\,456$		$-0.005\,687\,8$
		$y[x_1, x_2]$		$y[x_0, x_1, x_2, x_3]$	
		1.0585		$0.013\,468$	
x_2	$y[x_2]$		$y[x_1, x_2, x_3]$		
4	4.981 68		$-0.012\,648$		
		$y[x_2, x_3]$			
		1.0079			
x_3	$y[x_3]$				
6	6.997 52				

with intermediate steps:

$$y[x_0, x_1] = \frac{y[x_1] - y[x_0]}{x_1 - x_0} = \frac{2.8647 - 0}{2 - 0} = 1.4323$$

$$y[x_1, x_2] = \frac{y[x_2] - y[x_1]}{x_2 - x_1} = \frac{4.9817 - 2.8647}{4 - 2} = 1.0585$$

$$y[x_2, x_3] = \frac{y[x_3] - y[x_2]}{x_3 - x_2} = \frac{6.9975 - 4.9817}{6 - 4} = 1.0079$$

$$y[x_0, x_0, x_1] = \frac{y[x_0, x_1] - y[x_0, x_0]}{x_1 - x_0} = \frac{1.4323 - 2}{2 - 0} = -0.28383$$

$$y[x_0, x_1, x_2] = \frac{y[x_1, x_2] - y[x_0, x_1]}{x_2 - x_0} = \frac{1.0585 - 1.4323}{4 - 0} = -0.093\,456$$

$$y[x_1, x_2, x_3] = \frac{y[x_2, x_3] - y[x_1, x_2]}{x_3 - x_1} = \frac{1.0079 - 1.0585}{6 - 2} = -0.0126\,48$$

$$y[x_0, x_0, x_1, x_2] = \frac{y[x_0, x_1, x_2] - y[x_0, x_0, x_1]}{x_2 - x_0}$$
$$= \frac{-0.093\,456 + 0.283\,83}{4 - 0} = 0.047\,595$$

$$y[x_0, x_1, x_2, x_3] = \frac{y[x_1, x_2, x_3] - y[x_0, x_1, x_2]}{x_3 - x_0}$$
$$= \frac{-0.012\,648 + 0.093\,456}{6 - 0} = 0.013\,468$$

$$y[x_0, x_0, x_1, x_2, x_3] = \frac{y[x_0, x_1, x_2, x_3] - y[x_0, x_0, x_1, x_2]}{x_3 - x_0}$$
$$= \frac{0.013\,468 - 0.047\,595}{6 - 0} = -0.005\,687\,8.$$

Here, $y[x_0, x_0] = y_0' = 2$ is given in the problem statement, not calculated by finite differences.

The interpolation polynomial is then

$$P = 0 + (x - 0)\big(2 + (x - 0)\big(-0.283\,83 + (x - 2)\big(0.047\,595$$
$$+ (x - 4)\,(-0.005\,687\,8)\big)\big)\big),$$

and $P(1) = 1.6515$.

5.5   Discrete Fourier transform

The discrete Fourier transform is related to the interpolation problem

$$p(x) = \sum_{j=0}^{N-1} \alpha_j e^{2\pi i jx/T}, \tag{5.15}$$

$$p(x_k) = f_k, \tag{5.16}$$

for $k = 0, \ldots, N - 1$, or

$$p(x) = \sum_{j=0}^{N-1} \alpha_j \omega^j,$$ (5.17)

with

$$\omega(x) = e^{2\pi i x / T},$$

which emphasizes that it is just a special form of polynomial interpolation. The model assumes that x is periodic with period T.

If the support abscissas x_k are uniformly spaced, $x_k = kT/N$ for $k = 0, \ldots, N - 1$, then

$$\cos(2\pi i x_k / T) = \cos(2\pi i x_{N-k} / T)$$
$$\sin(2\pi i x_k / T) = -\sin(2\pi i x_{N-k} / T).$$

These identities, together with Euler's formula (A.23), make polynomial (5.15) equal to the polynomial

$$g(x) = \begin{cases} \frac{A_0}{2} + \sum_{j=1}^{(N-1)/2} \left[A_j \cos \frac{2\pi j x}{T} + B_j \sin \frac{2\pi j x}{T} \right] & N \text{ odd} \\ \frac{A_0}{2} + \sum_{j=1}^{(N/2)-1} \left[A_j \cos \frac{2\pi j x}{T} + B_j \sin \frac{2\pi j x}{T} \right] + \frac{A_{N/2}}{2} \cos \frac{2\pi N x}{2T} & N \text{ even} \end{cases}$$

(5.18)

at points $x = x_k$. The parameters α, A, B are related by

$$\alpha_0 = \frac{1}{2} A_0$$

$$\alpha_j = \begin{cases} \frac{1}{2}(A_j - i B_j) & j = 1, \ldots, (N/2) - 1 \\ \frac{1}{2}(A_{N-j} + i B_{N-j}) & j = (N/2) + 1, \ldots, N - 1 \end{cases}$$

$$\alpha_{N/2} = \frac{1}{2} A_{N/2},$$

when N is even, and

$$\alpha_0 = \frac{1}{2} A_0$$

$$\alpha_j = \begin{cases} \frac{1}{2}(A_j - i B_j) & j = 1, \ldots, (N - 1)/2 \\ \frac{1}{2}(A_{N-j} + i B_{N-j}) & j = (N + 1)/2, \ldots, N - 1, \end{cases}$$

when N is odd. The trigonometric function $g(x)$ is not equal to the complex phase function $p(x)$ everywhere: they are guaranteed equal only at the uniformly spaced support abscissas x_k.

In addition to the common assumption of uniformly spaced abscissas, it is also frequently assumed that N is a power of 2,

$$N = 2^m.$$

We will adopt both of these assumptions in the remainder of this section.

The first assumption permits a very compact representation of the solution. We write

$$\omega_j = \omega(x_j) = e^{2\pi i j/N}, \tag{5.19}$$

and the discrete phase factors ω_j have the following properties:

$$\begin{aligned} \omega_j^k &= e^{(2\pi i j/N)k} = e^{(2\pi i k/N)j} = \omega_k^j \\ \omega_j^{-k} &= e^{-(2\pi i j/N)k} = e^{(2\pi(-i)j/N)k} = \omega^{*k}_{\ j}, \end{aligned} \tag{5.20}$$

where ω^* is the complex conjugate of ω.

Because

$$e^{2\pi i} = 1,$$

it is true that $\omega_j^N = 1$ for any integer j. Also, $1 - \omega_j^N$ can be written (see (A.5))

$$1 - \omega_j^N = (1 - \omega_j) \sum_{k=0}^{N-1} \omega_j^k.$$

Now, since the left-hand side is zero, the right-hand side must also be zero, which implies that (a) $\omega_j = 1$ (i.e., $j = 0$); or that (b)

$$\sum_{k=0}^{N-1} \omega_j^k = 0, \tag{5.21}$$

for $j \neq 0$. Expression (5.21) is true for any nonzero integer $j \neq 0$ with ω_j defined by (5.19). Replace j with $(j - \ell)$, where $\ell \neq j$ is another arbitrary integer, and manipulate (5.21) in the manner of (5.20)

$$0 = \sum_{k=0}^{N-1} \omega_{j-\ell}^k = \sum_{k=0}^{N-1} e^{2\pi i(j-\ell)k/N} = \sum_{k=0}^{N-1} \omega_j^k \omega^{*k}_{\ \ell} = \sum_{k=0}^{N-1} \omega_k^j \omega^{*\ell}_{\ k}, \tag{5.22}$$

for $j \neq \ell$.

The significance of the orthogonality relation (5.22) is that the coefficients of the interpolation objective (5.16) may be determined from

$$\underbrace{\begin{pmatrix} \omega_0^0 & \omega_0^1 & \cdots & \omega_0^{N-1} \\ \omega_1^0 & \omega_1^1 & \cdots & \omega_1^{N-1} \\ \vdots & & & \vdots \\ \omega_{N-1}^0 & \omega_{N-1}^1 & \cdots & \omega_{N-1}^{N-1} \end{pmatrix}}_{\mathbf{A}} \begin{pmatrix} \alpha_0 \\ \alpha_1 \\ \vdots \\ \alpha_{N-1} \end{pmatrix} = \begin{pmatrix} f_0 \\ f_1 \\ \vdots \\ f_{N-1} \end{pmatrix},$$

and the inverse of the complex matrix $\mathbf{A}$ is

$$\mathbf{A}^{-1} = \frac{1}{N} \mathbf{A}^*, \tag{5.23}$$

because of the orthogonality relations. Therefore,

$$
\begin{pmatrix} \alpha_0 \\ \alpha_1 \\ \vdots \\ \alpha_{N-1} \end{pmatrix} = \frac{1}{N} \underbrace{\begin{pmatrix} \omega^{*0}_0 & \omega^{*1}_0 & \cdots & \omega^{*N-1}_0 \\ \omega^{*0}_1 & \omega^{*1}_1 & \cdots & \omega^{*N-1}_1 \\ \vdots & & & \vdots \\ \omega^{*0}_{N-1} & \omega^{*1}_{N-1} & \cdots & \omega^{*N-1}_{N-1} \end{pmatrix}}_{\mathbf{A}^{-1}} \begin{pmatrix} f_0 \\ f_1 \\ \vdots \\ f_{N-1} \end{pmatrix}. \tag{5.24}
$$

Equation (5.24) provides the coefficients α_j of the discrete Fourier transform problem by the simple multiplication of a matrix with a vector; $\mathcal{O}(N^2)$ operations.

As algorithms go, this is pretty simple. Yet, the cost is relatively high for some applications. With some rearrangement of these equations, a method requiring $\mathcal{O}(N \log_2 N)$ operations can be constructed. These faster algorithms are called *fast Fourier transforms* (FFT). The first such algorithm to be developed in recent times is the Cooley–Tukey algorithm [42], and that is the variation that will be described below. (Gauss discovered it first in 1805, though his work was not widely known [100].) The other principal variation was developed by W. M. Gentleman and G. Sande [83].

Equation (5.24) can be written

$$
\alpha(j) = \frac{1}{N} \sum_{k=0}^{N-1} f(k) e^{-2\pi i j k / N}. \tag{5.25}
$$

We assume $N = 2^m$, and write the index k in terms of the bits of $k_{m-1}, \ldots, k_0$ of the integer k:

$$
k = 2^{m-1} k_{m-1} + \cdots + 2^0 k_0.
$$

Then, using $f^{(0)}(k)$ for $f(k)$, (5.25) can be written

$$
\alpha(j) = \frac{1}{N} \sum_{k_0=0}^{1} e^{-2\pi i j 2^0 k_0 / N} \times \cdots
$$

$$
\times \underbrace{\left(\sum_{k_{m-2}=0}^{1} e^{-2\pi i j 2^{m-2} k_{m-2} / N} \underbrace{\left(\sum_{k_{m-1}=0}^{1} e^{-2\pi i j 2^{m-1} k_{m-1} / N} f^{(0)}(k) \right)}_{f^{(1)}} \right)}_{f^{(2)}}. \tag{5.26}
$$

The term $f^{(1)}$ can be expanded using the bits of index j:

$$
j = 2^{m-1} j_{m-1} + \cdots + 2^0 j_0.
$$

Now, because $N = 2^m$, this term will depend only on j_0:

$$
e^{-2\pi i j 2^{m-1} k_{m-1} / N} = e^{-\pi i j_0 k_{m-1}} \underbrace{e^{-2\pi i j_1 k_{m-1}}}_{1} \cdots \underbrace{e^{2^{m-1}\pi i j_{m-1} k_{m-1}}}_{1}.
$$

So, $f^{(1)}$ in (5.26) can be written

$$f^{(1)}(j_0, k_{m-2}, ..., k_0) = \sum_{k_{m-1}=0}^{1} f^{(0)}(k_{m-1}, k_{m-2}, ..., k_0)e^{-\pi i j_0 k_{m-1}},$$

with the relevant address bits explicit, and a recursion index in parenthesis as a superscript.

Equation (5.26) then becomes

$$\alpha(j) = \frac{1}{N} \sum_{k_0=0}^{1} e^{-2\pi i j 2^0 k_0/N} \times \cdots$$

$$\times \left(\underbrace{\sum_{k_{m-3}=0}^{1} e^{-2\pi i j 2^{m-3}k_{m-3}/N} \left(\underbrace{\sum_{k_{m-2}=0}^{1} e^{-2\pi i j 2^{m-2}k_{m-2}/N} f^{(1)}(j_0, k_{m-2}, ...)}_{f^{(2)}} \right)}_{f^{(3)}} \right).$$

The exponent in the $f^{(2)}$ term depends on j through j_0 and j_1 only, so we can write

$$f^{(2)}(j_0, j_1, k_{m-3}, ..., k_0) = \sum_{k_{m-2}=0}^{1} f^{(1)}(j_0, k_{m-2}, ..., k_0)e^{-\pi i(j_0+2j_1)k_{m-2}/2}.$$

Carrying on similarly, the expression (5.26) is determined to be a sequence of m steps,

(5.27)

$$f^{(p)}(j_0, ..., j_{p-1}, k_{m-1-p}, ..., k_0) = \sum_{k_{m-p}=0}^{1} f^{(p-1)}(j_0, ..., j_{p-2}, k_{m-p}, ..., k_0)w_{j(p)}^{k_{m-p}}$$

$$w_{j(p)} = \exp(-\pi i 2^{1-p}(2^{p-1}j_{p-1} + \cdots + j_0)), \quad (5.28)$$

for $p = 1, ..., m$. The recursion begins with starting values

$$f^{(0)}(k_{m-1}, ..., k_0) = f(x_k),$$

and the final result is

$$\alpha(j) = \frac{1}{N} f^{(m)}(j_0, ..., j_{m-1}).$$

There are $m = \log_2 N$ steps in the recursion, with N complex multiplications per step. The operation count is therefore $\mathcal{O}(N \log_2 N)$.

With some care, the algorithm requires only N complex numbers of storage. This is hinted at through the bit field addressing of the intermediate vectors $f^{(p)}$. Computing in place involves the complication that the bit fields of the starting vector, $(k_{m-1}, ..., k_0)$, and the bit fields of the ending vector, $(j_0, ..., j_{m-1})$, are reversed. The memory savings of computing in place easily justifies the modest complexity of performing these bit field manipulations.

The coefficient of j_{p-1} in the exponent of (5.28) is $-\pi i$, and $\exp(\pm \pi i) = -1$, therefore

$$w_{j(p)} = \begin{cases} \tilde{w}_{j(p)} & j_{p-1} = 0 \\ -\tilde{w}_{j(p)} & j_{p-1} = 1 \end{cases}$$

$$\tilde{w}_{j(p)} = \exp(-\pi i 2^{1-p}(2^{p-2} j_{p-2} + \cdots + j_0)),$$

and the sum (5.27) can be written

$$\underbrace{\begin{pmatrix} f^{(p)}(\dots, 0, \dots) \\ f^{(p)}(\dots, 1, \dots) \end{pmatrix}}_{\text{bit } j_{p-1} \text{ differs}} = \begin{pmatrix} 1 & 1 \\ 1 & -1 \end{pmatrix} \begin{pmatrix} 1 & \\ & \tilde{w}_{j(p)} \end{pmatrix} \underbrace{\begin{pmatrix} f^{(p-1)}(\dots, 0, \dots) \\ f^{(p-1)}(\dots, 1, \dots) \end{pmatrix}}_{\text{bit } k_{m-p} \text{ differs}}.$$

Each step of the FFT algorithm can therefore be represented as a matrix product $\mathbf{B}\mathbf{D}_p\mathbf{P}_p$ with

$$\mathbf{B} = \begin{pmatrix} 1 & 1 & & & & \\ 1 & -1 & & & & \\ & & \ddots & & & \\ & & & & 1 & 1 \\ & & & & 1 & -1 \end{pmatrix}$$

$$\mathbf{D}_p = \mathrm{diag}(1, \tilde{w}, 1, \tilde{w}, \dots),$$

and with $\mathbf{P}_p$ being a permutation. (This representation of the algorithm facilitates discussion, but is not how one would implement it. Do not use it.) The overall algorithm can be viewed as

$$\mathbf{f}^{(i)} = (\mathbf{B}\mathbf{D}_i\mathbf{P}_i)\mathbf{f}^{(i-1)} \quad \text{for } i = 1, \dots, m \tag{5.29a}$$

$$\alpha = \mathbf{f}^{(m)}/N. \tag{5.29b}$$

Each diagonal element of $\mathbf{D}_p$ is magnitude 1, so $\mathrm{cond}_2(\mathbf{D}_p) = 1$. Each block of block-diagonal matrix $\mathbf{B}_p$ has eigenvalues $\pm\sqrt{2}$. Since $\mathbf{B}_p$ is normal, all singular values of $\mathbf{B}_p$ are $\sqrt{2}$ and $\mathrm{cond}_2(\mathbf{B}_p) = 1$. Finally, any permutation $\mathbf{P}_p$ has $\mathrm{cond}_2(\mathbf{P}_p) = 1$. Therefore,

$$\mathrm{cond}(\mathbf{B}\mathbf{D}_i\mathbf{P}_i) = 1,$$

which means that error propagates like

$$\frac{\|\Delta\mathbf{f}^{(i)}\|_2}{\|\mathbf{f}^{(i)}\|_2} \leq \mathrm{cond}(\mathbf{B}\mathbf{D}_i\mathbf{P}_i)\frac{\|\Delta\mathbf{f}^{(i-1)}\|_2}{\|\mathbf{f}^{(i-1)}\|_2} = \frac{\|\Delta\mathbf{f}^{(i-1)}\|_2}{\|\mathbf{f}^{(i-1)}\|_2},$$

i.e., without amplification.

Also,

$$\mathbf{P}_i^{\mathsf{T}}\mathbf{P}_i = \mathbf{I}$$

$$\frac{1}{2}\mathbf{B}^{\mathsf{T}}\mathbf{B} = \mathbf{I}$$

$$\mathbf{D}_i^{\mathsf{T}}\mathbf{D}_i = \mathbf{I},$$

so

$$\|\mathbf{f}^{(i)}\|_2 = \sqrt{\mathbf{f}^{(i)\,H}\mathbf{f}^{(i)}} = \sqrt{\mathbf{f}^{(i-1)\,H}\mathbf{P}_i^{\mathsf{T}}\mathbf{D}_i^{H}\mathbf{B}^{\mathsf{T}}\mathbf{B}\mathbf{D}_i\mathbf{P}_i\mathbf{f}^{(i-1)}} = \sqrt{2}\|\mathbf{f}^{(i-1)}\|_2. \qquad (5.30)$$

Since step (5.29b) carries no error when N is a power of 2, we can conclude that

$$\|\boldsymbol{\alpha}\|_2 = 2^{-m/2}\|\mathbf{f}\|_2 = \frac{1}{\sqrt{N}}\|\mathbf{f}\|_2,$$

which is the discrete version of Parseval's theorem, also known as Rayleigh's energy theorem. This is also evident by examining (5.23).

Equation (5.30) is the key to understanding the numerical stability of the Cooley–Tukey algorithm. In each step of the algorithm there can be very large cancellation errors. However, cancellation in the calculation of a scalar, e.g., $x = a - b$, makes itself evident in the measure $|\Delta x|/|x|$ by the diminution of the denominator. In the present case, we are concerned with the error as measured by $\|\Delta \mathbf{f}\|_2/\|\mathbf{f}\|_2$ and (5.30) shows that the denominator grows, so the behavior in L^2 is quite different from scalar cancellation.

Example 5.6 Compute the discrete Fourier transform of the following array:

$$\mathbf{f}^{(0)} = \mathbf{f} = \begin{pmatrix} 1 \\ 2 \\ 3 \\ 0 \\ -1 \\ -3 \\ -2 \\ -1 \end{pmatrix}.$$

Solution

We will use the matrix representation (5.29) for illustrative purposes. Walking through the FFT algorithm, we begin with a permutation that pairs numbers by their least significant address bits:

$$\mathbf{P}_1 = \begin{array}{c} \\ 000 \\ 100 \\ 001 \\ 101 \\ 010 \\ 110 \\ 011 \\ 111 \end{array} \begin{array}{cccccccc} 000 & 001 & 010 & 011 & 100 & 101 & 110 & 111 \\ \left(1 \right. & & & & & & & \\ & & & & 1 & & & \\ & 1 & & & & & & \\ & & & & & 1 & & \\ & & 1 & & & & & \\ & & & & & & 1 & \\ & & & 1 & & & & \\ & & & & & & & \left. 1 \right) \end{array}.$$

The matrix **B** is the same for all steps,

$$
\mathbf{B} = \begin{pmatrix}
1 & 1 & & & & & & \\
1 & -1 & & & & & & \\
& & 1 & 1 & & & & \\
& & 1 & -1 & & & & \\
& & & & 1 & 1 & & \\
& & & & 1 & -1 & & \\
& & & & & & 1 & 1 \\
& & & & & & 1 & -1
\end{pmatrix}.
$$

The diagonal matrix **D** of phase factors depends on the "old j" address bits. In the first pass, there are no old j bits, so every phase factor is 1.

$$\mathbf{D}_1 = \mathbf{I}.$$

Multiplied out,

$$
\mathbf{f}^{(1)} = \mathbf{B}\mathbf{D}_1\mathbf{P}_1\mathbf{f}^{(0)} =
\begin{matrix}
0|00 \\ 1|00 \\ 0|01 \\ 1|01 \\ 0|10 \\ 1|10 \\ 0|11 \\ 1|11
\end{matrix}
\begin{pmatrix}
0 \\ 2 \\ -1 \\ 5 \\ 1 \\ 5 \\ -1 \\ 1
\end{pmatrix}.
$$

The address bits have been partitioned into the j and k components with j_0 taking on the old values of k_2, i.e., $j_0|k_1k_0$.

In the second step, we permute the address bits to pair values that differ only in k_1:

$$
\mathbf{P}_2 =
\begin{matrix}
0|00 \\ 0|10 \\ 0|01 \\ 0|11 \\ 1|00 \\ 1|10 \\ 1|01 \\ 1|11
\end{matrix}
\begin{array}{cccccccc}
0|00 & 1|00 & 0|01 & 1|01 & 0|10 & 1|10 & 0|11 & 1|11 \\
1 & & & & & & & \\
& & & & 1 & & & \\
& & 1 & & & & & \\
& & & & & & 1 & \\
& 1 & & & & & & \\
& & & & & 1 & & \\
& & & 1 & & & & \\
& & & & & & & 1
\end{array}.
$$

The phase factors $\tilde{w}$ depend on the j_0 bit. When $j_0 = 0$ the factor is 1, and when $j_0 = 1$ the factor is $\exp(-\pi i/2) = -i$:

$$\mathbf{D}_2 = \mathrm{diag}(1, 1, 1, 1, 1, -i, 1, -i).$$

Then,

$$\mathbf{f}^{(2)} = \mathbf{BD_2P_2f}^{(1)} = \begin{array}{c} 00|0 \\ 01|0 \\ 00|1 \\ 01|1 \\ 10|0 \\ 11|0 \\ 10|1 \\ 11|1 \end{array} \begin{pmatrix} 1 \\ -1 \\ -2 \\ 0 \\ 2-5i \\ 2+5i \\ 5-i \\ 5+i \end{pmatrix}.$$

The address bits have been repartitioned as $j_0 j_1 | k_0$.

In the third step, the permutation groups terms that differ in k_0 only:

$$\mathbf{P_3} = \begin{array}{c} \\ 00|0 \\ 00|1 \\ 10|0 \\ 10|1 \\ 01|0 \\ 01|1 \\ 11|0 \\ 11|1 \end{array} \begin{array}{cccccccc} 00|0 & 01|0 & 00|1 & 01|1 & 10|0 & 11|0 & 10|1 & 11|1 \\ \left(\begin{array}{cccccccc} 1 & & & & & & & \\ & & 1 & & & & & \\ & & & & 1 & & & \\ & & & & & & 1 & \\ & 1 & & & & & & \\ & & & 1 & & & & \\ & & & & & 1 & & \\ & & & & & & & 1 \end{array} \right) \end{array}.$$

The phase factors are 1, $\exp(-\pi i/4) = (1-i)/\sqrt{2}$, $\exp(-\pi i/2) = -i$, and $\exp(-3\pi i/4) = (-1-i)/\sqrt{2}$ for $j = 2j_1 + j_0 = 0, 1, 2, 3$, respectively:

$$\mathbf{D_3} = \mathrm{diag}\left(1, 1, 1, \frac{1-i}{\sqrt{2}}, 1, -i, 1, \frac{-1-i}{\sqrt{2}}\right),$$

$$\mathbf{f}^{(3)} = \mathbf{BD_3P_3f}^{(2)} = \begin{array}{c} 000 \\ 001 \\ 100 \\ 101 \\ 010 \\ 011 \\ 110 \\ 111 \end{array} \begin{pmatrix} -1 \\ 3 \\ 2(1+\sqrt{2}) - (5+3\sqrt{2})i \\ 2(1-\sqrt{2}) - (5-3\sqrt{2})i \\ -1 \\ -1 \\ 2(1-\sqrt{2}) + (5-3\sqrt{2})i \\ 2(1+\sqrt{2}) + (5+3\sqrt{2})i \end{pmatrix}.$$

The j address bits are in the reversed order, $j_0 j_1 j_2$. Permutation to the standard order

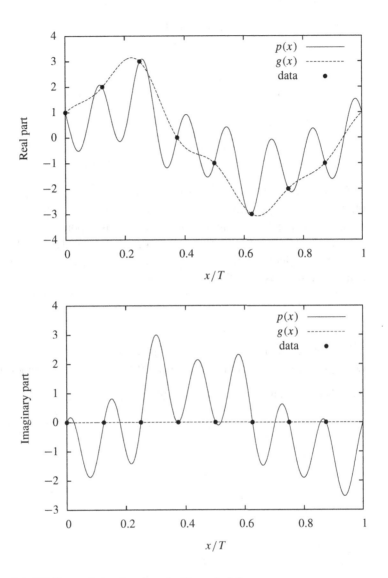

Figure 5.3 DFT interpolation functions for Example 5.6.

$j_2 j_1 j_0$, and division by $N = 8$, gives the final result

$$
\alpha = \begin{pmatrix}
-1/8 \\
(1 + \sqrt{2})/4 - (5 + 3\sqrt{2})i/8 \\
-1/8 \\
(1 - \sqrt{2})/4 + (5 - 3\sqrt{2})i/8 \\
3/8 \\
(1 - \sqrt{2})/4 - (5 - 3\sqrt{2})i/8 \\
-1/8 \\
(1 + \sqrt{2})/4 + (5 + 3\sqrt{2})i/8
\end{pmatrix}.
$$

In terms of the coefficients A and B of (5.18),

$$A_0 = 2\alpha_0 = -\frac{1}{4} \qquad\qquad A_4 = 2\alpha_4 = \frac{3}{4}.$$

$$A_1 = \alpha_1 + \alpha_7 = \frac{1+\sqrt{2}}{2} \qquad\qquad B_1 = i(\alpha_1 - \alpha_7) = \frac{5+3\sqrt{2}}{4}$$

$$A_2 = \alpha_2 + \alpha_6 = -\frac{1}{4} \qquad\qquad B_2 = i(\alpha_2 - \alpha_6) = 0$$

$$A_3 = \alpha_3 + \alpha_5 = \frac{1-\sqrt{2}}{2} \qquad\qquad B_3 = i(\alpha_3 - \alpha_5) = \frac{-5+3\sqrt{2}}{4}.$$

The original data and the interpolation functions $p(x)$ (5.15) and $g(x)$ (5.18) are plotted along with the given data in Figure 5.3. Note that while both functions match the given data, they differ significantly at intermediate points. Unless there is an a priori reason to use the complex form $p(x)$, don't use it. Note also that the original data was real, and the trigonometric interpolation function $g(x)$ is also everywhere real, but the corresponding phase function $p(x)$ is complex.

Problems

5.1 [40] Consider five equally spaced points $(-2h, y_{-2})$, $(-h, y_{-1})$, $(0, y_0)$, (h, y_1), $(2h, y_2)$. Use these to estimate the slope $y'(0)$ with order of accuracy h^4.

5.2 If (f, ϕ) and (g, γ) are DFT pairs:

$$f_k = \sum_{j=0}^{N-1} \phi_j \omega_k^j \qquad\qquad \phi_j = N^{-1} \sum_{k=0}^{N-1} f_k \omega^{*k}{}_j$$

$$g_k = \sum_{j=0}^{N-1} \gamma_j \omega_k^j \qquad\qquad \gamma_j = N^{-1} \sum_{k=0}^{N-1} g_k \omega^{*k}{}_j,$$

then

$$\underbrace{\sum_{p=0}^{N-1} f_p g_{k-p}}_{(f*g)_k} = \sum_{p,j,\ell=0}^{N-1} \phi_j \gamma_\ell \omega_p^j \omega_{k-p}^\ell = \sum_{j,\ell=0}^{N-1} \phi_j \gamma_\ell \omega_k^\ell \underbrace{\sum_{p=0}^{N-1} \omega_p^j \omega_p^{-\ell}}_{N\delta_{j\ell}} = N \sum_j \phi_j \gamma_j \omega_k^j,$$

i.e., $((f * g), N\phi\gamma)$ is a DFT pair, where $(f * g)$ is the "*discrete convolution*" of functions f and g (recall that g and f are periodic, so use $g_{k-p} = g_{N+k-p}$ if the index is negative).

(i) Compare the operation count of evaluating the convolution $(f * g)$ explicitly versus through the steps $f \to \phi$, $g \to \gamma$, $N\phi\gamma \to (f * g)$.

(ii) Use the FFT to compute the convolution of f and g:

$$f_i = \begin{cases} \frac{w_1^2 a_1}{4(i-c_1)^2+w_1^2} + \frac{w_2^2 a_2}{4(i-c_2)^2+w_2^2} & 0 \le i < 512 \\ 0 & 512 \le i < 1024 \end{cases}$$

$$g_i = \begin{cases} \frac{\exp(-i^2/(2\sigma^2))}{\sigma\sqrt{2\pi}} & 0 \le i < 512 \\ \frac{\exp(-(1024-i)^2/(2\sigma^2))}{\sigma\sqrt{2\pi}} & 512 \le i < 1024, \end{cases}$$

with $a_1 = 10$, $c_1 = 100$, $w_1 = 500$, $a_2 = 50$, $c_2 = 400$, $w_2 = 15$, and $\sigma = 10$.

Here f, composed of two Lorentzian peaks, approximates an absorption spectrum. g, consisting of a single Gaussian peak, approximates the response function of a grating spectrometer. The convolution $f * g$ then approximates an instrumental measurement of the true signal f. The signal f and response g are padded with many zeros to mask the periodicity inherent in the DFT treatment (because the underlying optical physics does not have this periodicity).

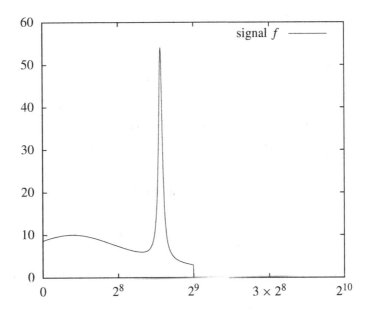

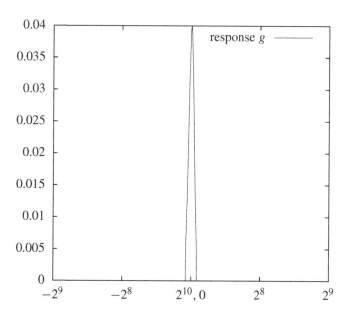

5.3 [204] The placement of support points has a significant impact on the interpolation error. With a change of variables such that $[a, b]$ maps to $[-1, 1]$, the smallest error for degree n polynomial interpolation comes from the placement of $n + 1$ support points at the zeros $\xi_k = \cos(\frac{(2k-1)\pi}{2(n+1)})$, $k = 1, ..., n + 1$, of Chebyshev polynomial $T_{n+1}(x)$ [36].

Demonstrate this by fitting

$$f(x) = \frac{1}{1 + 12x^2} \quad x \in [-1, 1]$$

to 11 support points:

(i) $x_i = -1 + 0.2i$, $i = 0, ..., 10$; and

(ii) $x_i = \cos((2i + 1)\pi/22)$, $i = 0, ..., 10$.

Note:

- as a function, $T_n(x) = \cos(n \cos^{-1} x)$
- as a polynomial, $T_n(x) = 2^{n-1}x^n +$ lower order terms.
- Chebyshev polynomials can be constructed numerically or analytically with the recurrence

$$T_0(x) = 1$$
$$T_1(x) = x$$
$$T_{k+1}(x) = 2x T_k(x) - T_{k-1}(x).$$

5.4 Is Horner's scheme as used to evaluate the Newton polynomial numerically stable?

5.5 (i) Rolle's theorem says $P_n^{(k)}(x) - f^{(k)}(x)$ has $n+1-k$ zeros, say $\zeta_0, ..., \zeta_{n-k}$, so $G(x) = P_n^{(k)}(x) - f^{(k)}(x) + K(x - \zeta_0) \cdots (x - \zeta_{n-k})$ has $n+2-k$ zeros. Complete this analysis to give a bound on the accuracy of derivatives of an interpolating polynomial. (ii) If h is the greatest distance between neighboring support abscissas, show

$$|P_n(x) - f(x)| \le \frac{\max |f^{(n+1)}|}{4(n+1)} h^{n+1}$$

$$|P_n^{(k)}(x) - f^{(k)}(x)| \le \frac{n! \max |f^{(n+1)}|}{(n+1-k)!(k-1)!} h^{n+1-k}, \quad k = 1, ..., n+1,$$

for $x \in I(x_0, ..., x_n)$, the interval spanned by the support abscissas.

5.6 [86] Show that the generic polynomial interpolation function of n support points

$$y(x) = \sum_{i=1}^{n} c_i(x) y(x_i)$$

is at best first-order accurate if all coefficients c_i are positive (i.e., such an interpolating function cannot interpolate exactly a quadratic or higher-degree polynomial function).

5.7 The error function

$$\text{erf}(x) = \frac{2}{\sqrt{\pi}} \int_0^x e^{-t^2} dt$$

does not have convenient analytical values, aside from

$$\text{erf}(0) = 0$$
$$\text{erf}(\infty) = 1,$$

so numerical integration, series expansion, or fitting is required to provide numerical values of erf for arbitrary x. However, derivatives of erf are easily constructed through the recursion, which begins by applying Leibniz' integral rule to the definition:

$$\text{erf}'(x) = \frac{2}{\sqrt{\pi}} e^{-x^2}$$
$$\text{erf}^{(n)}(x) = -2x \text{erf}^{(n-1)}(x) - 2(n-2)\text{erf}^{(n-2)}(x).$$

These derivatives are related to the Hermite polynomials,

$$\text{erf}^{(n)}(x) = (-1)^{n-1} \frac{2}{\sqrt{\pi}} H_{n-1}(x) e^{-x^2},$$

which possess the bound due to Cramér [45, 107]

$$|H_n(x)| \le K 2^{n/2} \sqrt{n!} e^{x^2/2},$$

where K is a constant less than 1.09. Using the result of only one numerical integration,

$$\text{erf}(2) = 0.995\,322\,265\,018\,952\,7,$$

construct a Hermite interpolation polynomial on $[0, 2]$ such that the approximation error is everywhere less than 5×10^{-3}.

5.8 The Black–Scholes model for the fluctuations in a stock price–time series holds that the rate of change of a price $\dot{p}$, divided by the price p, has a smooth part and a fluctuating random part possessing no correlation. Fluctuations in the time series $\dot{p}/p$ can be characterized the autocorrelation.

If $f \star g$ is the correlation of f and g, then $f \star f$ is the autocorrelation of f. The correlation is defined by

$$(f \star g)_n = \frac{1}{N-1} \sum_{i=0}^{N-1} f_i g_{n+i}^*,$$

and is easily computed using the discrete Fourier transform. If

$$f_i = \sum_{j=0}^{N-1} \phi_j \omega_i^j$$

$$g_i = \sum_{j=0}^{N-1} \gamma_j \omega_i^j,$$

then

$$\sum_{i=0}^{N-1} f_i g_{n+i}^* = \sum_{j,k=0}^{N-1} \phi_j \gamma_k^* \sum_{i=0}^{N-1} \omega_i^j \omega_{i+n}^{*j} = N \sum_{j,k=0}^{N-1} \phi_j \gamma_k^* \omega_n^{-j}$$

and $((N-1)(f \star g), N\phi\gamma^*)$ is a DFT pair.

(i)   Obtain a time series of stock market historical prices, e.g., 4096+1 adjusting closing values of the Dow Jones Industrial Average. One source for this kind of data is http://finance.yahoo.com. This series is p_i, where i labels the trading day. The quantity we wish to analyze is

$$r_i \equiv \frac{\dot{p}}{p} \approx \frac{p_{i+1} - p_i}{p_i}, \quad i = 0, ..., N-1.$$

(ii)   Assume that the normalized rate r is composed of a smoothly varying part s, and a fluctuation μ. Estimate s by computing the convolution of r with a boxcar filter, e.g.,

$$b_i = \frac{1}{21} \times \begin{cases} 1 & i \leq 10 \\ 1 & i \geq N - 10 \\ 0 & \text{otherwise.} \end{cases}$$

Then, $\mu \approx r - (r * b)$.

(iii)   Compute the autocorrelation normalized by the variance,

$$\frac{(\mu \star \mu)_i}{(\mu \star \mu)_0}.$$

If the stock market prices are governed by the Black–Scholes model with uncorrelated "noise," this quantity will resemble a discrete delta function.

6 Iterative methods and the roots of polynomials

The idea of iteration has already been encountered in discussing eigenvalues, and iterative approaches for linear systems. In this chapter, more general properties of iterative methods will be described, and the emphasis will largely be on root finding and polynomials. The first section will provide some definitions and basic concepts, then a variety of methods are presented in order of increasing order of convergence, culminating in the Newton–Raphson method.

The chapter closes with some polynomial-specific topics, and ideas for accelerating convergence.

6.1 Convergence and rates

A sequence of numbers

$$x_0, x_1, x_2, x_3, \ldots$$

is a *Cauchy sequence* [34] if for any positive real number ϵ there is an integer N such that

$$\|x_n - x_m\| < \epsilon,$$

for all $n \geq N$ and all $m \geq N$. A Cauchy sequence converges to a limit, given by $\lim_{n \to \infty} x_n$.

For example, the sequence

$$x_0 = 1, \ x_1 = \frac{1}{2}, \ x_2 = \frac{1}{3}, \ x_3 = \frac{1}{4}, \ \ldots$$

is a Cauchy sequence that converges to zero.

The sequence

$$x_0 = 1, \ x_1 = \cos(h), \ x_2 = \cos(2h), \ x_3 = \cos(3h), \ \ldots$$

is a Cauchy sequence only if h is a multiple of 2π, in which case each term in the sequence is 1. Otherwise, this sequence does not converge.

An abstract iterative method may be written

$$x^{(i+1)} = \Phi(x^{(i)}), \quad i \geq 0,$$

and iterative function Φ is said to converge for some starting point $x^{(0)}$ if the sequence

$$x^{(0)}, \ x^{(1)}, \ x^{(2)}, \ x^{(3)}, \ \ldots$$

generated by this function and starting point is a Cauchy sequence.

Example 6.1   Analyze the convergence of the sequence

$$x^{(0)} = 0$$
$$x^{(i+1)} = \sqrt{2 + x^{(i)}}. \tag{6.1}$$

Solution
This sequence is convergent, and $\lim_{n \to \infty} x^{(n)} = 2$. We can see by inspection that $0 \le x^{(i)} < x^{(i+1)} \le 2$, and this makes the sequence a Cauchy sequence: $(2 - x^{(i)}) \ge 0$ decreases with increasing i, so $|x^{(n)} - x^{(m)}| = |(2 - x^{(n)}) - (2 - x^{(m)})|$ also decreases to zero with increasing n, m. For any N, we can find a number $\epsilon(N)$ such that $|x^{(n)} - x^{(m)}| < \epsilon(N)$ for all $n, m \ge N$ (e.g., $\epsilon(N) = 2 - \sqrt{2 + x^{(N)}}$). That $\xi = 2$ is a fixed point follows from these inequalities, and from rearrangement of the statement $\xi = \Phi(\xi)$.

If we assume that Φ converges to ξ, then we can define the rate of convergence as follows. If the function converges, the distance $|x^{(i)} - \xi|$ must decrease with increasing iteration index i. Writing

$$|x^{(i+1)} - \xi| \le C|x^{(i)} - \xi|^p, \quad C \ge 0, \tag{6.2}$$

there are two possibilities. If $p = 1$ and $0 \le C < 1$, then the sequence generated by Φ will converge and we say Φ converges at first order (i.e., the *order of convergence* is 1). If $p > 1$ then for any $C \ge 0$ the sequence converges at order p.

We say a method is globally convergent with rate p if (6.2) is satisfied for any starting point $x^{(0)}$. Otherwise, the function is locally convergent.

To analyze local convergence, we can write a Taylor series of the function Φ:

$$x^{(i+1)} = \Phi(x^{(i)}) = \underbrace{\Phi(\xi)}_{\xi} + (x^{(i)} - \xi)\Phi'(\xi) + \frac{1}{2}(x^{(i)} - \xi)^2\Phi''(\xi) + \cdots$$

$$|x^{(i+1)} - \xi| = \left|(x^{(i)} - \xi)\Phi'(\xi) + \frac{1}{2}(x^{(i)} - \xi)^2\Phi''(\xi) + \cdots\right|$$

$$|x^{(i+1)} - \xi| \le |\Phi'(\xi)||x^{(i)} - \xi| + \left|\frac{1}{2}\Phi''(\xi)\right||x^{(i)} - \xi|^2 + \cdots \tag{6.3}$$

using the triangle inequality.

If $\Phi'(\xi) \ne 0$, then the first term on the right-hand side of (6.3) is not zero, and we can neglect the other right-hand side terms when $x^{(i)} - \xi$ is small enough. The result is

$$|x^{(i+1)} - \xi| \le |\Phi'(\xi)||x^{(i)} - \xi|, \tag{6.4}$$

and we conclude that Φ converges locally at first order provided $|\Phi'(\xi)| < 1$.

If $\Phi'(\xi) = 0$, but $\Phi''(\xi) \neq 0$, then retaining only the second term on the right-hand side of (6.3),

$$|x^{(i+1)} - \xi| \leq \left|\frac{1}{2}\Phi''(\xi)\right| |x^{(i)} - \xi|^2,$$

and we conclude that Φ converges locally at second order.

In these derivations we omitted terms $(x^{(i)} - \xi)^n$ of high order n, and this limits the validity of the results to a demonstration of *local* convergence.

A value $\xi = \Phi(\xi)$ is called a fixed point of the iteration.

Example 6.2   Find the order of convergence of the sequence (6.1).

Solution
The iteration of the previous example is $\Phi(x) = \sqrt{2 + x}$, so $\Phi'(x) = 1/(2\sqrt{2 + x})$ and $\Phi'(\xi) = 1/4$. Therefore, this iteration converges locally at first order.

There are a large number of iterative methods in common use, the most important of the general approaches being Newton–Raphson: a second-order scheme. A succession of lower-order methods will be examined first. With each of these methods, the same simple example of polynomial root finding is applied. Take note of the number of function evaluations required in each case to obtain the converged result. While the application focus here is on polynomial root finding, the methods are of course applicable to finding the zero of any sufficiently continuous function.

6.2   Bisection

If one has an ordered pair of values (a, b), such that $f(a)f(b) < 0$, then one or more roots of $f(x)$ must lie between a and b. The simplest possible algorithm for bounding such a root is bisection. To find an interval of width ϵ that bounds a root:

1. If $b - a < \epsilon$, then the root has been bounded to the desired accuracy. The best guess is $\xi = (b + a)/2$.
2. Otherwise, evaluate the function at the midpoint $c = (a + b)/2$. If $f(a)f(c) < 0$, then the new bracketing pair is $(a', b') = (a, c)$, otherwise $(a', b') = (c, b)$. Of course if $f(c) = 0$ exactly (unlikely) then the root has been found. Otherwise, redefine $(a, b) = (a', b')$ and go to step 1.

Example 6.3   Use bisection to seek a root of the function $f(x)$,

$$f(x) = (x - 1)(x - 10^{-3}),$$ (6.5)

with starting guesses $x_+^{(0)} = 10$ and $x_-^{(0)} = 0.5$.

Solution

Partial results of the sequence are given in Table 6.1. The order obtained is approximately 1, as shown in Figure 6.1 by plotting $|e_{n+1}| = |x^{(n+1)} - \xi|$ against $|e_n| = |x^{(n)} - \xi|$. According to (6.2), the points on this graph will lie on a line whose slope is the order p.

Table 6.1 Bisection solution to root of $(x - 1)(x - 10^{-3})$ with $a^{(0)} = 0.5$ and $b^{(0)} = 10$; Example 6.3.

n	$c^{(n)}$	$f(c^{(n)})$
1	5.250000000000000e+00	2.2308250000000001e+01
2	2.875000000000000e+00	5.3887499999999999e+00
3	1.687500000000000e+00	1.1594687500000000e+00
4	1.093750000000000e+00	1.0244531250000000e-01
5	7.968750000000000e-01	-1.6166210937500000e-01
⋮		
48	9.999999999998046e-01	-1.9520385347979901e-14
49	9.999999999999734e-01	-2.6618707052065505e-15
50	1.000000000000058e+00	5.7673865632405655e-15
51	1.000000000000016e+00	1.5527579290170075e-15
52	9.999999999999944e-01	-5.5455638809477148e-16

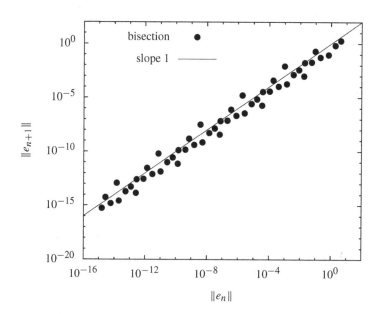

Figure 6.1 Order of convergence for bisection; Example 6.3.

6.3   Regula falsi

Regula falsi, or the *method of false position* [33], is another example of a bracketing method: its use requires two starting values a and b that bracket the root, i.e.,

$$f(a) > f(\xi) = 0$$
$$f(b) < f(\xi) = 0.$$

If we linearly interpolate between these guesses,

$$x^{(n)} = \frac{bf(a) - af(b)}{f(a) - f(b)},$$

we obtain an estimate of the root. Since a and b bracket the root, we also have that $x^{(n)}$ lies between a and b, and this guarantees convergence.

To make an iterative scheme of this method, we need to update the values a and b in such a way as to retain the bracketing property:

$$\begin{pmatrix} a \\ b \end{pmatrix} := \begin{cases} \begin{pmatrix} a \\ x^{(n)} \end{pmatrix} & \text{if } f(x^{(n)}) < 0 \\ \\ \begin{pmatrix} x^{(n)} \\ b \end{pmatrix} & \text{if } f(x^{(n)}) > 0. \end{cases}$$

One advantage of this method is that the derivatives of the function $f(x)$ are not required. Another advantage is that it is stable and guaranteed to converge if a single root is bracketed by the starting guesses. The disadvantages of the method are the need to obtain two starting guesses, and the potentially slow convergence.

Example 6.4   Use *regula falsi* to seek a root of the function $f(x)$ (6.5), with starting guesses $a = 10$ and $b = 0.5$.

Solution
Partial results of the sequence are given in Table 6.2. The order obtained is approximately 1, as shown in Figure 6.2.

Table 6.2: *Regula falsi* solution to root of $(x - 1)(x - 10^{-3})$ with $a^{(0)} = 10$ and $b^{(0)} = 0.5$; Example 6.4.

n	$x^{(n)}$	$f(x^{(n)})$
1	+5.2626592272870831e-01	-2.4883636722593816e-01
2	+5.5238974638303562e-01	-2.4680770422030418e-01
3	+5.7822972472898959e-01	-2.4345833989355545e-01
4	+6.0365026353729767e-01	-2.3886027313218597e-01
5	+6.2852467493174102e-01	-2.3310993260862198e-01
⋮		
324	+9.9999999999999845e-01	-1.4426394107092122e-15
325	+9.9999999999999856e-01	-1.3316171082466965e-15
326	+9.9999999999999878e-01	-1.1097893437561623e-15
327	+9.9999999999999889e-01	-9.8983881728143779e-16
328	+9.9999999999999900e-01	-8.8796157926562813e-16

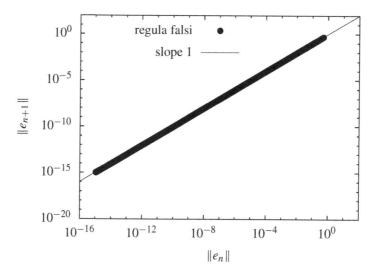

Figure 6.2: Order of convergence for *regula falsi*; Example 6.4.

In this particular example, *regula falsi* converged less rapidly than did bisection. Of course this is not always the case. However, it makes a couple of points about the pros and cons of these methods. Bisection is a very predictable method with guaranteed success, and a predictable outcome that the interval is halved each iteration. *Regula falsi* also guarantees convergence, but the interval is not guaranteed to diminish by as much as half with each iteration. If the function were linear, *regula falsi* would find the root in one iteration. This opportunity comes at the cost of the guaranteed rate of interval resolution. While the discussion in this chapter tends to emphasize the order of convergence p, it is not the whole story (see also [146, Section 8.5]). For the present specific problem, the coefficient C of (6.2) was over six times smaller for bisection than for *regula falsi*, and this led to a significant performance difference.

6.4   The secant method

The secant method is like *regula falsi*, but it sacrifices the bracketing property. Its origin is uncertain, though it is certainly ancient. The secant method was used in the fifteenth century by Parameśvara [176] (see [183]). The method proceeds as follows:

$$x^{(n+1)} = \frac{x^{(n-1)} f(x^{(n)}) - x^{(n)} f(x^{(n-1)})}{f(x^{(n)}) - f(x^{(n-1)})}. \tag{6.6}$$

This method has good convergence properties with an unexpected non-integer order of convergence. Equation (6.6) can be written in the form

$$
\begin{aligned}
(x^{(n+1)} - \xi) &= (x^{(n)} - \xi) - f(x^{(n)}) \frac{x^{(n)} - x^{(n-1)}}{f(x^{(n)}) - f(x^{(n-1)})} \\
&= (x^{(n)} - \xi) - \frac{f(x^{(n)})}{f[x^{(n)}, x^{(n-1)}]} \\
&= (x^{(n)} - \xi) \left(1 - \frac{f(x^{(n)}) - f(\xi)}{x^{(n)} - \xi} \frac{1}{f[x^{(n)}, x^{(n-1)}]} \right) \\
&= (x^{(n)} - \xi) \left(1 - \frac{f[x^{(n)}, \xi]}{f[x^{(n)}, x^{(n-1)}]} \right) \\
&= (x^{(n)} - \xi)(x^{(n-1)} - \xi) \left(\frac{f[x^{(n)}, x^{(n-1)}] - f[x^{(n)}, \xi]}{x^{(n-1)} - \xi} \frac{1}{f[x^{(n)}, x^{(n-1)}]} \right) \\
&= (x^{(n)} - \xi)(x^{(n-1)} - \xi) \frac{f[x^{(n)}, x^{(n-1)}, \xi]}{f[x^{(n)}, x^{(n-1)}]} \\
&= (x^{(n)} - \xi)(x^{(n-1)} - \xi) \frac{f''(\zeta_1)}{2 f'(\zeta_2)},
\end{aligned}
\tag{6.7}
$$

for some ζ_1 in the interval containing $x^{(n)}$, $x^{(n-1)}$, and ξ, and for some ζ_2 in the interval containing $x^{(n)}$ and $x^{(n-1)}$. Here, the terms $f[\cdots]$ are divided differences as defined in Chapter 5.

Now write

$$M = \frac{\sup f''}{2 \inf f'}$$

(sup is the *supremum*: the least upper bound, and inf is the *infimum*: the greatest lower bound) and write a scaled error

$$e^{(i)} = M |x^{(i)} - \xi|.$$

With these definitions, (6.7) may be written

$$e^{(i+1)} \leq e^{(i)} e^{(i-1)}. \tag{6.8}$$

Now hypothesize that

$$e^{(i)} \leq K^{p^i},$$

so (6.8) becomes

$$K^{p^{i+1}} \leq K^{p^i + p^{i-1}}.$$

Equality holds when

$$0 = p^2 - p - 1,$$

with positive root

$$p = \frac{1}{2}(1 + \sqrt{5}) \approx 1.618.$$

We conclude that the secant method converges superlinearly. This number $(\sqrt{5} + 1)/2$ is called the *golden ratio*.

The secant method retains the advantage that $f'(x)$ is not required. The two starting values do not need to bracket a root, but because we sacrifice the bracketing property we also sacrifice the assurance of convergence. A side effect of bracketing is the fact that the denominator $f(x_+) - f(x_-)$ does not have cancellation error. By sacrificing the bracketing property, the secant method may incur this cancellation error.

Example 6.5   Find a root of $f(x)$ (6.5) with the secant method, using starting guesses $x^{(0)} = 10$ and $x^{(1)} = 0.5$.

Solution
The results of the sequence are given in Table 6.3. The order obtained is approximately equal to the theoretical value, as shown in Figure 6.3.

Table 6.3: Secant solution to root of $(x - 1)(x - 10^{-3})$ with $x^{(0)} = 10$ and $x^{(1)} = 0.5$; Example 6.5.

n	$x^{(n)}$	$f(x^{(n)})$
2	+5.2626592272870831e-01	-2.4883636722593816e-01
3	+1.0374960937662511e+01	+9.7255478559422812e+01
4	+5.5140033335736582e-01	-2.4690940606410897e-01
5	+5.7627694986754130e-01	-2.4375810386877206e-01
6	+2.5005217525919123e+00	+3.7505867608408248e+00
7	+6.9370553462383866e-01	-2.1217187139071661e-01
8	+7.9044510768471543e-01	-1.6543208452969868e-01
9	+1.1328478078844779e+00	+1.5036350013630473e-01
10	+9.6981554600972641e-01	-2.9243168273592260e-02
11	+9.9636010535285280e-01	-3.6230059194576137e-03
12	+1.0001138323706684e+00	+1.1373149610647453e-04
13	+9.9999958377831788e-01	-4.1580528708736776e-07
14	+9.9999999995257838e-01	-4.7374090327706453e-11
15	+9.9999999999999989e-01	-6.5052130349130266e-19

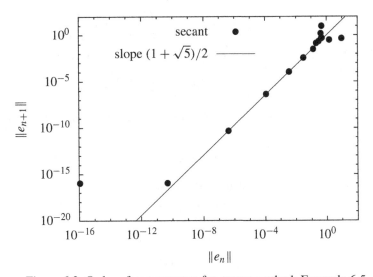

Figure 6.3: Order of convergence for secant method; Example 6.5.

Note that in Figure 6.3 the first few iterations do not lie on the line of slope 1.618 because we are not yet in the *asymptotic regime* (i.e., we are not in a regime where the assumptions of the *local* convergence rate analysis apply, because higher-order terms cannot be neglected). Note also that effective order near convergence is also not 1.618, because numerical errors dominate. The theory presented above did not account for numerical error.

6.5   Newton–Raphson

The Newton–Raphson iteration is designed to exploit the Taylor series analysis of local convergence described in Section 6.1. This method says that to determine a root of $f(x)$, we use the iteration

$$\Phi(x) = x - \frac{f(x)}{f'(x)}.$$

The method is attributed to Sir Isaac Newton and Joseph Raphson, who published versions of this algorithm in the seventeenth century, although an early version of the method specialized to computing square roots can be traced to first-century Greek scholar Heron of Alexandria. Raphson published the method in 1690 [187]. Newton developed it first around 1669 and published several versions of it; a history is presented by Ypma [247].

The Newton–Raphson method may be derived using the Taylor series in the following way:

$$f(x) = \underbrace{f(\xi)}_{0} + (x - \xi)f'(\xi) + \cdots$$

$$\xi \approx x - \frac{f(x)}{f'(\xi)} \approx x - \frac{f(x)}{f'(x)}.$$

If ξ is a simple root of $f(x)$, then $f'(\xi) \neq 0$, and clearly

$$\Phi(\xi) = \xi - \frac{f(\xi)}{f'(\xi)} = \xi,$$

so the root of $f()$ is a fixed point of the iteration. If $\mathbf{x}$ and $\mathbf{f}$ are vectors in $\mathbb{R}^n$, then we can write

$$\mathbf{x}^{(k+1)} = \mathbf{x}^{(k)} - \left(\nabla\mathbf{f}(\mathbf{x}^{(k)})\right)^{-1}\mathbf{f}(\mathbf{x}^{(k)}),$$

where $\nabla\mathbf{f}$ is an $n \times n$ matrix with elements $(\nabla\mathbf{f})_{ij} = \partial f_i/\partial x_j$.

Note that (6.7) implies that the secant method is equivalent to Newton–Raphson with the derivative estimated by finite differences using the previous two iteration values:

$$x^{(n+1)} = x^{(n)} - \frac{f(x^{(n)})}{f[x^{(n)}, x^{(n-1)}]} = x^{(n)} - \frac{f(x^{(n)})}{f'(\xi)},$$

for some $\xi \in [x^{(n)}, x^{(n-1)}]$.

The local order of convergence comes from the Taylor series analysis

$$\Phi(\xi) = \xi$$

$$\Phi'(\xi) = 1 - \frac{f'(\xi)}{f'(\xi)} + \frac{f(\xi)}{(f'(\xi))^2}f''(\xi) = 0 \quad \text{if } f'(\xi) \neq 0. \tag{6.9}$$

So, from the discussion of (6.3) we conclude that Newton–Raphson is a method of local second order if ξ is a simple root.

If ξ is not a simple root, then $f(x)$ must have the form

$$f(x) = (x - \xi)^m g(x), \tag{6.10}$$

where m is the multiplicity of the root, and $g(\xi) \neq 0$. The limit of f/f' as $x \to \xi$ is still zero, so ξ is still a fixed point of the iteration. Substituting (6.10) into (6.9) we obtain

$$\Phi'(x) = \frac{(x-\xi)^m g \left[m(m-1)(x-\xi)^{m-2}g + 2m(x-\xi)^{m-1}g' + (x-\xi)^m g'' \right]}{\left[m(x-\xi)^{m-1}g + (x-\xi)^m g' \right]^2}$$

$$= \frac{m(m-1)(x-\xi)^{2m-2}g^2 + 2m(x-\xi)^{2m-1}gg' + (x-\xi)^{2m}gg''}{m^2(x-\xi)^{2m-2}g^2 + 2m(x-\xi)^{2m-1}gg' + (x-\xi)^{2m}(g')^2}.$$

Φ' is indeterminate in the limit $x \to \xi$, and one could seek the limit by repeated application of L'Hôpital's rule. Or, by taking the lowest power of $(x-\xi)$ in both numerator and denominator (which is all L'Hôpital's rule accomplishes anyway),

$$\Phi'(\xi) = \frac{m(m-1)}{m^2} = 1 - \frac{1}{m}.$$

When the multiplicity m is greater than 1, $\Phi'(\xi)$ is not zero, but it is less than 1 in magnitude. We therefore obtain a locally convergent method of order 1.

Example 6.6   Seek numerically a root of the function (6.5) using Newton–Raphson, given a starting guess $x^{(0)} = 10$.

Solution
The results of the sequence, computed in double precision floating point, are given in Table 6.4. We expect second-order convergence since the roots of this function are simple. The order obtained is approximately 2, as shown in Figure 6.6.

Table 6.4: Newton–Raphson solution to root of $(x-1)(x-10^{-3})$ with $x^{(0)} = 10$; Example 6.6.

n	$x^{(n)}$	$f(x^{(n)})$
1	+5.2633822832780659e+00	+2.2435547394364086e+01
2	+2.9081332911779754e+00	+5.5471978147884906e+00
3	+1.7561310666044754e+00	+1.3271091254222926e+00
4	+1.2276680647248266e+00	+2.7927314435564971e-01
5	+1.0356401430494353e+00	+3.6874722702970195e-02
6	+1.0011868104206740e+00	+1.1870321292280904e-03
7	+1.0000014065868574e+00	+1.4051822491724635e-06
8	+1.0000000000019804e+00	+1.9785455825788167e-12
9	+9.9999999999999989e-01	-6.5052130349130266e-19

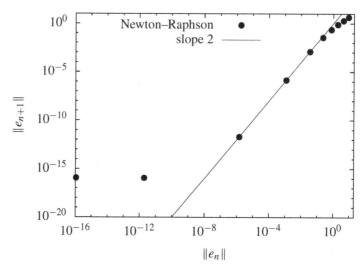

Figure 6.4: Order of convergence for Newton–Raphson; Example 6.6.

The set of initial values that will converge by Newton–Raphson to a given fixed point is difficult to assess a priori. There is a theorem by Kantorovich [126] which provides conditions under which points in a region will converge to a fixed point, but the conditions of the theorem require an intimate knowledge of the function being analyzed. Kantorovich's theorem can be summarized as follows. Let C be a convex region on which $\mathbf{f}(\mathbf{x})$ is continuously differentiable. Corresponding to this region is a Lipschitz condition:

$$\|\nabla\mathbf{f}(\mathbf{x}) - \nabla\mathbf{f}(\mathbf{y})\| \leq \gamma\|\mathbf{x} - \mathbf{y}\|,$$

for all $\mathbf{x}, \mathbf{y} \in C$. The initial value $\mathbf{x}^{(0)}$ is characterized by constants α and β:

$$\|(\nabla\mathbf{f})^{-1}(\mathbf{x}^{(0)})\mathbf{f}(\mathbf{x}^{(0)})\| \leq \alpha$$
$$\|(\nabla\mathbf{f})^{-1}(\mathbf{x}^{(0)})\| \leq \beta.$$

The constant h is defined by $h = \alpha\beta\gamma$, and define

$$r_{1,2} = \frac{\alpha}{h}\left[1 \mp \sqrt{1 - 2h}\right].$$

The region $S_{r_1}(\mathbf{x}^{(0)})$ is defined as the set of points within radius r_1 of point $\mathbf{x}^{(0)}$, and region $S_{r_2}(\mathbf{x}^{(0)})$ is defined similarly. If $h \leq 1/2$, and $S_{r_1}(\mathbf{x}^{(0)}) \subset C$, then all points $\mathbf{x}^{(k)}$ generated by Newton–Raphson will remain in $S_{r_1}(\mathbf{x}^{(0)})$, and will converge to the unique zero of $\mathbf{f}$ in $C \cap S_{r_2}(\mathbf{x}^{(0)})$.

The characterization of regions of convergence is very difficult. A complication is that even simple functions may have very complicated convergence behaviors. Consider the complex function $f = z^5 - 1$ [46]. A subset of the points that converge to $\xi = 1$ are shown (in black) in Figures 6.5 and 6.6.

Figure 6.5 Points in complex plane where Newton–Raphson applied to $f(z) = z^5 - 1$ converges to the root $\xi = 1 + 0i$ are shown in black. The figure is centered at $0.62 + 0.4464i$, and the width and height of the figure are 0.5.

Newton–Raphson has other well-known difficulties. For example, for no starting point other than $x^{(0)} = 0$ will Newton's method find the zero of $f(x) = x^{1/3}$. Another example is the function $f(x) = z^3 - 3z^2 + z - 1$ with starting value $x^{(0)} = 1$. For this problem, Newton–Raphson generates the non-convergent sequence $1, 0, 1, 0, \dots$.

For polynomials, the most useful assessment of the applicability Newton's method is a theorem of Smale [220]. He proves that there is a function $\alpha(x)$ of a single point x,

$$\alpha(x) = \left| \frac{f(x)}{f'(x)} \right| \sup_{k>1} \left| \frac{f^{(k)}(x)}{k! f'(x)} \right|^{1/(k-1)}, \tag{6.11}$$

and a constant $\alpha_0 \approx 0.130\,716\,94$, such that if $\alpha(x) < \alpha_0$, then the point x is an

Figure 6.6 $500\times$ expanded view of Figure 6.5.

approximate zero of $f(x)$, i.e., a point such that the sequence $x^{(0)} = x, x^{(1)}, x^{(2)}, \ldots$ generated by the Newton–Raphson method, will converge and will converge at second-order right from the start:

$$\left| x^{(k)} - x^{(k-1)} \right| \leq \left(\frac{1}{2} \right)^{2^{k-1}-1} \left| x^{(1)} - x^{(0)} \right|. \tag{6.12}$$

The theorem can also be used to determine in advance how many iterations are required to determine the root to a given accuracy. From the definition of approximate zero (6.12) one can write

$$\left| x^{(N)} - x^{(\ell)} \right| \leq \sum_{n=\ell+1}^{N} |x^{(n)} - x^{(n-1)}| \leq \left| x^{(1)} - x^{(0)} \right| \sum_{n=\ell+1}^{N} \left(\frac{1}{2} \right)^{2^{n-1}-1},$$

then take the limit $N \to \infty$, $x^{(N)} \to \xi$,

$$\left| \xi - x^{(\ell)} \right| \leq \left| x^{(1)} - x^{(0)} \right| \sum_{n=\ell+1}^{\infty} \left(\frac{1}{2} \right)^{2^{n-1}-1}$$

$$= \left(\frac{1}{2} \right)^{2^{\ell}-1} \left| x^{(1)} - x^{(0)} \right| \sum_{j=0}^{\infty} \left(\frac{1}{2} \right)^{2^{\ell}(2^{j}-1)}.$$

The sum on the right-hand side decreases with increasing ℓ, hence

$$\sum_{j=0}^{\infty} \left(\frac{1}{2} \right)^{2^{\ell}(2^{j}-1)} \leq \sum_{j=0}^{\infty} \left(\frac{1}{2} \right)^{(2^{j}-1)} \approx 1.632\,843\,018 < \frac{7}{4},$$

and all together one has

$$\left| \xi - x^{(\ell)} \right| \leq \left(\frac{1}{2} \right)^{2^{\ell}-1} \left| x^{(1)} - x^{(0)} \right| \frac{7}{4}. \tag{6.13}$$

An immediate application of (6.13) is the estimation of the number of iterations ℓ necessary to bound the root with accuracy ϵ:

$$\ell = 1 + \left\lceil \frac{1}{\ln 2} \ln \left(\frac{\ln \frac{7|x^{(1)}-x^{(0)}|}{4\epsilon}}{\ln 2} \right) \right\rceil.$$

Here $\lceil f \rceil$ is the *ceiling function*, which is the smallest integer greater than or equal to f.

In the special case $\ell = 0$ inequality (6.13) gives

$$\left| \xi - x^{(0)} \right| \leq \delta$$

$$\delta \equiv \frac{7}{4} \left| x^{(1)} - x^{(0)} \right| = \frac{7}{4} \left| \frac{f(x^{(0)})}{f'(x^{(0)})} \right|$$

$$\xi \in \left(x^{(0)} - \delta, \ x^{(0)} + \delta \right),$$

so when the conditions of Smale's theorem are satisfied, a bracket on the root is available using data from only one point.

Since it is difficult to guarantee convergence of Newton–Raphson, it is common to use the *modified Newton–Raphson method*,

$$\mathbf{x}^{(k+1)} = \mathbf{x}^{(k)} - \lambda \left(\nabla \mathbf{f}(\mathbf{x}^{(k)}) \right)^{-1} \mathbf{f}(\mathbf{x}^{(k)}), \tag{6.14}$$

where $\lambda = 1$ is the unmodified method.

The idea is that in the direction specified by $- \left(\nabla \mathbf{f}(\mathbf{x}^{(k)}) \right)^{-1} \mathbf{f}(\mathbf{x}^{(k)})$ some improvement in the scalar function $g(\mathbf{x}) = f(\mathbf{x})^{\mathrm{T}} \mathbf{f}(\mathbf{x})$ can be expected. This function is zero only where $\mathbf{f}$ is zero, and it has the Taylor series:

$$g(\mathbf{x} + \delta \mathbf{x}) = g(\mathbf{x}) + 2\mathbf{f}^{\mathrm{T}} (\nabla \mathbf{f}(\mathbf{x})) \delta \mathbf{x} + \text{h.o.t.} \tag{6.15}$$

(where h.o.t. stands for *higher-order terms*). With

$$\delta \mathbf{x} = -\lambda \left(\nabla \mathbf{f}(\mathbf{x}^{(k)}) \right)^{-1} \mathbf{f}(\mathbf{x}^{(k)})$$

one has

$$g(\lambda) \equiv g(\mathbf{x}^{(k)} + \delta \mathbf{x}) = g(\mathbf{x}^{(k)}) - 2\lambda \mathbf{f}(\mathbf{x}^{(k)})^{\mathrm{T}} (\nabla \mathbf{f}(\mathbf{x}^{(k)})) (\nabla \mathbf{f}(\mathbf{x}^{(k)}))^{-1} \mathbf{f}(\mathbf{x}^{(k)})$$

$$= g(\mathbf{x}^{(k)})(1 - 2\lambda),$$

so for some $\lambda > 0$ (small enough that the higher-order terms of Taylor series (6.15) can be neglected), iteration (6.14) will reduce g. Ultimately, a minimum of g will be determined, but this is not necessarily a zero of **f**. A bad enough starting guess is not compensated by this modification.

The question now arises as to how to estimate λ. Here is a simple method after [225]. Begin with $\lambda = 1$ (i.e., first try the unmodified Newton–Raphson method). If $g(\lambda)$ is not smaller than $g(0)$, then $\lambda := \lambda/2$. Repeat until some improvement in g is found.

6.6   Roots of a polynomial

If a polynomial $p(x)$ of degree n has real roots,

$$\xi_1 \geq \xi_2 \geq \xi_3 \geq \cdots \geq \xi_n$$

then it can be shown that the Newton–Raphson iteration sequence will converge monotonically to root ξ_1 if the starting point $x^{(0)}$ is greater than ξ_1.

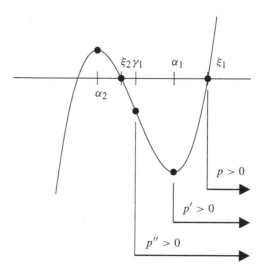

Figure 6.7 Why the greatest root can be approached monotonically.

To show this, assume $p(x^{(0)}) > 0$, and note that by Rolle's theorem the greatest zero of $p'(x)$, call it α_1, lies between ξ_1 and ξ_2 (see Figure 6.7). Then, $p'(x) > 0$ for $x > \alpha_1$. If the second zero of $p'(x)$ is $\alpha_2 \leq \alpha_1$, then by Rolle's theorem the greatest zero γ_1 of $p''(x)$ lies between α_1 and α_2, and $p''(x)$ is positive for all x greater than this zero.

Now consider the Taylor expansion of $p(x)$ including the Lagrange remainder formula, centered at the iteration $x^{(k)}$, which we assume to be larger than ξ_1,

$$0 = p(\xi_1) = p(x^{(k)}) + (\xi_1 - x^{(k)})p'(x^{(k)}) + \frac{1}{2}(\xi_1 - x^{(k)})^2 p''(\zeta),$$

for some ζ such that $\xi_1 \leq \zeta \leq x^{(k)}$.

According to the analysis of p, p', and p'' for $x > \xi_1$, we know that the Taylor remainder term is positive, and $p(x^{(k)}) > 0$ and $p'(x^{(k)}) > 0$.

$$p(x^{(k)}) + (\xi_1 - x^{(k)})p'(x^{(k)}) < 0$$

$$\frac{p(x^{(k)})}{p'(x^{(k)})} + (\xi_1 - x^{(k)}) < 0$$

$$\xi_1 < x^{(k)} - \frac{p(x^{(k)})}{p'(x^{(k)})} = x^{(k+1)},$$

and because $p(x^{(k)})$ and $p'(x^{(k)})$ are both positive for $x^{(k)} > \xi_1$, we have also

$$\xi_1 < x^{(k+1)} < x^{(k)},$$

which proves that the sequence converges and it converges monotonically (without overshoot) provided $x^{(0)} > \xi_1$.

To make use of this theorem we need some mechanism to construct a starting guess $x^{(0)}$ given the coefficients of the polynomial. One approach is to find a matrix whose characteristic equation is equal to the polynomial. The roots of the polynomial are then the eigenvalues of the matrix, and we can bound those eigenvalues with Gerschgorin's theorem.

A convenient matrix for this purpose is the $n \times n$ *Frobenius normal matrix*

$$\mathbf{F} = \begin{pmatrix} 0 & \cdots & \cdots & 0 & -a_0 \\ 1 & \ddots & & 0 & -a_1 \\ & \ddots & \ddots & \vdots & \vdots \\ & & \ddots & 0 & -a_{n-2} \\ 0 & & & 1 & -a_{n-1} \end{pmatrix}, \tag{6.16}$$

with characteristic equation

$$\psi(x) = \det(\mathbf{F} - x\mathbf{I}) = (-1)^n (a_0 + a_1 x + \cdots + a_{n-1}x^{n-1} + x^n). \tag{6.17}$$

Because of the association of the matrix (6.16) with the polynomial (6.17), the matrix is also known as the *companion matrix*. There is one Gerschgorin disk centered at 0 with radius $|a_0|$, $n - 2$ disks centered at 0 with radii $1 + |a_1|$, ..., $1 + |a_{n-2}|$, and one disk centered at $-a_{n-1}$ with radius 1. Therefore, all roots of $\psi(x)$ satisfy

$$|\xi| \leq \max\left(|a_0|, \ 1 + |a_1|, \ \ldots, \ 1 + |a_{n-2}|, \ 1 - a_{n-1}\right).$$

A different bound comes from analysis of $\mathbf{F}^{\mathrm{T}}$, which has the same eigenvalues as $\mathbf{F}$. From $\mathbf{F}^{\mathrm{T}}$ we determine that all roots of $\psi(x)$ satisfy

$$|\xi| \le \max\left(1, \; -a_{n-1} + \sum_{i=0}^{n-2} |a_i|\right).$$

Tighter bounds can be had by applying a similarity transform, as demonstrated in Example 3.2. Analyzing $\mathbf{DFD}^{-1}$, with $\mathbf{D} = \mathrm{diag}(1, \; a_1^{-1}, \; a_2^{-1}, \; ..., \; a_{n-1}^{-1})$ leads to

$$|\xi| \le \max\left(\left|\frac{a_0}{a_1}\right|, \; 2\left|\frac{a_1}{a_2}\right|, \; ..., \; 2\left|\frac{a_{n-2}}{a_{n-1}}\right|, \; 2\,|a_{n-1}|\right),$$

which is known as Kojima's bound [132]. A number of other bounds derived using Gerschgorin's theorem are described by Melman [161].

Once the maximum real root ξ_1 is determined the factor $(x - \xi_1)$ could be factored from the polynomial to give a new polynomial of lower degree:

$$p_1(x) = \frac{p(x)}{x - \xi_1}. \tag{6.18}$$

One could determine the coefficients of $p_1(x)$ by synthetic division. However, this *deflation* technique introduces numerical errors into p_1 which cause its roots to change. An alternative is to treat $p_1(x)$ using (6.18), with

$$p_1'(x) = \frac{p'(x)}{x - \xi_1} - \frac{p(x)}{(x - \xi_1)^2},$$

then find the roots of $p_1(x)$ using the Newton–Raphson method with a starting guess $x^{(0)} = \xi_1 - \epsilon$ for some small value ϵ such that $\xi_1 > \xi_1 - \epsilon \ge \xi_2$:

$$x^{(k+1)} = x^{(k)} - \frac{p_1(x^{(k)})}{p_1'(x^{(k)})} = x^{(k)} - \frac{p(x^{(k)})}{p'(x^{(k)}) - \frac{p(x^{(k)})}{x^{(k)} - \xi_1}}.$$

This alternative to deflation is called *zero suppression* and is due to Maehly [152]. Generalizing,

$$p_j(x) = \frac{p(x)}{\prod_{i=1}^{j}(x - \xi_i)}$$

$$p_j'(x) = \frac{p'(x)}{\prod_{i=1}^{j}(x - \xi_i)} - \frac{p(x)}{\prod_{i=1}^{j}(x - \xi_i)} \sum_{k=1}^{j} \frac{1}{x - \xi_k}$$

$$x^{(k+1)} = x^{(k)} - \frac{p(x^{(k)})}{p'(x^{(k)}) - p(x^{(k)}) \sum_{i=1}^{j} \frac{1}{x^{(k)} - \xi_i}},$$

where j is the number of roots that have already been determined, and we are seeking root ξ_{j+1} with a starting value $\xi_j - \epsilon$. This method assumes that the roots are simple, $\xi_1 > \xi_2 > \cdots > \xi_n$.

As mentioned in Section 1.6, the evaluation of polynomials is sensitive to perturbations in the coefficients. Let $p_n = a_0 + a_1 x + \cdots + a_n x^n$ be a polynomial of degree n, and let ξ be a root of multiplicity m, i.e., p_n could be written

$$p_n(x) = (x - \xi)^m h(x) \qquad h(\xi) \neq 0,$$

and $h(\xi)$ can be expressed in terms of the mth derivative of p_n,

$$m! h(\xi) = p_n^{(m)}(\xi).$$

Now perturb p_n by adding a small quantity, $\mu^s g(x)$, where $g(x)$ is a polynomial and μ^s scales the perturbation. The constant s will be specified momentarily. The root $\xi(\mu)$ of the modified polynomial $p + \mu^s g$ will obey

$$(\xi(\mu) - \xi)^m h(\xi(\mu)) + \mu^s g(\xi(\mu)) = 0,$$

or

$$(\xi(\mu) - \xi) h(\xi(\mu))^{1/m} = \mu^{s/m} [-g(\xi(\mu))]^{1/m}.$$

To find $\xi' = d\xi(\mu)/d\mu$, the change in the root with respect to the parameter μ, first differentiate this expression with respect to μ, then solve for ξ' in the limit $\mu \to 0$. Differentiation gives

$$\xi' h(\xi(\mu))^{1/m} + \frac{1}{n}(\xi(\mu) - \xi) h'(\xi(\mu)) h(\xi(\mu))^{(1-m)/m}$$

$$= \frac{s}{m}(\mu)^{(s-m)/m}[-g(\xi(\mu))]^{1/m} - \frac{1}{m}\mu^{s/m} g'(\xi(\mu)) \xi' [-g(\xi(\mu))]^{(1-m)/m}.$$

In order that the right-hand side be nonzero in the limit $\mu \to 0$, the parameter s must be equal to the multiplicity m. Then,

$$\xi' = \left(-\frac{g(\xi)}{h(\xi)}\right)^{1/m} = \left(-\frac{m! g(\xi)}{p_n^{(m)}(\xi)}\right)^{1/m},$$

and

$$\xi(\mu) \approx \xi + \mu\left(-\frac{m! g(\xi)}{p_n^{(m)}(\xi)}\right)^{1/m},$$

where p_n has a perturbation proportional to μ^m, or

$$\xi(\mu) \approx \xi + \mu^{1/m}\left(-\frac{m! g(\xi)}{p_n^{(m)}(\xi)}\right)^{1/m} \tag{6.19}$$

if p_n has a perturbation proportional to μ.

Now consider Wilkinson's polynomial of degree 20 [244],

$$p = \prod_{i=1}^{20}(x - i),$$

expressed in the expanded form $p = a_0 + a_1 x + \cdots + a_{20} x^{20}$. The roots are all simple, $m = 1$. Let the perturbation take the form

$$g = a_j x^j,$$

and let $\mu = \pm\epsilon$ (machine precision). Then, (6.19) will show the error in root $\xi_i = i$ due to a relative error ϵ in the single coefficient a_j. For this simple polynomial, the maximum root error occurs for $\xi = 14$ when coefficient a_{12} is perturbed:

$$a_{12} = 11\,310\,276\,995\,381$$

$$p'(14) = 4\,483\,454\,976\,000$$

$$\xi(\epsilon) = 14 - \mu\frac{11\,310\,276\,995\,381 \times 14^{12}}{4\,483\,454\,976\,000} = 14 - 1.4302 \times 10^{14}\mu.$$

If this perturbation is due to rounding in double precision, $|\mu| \lesssim 10^{-16}$, then the error in ξ amounts to a tenth of a percent error. But, under realistic circumstances the error could be disastrous. Evaluation of the polynomial by Horner's scheme carries an error equivalent to a relative error of 25ϵ on a_{12} (1.20), which increases the root error by a factor of 25. Evaluation of the polynomial in single precision increases the error by a factor of $2^{-24}/2^{-53} \approx 5.4 \times 10^8$. Combined, root ξ_{14} could be anywhere between -2.13×10^8 to $+2.13 \times 10^8$ – egregious. Note too that we are discussing the error in a single term, a_{12}. While the error associated with other terms is less, some are not less by much.

The point of this discussion is to show that the evaluation of some polynomials is very sensitive to the accuracy of the coefficients. Therefore, procedures like deflation, which introduce new errors to such coefficients, are risky. Zero suppression, on the other hand, does not modify the original representation, and does not require the expanded polynomial form. Of course not all polynomials are as sensitive as Wilkinson's example, and strategies do exist to mitigate the errors of deflation (see, e.g., [181, 242]). Nevertheless, the approach with least risk is zero suppression.

6.7   Newton–Raphson on the complex plane

A very interesting approach to root finding for polynomials exploits the geometric properties of the *Newton map*, e.g., Figures 6.5, 6.6, and 6.8, which show the mapping of a starting point in the complex plane to roots of a function. The boundaries separating *basins of attraction* of the roots is called the Julia set. The *immediate basin* of attraction of a root is the connected basin containing the root.

These maps are fascinating in their complexity, and possess so-called fractal structure (approximate self-similarity at all length scales: e.g., compare Figures 6.5 and 6.6). Yet, they also have a regularity to their structure that may be exploited to construct a root finding scheme. Hubbard, Schleicher, and Sutherland [114] derive a method that determines a set of points of cardinality $\approx 2.22d(\ln d)^2$ such that at least one point in the set will converge under the complex Newton–Raphson method to each of the d roots of a polynomial of degree d.

Amongst the properties that this method exploits are the facts that each immediate basin of attraction has at least one *channel* that extends to infinity, and the width of these channels is bounded. If m is the number of critical points in a given basin (number

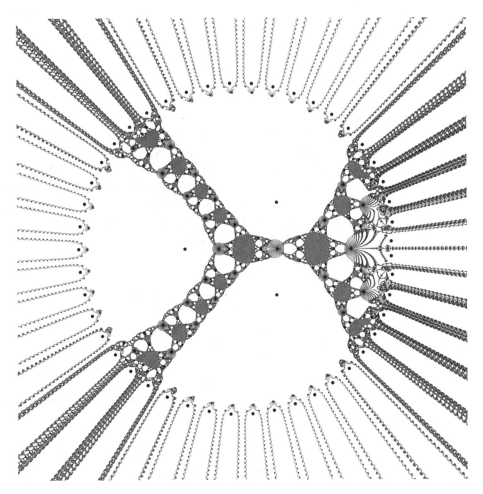

Figure 6.8 Basins of attraction for the polynomial $z^{50} + 8z^5 - \frac{80}{3}z^4 + 20z^3 - 2z + 1$, separated by Julia set (black lines). Roots are shown by black circles. Gray regions indicate a periodic orbit.

of zeros and inflection points), then the basin has m channels to infinity, and at least one of these channels has width greater than $\pi/\ln(m+1)$ radians.

The basic strategy is to create an array of starting points of sufficient density that at least one point will be in a channel associated with the immediate basin of each root. The points will be evenly spaced on s circles,

$$s = \lceil \alpha \ln d \rceil,$$

with n points per circle,

$$n = \lceil \beta d \ln d \rceil.$$

Here $\alpha = 0.266\,32$ and $\beta = 8.325\,47$ are parameters that approximately optimize the point count when the circles have radii

$$r_v = R \left(\frac{d-1}{d} \right)^{(2v-1)/4s} , \qquad v = 1, 2, ..., s,$$

and $R = (1 + \sqrt{2})\rho$, where ρ bounds the roots: $|\xi| < \rho$ for all roots ξ. This is a particular instance of the method of Hubbard, Schleicher, and Sutherland; the scheme as described in [114] offers considerable flexibility, including specialization in the case that the roots are known to be real. Note that when the polynomial coefficients are real, the roots are real or complex conjugate pairs, and this gives the Newton maps mirror symmetry about the real axis. In this important case, one need only discover half the roots, which means one need only use those starting points in the top half plane of the construction.

Example 6.7   Find the roots of

$$3x^5 - 10x^3 + 23x$$

by Newton's method, with the approach of Hubbard, Schleicher, and Sutherland. All roots lie in the circle $|z| < 1.7$.

Solution
The degree of the polynomial is five, so we have

$$s = \lceil \alpha \ln 5 \rceil = \lceil 0.428 \rceil = 1$$

circles of radius

$$r = (1 + \sqrt{2})1.7 \left(\frac{4}{5} \right)^{1/4} = 3.88.$$

On that circle, we will place

$$n = \lceil 5\beta \ln 5 \rceil = \lceil 66.996 \rceil = 67$$

evenly spaced starting points. This collection of 67 points is sure to include at least one in the immediate basin of each root. See Figure 6.9.

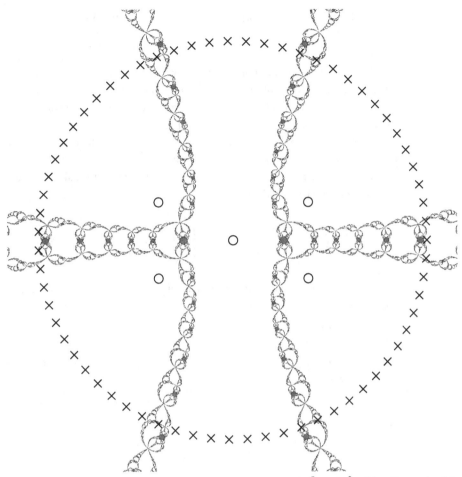

Figure 6.9: The Basins of attraction for the polynomial $3x^5 - 10x^3 + 23x$, Example 6.7, are separated by the Julia set (black lines). Roots are shown by open black circles. Gray regions indicate a periodic orbit. Crosses mark Newton–Raphson starting points by the method of Hubbard, Schleicher, and Sutherland.

6.8 Bairstow's method

Even real polynomials can have complex roots. When these show up, they always occur in complex conjugate pairs, e.g., quadratic factors

$$(x - \xi)(x - \xi^*) = x^2 - 2x\Re(\xi) + |\xi|^2. \qquad (6.20)$$

Bairstow's method [5] finds roots by pairs, and so can find complex roots in real polynomials using only real math.

The idea is to seek a factorization like (6.20) for polynomial p of degree n:

$$p(x) = p_1(x)(x^2 - rx - q) + ax + b, \tag{6.21}$$

where p_1 is a polynomial of degree $n - 2$, $(x^2 - rx - q)$ is a quadratic term, and $ax + b$ is a remainder. If the two solutions of $(x^2 - rx - q) = 0$ are roots of p, then the remainder $ax + b$ is zero. For any choice of r, q, one could do synthetic division of $p(x)$ to divide the factor $(x^2 - rx - q)$ with remainder $ax + b$. Therefore, we can consider a, b to be functions of r, q. Likewise p_1 depends on both r and q. The original polynomial $p(x)$ does not depend on r or q – only its representation as (6.21) depends on these numbers.

Bairstow's method adjusts the parameters r, q in order to make a, b be zero. This is done by the Newton–Raphson method, i.e.,

$$\begin{pmatrix} r \\ q \end{pmatrix}^{(k+1)} = \begin{pmatrix} r \\ q \end{pmatrix}^{(k)} - \left[\left(\begin{matrix} \frac{\partial a}{\partial r} & \frac{\partial a}{\partial q} \\ \frac{\partial b}{\partial r} & \frac{\partial b}{\partial q} \end{matrix} \right)^{(k)} \right]^{-1} \begin{pmatrix} a(r, q) \\ b(r, q) \end{pmatrix}^{(k)}. \tag{6.22}$$

The trick now is to determine the partial derivatives that appear in (6.22). We get these by differentiating (6.21):

$$\frac{\partial p(x)}{\partial r} = 0 = -xp_1(x) + (x^2 - rx - q)\frac{\partial p_1(x)}{\partial r} + x\frac{\partial a}{\partial r} + \frac{\partial b}{\partial r} \tag{6.23a}$$

$$\frac{\partial p(x)}{\partial q} = 0 = -p_1(x) + (x^2 - rx - q)\frac{\partial p_1(x)}{\partial q} + x\frac{\partial a}{\partial q} + \frac{\partial b}{\partial q}. \tag{6.23b}$$

Let the roots of $(x^2 - rx - q)$ be labeled x_0 and x_1. If we evaluate (6.23a) and (6.23b) at $x_i = x_0, x_1$:

$$0 = -x_i p_1(x_i) + x_i\frac{\partial a}{\partial r} + \frac{\partial b}{\partial r}$$
$$0 = -p_1(x_i) + x_i\frac{\partial a}{\partial q} + \frac{\partial b}{\partial q}. \tag{6.24}$$

If one tried to factor p_1 in a manner analogous to the factorization (6.21), then

$$p_1(x) = p_2(x)(x^2 - rx - q) + \alpha x + \beta, \tag{6.25}$$

and for $x_i = x_0, x_1$:

$$p_1(x_i) = \alpha x_i + \beta. \tag{6.26}$$

Now substitute (6.26) into (6.24). For $x_i = x_0, x_1$:

$$0 = -x_i(\alpha x_i + \beta) + x_i\frac{\partial a}{\partial r} + \frac{\partial b}{\partial r}$$
$$= -(\alpha r + \beta)x_i - \alpha q + x_i\frac{\partial a}{\partial r} + \frac{\partial b}{\partial r} \tag{6.27a}$$

$$0 = -(\alpha x_i + \beta) + x_i\frac{\partial a}{\partial q} + \frac{\partial b}{\partial q}, \tag{6.27b}$$

where $x^2 - rx - q = 0$ is used to eliminate x_i^2 in (6.27a).

Comparing powers of x_i in (6.27b) and then in (6.27a) one finds

$$\frac{\partial a}{\partial q} = \alpha$$

$$\frac{\partial b}{\partial q} = \beta$$

$$\frac{\partial a}{\partial r} = \alpha r + \beta \qquad\qquad (6.28)$$

$$\frac{\partial b}{\partial r} = \alpha q.$$

The synthetic division has to be done twice per iteration, since p_1 and p_2 and their remainders are functions of r and q. This is most easily done by working backwards. If $p(x)$ is of degree n then $p_1(x)$ is of degree $n - 2$. Write the series for $p_1(x)$ as

$$p_1(x) = d_0 + d_1 x + d_2 x^2 + \cdots + d_{n-2} x^{n-2},$$

then multiply by $(x^2 - rx - q)$ to give

$$p_1(x)(x^2 - rx - q) = - d_0 q - (d_1 q + r d_0)x$$
$$- (d_2 q + r d_1 - d_0)x^2 \ldots - (d_{n-2} q + r d_{n-3} - d_{n-4})x^{n-2}$$
$$+ (d_{n-3} - r d_{n-2})x^{n-1} + d_{n-2} x^n,$$

and compare term by term with the series

$$p(x) = c_0 + c_1 x + c_2 x^2 + \cdots + c_n x^n,$$

to give

$$d_{n-2} = c_n$$
$$d_{n-3} = c_{n-1} + r d_{n-2}$$

$$\vdots$$

$$d_{i-2} = c_i + r d_{i-1} + q d_i$$

$$\vdots$$

$$d_1 = c_3 + r d_2 + q d_3$$
$$d_0 = c_2 + r d_1 + q d_2$$
$$a = c_1 + q d_1 + r d_0$$
$$b = c_0 + q d_0.$$

The same procedure, again, determines α and β by constructing p_2 and its remainder.

Example 6.8   Use Bairstow's method to factor

$$p = 24 - 18x + 131x^2 - 96x^3 + 56x^4 - 30x^5 + 5x^6.$$

Solution

For the first root pair, begin with $r^{(0)} = q^{(0)} = 0$. The factorization (6.21) gives

$$p = \underbrace{(131 - 96x + 56x^2 - 30x^3 + 5x^4)}_{p_1} \underbrace{(x^2)}_{x^2 - rx - q} + \underbrace{(-18)}_{a} x + \underbrace{(24)}_{b}.$$

Since $a, b \neq 0$, the guess $(r, q)^{(0)}$ does not give a quadratic factor, so some iteration is required. If p_1 is factored in the same manner, per (6.25), then

$$p_1 = \underbrace{(56 - 30x + 5x^2)}_{p_2} \underbrace{(x^2)}_{x^2 - rx - q} + \underbrace{(-96)}_{\alpha} x + \underbrace{(131)}_{\beta}.$$

With r, q, α, and β known, the derivatives (6.28) can be determined and the Newton–Raphson step (6.22) can be taken:

$$\begin{pmatrix} r \\ q \end{pmatrix}^{(1)} = \begin{pmatrix} 0 \\ 0 \end{pmatrix} - \begin{pmatrix} 131 & -96 \\ 0 & 131 \end{pmatrix}^{-1} \begin{pmatrix} -18 \\ 24 \end{pmatrix} = \begin{pmatrix} 0.031\,467 \\ -0.183\,21 \end{pmatrix}.$$

Now, with the new r, q pair a, b and p_1 are computed. However, a, b are not zero, so p_1 is factored giving α, β and p_2. Then r, q, α, β permit determination of a new r, q by the Newton–Raphson step. This procedure continues until such time as $a, b = 0$ (or, better, $\mathcal{O}(\epsilon)$), when

$$r^{(4)} = 1.0618 \times 10^{-17}$$

$$q^{(4)} = -0.200\,00,$$

giving complex roots

$$x_1 = 5.3091 \times 10^{-18} + 0.447\,21i$$

$$x_2 = 5.3091 \times 10^{-18} - 0.447\,21i,$$

and $p_1 = 120 - 90x + 55x^2 - 30x^3 + 5x^4$.

Since p_1 has degree greater than two, we let $p := p_1$ and repeat the root-finding sequence. Beginning again with $(q, r)^{(0)} = (0, 0)$, six iterations give

$$r^{(6)} = 9.4005 \times 10^{-17}$$

$$q^{(6)} = -3.0000,$$

which gives the complex roots

$$x_3 = 4.7003 \times 10^{-17} - 1.7321i$$

$$x_4 = 4.7003 \times 10^{-17} + 1.7321i,$$

and $p_1 = 40 - 30x + 5x^2$.

This last factor is degree two and can be solved immediately to give

$$x_5 = 4.0000$$

$$x_6 = 2.0000.$$

6.9 Improving convergence

The sequence $\{x^{(0)}, x^{(1)}, x^{(2)}, x^{(3)}, ...\}$ may converge, but slowly. One could analyze this sequence to anticipate the converged result $\xi = x^{(\infty)}$, thereby accelerating the convergence. One approach to this problem is due to A. Aitken [3]. We hypothesize the geometric convergence model (i.e., first-order: see (6.4))

$$x^{(i+1)} - \xi = k(x^{(i)} - \xi),$$

so also

$$x^{(i+2)} - \xi = k(x^{(i+1)} - \xi).$$

These two equations may be solved to give k and ξ:

$$
\begin{aligned}
k &= \frac{x^{(i+2)} - x^{(i+1)}}{x^{(i+1)} - x^{(i)}} \\
\xi &= x^{(i)} - \frac{(x^{(i+1)} - x^{(i)})^2}{x^{(i+2)} - 2x^{(i+1)} + x^{(i)}}.
\end{aligned}
\tag{6.29}
$$

This is called *Aitken's Δ^2 method*.

Example 6.9 Apply Aitken's method to the first-order iterative method

$$x^{(0)} = 0.5$$
$$x^{(k)} = \cos(x^{(k)})$$

to find the solution $\xi = \cos \xi$.

Solution
The sequence $\{x^{(k)}\}$ and the sequence $\{y^{(k)}\}$ derived from sequence $\{x^{(k)}\}$ using Aitken's method are shown in Table 6.5. The convergence is clearly improved.

Table 6.5: Aitken's method; Example 6.9.

k	$x^{(k)}$	$|x^{(k)} - \cos(x^{(k)})|$	$y^{(k)}$	$|y^{(k)} - \cos(y^{(k)})|$
1	0.5000000000	3.776e-01	0.7313851864	1.286e-02
2	0.8775825619	2.386e-01	0.7360866917	5.015e-03
3	0.6390124942	1.637e-01	0.7376528714	2.396e-03
4	0.8026851007	1.079e-01	0.7384692209	1.031e-03
5	0.6947780268	7.342e-02	0.7387980652	4.804e-04
$\vdots$				
25	0.7390690012	2.700e-05	0.7390851332	6.478e-11

Problems

6.1 For what starting value $x^{(0)}$ will Newton's method, applied to

$$f(x) = x^2 - a, \quad a > 0,$$

converge quadratically to the root $\xi = \sqrt{a}$?

6.2 Use a Taylor series with Lagrange remainder (A.7) to show

$$(\xi - x^{(1)}) = -(\xi - x^{(0)})^2 \frac{f''(\eta)}{2f'(x^{(0)})}$$

for Newton's method, where η is in the interval $I[\xi, x^{(0)}]$. Use this with Newton's method for

$$f(x) = x^2 - a, \quad a > 0,$$

to find values $x^{(0)}$ such that Newton's method will converge monotonically. That is, so

$$x^{(0)} - \xi \geq x^{(1)} - \xi \geq \cdots \geq x^{(n)} - \xi \geq \cdots \geq 0$$

or

$$x^{(0)} - \xi \leq x^{(1)} - \xi \leq \cdots \leq x^{(n)} - \xi \leq \cdots \leq 0.$$

6.3 Treating the vapor phase as ideal, a simplified thermodynamic model for vapor–liquid equilibrium in the system water–ethanol is [4] :

$$\tau_{12} = \frac{\Delta\gamma_{12}}{RT}$$

$$\tau_{21} = \frac{\Delta\gamma_{21}}{RT}$$

$$G_{12} = \exp(-\alpha_{12}\tau_{12})$$

$$G_{21} = \exp(-\alpha_{21}\tau_{21})$$

$$\gamma_1 = \exp\left(x_2^2 \left[\frac{\tau_{21}G_{21}^2}{(x_1 + x_2 G_{21})^2} + \frac{\tau_{12}G_{12}}{(x_2 + x_1 G_{12})^2} \right]\right)$$

$$\gamma_2 = \exp\left(x_1^2 \left[\frac{\tau_{12}G_{12}^2}{(x_2 + x_1 G_{12})^2} + \frac{\tau_{21}G_{21}}{(x_1 + x_2 G_{21})^2} \right]\right)$$

$$P_1^{\text{sat}} = 10^{A_1 - B_1/(T + C_1)}$$

$$P_2^{\text{sat}} = 10^{A_2 - B_2/(T + C_2)}$$

$$y_1 = \frac{\gamma_1 x_1 P_1^{\text{sat}}}{P}$$

$$y_2 = \frac{\gamma_2 x_2 P_2^{\text{sat}}}{P},$$

with parameters

R	8.314 472 J/K mol
P	101.325 kPa
A_1, B_1, C_1	7.168 79, 1552.601, -50.731
A_2, B_2, C_2	7.232 55, 1750.286, -38.15
$\alpha_{12} = \alpha_{21}$	-0.351
$\Delta\gamma_{12}$	3159.86 J/mol
$\Delta\gamma_{21}$	-618.37 J/mol .

Here x_1 is the mole fraction of ethanol in the liquid, $x_2 = 1 - x_1$ is the mole fraction of water in the liquid, and y_1 and y_2 are the respective mole fractions in the vapor phase, and they are subject to a constraint $y_1 + y_2 = 1$.

For a choice of x_1, and a guess of T, one can solve the given equations in the order presented to calculate y_1, y_2, and the error $y_1 + y_2 - 1$. If this error is not zero, a different temperature guess T is required.

(i) Develop an iteration to solve this system of equations for any $0 < x_1 < 1$.

(ii) Find the azeotrope: the composition that cannot be changed by distillation ($x_1 = y_1$ and $x_2 = y_2$).

6.4 What point on the curve $y = \tan(x)$, with $x \in [-\pi/2, \pi/2]$, is closest to the point $\bar{x} = -1/2$, $\bar{y} = 1$?

6.5 On the following resistor network

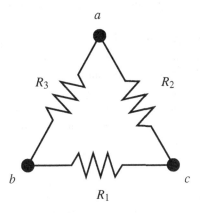

the resistance across points a and b, R_{ab} is given by

$$\frac{1}{R_{ab}} = \frac{1}{R_3} + \frac{1}{R_1 + R_2}.$$

The following resistances are measured: $R_{ab} = 116.68$, $R_{ac} = 123.81$, and $R_{bc} = 107.14$. What are R_1, R_2, and R_3?

7 Optimization

The goal of optimization is to find a vector solution $\mathbf{x}$ that minimizes some scalar function $f(\mathbf{x})$. We will assume that the function is at least C^2 (i.e., is continuous, has continuous derivatives and has continuous second derivatives). The solution space may be subject to constraints, e.g., the positivity of some elements of $\mathbf{x}$.

We have seen already one method, conjugate gradients (Section 4.1), which assumes a specific quadratic function f with a minimum that solves the linear equation $\mathbf{Ax} = \mathbf{b}$. We are now interested in more complicated functions, and for many such functions the solution is notoriously difficult. For example, the function may possess any number of local minima. In the neighborhood of a local minimum, quasi-Newton type methods will converge to the local minimum, which may be distant from the desired global minimum. None of the approaches described here can cope with that problem – the success of the methods relies on smoothness of the function $f(\mathbf{x})$ and the quality of the starting guess $\mathbf{x}^{(0)}$.

We will examine separately the case of nonlinear functions $f(\mathbf{x})$ without constraints (variable metric methods), and linear constraints applied to linear functions (linear programming). Methods for constrained nonlinear problems (nonlinear programming) are built on these simpler methods.

The variable metric methods for n-dimensional problems rely on accurate and stable methods for simpler one-dimensional (1D) problems. The problem of finding a local minimum in 1D is broken down into two parts. Problem I is called bracketing. Given

$$f'(0) < 0,$$

find numbers λ_- and λ_+ such that $f(\lambda)$ is a minimum for some $\lambda_- \leq \lambda \leq \lambda_+$. The pair (λ_-, λ_+) is called the bracket. Problem II is interval refinement. Given brackets (λ_-, λ_+), find the optimum λ that minimizes the function in this interval. These 1D methods have application by themselves, but we will limit our interest here to conditions required to support the variable metric algorithms.

7.1 1D: Bracketing

If an ordered triplet of points (x_a, x_b, x_c) can be found with $f(x_b) < f(x_a)$ and $f(x_b) < f(x_c)$, then (x_a, x_c) will be a bracket. The proof of this uses divided differences: $(f(x_b) - f(x_a))/(x_b - x_a)$ is negative, and is the derivative $f'(\xi)$ for some

$\xi \in [x_a, x_b]$; $(f(x_c) - f(x_b))/(x_c - x_b)$ is positive, and is the derivative $f'(\zeta)$ for some $\zeta \in [x_b, x_c]$. So, somewhere between ξ and ζ, f' must be zero – a local extrema. Since $f(x_b)$ is smaller than the bracketing points, the local extrema is a minimum.

Here is an algorithm for finding such a bracket:

1. Begin with $\lambda_- = 0$ and $\lambda_+ = h$, with h a (problem-dependent) constant.
2. If $f(\lambda_+) > f(\lambda_-)$ then (λ_-, λ_+) is a bracket, assuming $f'(0) < 0$.
3. Otherwise, let $\lambda_0 = \lambda_+$, and redefine

$$\lambda_+ = \lambda_0 + \Gamma(\lambda_0 - \lambda_-),$$

   with $\Gamma = (1 + \sqrt{5})/2 \approx 1.618$ (the golden ratio).
4. If $f(\lambda_0) < f(\lambda_-)$ and $f(\lambda_0) < f(\lambda_+)$, then (λ_-, λ_+) is a bracket. Otherwise, go to step 3.

Note that $\lambda_- = 0$ was never adjusted.

If bracketing is the final objective, the constant Γ can be anything larger than 1. The particular value employed here is chosen for compatibility with the golden section method of interval refinement.

7.2   1D: Refinement by interpolation

Now, given $f_0 = f(0)$, $f_0' = f'(0)$, and $f_1 = f(\lambda_+)$, we have enough information to construct a quadratic estimate of the function in the interval $(0, \lambda_+)$ using Hermite interpolation:

$$
\begin{array}{ll}
0 & f_0 \\
 & \quad f_0' \\
0 & f_0 \qquad\qquad\qquad (f_1 - f_0 - \lambda_+ f_0')/\lambda_+^2 \\
 & \quad (f_1 - f_0)/\lambda_+ \\
\lambda_+ & f_1
\end{array}
$$

$$f(\lambda) = f_0 + \lambda \left[f_0' + \lambda \frac{(f_1 - f_0 - \lambda_+ f_0')}{\lambda_+^2} \right],$$

with minimum at

$$\lambda = \frac{f_0' \lambda_+^2}{2(f_0 + f_0'\lambda_+ - f_1)}.$$

If the gradient $f'(\lambda_+)$ can be measured, cubic interpolation can be used [51]. In that case, Hermite interpolation again gives

$$0 \quad f_0$$

$$f_0'$$

$$0 \quad f_0 \qquad \frac{f_1 - f_0 - \lambda_+ f_0'}{\lambda_+^2}$$

$$\frac{f_1 - f_0}{\lambda_+} \qquad\qquad \frac{\lambda_+(f_1' + f_0') - 2(f_1 - f_0)}{\lambda_+^3}$$

$$\lambda_+ \quad f_1 \qquad \frac{\lambda_+ f_1' - f_1 + f_0}{\lambda_+^2}$$

$$f_1'$$

$$\lambda_+ \quad f_1$$

$$f(\lambda) = f_0 + \lambda \left[f_0' + \lambda \left\{ \frac{f_1 - f_0 - \lambda_+ f_0'}{\lambda_+^2} \right. \right. \tag{7.1}$$

$$\left. \left. + (\lambda - \lambda_+) \left(\frac{\lambda_+(f_1' + f_0') - 2(f_1 - f_0)}{\lambda_+^3} \right) \right\} \right]$$

$$f'(\lambda) = 3\frac{\lambda_+(f_0' + f_1') + 2(f_0 - f_1)}{\lambda_+^3}\lambda^2 - 2\frac{3(f_0 - f_1) + \lambda_+(f_1' + 2f_0')}{\lambda_+^2}\lambda + f_0'.$$

With $f_0' < 0$ and (f_0, f_1) a bracket, this quadratic has two real roots, and the smallest positive root will be the minimum of the cubic interpolating polynomial.

When the function is locally quadratic or cubic, these polynomial estimates may be perfectly adequate. The convergence properties of the variable metric method relies on "exact" 1D minimization of arbitrary functions, however, so some iterative approach is often used.

7.3   1D: Refinement by golden section search

A simple scheme that guarantees that λ is found to some accuracy ϵ is the golden section search [129]. It systematically shrinks the bracket, while maintaining an ordered triplet of points.

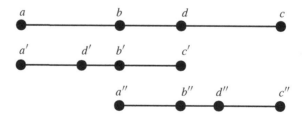

Figure 7.1 Self-similarity of golden section search points.

We begin with an ordered triplet of points (a, b, c) that bracket the minimum, where b has some definite relative position in the interval (a, c), to be determined.

Next, a fourth point d will be introduced somewhere in (a, c). Let's assume for now the ordering (a, b, d, c), as in Figure 7.1. If $f(b) < f(d)$, then $(a', b', c') = (a, b, d)$ is a new ordered triplet of points on a smaller interval. However, if $f(b) > f(d)$, then $(a'', b'', c'') = (b, d, c)$ is a new ordered triplet.

By maintaining b, d in a special relative position within the interval, we have the property (i) that the size of the new interval is always the same $c' - a' = c'' - a''$, and (ii) the new interior point b' or b'' is in a special relative position such that self-similarity can be maintained with successive iterations.

With reference to Figure 7.1, the special position requirement is

$$\frac{d-a}{c-a} = \frac{c-b}{c-a} \quad \text{(symmetry property)}$$

$$\underbrace{\frac{d-a}{c-a}}_{\alpha} = \frac{b'-a'}{c'-a'} \quad \text{(self-similarity property)}$$

$$\alpha = \frac{b-a}{d-a} \quad \text{(because } (a', b', c') = (a, b, d))$$

$$= \frac{(c-a)-(c-b)}{d-a} = \frac{(c-a)-(d-a)}{d-a} = \frac{1}{\alpha} - 1$$

$$\alpha = \frac{\sqrt{5}-1}{2} = \Gamma - 1,$$

where Γ is the golden ratio (also, the order of convergence of the secant method).

The algorithm is then:

1. Begin with bracket $(a, c) = (\lambda_-, \lambda_+)$ and choose $b = a + (\Gamma - 1)(c - a)$.
2. If $c - a \le \epsilon$, then the bracket is tight enough, and $\lambda = b$ is the best value. Stop.
3. If $|c - b| > |b - a|$, let

$$d = b + (2 - \Gamma)(c - b),$$

and the ordered set is (a, b, d, c). Otherwise, let

$$d = b - (2 - \Gamma)(b - a),$$

and the ordered set is (a, d, b, c).
4. Evaluate $f(d)$, and select the appropriate ordered triplet from the available four points. Go to step 2.

7.4   nD: Variable metric methods

When $f(\mathbf{x})$ is a minimum, the gradient at $\mathbf{x}$ is zero:

$$\mathbf{g}(\mathbf{x}) = \nabla_{\mathbf{x}} f(\mathbf{x})$$

$$f(\boldsymbol{\xi}) \text{ a minimum} \quad \Longrightarrow \quad \mathbf{g}(\boldsymbol{\xi}) = \mathbf{0},$$

where $\implies$ means "implies." In principle, a vector version of Newton–Raphson could be employed to find the zero of the gradient:

$$\mathbf{x}^{(k+1)} = \mathbf{x}^{(k)} - \mathbf{G}(\mathbf{x}^{(k)})^{-1}\mathbf{g}(\mathbf{x}^{(k)}),$$

where $\mathbf{G}$ is the *Hessian*, or matrix of second derivatives of the function f:

$$\mathbf{G} = \nabla_{\mathbf{x}}\mathbf{g}(\mathbf{x})$$

$$G_{\alpha\beta} = \frac{\partial g_\alpha}{\partial x_\beta} = \frac{\partial^2 f}{\partial x_\alpha \partial x_\beta}.$$

Note that if $\mathbf{G}$ were constant, then f would be quadratic

$$f(\mathbf{x}) = f(\boldsymbol{\xi}) + \frac{1}{2}(\mathbf{x} - \boldsymbol{\xi})^{\mathrm{T}}\mathbf{G}(\mathbf{x} - \boldsymbol{\xi}), \tag{7.2}$$

and $\mathbf{G}$ will be symmetric positive definite if f has a unique minimum. This is the Taylor series of $f(\mathbf{x})$, the linear term $\mathbf{g}^{\mathrm{T}}(\boldsymbol{\xi})(\mathbf{x}-\boldsymbol{\xi})$ being zero because $\mathbf{g}(\boldsymbol{\xi}) = 0$ at the minimum. The gradient of f is

$$\mathbf{g}(\mathbf{x}) = \mathbf{G}(\mathbf{x} - \boldsymbol{\xi}),$$

and

$$\mathbf{x} - \boldsymbol{\xi} = \mathbf{G}^{-1}\mathbf{g}(\mathbf{x}). \tag{7.3}$$

Simplifying (7.2) with (7.3),

$$f(\mathbf{x}) = f(\boldsymbol{\xi}) + \frac{1}{2}\mathbf{g}(\mathbf{x})^{\mathrm{T}}\mathbf{G}^{-1}\mathbf{g}(\mathbf{x}). \tag{7.4}$$

The significance of this equation is that $\mathbf{G}^{-1}$ is a metric tensor: a sort of $n \times n$ matrix that endows the $\mathbb{R}^n$ space of $\mathbf{g}$ with a measure of distance, in units of f. The method to be developed in this section evolves an estimate $\mathbf{H}$ of the inverse Hessian, hence the name "variable metric method."

There are two difficulties with the Newton–Raphson approach. First, it is not guaranteed to converge for any starting value. The modified Newton method should be used, and even then a bad starting guess (i.e., a starting guess where Φ is not a contractive mapping) will fail to converge. Second, it may be impractical to compute $\mathbf{G}$. Finite differences can be used to approximate $\mathbf{G}$, but this requires $\mathcal{O}(n^2)$ evaluations of $f(\mathbf{x})$, and it degrades the performance of the method.

Variable metric methods are designed to approximate $\mathbf{G}^{-1}$ thereby eliminating one difficulty of the quasi-Newton approach. With $\mathbf{H}$ being the approximation to $\mathbf{G}^{-1}$, the algorithm as described by W. C. Davidon [51] is

Ready: Use an estimate of $\mathbf{G}^{-1}$ to determine a search direction $\mathbf{p}^{(k)}$,

$$\mathbf{p}^{(k)} = -\mathbf{H}^{(k)}\mathbf{g}^{(k)} \tag{7.5}$$

and bracket an interval in this direction such that

$$f(\mathbf{x}^{(k)} + \lambda\mathbf{p}^{(k)}) < f(\mathbf{x}^{(k)})$$

has a minimum for some λ in the interval (λ_-, λ_+). Note that using (7.4) we

Aim:

have $\delta f = -\mathbf{g}^{(k)\mathrm{T}}\mathbf{p}^{(k)}/2$ as an estimate of the improvement in f to be had. If δf is small enough, consider the method converged and stop.

Aim: Find the value of $\lambda \in (\lambda_-, \lambda_+)$ that gives the smallest value of $f(\mathbf{x}^{(k+1)} = \mathbf{x}^{(k)} + \lambda\mathbf{p}^{(k)})$.

Fire: Evaluate $f(\mathbf{x}^{(k+1)})$ and the gradient $\mathbf{g}(\mathbf{x}^{(k+1)})$.

Davidon uses cubic interpolation for *aim*, and he uses $f(\mathbf{x}^{(k+1)})$ and $\mathbf{g}(\mathbf{x}^{(k+1)})$ to assess whether the assumptions underlying the use of cubic interpolation were justified. If not, $f(\mathbf{x}^{(k+1)}), \mathbf{g}(\mathbf{x}^{(k+1)})$ together with one of the bracketing points from *ready* can be used to correct the interpolation result.

Dress: Use the search information, $\mathbf{p}^{(k)}$ and λ, to improve the estimate of $\mathbf{H}$.

Stuff: Assess how well the function has been minimized, and assess the accuracy of the estimate of $\mathbf{H}$.

Davidon tests whether $\mathbf{H} \approx \mathbf{G}(\xi)^{-1}$ by employing property (7.4) to see if a random perturbation $\delta\mathbf{x}$ changes the function by an expected amount. Note that while (7.4) is exact for a quadratic function, is it also accurate to $\mathcal{O}|\mathbf{x} - \xi|^3$ for the general function.

Note that if $\mathbf{H}$ were the inverse Hessian ($\mathbf{H} = \mathbf{G}^{-1}$), then the search direction (7.5) would be equal to the Newton–Raphson search direction and, if $\mathbf{H}$ were the identity matrix $\mathbf{I}$, then this would be the method of steepest descents. Note also that the bracketing (λ_-, λ_+) and subsequent line search are essentially the modified Newton method embellishment.

Davidon's breakthrough is the updating of the estimate $\mathbf{H}$ so that the method begins with steepest descents, $\mathbf{H}^{(0)} = \mathbf{I}$, and progresses to become the inverse Hessian, $\lim_{k\to\infty} \mathbf{H}^{(k)} = \mathbf{G}^{-1}$, without ever having to evaluate explicitly the second derivatives of $f(\mathbf{x})$. The idea is to require

$$\mathbf{H}^{(k+1)}(\mathbf{g}^{(k+1)} - \mathbf{g}^{(k)}) = (\mathbf{x}^{(k+1)} - \mathbf{x}^{(k)}), \tag{7.6}$$

with $\mathbf{H}$ carrying forward as much information from previous iterations as possible. This is accomplished with the algorithm, simplified by Fletcher and Powell [66]

$$\sigma^{(k)} = \mathbf{x}^{(k+1)} - \mathbf{x}^{(k)}$$
$$\mathbf{y}^{(k)} = \mathbf{g}^{(k+1)} - \mathbf{g}^{(k)}$$
$$\mathbf{H}^{(k+1)} = \mathbf{H}^{(k)} + \frac{\sigma^{(k)}\sigma^{(k)\mathrm{T}}}{\sigma^{(k)\mathrm{T}}\mathbf{y}^{(k)}} - \frac{\mathbf{H}^{(k)}\mathbf{y}^{(k)}\mathbf{y}^{(k)\mathrm{T}}\mathbf{H}^{(k)}}{\mathbf{y}^{(k)\mathrm{T}}\mathbf{H}^{(k)}\mathbf{y}^{(k)}}. \tag{7.7}$$

That (7.7) satisfies (7.6) is easily shown by multiplying (7.7) by $\mathbf{y}^{(k)}$ on the right. Note that $\mathbf{G}$ is symmetric, since the order of differentiation does not matter for an analytic function. With $\mathbf{H}^{(0)} = \mathbf{I}$ being symmetric, the update (7.7) is symmetric.

The updating formula (7.7) is not unique, for example the rank-1 update

$$\mathbf{H}^{(k+1)} = \mathbf{H}^{(k)} + \frac{(\sigma^{(k)} - \mathbf{H}^{(k)}\mathbf{y}^{(k)})(\sigma^{(k)} - \mathbf{H}^{(k)}\mathbf{y}^{(k)})^{\mathrm{T}}}{\mathbf{y}^{(k)\mathrm{T}}(\sigma^{(k)} - \mathbf{H}^{(k)}\mathbf{y}^{(k)})}$$

satisfies (7.6), but is impractical for computation since it is unbounded and is not positive definite. This lack of uniqueness provides an opportunity to find an update with optimal behavior, and a number of researchers were engaged in this search in the late 1960s and early '70s. A family of successful methods emerged, which may be summarized as follows [173]:

$$\mathbf{H}^{(k+1)} = \gamma^{(k)} \mathbf{H}^{(k)} + \left(1 + \gamma^{(k)} \beta^{(k)} \frac{\mathbf{y}^{(k)\mathrm{T}} \mathbf{H}^{(k)} \mathbf{y}^{(k)}}{\sigma^{(k)\mathrm{T}} \mathbf{y}^{(k)}} \right) \frac{\sigma^{(k)} \sigma^{(k)\mathrm{T}}}{\sigma^{(k)\mathrm{T}} \mathbf{y}^{(k)}} \tag{7.8}$$

$$- \gamma^{(k)} (1 - \beta^{(k)}) \frac{\mathbf{H}^{(k)} \mathbf{y}^{(k)} \mathbf{y}^{(k)\mathrm{T}} \mathbf{H}^{(k)}}{\mathbf{y}^{(k)\mathrm{T}} \mathbf{H}^{(k)} \mathbf{y}^{(k)}}$$

$$- \gamma^{(k)} \beta^{(k)} \frac{(\sigma^{(k)} \mathbf{y}^{(k)\mathrm{T}} \mathbf{H}^{(k)} + \mathbf{H}^{(k)} \mathbf{y}^{(k)} \sigma^{(k)\mathrm{T}})}{\sigma^{(k)\mathrm{T}} \mathbf{y}^{(k)}}.$$

Special cases of this update are:

1. $\gamma = 1, \beta = 0$: The approach of Davidon and of Fletcher and Powell [66], a.k.a. DFP.
2. $\gamma = 1, \beta = 1$: The BFGS method of Broyden [28], Fletcher [65], Goldfarb [87], and Shanno [215].

This update formula has some nice properties, at least for quadratic functions

$$f(\mathbf{x}) = \frac{1}{2} \mathbf{x}^{\mathrm{T}} \mathbf{G} \mathbf{x} + \mathbf{b}^{\mathrm{T}} \mathbf{x} + c, \tag{7.9}$$

with symmetric positive definite $\mathbf{G}$, and with exact line searches. Fletcher and Powell [66] show that assuming these properties the variable metric method is a conjugate gradient method.

Using model (7.9),

$$\mathbf{g} = \mathbf{G} \mathbf{x} + \mathbf{b}, \tag{7.10}$$

and (7.10) gives

$$\mathbf{y}^{(k)} = \mathbf{G} \sigma^{(k)}. \tag{7.11}$$

First, with exact line searches

$$0 = \mathbf{g}^{(k+1)} \cdot \sigma^{(k)} \propto \mathbf{g}^{(k+1)} \cdot \mathbf{p}^{(k)} \propto \mathbf{g}^{(k+1)} \mathbf{H}^{(k)} \mathbf{g}^{(k)}. \tag{7.12}$$

If this were not the case, then at $\mathbf{x}^{(k+1)}$ there would be a component of the gradient $\mathbf{g}(\mathbf{x}^{(k+1)})$ in the search direction $\mathbf{p}^{(k)}$, so $f(\mathbf{x}^{(k+1)})$ cannot be a minimum on that line.

Second, condition (7.6) with (7.11) gives

$$\mathbf{H}^{(k)} \mathbf{G} \sigma^{(k-1)} = \sigma^{(k-1)}.$$

The Fletcher and Powell result is a proof by induction. One has that

$$\sigma^{(i)\mathrm{T}} \mathbf{G} \sigma^{(j)} = 0 \qquad\qquad 0 \le i < j < k \tag{7.13}$$

$$\mathbf{H}^{(k)} \mathbf{G} \sigma^{(i)} = \sigma^{(i)} \qquad\qquad 0 \le i < k \tag{7.14}$$

is true for $k = 1$, and it will be shown that these are true for $k + 1$.

$$\mathbf{g}^{(k)} = \mathbf{b} + \mathbf{G}\mathbf{x}^{(k)}$$

$$\mathbf{g}^{(k)} = \mathbf{b} + \mathbf{G}\mathbf{x}^{(k-1)} + \mathbf{G}\sigma^{(k-1)}$$

$$= \mathbf{b} + \mathbf{G}\mathbf{x}^{(i+1)} + \mathbf{G}\sigma^{(i+1)} + \cdots + \mathbf{G}\sigma^{(k-1)}$$

$$= \mathbf{g}^{(i+1)} + \mathbf{G}\sigma^{(i+1)} + \cdots + \mathbf{G}\sigma^{(k-1)}.$$

Then, using (7.13) and (7.12),

$$\sigma^{(i)\mathrm{T}}\mathbf{g}^{(k)} = \underbrace{\sigma^{(i)\mathrm{T}}\mathbf{g}^{(i+1)}}_{0 \text{ by (7.12)}} + \sum_{j=i+1}^{k-1} \underbrace{\sigma^{(i)\mathrm{T}}\mathbf{G}\sigma^{(j)}}_{0 \text{ by (7.13)}}.$$

And, combining this result with (7.14) gives

$$0 = \sigma^{(i)\mathrm{T}}\mathbf{G}\mathbf{H}^{(k)}\mathbf{g}^{(k)} = -\sigma^{(i)}\mathbf{G}\mathbf{p}^{(k)} \propto \sigma^{(i)\mathrm{T}}\mathbf{G}\sigma^{(k)},$$

for $i < k$, using (7.5) and $\sigma^{(k)} = \lambda\mathbf{p}^{(k)}$. This establishes that search directions $\mathbf{p} \propto \sigma$ are $\mathbf{G}$-conjugate, and verifies by induction (7.13).

Now, (7.11), together with (7.14) and the conjugacy result, gives

$$\mathbf{y}^{(k)\mathrm{T}}\mathbf{H}^{(k)}\mathbf{G}\sigma^{(i)} = \mathbf{y}^{(k)\mathrm{T}}\sigma^{(i)} = \sigma^{(k)\mathrm{T}}\mathbf{G}\sigma^{(i)} = 0, \qquad (7.15)$$

for $i < k$. Multiplication of (7.8) from the right by $\mathbf{G}\sigma^{(i)}$, and simplification with (7.15) and (7.14) then gives

$$\mathbf{H}^{(k+1)}\mathbf{G}\sigma^{(i)} = \mathbf{H}^{(k)}\mathbf{G}\sigma^{(i)} = \sigma^{(i)},$$

for $i < k$, which completes the proof by induction.

As a conjugate gradient method, the variable metric method (with exact line searches, a quadratic function, and exact math) will locate the minimum in at most n iterations. The conjugacy result (7.13) verifies that all n search directions $\sigma^{(i)}$ are independent, and result (7.14) says that matrix $\mathbf{H}^{(n)}\mathbf{G}$ has n independent eigenvectors with eigenvalue 1. This requires that $\mathbf{H}^{(n)}\mathbf{G} = \mathbf{I}$: the approximate inverse Hessian evolves to be the exact inverse Hessian after n steps.

Broyden [28] shows that when $\gamma = 1$ and $\mathbf{H}^{(0)}$ is symmetric positive definite, $\mathbf{H}^{(k)}$ will be symmetric positive definite for all $k > 0$ if $\beta > 0$. Oren and Luenberger [173] extend this result for $\gamma \neq 1$. The significance of the symmetric positive definite property is that it guarantees that the search direction $\mathbf{p}$ will always point in the direction that minimizes the function. The significance of the parameter γ is that it effectively rescales the problem, and this property may be used to improve the condition number, and thereby improve the convergence properties of the algorithm. Oren [172] and Oren and Spedicato [174] propose a number of rules for selecting β and γ at each step. Perhaps the simplest of these is

$$\gamma, \beta = \begin{cases} \epsilon/s, 0 & \text{if } \epsilon/s \leq 1 \\ s/\tau, 1 & \text{if } s/\tau \geq 1 \\ 1, s(\epsilon - s)/(\epsilon\tau - s^2) & \text{if } s/\tau \leq 1 \leq \epsilon/s, \end{cases}$$

where

$$s = \mathbf{y}^{(k)\mathrm{T}} \boldsymbol{\sigma}^{(k)}$$

$$\tau = \mathbf{y}^{(k)\mathrm{T}} \mathbf{H}^{(k)} \mathbf{y}^{(k)}$$

$$\epsilon = \boldsymbol{\sigma}^{(k)\mathrm{T}} (\mathbf{H}^{(k)})^{-1} \boldsymbol{\sigma}^{(k)}.$$

This prescription appears expensive, since one requires a vector $\mathbf{z}$ determined by $\mathbf{Hz} = \boldsymbol{\sigma}$. However, Oren and Spedicato show that the matrix $\mathbf{H}^{-1}$ can be obtained by a simple procedure along with the update of $\mathbf{H}$. Writing (7.8) as

$$\mathbf{H}^{(k+1)} = \Psi(\mathbf{H}^{(k)}, \mathbf{y}^{(k)}, \boldsymbol{\sigma}^{(k)}, \beta^{(k)}, \gamma^{(k)}),$$

then

$$\Psi(\mathbf{H}^{(k)}, \mathbf{y}^{(k)}, \boldsymbol{\sigma}^{(k)}, \beta^{(k)}, \gamma^{(k)})^{-1} = \Psi((\mathbf{H}^{(k)})^{-1}, \boldsymbol{\sigma}^{(k)}, \mathbf{y}^{(k)}, \bar{\beta}^{(k)}, 1/\gamma^{(k)}),$$

where

$$\bar{\beta} = \frac{s^2(1 - \beta)}{s^2(1 - \beta) + \beta \epsilon \tau}.$$

So, with $\mathbf{H}^{(0)} = (\mathbf{H}^{(0)})^{-1} = \mathbf{I}$, the matrix $\mathbf{H}^{-1}$ is easily estimated at each iteration, facilitating use of the γ scaling capability.

Note that when $\gamma \neq 1$ the matrix $\mathbf{H}$ does *not* evolve into $\mathbf{G}^{-1}$ for a quadratic function, in n steps. In fact, n consecutive steps with $\gamma^{(k)} = 1$ are required for this correspondence to occur.

Regarding variable metric methods generally, there are differing opinions regarding the need to restart. For some problems, convergence is slow until $\mathbf{H}$ becomes a good approximation to $\mathbf{G}^{-1}$ in $\mathcal{O}(n)$ iterations, and then convergence is rapid. This argues that restarting should not occur unless absolutely necessary. Other problems seem to fare best when the method is restarted periodically, as with the conjugate gradient method.

Example 7.1   Starting with $\mathbf{x}^{(0)} = (-1.2, 1.0)^{\mathrm{T}}$, find the minimum of "Rosenbrock's banana" [201]

$$f(\mathbf{x}) = 100(x_2 - x_1^2)^2 + (1 - x_1)^2$$

with the BFGS method (note $\boldsymbol{\xi} = (1, 1)^{\mathrm{T}}$).

Solution
There is a great deal of flexibility in the details of a BFGS implementation. In this example bracketing with rather tight interval refinement is employed in order to emulate an exact line search. This is not the most efficient choice – the number of VMM iterations may be reduced, but at the expense of a number of function evaluations in the line search.

The sequence of search points so obtained is shown in Figure 7.2. The method immediately locates the valley of the banana, then follows it to the minimum. For a method that initially contains no information of the Hessian, the method is remarkably effective.

On closer observation, it can be seen that some search directions appear poorly chosen. Qualitatively, this behavior is a consequence of $\mathbf{G}^{-1}(\mathbf{x})$ changing along the path, with $\mathbf{H}$ not quite keeping up – it is always $\mathcal{O}(n)$ steps out of date.

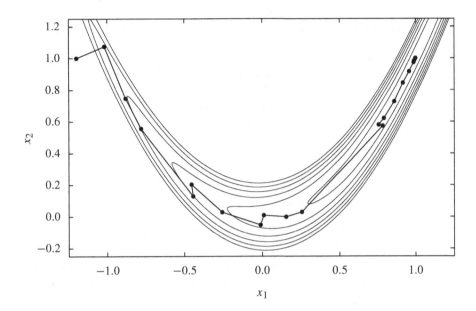

Figure 7.2 Rosenbrock's banana and the search path of the BFGS method; Example 7.1.

For the problem of Example 7.1, and with the particular line search implementation used, the DFP method takes 50% more iterations than does BFGS (32 versus 21), and Oren–Spedicato is somewhere in between (25).

7.5   Linear programming

Linear programming is an important application of solvers for $\mathbf{Ax} = \mathbf{b}$ to systems of linear equations with constraints. The basic idea will be described by way of an example, then a more formal statement of the algorithm will be presented. Linear programming is very important for financially oriented problems, and there is a vast literature with specialized language. It is frequently said that the simplex method was created by G. B. Danzig in 1947 [49], although L. V. Kantorovich appears to have discovered it a few years earlier [125]. It was estimated at one time that as much as 25% of all scientific computation was devoted to this method [108]. It is hard to overstate how important this technique is to engineering and economics.

Example 7.2   Maximize your likely profit by investing $10 000. You have three investment choices: (1) a municipal bond that returns 6% per year, (2) a bank certificate of deposit (CD) that returns 4% per year, (3) and a stock fund that has an expected appreciation of 8% per year. To minimize the risk of the stock, you will not invest more than $2000 that way. To satisfy your sense of social responsibility you wish your municipal fund investment to be at least four times the size of your bank CD investment.

Solution

As a system of inequalities we have:

$$\text{maximize} \quad 0.06x_1 + 0.04x_2 + 0.08x_3$$
$$x_1 + x_2 + x_3 = 10\,000$$
$$x_3 \leq 2000$$
$$x_1 \geq 4x_2,$$

and you cannot have a negative investment:

$$x_1, x_2, x_3 \geq 0.$$

The first rearrangement of these equations will be to introduce a variable for the objective function (the profit to be maximized), and *slack* variables that turn the inequality constraints into equalities.

$$\text{maximize} \quad x_4$$
$$0.06x_1 + 0.04x_2 + 0.08x_3 - x_4 = 0$$
$$x_1 + x_2 + x_3 = 10\,000$$
$$x_3 + x_5 = 2000$$
$$-x_1 + 4x_2 + x_6 = 0$$
$$x_1, x_2, x_3, x_5, x_6 \geq 0.$$

This formulation of the problem is called the *standard form*.

Variables x_1, x_2, x_3, x_5, and x_6 are *restricted* variables: they must be positive. We write $I = \{1, 2, 3, 5, 6\}$ for the set of restricted variables. In total there are six variables and four equalities to be satisfied. We will partition the six variables into two lists: list J will consist of the four *basic* variables – those to be determined by the linear equations, and list K will consist of the two *nonbasic* variables – those whose value will be taken as zero.

One could pick any four variables for membership in J, but not all choices make sense. If a variable is in J and in I then its computed value must be positive. A choice J that satisfies this condition for all members is called a *feasible basis*.

To start the calculation one must have a feasible basis. There is a systematic way to construct these, but for now let us assume $J^{(0)} = \{1, 2, 4, 5\}$ and $K^{(0)} = \{3, 6\}$.

The linear equations can be written in the following form:

$$
\underbrace{\begin{pmatrix} 0.06 & 0.04 & -1 & 0 \\ 1 & 1 & 0 & 0 \\ 0 & 0 & 0 & 1 \\ -1 & 4 & 0 & 0 \end{pmatrix}}_{\mathbf{A}_J} \underbrace{\begin{pmatrix} x_1 \\ x_2 \\ x_4 \\ x_5 \end{pmatrix}}_{\mathbf{x}_J} + \underbrace{\begin{pmatrix} 0.08 & 0 \\ 1 & 0 \\ 1 & 0 \\ 0 & 1 \end{pmatrix}}_{\mathbf{A}_K} \underbrace{\begin{pmatrix} x_3 \\ x_6 \end{pmatrix}}_{\mathbf{x}_K} = \underbrace{\begin{pmatrix} 0 \\ 10\,000 \\ 2000 \\ 0 \end{pmatrix}}_{\mathbf{b}}
$$

$$
\begin{pmatrix} x_1 \\ x_2 \\ x_4 \\ x_5 \end{pmatrix} = \underbrace{\begin{pmatrix} 8000 \\ 2000 \\ 560 \\ 2000 \end{pmatrix}}_{\mathbf{A}_J^{-1}\mathbf{b}} + \underbrace{\begin{pmatrix} -0.8 & 0.2 \\ -0.2 & -0.2 \\ 0.024 & 0.004 \\ -1 & 0 \end{pmatrix}}_{-\mathbf{A}_J^{-1}\mathbf{A}_K} \begin{pmatrix} x_3 \\ x_6 \end{pmatrix}.
$$

(7.16)

The interpretation of this is that a profit of \$560 per year will result from an investment of \$8000 in bonds and \$2000 in CDs. From the coefficients of the matrix in the second line, it is evident that increasing nonbasic variable x_3 while holding constant x_6 will improve profits x_4 (because coefficient 0.024 is positive), and also increasing nonbasic variable x_6 while holding constant x_3 will improve profits x_4 (because coefficient 0.004 is positive). The larger coefficient is indicative of a faster payoff, so let's consider the consequence of increasing x_3.

One cannot increase x_3 arbitrarily because increasing x_3 will decrease restricted variables x_1, x_2, and x_5:

$$
\begin{aligned}
x_1 &= 8000 - 0.8x_3 & \Longrightarrow & & x_3 &\le 10\,000 \\
x_2 &= 2000 - 0.2x_3 & \Longrightarrow & & x_3 &\le 10\,000 \\
x_5 &= 2000 - x_3 & \Longrightarrow & & x_3 &\le 2000.
\end{aligned}
$$

When x_3 is increased to 2000, x_5 will decrease to zero. We pick the first variable to reach zero, and exchange it for x_3 on the nonbasic list. Therefore we obtain a new nonbasic set $K^{(1)} = \{5, 6\}$ with the new basic variable set $J^{(1)} = \{1, 2, 4, 3\}$:

$$
\begin{pmatrix} 0.06 & 0.04 & -1 & 0.08 \\ 1 & 1 & 0 & 1 \\ 0 & 0 & 0 & 1 \\ -1 & 4 & 0 & 0 \end{pmatrix} \begin{pmatrix} x_1 \\ x_2 \\ x_4 \\ x_3 \end{pmatrix} + \begin{pmatrix} 0 & 0 \\ 0 & 0 \\ 1 & 0 \\ 0 & 1 \end{pmatrix} \begin{pmatrix} x_5 \\ x_6 \end{pmatrix} = \begin{pmatrix} 0 \\ 10\,000 \\ 2000 \\ 0 \end{pmatrix}
$$

$$
\begin{pmatrix} x_1 \\ x_2 \\ x_4 \\ x_3 \end{pmatrix} = \begin{pmatrix} 6400 \\ 1600 \\ 608 \\ 2000 \end{pmatrix} + \begin{pmatrix} 0.8 & 0.2 \\ 0.2 & -0.2 \\ -0.024 & 0.004 \\ -1 & 0 \end{pmatrix} \begin{pmatrix} x_5 \\ x_6 \end{pmatrix}.
$$

(7.17)

Increasing x_5 will make the profit lower (because -0.024 is negative), but increasing x_6 will improve the profit (because 0.004 is positive). This means we will make x_6 a basic variable in the next iteration. Clearly x_2 will be the new nonbasic variable since it is the only restricted variable with a corresponding negative coefficient (-0.2) in the matrix $(-\mathbf{A}_J^{-1}\mathbf{A}_K)$. The next step proceeds with $J^{(2)} = \{1, 6, 4, 3\}$ and $K^{(2)} = \{5, 2\}$:

$$
\begin{pmatrix} 0.06 & 0 & -1 & 0.08 \\ 1 & 0 & 0 & 1 \\ 0 & 0 & 0 & 1 \\ -1 & 1 & 0 & 0 \end{pmatrix} \begin{pmatrix} x_1 \\ x_6 \\ x_4 \\ x_3 \end{pmatrix} + \begin{pmatrix} 0 & 0.04 \\ 0 & 1 \\ 1 & 0 \\ 0 & 4 \end{pmatrix} \begin{pmatrix} x_5 \\ x_2 \end{pmatrix} = \begin{pmatrix} 0 \\ 10000 \\ 2000 \\ 0 \end{pmatrix}
$$

$$
\begin{pmatrix} x_1 \\ x_6 \\ x_4 \\ x_3 \end{pmatrix} = \begin{pmatrix} 8000 \\ 8000 \\ 640 \\ 2000 \end{pmatrix} + \begin{pmatrix} 1 & -1 \\ 1 & -5 \\ -0.02 & -0.02 \\ -1 & 0 \end{pmatrix} \begin{pmatrix} x_5 \\ x_2 \end{pmatrix}.
$$

Any increase in restricted variable x_5 will reduce the profit x_4. Likewise, increasing x_2 will reduce the profit. We have achieved the final solution: \$640 profit with \$8000 in bonds, \$0 in CDs (because x_2 is a nonbasic variable), and \$2000 in stocks.

For this simple problem a graphical solution can be developed (Figure 7.3). The inequality constraints define a convex polygon where feasible solutions exist (a *simplex*). Since the objective function x_4 is linear, the maximum value of the objective function must lie at a vertex of the polygon. Our algorithm began at vertex (a), and moved to neighboring vertex (b) that increased the objective function. The algorithm terminated at vertex (c) when there was no improvement possible.

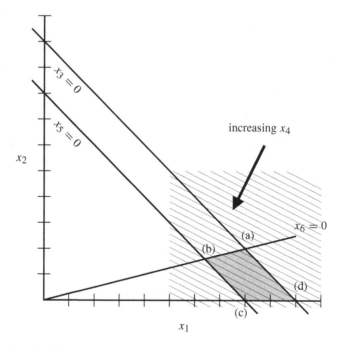

Figure 7.3 Graph of linear programming problem; Example 7.2. Vertex (a) corresponds to $J = \{1, 2, 4, 5\}$, $K = \{3, 6\}$; (b) to $J = \{1, 2, 3, 4\}$, $K = \{5, 6\}$; (c) to $J = \{1, 3, 4, 6\}$, $K = \{2, 5\}$; and (d) to $J = \{1, 4, 5, 6\}$, $K = \{2, 3\}$.

Not every linear programming problem has a unique solution. If the contours of constant x_4 were parallel to a bounding surface of the feasible region, then the circumstance would occur where more than one solution is possible with the same value of the objective solution. It is also possible to compose a problem that has inconsistent constraints. In this case there is no feasible region, and no solution exists. Finally, it could occur that the inequality constraints do not define a convex surface that bound the optimum. This is another failure mode that should be checked.

The algorithm used here is called "phase II of the simplex method." Phase I concerns the determination of the first feasible basis. The algorithm for phase II can be stated as follows. Given a feasible basis J, a nonbasic set K, and a partitioned matrix $\mathbf{A} = \mathbf{A}_J \cup \mathbf{A}_K$, one step of phase II of the simplex method is:

1. Solve $\mathbf{A}_J \mathbf{x}_J = \mathbf{b}$ for $\mathbf{x}_J$ – the feasible solution corresponding to basis J.
2. If the pth element of J is the variable to be maximized, determine $\mathbf{r} = -\mathbf{e}_p^T \mathbf{A}_J^{-1} \mathbf{A}_K$. This is the row of $-\mathbf{A}_J^{-1} \mathbf{A}_K$ that corresponds to the objective function.
3. Choose an index k such that

$$r_k > 0 \quad \text{and} \quad K_k \in I$$

or

$$r_k \neq 0 \quad \text{and} \quad K_k \notin I,$$

and define $\sigma = \text{sign}(r_k)$.

If there is no such index k then the feasible solution $\mathbf{x}_J$ is as good as it's going to get. Terminate the algorithm.

4. If $\sigma \mathbf{e}_j^T \mathbf{A}_J^{-1} \mathbf{A}_k \mathbf{e}_k \leq 0$ for all $J_j \in I$ then there is no finite optimum: the feasible region is not bounded in the direction of the optimum. Terminate.
5. Find the index j such that $J_j \in I$, and such that

$$\frac{\mathbf{e}_j^T \mathbf{x}_J}{\sigma \mathbf{e}_j^T \mathbf{A}_J^{-1} \mathbf{A}_K \mathbf{e}_k} > 0$$

is a minimum. This variable J_j is then the new nonbasic variable:

$$J' = (J \setminus J_j) \cup K_k$$

and

$$K' = (K \setminus K_k) \cup J_j$$

define the new basic and nonbasic variables.

Phase I of the simplex method consists of constructing a specially engineered problem. We augment the given problem with a number of *artificial variables*, and with an artificial objective function. This new problem is constructed so that (i) a feasible basis is trivial to construct, and (ii) when the artificial objective is optimized, all artificial variables will ideally be nonbasic. Then, once the solution to this artificial problem is discovered, its optimal basis will be a feasible basis of the problem we wanted to solve in the first place.

Let us assume a general problem

$$\text{maximize} \quad c_1 x_1 + c_2 x_2 + \cdots + c_n x_n$$
$$a_{j1} x_1 + \cdots + a_{jn} x_n = b_j \geq 0, \quad j = 1, ..., m. \tag{7.18}$$

This is a problem in the standard form, with the additional condition that all constants b_j be positive, which only requires that one multiply the equations by -1 as necessary.

We augment this system by addition of artificial variables $x_{n+1}, ..., x_{n+m}$ – one new variable per equation. And, we introduce a new objective function:

$$\text{maximize} \quad x_{n+m+1}$$

$$x_{n+1} + \cdots + x_{n+m+1} = 2 \sum_{j=1}^{m} b_j$$

$$a_{j1} x_1 + \cdots + a_{jn} x_n + x_{n+j} = b_j \geq 0, \quad j = 1, ..., m.$$

The artificial variables $x_{n+1}, ..., x_{n+m}$ are restricted. For this modified system, a feasible basis consists of $J = \{x_{n+1}, ..., x_{n+m}, x_{n+m+1}\}$.

The next step is to perform phase II operations on this system. There are three possible outcomes:

1. An optimal solution is found with $x_{n+m+1} = 2 \sum b_j$ and each of the artificial variables $x_{n+1}, ..., x_{n+m}$ are nonbasic.

   If this occurs, then the basic variables are x_{n+m+1}, and m variables from the original problem (7.18). Phase II may proceed with these m basic variables, together with the appropriate objective function variable.

2. It can occur that an optimal solution is found with $x_{n+m+1} < 2 \sum b_j$, and this must imply that the original problem was not feasible.

3. Finally, it is possible that the objective is optimized with $x_{n+m+1} = 2 \sum b_j$, but some of the artificial variables are basic although having basic solutions zero (see Problem 7.2).

In the latter case, one "lives with" the basic artificial variables, but takes measures to ensure that they remain zero. To simplify the presentation, assume that artificial variables $x_{n+1}, ..., x_{n+\ell}$ are basic. The system we will now examine is this:

$$\text{maximize} \quad x_{n+\ell+2}$$
$$c_1 x_1 + \cdots + c_n x_n - x_{n+\ell+2} = 0$$
$$x_{n+1} + \cdots + x_{n+\ell} + x_{n+\ell+1} = 0 \tag{7.19}$$
$$a_{j1} x_1 + \cdots + a_{jn} x_n + x_{n+j} = b_j, \quad j = 1, ..., \ell$$
$$a_{k1} x_1 + \cdots + a_{kn} x_n = b_k, \quad j = \ell + 1, ..., m,$$

where the new artificial variable $x_{n+\ell+1}$ is restricted. With all variables in (7.19) restricted, these artificial variables are constrained to remain zero. The phase II initial feasible basis will consist of the phase I feasible basis, omitting the phase I objective

(x_{n+m+1}), and adding the artificial variable $(x_{n+\ell+1})$ and the new objective variable $(x_{n+\ell+2})$. The purpose of variable $x_{n+\ell+1}$ is to provide an extra feasible basis member, as required by the addition of the the new constraint equation $x_{n+1}+\cdots x_{n+\ell}+x_{n+\ell+1}=0$.

Example 7.3 Perform phase I of the simplex method for the previous example.

Solution
The system to investigate will be

$$\text{maximize} \quad x_{10}$$
$$x_7 + x_8 + x_9 + x_{10} = 24\,000$$
$$x_1 + x_2 + x_3 + x_7 = 10\,000$$
$$x_3 + x_5 + x_8 = 2000$$
$$-x_1 + 4x_2 + x_6 + x_9 = 0$$
$$x_1, x_2, x_3, x_5, x_6, x_7, x_8, x_9 \geq 0,$$

with feasible basis $J^{(0)} = \{7, 8, 9, 10\}$ and nonbasic variables $K^{(0)} = \{1, 2, 3, 5, 6\}$. Now perform the operations of phase II:

$$\begin{pmatrix} 1 & 1 & 1 & 1 \\ 1 & 0 & 0 & 0 \\ 0 & 1 & 0 & 0 \\ 0 & 0 & 1 & 0 \end{pmatrix}\begin{pmatrix} x_7 \\ x_8 \\ x_9 \\ x_{10} \end{pmatrix} + \begin{pmatrix} 0 & 0 & 0 & 0 & 0 \\ 1 & 1 & 1 & 0 & 0 \\ 0 & 0 & 1 & 1 & 0 \\ -1 & 4 & 0 & 0 & 1 \end{pmatrix}\begin{pmatrix} x_1 \\ x_2 \\ x_3 \\ x_5 \\ x_6 \end{pmatrix} = \begin{pmatrix} 24\,000 \\ 10\,000 \\ 2000 \\ 0 \end{pmatrix}$$

$$\begin{pmatrix} x_7 \\ x_8 \\ x_9 \\ x_{10} \end{pmatrix} = \begin{pmatrix} 10\,000 \\ 2000 \\ 0 \\ 12\,000 \end{pmatrix} + \begin{pmatrix} -1 & -1 & -1 & 0 & 0 \\ 0 & 0 & -1 & -1 & 0 \\ 1 & -4 & 0 & 0 & -1 \\ 0 & 5 & 2 & 1 & 1 \end{pmatrix}\begin{pmatrix} x_1 \\ x_2 \\ x_3 \\ x_5 \\ x_6 \end{pmatrix}.$$

Choose x_2 as the variable to go from the K list to the J list. To find the new K variable, analyze

$$x_7 = 10\,000 - x_2 \qquad \Longrightarrow \qquad x_2 \leq 10\,000$$
$$x_8 = 20\,000 + 0x_2 \qquad \Longrightarrow \qquad x_2 \quad \text{not bounded}$$
$$x_9 = 0 - 4x_2 \qquad \Longrightarrow \qquad x_2 \leq 0,$$

so clearly x_9 is the new K variable: $J^{(1)} = \{7, 8, 2, 10\}$ and $K^{(1)} = \{1, 9, 3, 5, 6\}$;

$$\begin{pmatrix} 1 & 1 & 0 & 1 \\ 1 & 0 & 1 & 0 \\ 0 & 1 & 0 & 0 \\ 0 & 0 & 4 & 0 \end{pmatrix}\begin{pmatrix} x_7 \\ x_8 \\ x_2 \\ x_{10} \end{pmatrix} + \begin{pmatrix} 0 & 1 & 0 & 0 & 0 \\ 1 & 0 & 1 & 0 & 0 \\ 0 & 0 & 1 & 1 & 0 \\ -1 & 1 & 0 & 0 & 1 \end{pmatrix}\begin{pmatrix} x_1 \\ x_9 \\ x_3 \\ x_5 \\ x_6 \end{pmatrix} = \begin{pmatrix} 24\,000 \\ 10\,000 \\ 2000 \\ 0 \end{pmatrix}$$

$$\begin{pmatrix} x_7 \\ x_8 \\ x_2 \\ x_{10} \end{pmatrix} = \begin{pmatrix} 10\,000 \\ 2000 \\ 0 \\ 12\,000 \end{pmatrix} + \begin{pmatrix} -5/4 & 1/4 & -1 & 0 & 1/4 \\ 0 & 0 & -1 & -1 & 0 \\ 1/4 & -1/4 & 0 & 0 & -1/4 \\ 5/4 & -5/4 & 2 & 1 & -1/4 \end{pmatrix} \begin{pmatrix} x_1 \\ x_9 \\ x_3 \\ x_5 \\ x_6 \end{pmatrix}.$$

Now pick x_3 as the new J variable:

$$x_7 = 10\,000 - x_3 \qquad \Longrightarrow \qquad x_3 \le 10\,000$$

$$x_8 = 2000 - x_3 \qquad \Longrightarrow \qquad x_3 \le 2000$$

$$x_2 = 0 - 0x_3 \qquad \Longrightarrow \qquad x_3 \quad \text{not bounded,}$$

so x_8 is the new nonbasic variable: $J^{(2)} = \{7, 3, 2, 10\}$ and $K^{(2)} = \{1, 9, 8, 5, 6\}$;

$$\begin{pmatrix} 1 & 0 & 0 & 1 \\ 1 & 1 & 1 & 0 \\ 0 & 1 & 0 & 0 \\ 0 & 0 & 4 & 0 \end{pmatrix} \begin{pmatrix} x_7 \\ x_3 \\ x_2 \\ x_{10} \end{pmatrix} + \begin{pmatrix} 0 & 1 & 1 & 0 & 0 \\ 1 & 0 & 0 & 0 & 0 \\ 0 & 0 & 1 & 1 & 0 \\ -1 & 1 & 0 & 0 & 1 \end{pmatrix} \begin{pmatrix} x_1 \\ x_9 \\ x_8 \\ x_5 \\ x_6 \end{pmatrix} = \begin{pmatrix} 24\,000 \\ 10\,000 \\ 2000 \\ 0 \end{pmatrix}$$

$$\begin{pmatrix} x_7 \\ x_3 \\ x_2 \\ x_{10} \end{pmatrix} = \begin{pmatrix} 8000 \\ 2000 \\ 0 \\ 16\,000 \end{pmatrix} + \begin{pmatrix} -5/4 & 1/4 & 1 & 1 & 1/4 \\ 0 & 0 & -1 & -1 & 0 \\ 1/4 & -1/4 & 0 & 0 & -1/4 \\ 5/4 & -5/4 & -2 & -1 & -1/4 \end{pmatrix} \begin{pmatrix} x_1 \\ x_9 \\ x_8 \\ x_5 \\ x_6 \end{pmatrix}.$$

Now x_1 is the only variable that improves the objective function

$$x_7 = 8000 - \frac{5}{4}x_1 \qquad \Longrightarrow \qquad x_1 \le 6400$$

$$x_3 = 2000 + 0x_1 \qquad \Longrightarrow \qquad x_1 \quad \text{not bounded}$$

$$x_2 = 0 + \frac{1}{4}x_1 \qquad \Longrightarrow \qquad x_1 \quad \text{not bounded,}$$

so x_7 is the new nonbasic variable: $J^{(3)} = \{1, 3, 2, 10\}$ and $K^{(3)} = \{7, 9, 8, 5, 6\}$;

$$\begin{pmatrix} 0 & 0 & 0 & 1 \\ 1 & 1 & 1 & 0 \\ 0 & 1 & 0 & 0 \\ -1 & 0 & 4 & 0 \end{pmatrix} \begin{pmatrix} x_1 \\ x_3 \\ x_2 \\ x_{10} \end{pmatrix} + \begin{pmatrix} 1 & 1 & 1 & 0 & 0 \\ 1 & 0 & 0 & 0 & 0 \\ 0 & 0 & 1 & 1 & 0 \\ 0 & 1 & 0 & 0 & 1 \end{pmatrix} \begin{pmatrix} x_7 \\ x_9 \\ x_8 \\ x_5 \\ x_6 \end{pmatrix} = \begin{pmatrix} 24\,000 \\ 10\,000 \\ 2000 \\ 0 \end{pmatrix}$$

$$\begin{pmatrix} x_1 \\ x_3 \\ x_2 \\ x_{10} \end{pmatrix} = \begin{pmatrix} 6400 \\ 2000 \\ 1600 \\ 24\,000 \end{pmatrix} + \begin{pmatrix} -4/5 & 1/5 & 4/5 & 4/5 & 1/5 \\ 0 & 0 & -1 & -1 & 0 \\ -1/5 & -1/5 & 1/5 & 1/5 & -1/5 \\ -1 & -1 & -1 & 0 & 0 \end{pmatrix} \begin{pmatrix} x_7 \\ x_9 \\ x_8 \\ x_5 \\ x_6 \end{pmatrix}.$$

Artificial objective x_{10} is now optimized, and all other artificial variables are nonbasic. We can create a phase II calculation of the original problem with J consisting of

x_1, x_2, x_3 together with the objective function of the original problem (this is vertex (b) of Figure 7.3).

The simplex method as described above involves the equivalent of one matrix inversion per step (or, more appropriately, one matrix decomposition, QR or LR, followed by several back substitutions). This is what formally happens, but the mechanics can be significantly simplified. To be specific, the inverse of $\mathbf{A}_{J(k)}$ and the inverse of $\mathbf{A}_{J(k+1)}$ are very closely related, since only one column of the matrix $\mathbf{A}_J$ changed during the step. The idea here is to *modify* the matrices, versus reconstruct them through a matrix decomposition.

Begin with the formal solution at step k,

$$\underbrace{\mathbf{A}_J^{-1}\mathbf{A}_J}_{\mathbf{I}}\mathbf{x}_K = \underbrace{\mathbf{A}_J^{-1}\mathbf{b}}_{\mathbf{b}_J} - \mathbf{A}_J^{-1}\mathbf{A}_K\mathbf{x}_K,$$

and label the columns of $\mathbf{A}_J$ and $\mathbf{A}_K$ to identify $\mathbf{a}_r$ as the column corresponding to the basic variable that will become nonbasic, and $\mathbf{a}_s$ as the column associated with the nonbasic variable that will become basic:

$$\mathbf{A}_J^{-1}\begin{pmatrix}\mathbf{a}_1 & \cdots & \mathbf{a}_{r-1} & \mathbf{a}_r & \mathbf{a}_{r+1} & \cdots & \mathbf{a}_n\end{pmatrix}\mathbf{x}_J$$
$$= \mathbf{A}_J^{-1}\mathbf{b} - \mathbf{A}_J^{-1}\begin{pmatrix}\mathbf{a}_{n+1} & \cdots & \mathbf{a}_{s-1} & \mathbf{a}_s & \mathbf{a}_{s+1} & \cdots & \mathbf{a}_m\end{pmatrix}\mathbf{x}_K.$$

Now, symbolically swap $\mathbf{a}_s$ and $\mathbf{a}_r$ to move to the next solution. The following equation holds,

$$\underbrace{\mathbf{A}_J^{-1}\begin{pmatrix}\cdots & \mathbf{a}_{r-1} & \mathbf{a}_s & \mathbf{a}_{r+1} & \cdots\end{pmatrix}}_{\mathbf{M}}\mathbf{x}_J' \qquad (7.20)$$
$$= \mathbf{A}_J^{-1}\mathbf{b} - \mathbf{A}_J^{-1}\begin{pmatrix}\cdots & \mathbf{a}_{s-1} & \mathbf{a}_r & \mathbf{a}_{s+1} & \cdots\end{pmatrix}\mathbf{x}_K',$$

but the matrix $\mathbf{M}$ is not the identity matrix $\mathbf{I}$. Instead, it will be a *Frobenius matrix*

$$\mathbf{M} = \begin{pmatrix} 1 & & & m_1 & & & \\ & \ddots & & \vdots & & & \\ & & 1 & m_{r-1} & & & \\ & & & m_r & & & \\ & & & m_{r+1} & 1 & & \\ & & & \vdots & & \ddots & \\ & & & m_n & & & 1 \end{pmatrix}$$
$$m_i = (\mathbf{A}_J^{-1})_{ij}(\mathbf{a}_r)_j,$$

and this Frobenius matrix has a simple inverse:

$$
\mathbf{M}^{-1} = \begin{pmatrix}
1 & & & & -m_1/m_r & & \\
& \ddots & & & \vdots & & \\
& & 1 & & -m_{r-1}/m_r & & \\
& & & & 1/m_r & & \\
& & & & -m_{r+1}/m_r & 1 & \\
& & & & \vdots & & \ddots \\
& & & & -m_n/m_r & & 1
\end{pmatrix}.
$$

So, to make the matrix on the left-hand side of (7.20) into the identity matrix, we multiply both sides of the equation by $\mathbf{M}^{-1}$:

$$
\mathbf{x}'_J = \underbrace{\mathbf{M}^{-1}\mathbf{b}_J}_{\mathbf{b}'_J} - \underbrace{\mathbf{M}^{-1}\overbrace{\mathbf{A}_J^{-1}}^{\mathbf{A}'_J{}^{-1}}\mathbf{A}'_K}_{\mathbf{A}'_J{}^{-1}\mathbf{A}'_K}\,\mathbf{x}'_K.
$$

The key point here is that $\mathbf{M}$ and $\mathbf{M}^{-1}$ have very simple structures, and these matrices do not actually need to be made – the elements of the matrices can be easily computed, then applied to stored values of vector $\mathbf{b}_J{}^{(k)}$ and matrix $(\mathbf{A}_J^{-1}\mathbf{A}_K)^{(k)}$ to give values for these quantities for iteration $k + 1$. In fact, the values m_i are already known: vector $\mathbf{m}$ is the column of $(\mathbf{A}_J^{-1}\mathbf{A}_K)$ corresponding to variable s. Also, the new r column of matrix $(\mathbf{A}_J^{-1}\mathbf{A}'_K)$ is $\mathbf{e}_r$: the rth column of $(\mathbf{A}_J^{-1}\mathbf{A}_J)$.

The phase I algorithm described here begins with a trivial matrix $\mathbf{A}_J$ – the transpose of an easily inverted Frobenius matrix. So, the entirety of phase I does not require a genuine matrix inversion. Reformulation of the phase I output to a phase II problem can also be accomplished with matrix modification ideas, but with considerably greater complexity. Since this reformulation happens just once, use QR or LR.

A high level summary of the solution method after columns r and s have been chosen is this:

- Make $\mathbf{m}$ by copying the s column of $(\mathbf{A}_J^{-1}\mathbf{A}_K)^{(k)}$.
- Replace the s column of $(\mathbf{A}_J^{-1}\mathbf{A}_K)$ with $\mathbf{e}_r$.
- "Multiply" $\mathbf{M}^{-1}$ with $\mathbf{b}_J$ and $(\mathbf{A}_J^{-1}\mathbf{A}_K)$ without really making matrix $\mathbf{M}^{-1}$; e.g.,

$$
(\mathbf{b}_J)_r := (\mathbf{b}_J)_r/m_r,
$$

then, for all $i \neq r$:

$$
(\mathbf{b}_J)_i := (\mathbf{b}_J)_i - m_i(\mathbf{b}_J)_r.
$$

The "multiplication" of $\mathbf{M}^{-1}$ with $(\mathbf{A}_J^{-1}\mathbf{A}_K)$ follows this algorithm, treating each column analogously with the treatment of vector $\mathbf{b}_J$.

Example 7.4   Use the matrix modification algorithm to show the transition from solution (7.16) to solution (7.17).

Solution

The vector $\mathbf{b}^{(1)}$ and matrix $(-\mathbf{A}_J^{-1}\mathbf{A}_K)^{(1)}$ can be written

$$\left(\mathbf{b}\,\middle|-\mathbf{A}_J^{-1}\mathbf{A}_K\right)^{(1)} = \begin{pmatrix} 8000 & -0.8 & 0.2 \\ 2000 & -0.2 & -0.2 \\ 560 & 0.024 & 0.004 \\ 2000 & -1 & 0 \end{pmatrix}.$$

We are exchanging x_5, the fourth entry of $\mathbf{x}_J$, with x_3, the first entry of $\mathbf{x}_K$. Thus,

$$\mathbf{m} = \begin{pmatrix} 0.8 \\ 0.2 \\ -0.024 \\ 1 \end{pmatrix},$$

and $\mathbf{e}_s = \mathbf{e}_4$. Replace the r column of $(-\mathbf{A}_J^{-1}\mathbf{A}_K)$ with $-\mathbf{e}_4$,

$$\begin{pmatrix} 8000 & 0 & 0.2 \\ 2000 & 0 & -0.2 \\ 560 & 0 & 0.004 \\ 2000 & -1 & 0 \end{pmatrix},$$

divide the fourth row by $m_4 = 1$ (trivial), then subtract m_i times the fourth row from each row $i \neq 4$:

$$\begin{pmatrix} 6400 & 0.8 & 0.2 \\ 1600 & 0.2 & -0.2 \\ 608 & -0.024 & 0.004 \\ 2000 & -1 & 0 \end{pmatrix} = \left(\mathbf{b}\,\middle|-\mathbf{A}_J^{-1}\mathbf{A}_K\right)^{(2)}.$$

7.6   Quadratic programming

The ideas developed in linear programming and variable metric methods can be combined to seek constrained solutions to general nonlinear optimization problems. An elegant solution is available in the special case that the function to be optimized is strictly quadratic and symmetric positive definite. That approach, due to Goldfarb and Idnani [88, 89], will be presented after first explaining some of the underlying theory.

7.6.1   Lagrange multipliers

The method of Lagrange multipliers (named for Lagrange, e.g., [139, Chapters XI–XIII], but first discovered by Euler 50 years previously [60]; see also [74]) is a technique that augments the function to be optimized with artificial variables in such a way that an extrema or inflection point of the augmented function is a solution to a constrained system. The extension of the method of Lagrange multipliers to inequality constraints is due to Karush [127] though often attributed to Kuhn and Tucker [135]. Consider for example the following quadratic function, where we will assume that matrix $\mathbf{G}$ is symmetric positive definite:

$$\text{minimize} \quad f(\mathbf{x}) = \frac{1}{2}\mathbf{x}^{\mathrm{T}}\mathbf{G}\mathbf{x} + \mathbf{b}^{\mathrm{T}}\mathbf{x} \tag{7.21}$$

$$\mathbf{x} \in \mathbb{R}^n.$$

We will seek the minimum of $f(\mathbf{x})$ subject to linear equality and inequality constraints,

$$\mathbf{h} = \mathbf{C}^{\mathrm{T}}\mathbf{x} + \mathbf{c} \tag{7.22}$$

$$\mathbf{h}_i(\mathbf{x}) \geq 0 \text{ if } i \in \bar{\mathbb{I}}$$

$$\mathbf{h}_i(\mathbf{x}) = 0 \text{ if } i \notin \bar{\mathbb{I}}.$$

Here $\mathbf{h}, \mathbf{c} \in \mathbb{R}^m$, $\mathbf{C} \in \mathbb{R}^{n \times m}$, and $\bar{\mathbb{I}}$ is the subset of the indices $1, ..., m$ corresponding to inequality constraints.

The augmented *Lagrangian* function L is

$$L(\mathbf{x}, \boldsymbol{\lambda}) = f(\mathbf{x}) - \boldsymbol{\lambda}^{\mathrm{T}}\mathbf{h}(\mathbf{x}), \tag{7.23}$$

where the augmented variables $\boldsymbol{\lambda}$ are called *Lagrange multipliers*. According to Karush, Kuhn, and Tucker, the solution to (7.21), subject to (7.22), is the saddle point $\{\mathbf{x}^0, \boldsymbol{\lambda}^0\}$ of L (7.23)

$$\min_{\mathbf{x} \in \mathbb{R}^n} L(\mathbf{x}, \boldsymbol{\lambda}) \leq \min_{\mathbf{x} \in \mathbb{R}^n} \max_{\boldsymbol{\lambda}} L(\mathbf{x}, \boldsymbol{\lambda}) \leq \max_{\boldsymbol{\lambda}} L(\mathbf{x}, \boldsymbol{\lambda}). \tag{7.24}$$

The multipliers λ_i are strictly positive if $i \in \bar{\mathbb{I}}$, but unconstrained otherwise.

Why does this work? Let

$$D(\boldsymbol{\lambda}|\mathbf{x}^0) = L(\mathbf{x}^0, \boldsymbol{\lambda}) = f(\mathbf{x}^0) - \boldsymbol{\lambda}^{\mathrm{T}}\mathbf{h}(\mathbf{x}^0)$$

be the *dual function*, a function of the Lagrange multipliers with $\mathbf{x}^0$ a parameter. The maximum with respect to an unconstrained parameter λ_i, $i \notin \bar{\mathbb{I}}$, occurs either (i) when $\mathbf{h} \neq 0$, and $-\lambda_i \mathbf{h} \to \infty$, or (ii) when

$$\frac{\partial}{\partial \lambda_i} D(\boldsymbol{\lambda}|\mathbf{x}^0) = 0 = -\mathbf{h}(\mathbf{x}^0),$$

and the equality constraint is satisfied.

The maximum with respect to a constrained parameter λ_i, $i \in \bar{\mathbb{I}}$, likewise corresponds to the unbounded possibility (i) $\mathbf{h} < 0$, $-\lambda_i \mathbf{h} \to \infty$; or the bounded possibilities (iia)

$$\frac{\partial}{\partial \lambda_i} D(\boldsymbol{\lambda}|\mathbf{x}^0) = 0 = -h_i(\mathbf{x}),$$

or (iib)

$$\lambda_i = 0 \quad \text{if} \quad h_i(\mathbf{x}) > 0.$$

It follows that the maximum of the dual function is unbounded (∞) when the constraints are violated, but is bounded with value $f(\mathbf{x}^0)$ when the constraints are obeyed. The saddle point of (7.24) can be written

$$\min_{\mathbf{x}\in\mathbb{R}^n} \max_{\lambda} L(\mathbf{x}, \lambda) = \min_{\mathbf{x}\in\mathbb{R}^n} \begin{cases} f(\mathbf{x}) & \text{if all constraints obeyed} \\ \infty & \text{if any constraints are violated.} \end{cases}$$

This minimum is achieved only when the constraints are obeyed, in which case the smallest value of $f(\mathbf{x})$ is found that is consistent with the given constraints.

Now consider minimization with respect to $\mathbf{x}$. If the constraints are violated, then L may be unbounded: $-\lambda^T\mathbf{h} \rightarrow -\infty$. But, if the constraints are obeyed then the minimum will be bounded (because $\mathbf{G}$ is symmetric positive definite), with

$$\nabla_\mathbf{x} L(\mathbf{x}, \lambda) = \nabla_\mathbf{x} f(\mathbf{x}) - \mathbf{C}\lambda = 0 \tag{7.25}$$

and

$$\lambda = \mathbf{C}^\dagger \nabla_\mathbf{x} f(\mathbf{x}).$$

A *pseudoinverse* of the constraint matrices $\mathbf{C}$ determines the Lagrange multipliers (a pseudoinverse is the generalization of a matrix inverse to nonsquare matrices, and its relation to Gaussian elimination, Householder, and SVD, will be discussed in Chapter 8).

The saddle point is the intersection of the bounded minimum of the $\mathbf{x}$ problem, and the bounded maximum of the λ problem: the finite solution that minimizes $f(\mathbf{x})$ subject to all applicable constraints.

An algorithm that takes full advantage of this Lagrange function is the dual method of Goldfarb and Idnani [88, 89]. The basic idea is a combination of simplex-like progression through the constraints, with direct solution of linear systems (7.25).

7.6.2 The Goldfarb–Idnani dual algorithm

The Goldfarb–Idnani algorithm is based on a couple of key observations. Here let $\mathbf{n}$ be a particular column of the constraint matrix $\mathbf{C}$, and let $\mathbf{N}$ be a matrix composed of select columns of $\mathbf{C}$. $\mathbf{g}$ will stand for the gradient, nominally $\nabla_\mathbf{x} f(\mathbf{x})$.

I. When $\mathbf{n}_i^T\mathbf{x} > -c_i$, $i \in \mathbb{I}$, the constraint is inactive. The constrained optimum solution with, or without, the particular constraint would be the same. Accordingly, we will at any one time deal with a set of constraints $\mathbb{A}$, the "active set," for which $\mathbf{n}_i^T\mathbf{x} = -c_i$. $\mathbf{N}$ will be the constraint matrix composed of columns of $\mathbf{C}$ from the active set.

II. The least squares problem

$$\mathbf{N}\lambda = \mathbf{g}$$

is not unique. One can write $\lambda = \mathbf{N}^\dagger\mathbf{g}$, but the pseudoinverse $\mathbf{N}^\dagger$ is not unique. For example, $\mathbf{N}^\dagger = (\mathbf{N}^T\mathbf{W}\mathbf{N})^{-1}\mathbf{N}^T\mathbf{W}$ for any invertible $\mathbf{W}$. If

$$\mathbf{N}^\dagger = (\mathbf{N}^T\mathbf{G}^{-1}\mathbf{N})^{-1}\mathbf{N}^T\mathbf{G}^{-1}, \tag{7.26}$$

then $\lambda(\mathbf{x})$ is independent of $\mathbf{x}$:

$$\lambda = \mathbf{N}^\dagger \mathbf{g} \tag{7.27}$$
$$= (\mathbf{N}^T \mathbf{G}^{-1} \mathbf{N})^{-1} \mathbf{N}^T \mathbf{G}^{-1} (\mathbf{G}\mathbf{x} + \mathbf{b})$$
$$= -(\mathbf{N}^T \mathbf{G}^{-1} \mathbf{N})^{-1} \left(\mathbf{c}' + \mathbf{N}^T \mathbf{x}_0 \right),$$

where $\mathbf{x}_0 = -\mathbf{G}^{-1}\mathbf{b}$ is the solution to the unconstrained minimization problem, and $\mathbf{c}'$ is the subset of the vector $\mathbf{c}$ corresponding to the active constraints $\mathbb{A}$.

III. For a given choice of $\mathbf{N}^\dagger$, the effect of active constraints is to modify the Hessian of the system. The direction of steepest descents at some point $\tilde{\mathbf{x}}$ is $-\mathbf{g}(\tilde{\mathbf{x}})$: the minimum of the unconstrained quadratic $f(\mathbf{x})$ is

$$\mathbf{x} = \tilde{\mathbf{x}} - \mathbf{G}^{-1}\mathbf{g}(\tilde{\mathbf{x}}).$$

However, in the presence of constraints the direction $\mathbf{g}$ must be modified to be orthogonal to the constraints $\mathbf{N}$. The direction of steepest descents, made orthogonal to the constraints is

$$\mathbf{g}_\perp(\tilde{\mathbf{x}}) = (\mathbf{I} - \mathbf{N}\mathbf{N}^\dagger)\mathbf{g}(\tilde{\mathbf{x}}), \tag{7.28}$$

in the sense that

$$\mathbf{N}^T\mathbf{x} = \mathbf{N}^T\tilde{\mathbf{x}} - \mathbf{N}^T\mathbf{G}^{-1}(\mathbf{I} - \mathbf{N}\mathbf{N}^\dagger)\mathbf{g}(\tilde{\mathbf{x}}) = \mathbf{N}^T\tilde{\mathbf{x}}$$

so active constraints are unaffected: $\mathbf{h}'(\mathbf{x}) = \mathbf{N}^T\mathbf{x} + \mathbf{c}' = \mathbf{h}'(\tilde{\mathbf{x}})$. (Compare (7.28) to Gram–Schmidt orthogonalization, (3.12).)

We can define the constrained inverse Hessian

$$\mathbf{H} = \mathbf{G}^{-1}(\mathbf{I} - \mathbf{N}\mathbf{N}^\dagger),$$

so that

$$\mathbf{x} = \tilde{\mathbf{x}} - \mathbf{H}\mathbf{g}(\tilde{\mathbf{x}}) \tag{7.29}$$

gives the constrained minimum (assuming the active constraint set $\mathbb{A}$ is unchanged on the line from $\tilde{\mathbf{x}}$ to $\mathbf{x}$).

IV. The solution to the problem defined by constrains $\mathbb{A} \cup \{p\}$ can be generated simply from the solution to the problem defined by $\mathbb{A}$ (here p is the index of a new constraint to be added to the active set $\mathbb{A}$). Thus, beginning with $\mathbb{A} = \emptyset$ one can systematically obtain the solution of the fully constrained problem. Herein lies the chief advantage of the Goldfarb–Idnani method: there is no analog to the simplex phase I. The method may be started with the solution $\mathbf{x}_0$ to the unconstrained problem, with $\mathbb{A} = \emptyset$.

With these fact established, consider how the solution is evolved upon addition of new constraints. Let $\{\mathbf{x}, \lambda, \mathbb{A}\}$ define a solution, and let p be an index such that $p \notin \mathbb{A}$ for which the constraint is violated: $p \in \bar{\mathbb{I}}$ with $h_p(\mathbf{x}) < 0$; or $p \notin \bar{\mathbb{I}}$ and $h_p(\mathbf{x}) \neq 0$. Let $\mathbf{n}_p$ be the corresponding column of matrix $\mathbf{C}$.

One can view introduction of constraint p as the introduction of a gradient proportional to $\mathbf{n}_p$:

$$\nabla_\mathbf{x} L \rightarrow (\mathbf{g}(\mathbf{x}) - \mathbf{n}_p\lambda_p) - \mathbf{N}\lambda.$$

This suggests that λ will be changed in proportion to

$$\delta\lambda = -\mathbf{N}^\dagger \mathbf{n}_p \qquad (7.30)$$

(cf. (7.27)), and that $\mathbf{x}$ will be changed in proportion to

$$\delta\mathbf{x} = \mathbf{H}\mathbf{n}_p \qquad (7.31)$$

(cf. (7.29)). In both cases the constant of proportionality is λ_p. These changes to the primary and dual variables are independent because λ is independent of $\mathbf{x}$ with choice (7.26).

If $i \in \mathbb{A} \cap \bar{\mathbb{I}}$, so $\lambda_i \in \mathbb{R}_+$, then

$$\lambda_i + \lambda_p \delta\lambda_i > 0$$

places a constraint on λ_p:

$$\lambda_p \gtrless -\lambda_i / \delta\lambda_i \qquad \text{if } \delta\lambda_i \gtrless 0 \quad \forall \quad i \in \mathbb{A} \cap \bar{\mathbb{I}}. \qquad (7.32)$$

Also,

$$\lambda_p > 0 \quad \text{if } p \in \bar{\mathbb{I}}. \qquad (7.33)$$

The fact that p corresponds to a violated constraint also poses a constraint. For the new constraint p to be satisfied,

$$h_p = \mathbf{n}_p^T(\mathbf{x} + \lambda_p \delta\mathbf{x}) + c_p = 0$$
$$\lambda_p = -\frac{c_p + \mathbf{n}_p^T \mathbf{x}}{\mathbf{n}_p^T \mathbf{H}\mathbf{n}_p}. \qquad (7.34)$$

If all three conditions (7.32), (7.33), and (7.34) are satisfied for some finite λ_p, then a solution λ_p has been determined. The solution $\mathbf{x}$ and λ is updated:

$$\mathbf{x} := \mathbf{x} + \lambda_p \mathbf{H}\mathbf{n}_p$$
$$\lambda := \lambda - \lambda_p \mathbf{N}^\dagger \mathbf{n}_p,$$

one appends p to the active set $\mathbb{A}$, appends λ_p to the vector λ, appends $\mathbf{n}_p$ to the matrix $\mathbf{N}$, and recomputes $\mathbf{N}^\dagger$ and $\mathbf{H}$. If any constraints remain that are violated, this procedure is repeated.

It is not always possible to satisfy all conditions in a given step, so the simple stepping algorithm described in the previous paragraph must be modified. To simplify the analysis, consider first all equality constraints prior to consideration of any inequality constraints.

With this special ordering, one can ignore conditions (7.32) and (7.33) as equality constraints are added to the system. If the equality constraints are independent, then $\delta\mathbf{x}$ (7.30) will not be zero, so the remaining condition (7.34) is easily satisfied. If $\delta\mathbf{x} = 0$ then the problem was inconsistent and no solution was possible. Once all equality constraints are added to the system, they remain active.

Let

$$
t_d = \begin{cases} \min_j \frac{-\lambda_j}{\delta\lambda_j} & \text{over all } j \in \mathbb{A} \cap \bar{\mathbb{I}} \text{ such that } \delta\lambda_j < 0 \\ \infty & \text{if no such term,} \end{cases}
\tag{7.35}
$$

where t_d is the maximum dual step size before some active inequality constraint becomes infeasible. Then, conditions (7.32) and (7.33) become

$$
0 < \lambda_p < t_d.
$$

If $\lambda_p \geq t_d$ is indicated, then one of the active inequality constraints becomes infeasible. Let this constraint be ℓ: $-\lambda_\ell / \delta\lambda_\ell = t_d$.

And, let

$$
t_p = \begin{cases} -\dfrac{h_p(\mathbf{x})}{\delta\mathbf{x}\cdot\mathbf{n}_p} & \text{if } \|\delta\mathbf{x}\| \neq 0, \\ \infty & \text{otherwise,} \end{cases}
\tag{7.36}
$$

where t_p is the maximum primal space step size. If $\lambda_p = t_p$ then new inequality constraint p will be satisfied. Condition (7.34) is

$$
\lambda_p = t_p.
$$

With this new notation the special cases are more easily described. Begin with $\lambda_p = 0$, and commence the analysis with computation of $\delta\mathbf{x}$ and $\delta\lambda$.

(i) If $t_p = t_d = \infty$ then the problem is infeasible. No solution exists.

(ii) If $t_p = \infty$, with t_d finite, then constraint ℓ will become infeasible before constraint p can be satisfied, if at all. A partial step in the dual space is suggested,

$$
\lambda_p := t_d,
$$

and infeasible constraint ℓ should be removed from $\mathbb{A}$, and $\mathbf{H}$, $\mathbf{N}$, and $\mathbf{N}^\dagger$ should be updated accordingly. One then reevaluates $\delta\mathbf{x}$ and $\delta\lambda$, and seeks again to satisfy all conditions.

(iii) Otherwise, a step can be made in both primal and dual spaces:

$$
t = \min(t_d, t_p)
$$
$$
\mathbf{x} := \mathbf{x} + t\delta\mathbf{x}
$$
$$
\lambda := \lambda + t\delta\lambda.
$$
$$
\lambda_p := t.
$$

(iiia) If $t_p \leq t_d$, then all conditions have been satisfied (a full step has been taken). One appends $\mathbf{n}_p$ to $\mathbf{N}$, recomputes $\mathbf{H}$ and $\mathbf{N}^\dagger$, and proceeds to consider again whether inequality constraints are satisfied. (iiib) However, if $t_p > t_d$, then this partial step has not yet satisfied condition (7.34). As with case (ii), one removes constraint ℓ, modifies $\mathbb{A}$, $\mathbf{H}$, $\mathbf{N}$, and $\mathbf{N}^\dagger$ accordingly, reevaluates $\delta\mathbf{x}$ and $\delta\lambda$, and seeks once more to satisfy all conditions.

Example 7.5   [88] Minimize

$$f(\mathbf{x}) = 6x_1 + 2(x_1^2 - x_1x_2 + x_2^2)$$

subject to

$$x_1 \geq 0$$
$$x_2 \geq 0$$
$$x_1 + x_2 \geq 2.$$

Solution

We first put the question in the form of (7.21) and (7.22):

$$f(\mathbf{x}) = \frac{1}{2}\mathbf{x}^T \begin{pmatrix} 4 & -2 \\ -2 & 4 \end{pmatrix} \mathbf{x} + \begin{pmatrix} 6 & 0 \end{pmatrix} \mathbf{x}$$

$$\mathbf{h} = \begin{pmatrix} 1 & 0 \\ 0 & 1 \\ 1 & 1 \end{pmatrix} \mathbf{x} + \begin{pmatrix} 0 \\ 0 \\ -2 \end{pmatrix} \geq 0. \quad \text{\tiny ●}$$

The calculation begins with $\mathbb{A} = \emptyset$, and

$$\mathbf{x} = -\mathbf{G}^{-1}\mathbf{b} = -\begin{pmatrix} 1/3 & 1/6 \\ 1/6 & 1/3 \end{pmatrix}\begin{pmatrix} 6 \\ 0 \end{pmatrix} = \begin{pmatrix} -2 \\ -1 \end{pmatrix},$$

and all three constraints are violated. Consider the first constraint, $p = 1$, $\mathbf{n}_1 = (1\ 0)^T$. $\mathbf{H} = \mathbf{G}^{-1}$ at this stage, and $\delta\mathbf{x} = (1/3\ 1/6)^T$ from (7.31). The dual step is $t_d = \infty$ since there exist no active constraints, and the primal step is $t_p = 6$ from (7.36). Since $t_p < t_d$ we take a full step:

$$\mathbb{A} = \{1\}$$
$$\lambda = (6)$$
$$\mathbf{x} := \begin{pmatrix} 0 \\ 0 \end{pmatrix}$$
$$\mathbf{N} = \begin{pmatrix} 1 \\ 0 \end{pmatrix}$$
$$\mathbf{N}^\dagger = \left[\begin{pmatrix} 1 & 0 \end{pmatrix}\begin{pmatrix} 1/3 & 1/6 \\ 1/6 & 1/3 \end{pmatrix}\begin{pmatrix} 1 \\ 0 \end{pmatrix}\right]^{-1}\begin{pmatrix} 1 & 0 \end{pmatrix}\begin{pmatrix} 1/3 & 1/6 \\ 1/6 & 1/3 \end{pmatrix} = \begin{pmatrix} 1 & \frac{1}{2} \end{pmatrix}$$
$$\mathbf{H} = \begin{pmatrix} 1/3 & 1/6 \\ 1/6 & 1/3 \end{pmatrix}\left[\begin{pmatrix} 1 & 0 \\ 0 & 1 \end{pmatrix} - \begin{pmatrix} 1 & 1/2 \\ 0 & 0 \end{pmatrix}\right] = \begin{pmatrix} 0 & 0 \\ 0 & 1/4 \end{pmatrix}.$$

This step satisfied the first constraint, and coincidentally satisfied also the second constraint, but the third remains violated with $h_3 = -2$. We must therefore choose $p = 3$, and compute $\delta\mathbf{x} = (0\ 1/4)^T$ from (7.31) and $\delta\lambda = (-3/2)$ from (7.30). The primal step size is 8 from (7.36) and the dual step size is 4 from (7.35). Since $t_d < t_p$, we take a partial step

$$\mathbf{x} := \begin{pmatrix} 0 \\ 1 \end{pmatrix}$$

$$\lambda := (0),$$

then eliminate constraint $\ell = 1$ from the system.

Again, we have $\mathbb{A} = \emptyset$ and $\mathbf{H} = \mathbf{G}^{-1}$, but only constraint $p = 3$ remains violated, but now with $h_3 = -1$. We compute $\delta\mathbf{x} = (1/2 \ 1/2)^{\mathrm{T}}$ and $t_p = 1$. Since there are no active constraints, $t_d = \infty$ and we take a full step:

$$\mathbf{x} = \begin{pmatrix} 1/2 \\ 3/2 \end{pmatrix}.$$

Now all constraints are satisfied so the problem is solved with $f = 13/2$.

Contours of the unconstrained function are shown together with the optimization trajectory in Figure 7.4. The shaded region is the feasible region in which all constraints are satisfied.

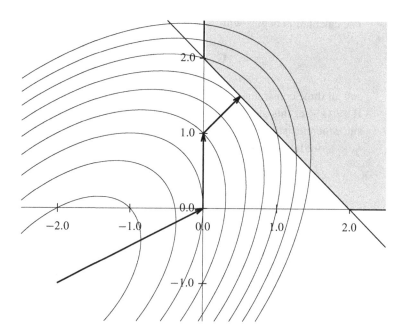

Figure 7.4 Solution path for Example 7.5.

An efficient implementation of this algorithm makes use of matrix modification techniques. Since $\mathbf{G}$ is symmetric positive definite, it has a Cholesky decomposition:

$$\mathbf{G} = \mathbf{L}\mathbf{L}^\mathsf{T}.$$

Consider $\mathbf{N}^\dagger$,

$$\mathbf{N}^\dagger = (\mathbf{N}^\mathsf{T}\mathbf{G}^{-1}\mathbf{N})^{-1}\mathbf{N}^\mathsf{T}\mathbf{G}^{-1}$$
$$= (\mathbf{N}^\mathsf{T}\mathbf{L}^{-\mathsf{T}}\mathbf{L}^{-1}\mathbf{N})^{-1}\mathbf{N}^\mathsf{T}\mathbf{L}^{-\mathsf{T}}\mathbf{L}^{-1},$$

and introduce the Householder decomposition of $\mathbf{L}^{-1}\mathbf{N} = \mathbf{Q}\mathbf{R}$,

$$\mathbf{N}^\dagger = (\mathbf{R}^\mathsf{T}\mathbf{R})^{-1}\mathbf{R}^\mathsf{T}\mathbf{Q}^\mathsf{T}\mathbf{L}^{-1}$$
$$= \mathbf{R}^{-1}\mathbf{Q}^\mathsf{T}\mathbf{L}^{-1}.$$

Then, define

$$\mathbf{J} = \mathbf{L}^{-\mathsf{T}}\mathbf{Q} = (\mathbf{Q}^\mathsf{T}\mathbf{L}^{-1})^\mathsf{T}$$
$$= \begin{pmatrix} \mathbf{J}_1 & | & \mathbf{J}_2 \end{pmatrix} = \begin{pmatrix} \mathbf{L}^{-\mathsf{T}}\mathbf{Q}_1 & | & \mathbf{L}^{-\mathsf{T}}\mathbf{Q}_2 \end{pmatrix},$$

so

$$\mathbf{N}^\dagger = \mathbf{R}^{-1}\mathbf{J}_1^\mathsf{T}.$$

For $\mathbf{H}$ one has

$$\mathbf{H} = \mathbf{G}^{-1}(\mathbf{I} - \mathbf{N}\mathbf{N}^\dagger)$$

$$= \mathbf{L}^{-\mathsf{T}}\mathbf{L}^{-1} - \underbrace{\mathbf{L}^{-\mathsf{T}}\overbrace{\mathbf{L}^{-1}\mathbf{N}\,\mathbf{R}^{-1}}^{\mathbf{QR}}\mathbf{J}_1^\mathsf{T}}_{\mathbf{J}}$$

$$= \mathbf{J}\mathbf{J}^\mathsf{T} - \mathbf{J}\mathbf{J}_1^\mathsf{T} = \begin{pmatrix} \mathbf{J}_1 & \mathbf{J}_2 \end{pmatrix}\begin{pmatrix} 0 \\ \mathbf{J}_2 \end{pmatrix}$$

$$\mathbf{H} = \mathbf{J}_2\mathbf{J}_2^\mathsf{T}.$$

Through these manipulations, it is apparent that $\mathbf{J}$ and $\mathbf{R}$ are the two quantities required, both related to the $\mathbf{QR}$ decomposition of $\mathbf{L}^{-1}\mathbf{N}$. The matrix manipulation techniques that are required are those associated with the addition or subtraction of a column of $\mathbf{N}$.

7.6.3 Matrix modification for QR

Here we consider matrix modification for the Householder decomposition of $\mathbf{A} = \mathbf{QR}$ (with $\mathbf{A}, \mathbf{R} \in \mathbb{R}^{n\times m}$, $\mathbf{Q} \in \mathbb{R}^{n\times n}$) upon deletion and addition of a column of the matrix $\mathbf{A}$.

First, consider the matrix $\tilde{\mathbf{A}}$ obtained from $\mathbf{A}$ by deletion of the ith column. Let $\mathbf{R}'$ be the matrix resulting from the product $\mathbf{Q}^T\tilde{\mathbf{A}}$:

$$\mathbf{Q}^T\tilde{\mathbf{A}} = \mathbf{R}' = \begin{pmatrix} & & \\ & & \\ & & \end{pmatrix},$$

i.e., a Hessenberg matrix for which the subdiagonal commences at column i. The solution is to shorten this subdiagonal one row at a time by application of a succession of Givens reflection matrices:

$$\underbrace{\mathbf{G}_{m-2,m-1}...\mathbf{G}_{i,i+1}\mathbf{Q}^T}_{\tilde{\mathbf{Q}}^T}\tilde{\mathbf{A}} = \underbrace{\mathbf{G}_{m-1,m}...\mathbf{G}_{i,i+1}\mathbf{R}'}_{\tilde{\mathbf{R}}\in\mathbb{R}^{n\times(m-1)}} = \begin{pmatrix} & & \\ & & \\ & & 0 \end{pmatrix}.$$

Now consider the case that $\tilde{\mathbf{A}}$ is derived from $\mathbf{A}$ by adding a column at the right.

$$\mathbf{Q}^T\tilde{\mathbf{A}} = \mathbf{R}' = \begin{pmatrix} & & \\ & & \\ & & \end{pmatrix}.$$

A single Householder matrix will bring the last column of the right-hand side into the right triangular form:

$$\underbrace{\hat{\mathbf{Q}}\mathbf{Q}^T}_{\tilde{\mathbf{Q}}^T}\tilde{\mathbf{A}} = \underbrace{\hat{\mathbf{Q}}\mathbf{R}'}_{\tilde{\mathbf{R}}} = \begin{pmatrix} & & \\ & & \\ & & \end{pmatrix},$$

with

$$\hat{Q} = \left(\begin{array}{c|c} \mathbf{I}_{m \times m} & \\ \hline & \mathbf{I} - 2\mathbf{w}\mathbf{w}^T \end{array} \right).$$

In the Goldfarb–Idnani method, the matrix $\mathbf{R}$ is stored directly, as is $\mathbf{J} = \mathbf{L}^{-1}\mathbf{Q}$. The modification of $\mathbf{J}$ corresponding to elimination of column i is therefore

$$\tilde{\mathbf{J}} = \mathbf{J}\mathbf{G}_{i,i+1}^T \cdots \mathbf{G}_{m-2,m-1}^T,$$

and for addition of a column it is

$$\tilde{\mathbf{J}} = \mathbf{J}\hat{\mathbf{Q}},$$

where in both cases the modification factors $(\mathbf{G}, \mathbf{w})$ are motivated by the transformation $\mathbf{R} \to \mathbf{R}'$ accompanying the column change in $\mathbf{A} = \mathbf{L}^{-1}\mathbf{N}$. In the particular case that constraint $\mathbf{n}_p$ is being added, the new column of $\mathbf{R}'$ is given by $\mathbf{J}^T\mathbf{n}_p$.

Problems

7.1 [50] Find a combination of alloys such that the mixture has a composition of 30% lead (Pb), 30% zinc (Zn), and 40% tin (Sn), and such that the cost is minimized.

| | alloy | | | | | | | | | mix |
	A	B	C	D	E	F	G	H	I	
% Pb	10	10	40	60	30	30	30	50	20	30
% Zn	10	30	50	30	30	40	20	40	30	30
% Sn	80	60	10	10	40	30	50	10	50	40
cost	$4.1	$4.3	$5.8	$6.0	$7.6	$7.5	$7.3	$6.9	$7.3	min.

7.2 Use phase I of the simplex method to find a feasible basis for the following problem:

$$\text{maximize } x_3$$
$$x_0 - x_1 = 0$$
$$x_1 - x_0 = 0$$
$$x_0 + 2x_1 + x_2 + x_3 = 3$$
$$x_0, x_1, x_2 \geq 0$$

(the dependence of the first two equations is intentional).

7.3 [66] Starting at $\mathbf{x} = (-1, 0, 0)$ find the minimum of the following function:

$$f(x_1, x_2, x_3) = 100 \left\{ [x_3 - 10\theta(x_1, x_2)]^2 + [r(x_1, x_2) - 1]^2 \right\} + x_3^2$$

$$\theta(x_1, x_2) = \frac{1}{2\pi} \begin{cases} \arctan(x_2/x_1) & x_1 > 0 \\ \pi + \arctan(x_2/x_1) & x_1 < 0 \end{cases}$$

$$r(x_1, x_2) = \sqrt{x_1^2 + x_2^2}.$$

7.4 [162] Find the minimum of

$$f = (x_1 - x_2)^2 + (x_2 + x_3 - 2)^2 + (x_4 - 1)^2 + (x_5 - 1)^2$$

subject to the linear constraints

$$x_1 + 3x_2 = 0$$
$$x_3 + x_4 - 2x_5 = 0$$
$$x_2 - x_5 = 0.$$

7.5 [113] Plan farming activities on 600 acres to maximize profit. The expected revenue (in dollars per acre of crop), and the resources required (water in acre-feet, and labor in hours, per acre of crop) are given in the following table. In addition, you cannot sell more than 6000 tons of tomato, with 33.25 tons expected per acre.

	alfalfa	wheat	corn	tomato	limit
revenue	121	160	135	825	
water	4.5	2.5	3.5	3.25	1800
labor	6.0	4.2	5.6	14.0	5000

8 Data fitting

Data fitting can be viewed as a generalization of polynomial interpolation to the case where we have more data than is needed to construct a polynomial of specified degree.

C. F. Gauss claims to have first developed solutions to the least squares problem, and both Gaussian elimination and the Gauss–Seidel iterative method were developed to solve these problems [52, 79]. In fact, interest in least squares by Galileo predates Gauss by over 200 years – a comprehensive history and analysis is given by Harter [97]. In addition to Gauss' contributions, the Jacobi iterative method [118] and the Cholesky decomposition method [13] were developed to solve least squares problems. Clearly, the least squares problem was (and continues to be) a problem of considerable importance. All these methods were applied to the normal equations, which represent an overdetermined system as a square and symmetric positive definite matrix. Despite the astounding historical importance of the normal equations, the argument will be made that you should not ever use them. Extensions of least squares to nonlinear problems, and linear problems with normal error, are described.

Least squares refers to a best fit in the L^2 norm, and this is by far the most commonly used norm. However, other norms are important for certain applications. Covariance-weighting leads to minimization in the Mahalanobis norm. L^1 is commonly used in financial modeling, and L^∞ may be most suitable when the underlying error distribution is uniform, versus normal. The history for these alternative norms is also quite rich. In 1824 Fourier [70] described fits in the L^1 and L^∞ norms using a geometric approach reminiscent to linear programming, which is the approach adopted here.

8.1 Least squares

Suppose we have a set of n data points (x_1, y_1), ..., (x_n, y_n) and we wish to find the line

$$y(x) = mx + b$$

that represents this data in the best way. The method used will obviously depend on what we mean by "best." A very common choice is "least squares," which means that we want to find parameters m and b in order that the error function

$$E^2 = \sum_{i=1}^{n} (y_i - y(x_i))^2 \tag{8.1}$$

is as small as possible. The minimum of E^2 will occur when

$$\frac{\partial E^2}{\partial m} = 0 = -2 \sum_{i=1}^{n} (y_i - y(x_i)) x_i$$

$$= -2 \sum_{i=1}^{n} (y_i x_i) + 2m \sum_{i=1}^{n} x_i^2 + 2b \sum_{i=1}^{n} x_i$$

$$\frac{\partial E^2}{\partial b} = 0 = -2 \sum_{i=1}^{n} (y_i - y(x_i))$$

$$= -2 \sum_{i=1}^{n} y_i + 2m \sum_{i=1}^{n} x_i + 2b \sum_{i=1}^{n} 1.$$

This gives two equations for the two unknowns m, b, and the solution can be determined

$$\begin{pmatrix} \sum x_i^2 & \sum x_i \\ \sum x_i & n \end{pmatrix} \begin{pmatrix} m \\ b \end{pmatrix} = \begin{pmatrix} \sum x_i y_i \\ \sum y_i \end{pmatrix}$$

$$m = \frac{n \left(\sum x_i y_i \right) - \left(\sum x_i \right) \left(\sum y_i \right)}{n \left(\sum x_i^2 \right) - \left(\sum x_i \right)^2}$$

$$b = \frac{\left(\sum x_i^2 \right) \left(\sum y_i \right) - \left(\sum x_i \right) \left(\sum x_i y_i \right)}{n \left(\sum x_i^2 \right) - \left(\sum x_i \right)^2}.$$

These are called the *normal equations*.

Another way to look at this same problem is

$$\underbrace{\begin{pmatrix} x_1 & 1 \\ \vdots & \vdots \\ x_n & 1 \end{pmatrix}}_{\mathbf{A}} \underbrace{\begin{pmatrix} m \\ b \end{pmatrix}}_{\mathbf{x}} = \underbrace{\begin{pmatrix} y_1 \\ \vdots \\ y_n \end{pmatrix}}_{\mathbf{b}}, \tag{8.2}$$

so the goal is really the solution of $\mathbf{Ax} = \mathbf{b}$, but where now $\mathbf{A}$ is not square.

We can make the system square by multiplying through by $\mathbf{A}^{\mathrm{T}}$:

$$\underbrace{\begin{pmatrix} \sum x_i^2 & \sum x_i \\ \sum x_i & n \end{pmatrix}}_{\mathbf{A}^{\mathrm{T}}\mathbf{A}} \mathbf{x} = \underbrace{\begin{pmatrix} \sum x_i y_i \\ \sum y_i \end{pmatrix}}_{\mathbf{A}^{\mathrm{T}}\mathbf{b}},$$

which clearly results in the normal equations again. This approach is appealing because of its intuitive feel and because it is simple, but don't ever do this. The reason is that the condition number of $\mathbf{A}^{\mathrm{T}}\mathbf{A}$ is the square of the condition number of $\mathbf{A}$, and so the numerical errors associated with the normal equations are larger than they have to be.

The best approach is to use a well-conditioned method like QR. As previously noted, the Householder reduction method works perfectly well for non-square matrices. In the present case

$$Ax = Q \underbrace{\begin{pmatrix} * & * \\ 0 & * \\ 0 & 0 \\ \vdots & \vdots \\ 0 & 0 \end{pmatrix}}_{R} x = b,$$

so x can be determined through back substitution after calculating R and $Q^T b$.

Mathematically, the QR result is the same as the normal equations result. The difference lies only in numerical error. To show the equivalence, construct the normal equation solution using the QR form of the matrix

$$A^T A x = A^T b$$
$$R^T \underbrace{Q^T Q}_{I} R x = R^T Q^T b \tag{8.3}$$
$$R^T \left(R x - Q^T b \right) = 0.$$

If A has size $n \times m$, then the first m rows of the equation

$$R x = Q^T b \tag{8.4}$$

hold. The subsequent $n - m$ rows do not satisfy (8.4), That part of $Q^T b$ lies in the null space of R, which is the null space of A.

Another approach is to use singular value decomposition:

$$Ax = U \Sigma V^T x = b \tag{8.5a}$$
$$\Sigma y = U^T b \tag{8.5b}$$
$$x = V y. \tag{8.5c}$$

Here, one first solves (8.5b) for y, and then finds x from (8.5c). In solving (8.5b) Σ will have the form

$$\Sigma = \begin{pmatrix} \sigma_1 & 0 \\ 0 & \sigma_2 \\ 0 & 0 \\ \vdots & \vdots \\ 0 & 0 \end{pmatrix}.$$

Again, the SVD result is mathematically the same as the normal equations result. And, again, we show this by writing the normal equations using the SVD decomposition of the matrix

$$\mathbf{A}^T\mathbf{A}\mathbf{x} = \mathbf{A}^T\mathbf{b}$$
$$\mathbf{V}\boldsymbol{\Sigma}^T\mathbf{U}^T\mathbf{U}\boldsymbol{\Sigma}\mathbf{V}^T\mathbf{x} = \mathbf{V}\boldsymbol{\Sigma}^T\mathbf{U}^T\mathbf{b}$$
$$\boldsymbol{\Sigma}^T\boldsymbol{\Sigma}\mathbf{V}^T\mathbf{x} = \boldsymbol{\Sigma}^T\mathbf{U}^T\mathbf{b}$$
$$\boldsymbol{\Sigma}^T(\boldsymbol{\Sigma}\mathbf{V}^T\mathbf{x} - \mathbf{U}^T\mathbf{b}) = 0,$$

and for those rows corresponding to nonzero singular values σ_i,

$$\boldsymbol{\Sigma}\mathbf{V}^T\mathbf{x} = \mathbf{U}^T\mathbf{b},$$

which is the SVD solution (8.5).

From a computational complexity point of view, QR is superior to SVD for this type of problem. There are simply fewer operations to perform with QR. Nonetheless, for very poorly conditioned problems SVD is a uniquely capable technique (see, e.g., Example 3.8).

Each of the methods described above for the least squares problem can be discussed in terms of the *pseudoinverse* $\mathbf{A}^\dagger$:

$$\mathbf{A}\mathbf{x} = \mathbf{b}, \quad \mathbf{A} \in \mathbb{R}^{n \times m}, \quad n > m$$
$$\mathbf{x} = \mathbf{A}^\dagger\mathbf{b}$$
$$\mathbf{A}^\dagger = \begin{cases} (\mathbf{A}^T\mathbf{A})^{-1}\mathbf{A}^T & \text{normal equations} \\ \mathbf{R}^{-1}\mathbf{Q}^T & \text{QR} \\ \mathbf{V}\boldsymbol{\Sigma}^{-1}\mathbf{U}^T & \text{SVD.} \end{cases}$$

Here it is understood that $\mathbf{R}$ and $\boldsymbol{\Sigma}$ are the upper square partitions of those matrices. The idea of the pseudoinverse was developed by E. H. Moore and R. Penrose [165, 180], and it is frequently called the *Moore–Penrose pseudoinverse*.

This discussion has focused on elimination methods for the least squares problem. Recall that the iterative conjugate gradient algorithm (4.16) also solves $\mathbf{A}^T\mathbf{A}\mathbf{x} = \mathbf{A}^T\mathbf{b}$ without forming $\mathbf{A}^T\mathbf{A}$, and therefore has similarly good numerical properties.

It is worth repeating a key result of this discussion: the solution of $\mathbf{A}\mathbf{x} = \mathbf{b}$ by $\mathbf{x} = \mathbf{A}^\dagger\mathbf{b}$, is equivalent to minimizing $\|\mathbf{r}\|_2$ with respect to $\mathbf{x}$ where $\mathbf{r} = \mathbf{b} - \mathbf{A}\mathbf{x}$. For system (8.2), $\|\mathbf{r}\|_2^2$ is given by (8.1). Another way to view it is by writing

$$\mathbf{r} = \mathbf{b} - \mathbf{A}\mathbf{x} = (\mathbf{I} - \mathbf{A}(\mathbf{A}^T\mathbf{A})^{-1}\mathbf{A}^T)\mathbf{b},$$

which is a Gram–Schmidt orthogonalization of $\mathbf{b}$ with respect to the column space of $\mathbf{A}$ (cf. (3.12) and (7.28)). Thus, the residual is just the component of $\mathbf{b}$ that is orthogonal to the column space of $\mathbf{A}$ [227]. We can use this observation to map a quadratic function (e.g., (8.1)) to an equivalent linear system (e.g., (8.2)), and this will be exploited below in Sections 8.3, 8.4, and 8.5.

8.2   An application to the Taylor series

On the one hand, any data (x_i, y_i) may be fit to a polynomial model

$$y = a_0 + a_1 x + a_2 \frac{x^2}{2} + \cdots$$

using least squares. On the other hand, if y is a function of x, $y = f(x)$, and if that function possesses sufficient derivatives, one has the Taylor series

$$y = f(0) + x f'(0) + x^2 \frac{f''(0)}{2} + \cdots .$$

Intuition suggest that the fit parameters determine the derivatives,

$$a_j \approx f^{(j)}(0), \tag{8.6}$$

and this is in fact true. To assess the accuracy of this approximation it is useful to separate the relative orientation of the support abscissas (the *stencil*), and the length scale associated with that orientation.

To illustrate this point, let's consider a specific example. We will sample a function $f(x)$ at certain discrete points.

x_i	y_i
$x_0 - 2h$	$f(x_0 - 2h)$
$x_0 - 1h$	$f(x_0 - h)$
x_0	$f(x_0)$
$x_0 + 1h$	$f(x_0 + h)$
$x_0 + 2h$	$f(x_0 + 2h)$

The stencil in this case is a regularly spaced linear array of points centered about the point x_0 where we wish to determine the Taylor coefficients. The length scale associated with the stencil is the separation h.

Since we have five data points, we might hope to estimate the first five parameters of the Taylor series. The equations to be solved are of the form

$$y_i = f^{(0)}(x_0) + (x_i - x_0) f^{(1)}(x_0) + \cdots + (x_i - x_0)^4 \frac{f^{(4)}(x_0)}{4!} + (x_i - x_0)^5 \frac{f^{(5)}(\xi_i)}{5!},$$

where the last term is the Lagrange remainder. Combining expansions of this form for each of the support points we obtain the linear identity

$$
\begin{pmatrix}
1 & -2h & 2h^2 & -\frac{4h^3}{3} & \frac{2h^4}{3} \\
1 & -h & \frac{h^2}{2} & -\frac{h^3}{6} & \frac{h^4}{24} \\
1 & 0 & 0 & 0 & 0 \\
1 & h & \frac{h^2}{2} & \frac{h^3}{6} & \frac{h^4}{24} \\
1 & 2h & 2h^2 & \frac{4h^3}{3} & \frac{2h^4}{3}
\end{pmatrix}
\begin{pmatrix}
f^{(0)}(x_0) \\
f^{(1)}(x_0) \\
f^{(2)}(x_0) \\
f^{(3)}(x_0) \\
f^{(4)}(x_0)
\end{pmatrix}
=
\begin{pmatrix}
y(x_0 - 2h) \\
y(x_0 - h) \\
y(x_0) \\
y(x_0 + h) \\
y(x_0 + 2h)
\end{pmatrix}
- h^5
\begin{pmatrix}
-\frac{4}{15} f^{(5)}(\xi_1) \\
-\frac{1}{120} f^{(5)}(\xi_2) \\
0 \\
\frac{1}{120} f^{(5)}(\xi_4) \\
\frac{4}{15} f^{(5)}(\xi_5)
\end{pmatrix},
$$

$$\tag{8.7}$$

or $\mathbf{Af} = \mathbf{y} - h^5\mathbf{r}$, This equation cannot be solved exactly since we don't know the parameters ξ, but we can ignore them to give the approximation $\mathbf{Aa} = \mathbf{y}$. Symbolically, one has

$$\mathbf{a} = \mathbf{A}^{-1}\mathbf{y},$$

which may be compared with

$$\mathbf{f} = \mathbf{A}^{-1}\mathbf{y} - h^5\mathbf{A}^{-1}\mathbf{r}$$

to give an expression for the error associated with approximation (8.6):

$$\mathbf{a} - \mathbf{f} = h^5\mathbf{A}^{-1}\mathbf{r}. \tag{8.8}$$

The matrix $\mathbf{A}$ depends on h in a clear way: the ith column (counting from zero) is proportional to h^i. If we take out this explicit dependence,

$$\mathbf{A} = \bar{\mathbf{A}}\operatorname{diag}\left(1, h, h^2, h^3, h^4\right)$$

$$\bar{\mathbf{A}} = \begin{pmatrix} 1 & -2 & 2 & -\frac{4}{3} & \frac{2}{3} \\ 1 & -1 & \frac{1}{2} & -\frac{1}{6} & \frac{1}{24} \\ 1 & 0 & 0 & 0 & 0 \\ 1 & 1 & \frac{1}{2} & \frac{1}{6} & \frac{1}{24} \\ 1 & 2 & 2 & \frac{4}{3} & \frac{2}{3} \end{pmatrix},$$

then (8.8) becomes

$$\mathbf{a} - \mathbf{f} = \operatorname{diag}\left(h^5, h^4, h^3, h^2, h\right)\bar{\mathbf{A}}^{-1}\mathbf{r}. \tag{8.9}$$

The vector $\mathbf{r}$ depends on h in a more subtle way: each element is proportional to a value of $f^{(5)}(\xi)$ for some ξ in an interval that depends on h (e.g., $x_0 - 2h \le \xi_1 \le x_0$). That sensitivity can be eliminated by defining a constant M, say, with

$$\max_x |f^{(5)}| \le M,$$

so

$$|\mathbf{r}| \le M\bar{\mathbf{r}},$$

where $\bar{\mathbf{r}}$ depends on the shape of the stencil but not the scale h. Combined with (8.9), we have

$$|\mathbf{a} - \mathbf{f}| \le \operatorname{diag}(h^5, h^4, h^3, h^2, h)M|\bar{\mathbf{A}}^{-1}\bar{\mathbf{r}}|.$$

In this example the important result is

$$a_0 - f^{(0)}(x_0) = \mathcal{O}(h^5)$$
$$a_1 - f^{(1)}(x_0) = \mathcal{O}(h^4)$$
$$a_2 - f^{(2)}(x_0) = \mathcal{O}(h^3)$$
$$a_3 - f^{(3)}(x_0) = \mathcal{O}(h^2)$$
$$a_4 - f^{(4)}(x_0) = \mathcal{O}(h).$$

The details of the stencil ($\bar{\mathbf{A}}$ and $\bar{\mathbf{r}}$) and the magnitude of the remainder derivative (M) are coefficients buried in the $\mathcal{O}$ notation.

This analysis can be carried over to overdetermined stencils by substituting $\mathbf{A}^\dagger$ for $\mathbf{A}^{-1}$, and is easily adapted to multivariate functions. However, it is not stable for high order derivatives because the matrix (8.7) can be poorly conditioned. Note its resemblance to the Vandermonde matrix (cf. (2.43)).

8.3 Data with experimental error

Suppose our data consists of points (x_i, y_i) and also the variance and covariance of the y data, σ_i^2 and $\text{cov}(y_i, y_j)$. We will assume in this section that the abscissas x are known with certainty.

The underlying assumption is that the data y_i is random – a number drawn from a Gaussian normal distribution with mean μ_i, and standard deviation σ_i. In this circumstance, the probability of observing a measurement in the interval $(y_i, y_i + dy_i)$ is

$$P(y_i) = \frac{1}{\sigma_i \sqrt{2\pi}} e^{-(y_i - \mu_i)^2/(2\sigma_i^2)}, \tag{8.10}$$

and the probability of y_i lying in some interval is 1,

$$\int_{-\infty}^{+\infty} P(y_i) dy_i = 1.$$

P is a *probability density function*.

We will say that variable y_i is distributed according to (8.10), or

$$y_i = \mathcal{N}(\mu_i, \sigma_i^2), \tag{8.11}$$

if it has mean μ_i and variance σ_i^2. Note that

$$\mathcal{N}(\mu, \sigma^2)$$
$$\mu + \mathcal{N}(0, \sigma^2),$$

and

$$\mu + \sigma \mathcal{N}(0, 1)$$

are equivalent.

The expectation of y, written $\mathbb{E}(y)$, is the average of a large number of samples. If $y = \mathcal{N}(\mu, \sigma^2)$ then $\mathbb{E}(y) = \mu$:

$$\mathbb{E}(y) = \int_{-\infty}^{+\infty} y P(y) dy$$

$$y = \sigma z + \mu$$

$$\mathbb{E}(y) = \int_{-\infty}^{+\infty} (\sigma z + \mu) dz \frac{1}{\sqrt{2\pi}} \exp(-z^2/2) = \mu,$$

using the fact that $\int z \exp(-z^2/2)dz = 0$ because the integrand is odd.

The variance of a number is given by the expectation $\sigma^2 = \mathbb{E}([y - \mu]^2) = \mathbb{E}(y^2) - \mu^2$:

$$
\begin{aligned}
\mathbb{E}([y - \mu]^2) &= \int_{-\infty}^{+\infty} (y - \mu)^2 P(y)dy \\
&= \frac{1}{\sqrt{2\pi}} \int_{-\infty}^{+\infty} \sigma^2 z^2 \exp(-z^2/2)dz \\
&= -2\sigma^2 \lim_{s \to 1} \frac{d}{ds} \int_{-\infty}^{+\infty} dz \frac{1}{\sqrt{2\pi}} \exp(-sz^2/2) \\
&= -2\sigma^2 \lim_{s \to 1} \frac{d}{ds} s^{-1/2} \\
&= \sigma^2.
\end{aligned}
\tag{8.12}
$$

If y_i and y_j, $j \neq i$, are normally distributed and independent then $\mathbb{E}([y_i - \mu_i][y_j - \mu_j]) = 0$. In the more general case, there may be a covariance $\text{cov}(y_i, y_j) = \mathbb{E}([y_i - \mu_i][y_j - \mu_j]) = \mathbb{E}(y_i y_j) - \mu_i \mu_j$.

For the collection of measured data, we assume an analogous probability density function

$$
P(\mathbf{y}) = \frac{1}{\sqrt{(2\pi)^n \det(\mathbf{\Sigma})}} \exp\left[-\frac{1}{2}(\mathbf{y} - \mathbf{y}_m)^\mathsf{T} \mathbf{\Sigma}^{-1}(\mathbf{y} - \mathbf{y}_m)\right],
$$

where $\mathbf{y}$ are the experimental measurements, $\mathbf{y}_m(x)$ is the model to which we will fit the data, and $\mathbf{\Sigma}$ will be shown to be the symmetric positive definite covariance matrix

$$
\Sigma_{ij} = \text{cov}(y_i, y_j),
$$

with the understanding $\text{cov}(y_i, y_i) = \sigma_i^2$.

To see that this is a probability density, first note that $P(\mathbf{y}) \geq 0$. Second, the sum of all probabilities is one,

$$
\mathbf{y} = \mathbf{\Sigma}^{1/2}\mathbf{z} + \mathbf{y}_m
$$
$$
dy^n = \det(\mathbf{\Sigma}^{1/2})dz^n = \det(\mathbf{\Sigma})^{1/2}dz^n
$$
$$
\int_{-\infty}^{+\infty} \cdots \int dy^n P(\mathbf{y}) = \frac{1}{\sqrt{(2\pi)^n}} \int_{-\infty}^{+\infty} \cdots \int dz^n \exp\left(-\frac{1}{2}\mathbf{z}^\mathsf{T}\mathbf{z}\right)
$$
$$
= \prod_{i=1}^{n} \int_{-\infty}^{+\infty} dz_i \frac{1}{\sqrt{2\pi}} \exp(-z_i^2/2) = 1.
$$

The transformed variables $\mathbf{z}$ are $\mathcal{N}(0, 1)$ normally distributed and independent; $\mathbb{E}(z_i) = 0$, and $\mathbb{E}(z_i z_j) = \delta_{ij}$ Third, this combined model embodies variances consistent with

(8.10), and covariances consistent with expectation:

$$y_i - y_{m,i} = \sum_k \Sigma_{ik}^{1/2} z_k$$

$$(y_i - y_{m,i})(y_j - y_{m,j}) = \sum_{k\ell} \Sigma_{ik}^{1/2} \Sigma_{j\ell}^{1/2} z_k z_\ell$$

$$\text{cov}(y_i, y_j) = \mathbb{E}[(y_i - y_{m,i})(y_j - y_{m,j})]$$

$$= \int_{-\infty}^{+\infty} \cdots \int dx^n (y_i - y_{m,i})(y_j - y_{m,j}) P(\mathbf{y})$$

$$= \frac{1}{\sqrt{(2\pi)^n}} \int_{-\infty}^{+\infty} \cdots \int dz^n \sum_{k\ell} \Sigma_{ik}^{1/2} \Sigma_{j\ell}^{1/2} z_k z_\ell \exp\left(-\frac{1}{2}\mathbf{z}^{\mathsf{T}}\mathbf{z}\right)$$

$$= \sum_{k\ell} \Sigma_{ik}^{1/2} \Sigma_{j\ell}^{1/2} \mathbb{E}(z_k z_\ell) = \sum_k \Sigma_{ik}^{1/2} \Sigma_{jk}^{1/2} = \sum_k \Sigma_{ik}^{1/2} \Sigma_{kj}^{1/2}$$

$$= \Sigma_{ij},$$

using the fact that $\Sigma^{1/2}$ is symmetric positive definite when Σ is symmetric positive definite. Therefore, the matrix Σ is the matrix of covariances. The multivariate extension of notation $\mathcal{N}(\mu, \sigma^2)$ is $\mathcal{N}(\boldsymbol{\mu}, \boldsymbol{\Sigma})$.

If the expectation is that our data truly obeys the model $y_m(x)$, but for normally distributed error, then the most probable state will minimize

$$D_M^2 = \chi^2 = \sum_{i,j=1}^n (y_i - y_m(x_i)) \Sigma_{ij}^{-1} (y_j - y_m(x_j)),$$

since $P(\mathbf{y}) \propto \exp(-\chi^2/2)$. If Σ were replaced with $\mathbf{I}$, this reduces to the least squares problem of Section 8.1. Formally, the data fitting problem seeks to maximize $L(\mathbf{y}_m \mid \mathbf{y})$, the likelihood of model $\mathbf{y}_m$ given data $\mathbf{y}$, which is the *dual* of the probability density function $P(\mathbf{y} \mid \mathbf{y}_m)$. D_M is known as the Mahalanobis distance [153], when the inverse of the covariance matrix is viewed as a metric tensor (cf. (7.4)). D_M can also be viewed as a particular norm, the Mahalanobis norm $\|\mathbf{y} - \mathbf{y}_m\|_{\Sigma^{-1}}$ (cf. (4.17)).

By analogy to the least squares problem, we can see that χ^2 is just the L^2 norm of the residual

$$\mathbf{r} = \Sigma^{-1/2}\mathbf{y}_m - \Sigma^{-1/2}\mathbf{y},$$

and this residual is minimized by solving the overdetermined equations

$$\Sigma^{-1/2} \begin{pmatrix} x_1 & 1 \\ \vdots & \vdots \\ x_n & 1 \end{pmatrix} \begin{pmatrix} m \\ b \end{pmatrix} = \Sigma^{-1/2} \begin{pmatrix} y_1 \\ \vdots \\ y_n \end{pmatrix},$$

if $y_m(x)$ is a line, which is just a $\Sigma^{-1/2}$-weighted version of (8.2).

As before this is best solved by QR or SVD. To compute $\Sigma^{-1/2}$, use the eigenvalue decomposition (QR for symmetric positive definite matrices), or consider the singular value decomposition $\mathbf{\Sigma} = \mathbf{UDV}^T$. Since $\mathbf{\Sigma}$ is symmetric positive definite, $\mathbf{V} = \mathbf{U}$, and the singular value decomposition is just the eigenvalue decomposition. Thus, $\mathbf{\Sigma}^{-1/2} = \mathbf{UD}^{-1/2}\mathbf{V}^T$. The decomposition

$$\mathbf{\Sigma} = \mathbf{UD}^2\mathbf{U}^T, \tag{8.13}$$

with $\mathbf{\Sigma}$ symmetric positive definite, $\mathbf{U}$ unitary, and $\mathbf{D}$ diagonal, is called the Löwdin decomposition.

Since the slope m and intercept b depend on the ordinates y, which in turn are subject to error with standard deviation σ, it follows that m and b should have some distribution characterized by standard deviations. Let's write

$$\mu_m = m(\boldsymbol{\mu}, \mathbf{x}) \tag{8.14}$$

$$\mu_b = b(\boldsymbol{\mu}, \mathbf{x}) \tag{8.15}$$

to signify that m, b are functions of y, and that the mean value of m, μ_m, is a function of the mean values of the ordinates μ_i and the abscissas x_i.

Now, perform an error analysis on m, b – use the Taylor series, keeping only the leading terms:

$$\mu_m + \sigma_m \mathcal{N}_m(0, 1) = m(\boldsymbol{\mu}, \mathbf{x}) + \sum_i \frac{\partial m}{\partial y_i} \mathcal{N}_i(\mathbf{0}, \mathbf{\Sigma})$$

$$\mu_b + \sigma_b \mathcal{N}_b(0, 1) = b(\boldsymbol{\mu}, \mathbf{x}) + \sum_i \frac{\partial b}{\partial y_i} \mathcal{N}_i(\mathbf{0}, \mathbf{\Sigma});$$

then subtract (8.14) and (8.15), respectively, to give

$$\sigma_m \mathcal{N}_m(0, 1) = \sum_i \frac{\partial m}{\partial y_i} \mathcal{N}_i(\mathbf{0}, \mathbf{\Sigma}) \tag{8.16a}$$

$$\sigma_b \mathcal{N}_b(0, 1) = \sum_i \frac{\partial b}{\partial y_i} \mathcal{N}_i(\mathbf{0}, \mathbf{\Sigma}). \tag{8.16b}$$

If we take the expectation of (8.16a) we get $0 = 0$, but if we square both sides first, we get

$$\mathbb{E}\sigma_m^2 \mathcal{N}_m(0, 1)^2 = \sum_{i=1}^n \sum_{j=1}^n \frac{\partial m}{\partial y_i} \frac{\partial m}{\partial y_j} \mathbb{E} \mathcal{N}_i(\mathbf{0}, \mathbf{\Sigma}) \mathcal{N}_j(\mathbf{0}, \mathbf{\Sigma})$$

$$\sigma_m^2 = \sum_{i=1}^n \sum_{j=1}^n \frac{\partial m}{\partial y_i} \frac{\partial m}{\partial y_j} \operatorname{cov}(y_i, y_j).$$

Likewise, for the intercept we have

$$\sigma_b^2 = \sum_{i=1}^n \sum_{j=1}^n \frac{\partial b}{\partial y_i} \frac{\partial b}{\partial y_j} \operatorname{cov}(y_i, y_j).$$

Since m and b depend on the same data y, they should be expected to have correlated errors. To estimate the covariance, multiply (8.16a) and (8.16b), and take the expectation of the product

$$\text{cov}(m, b) = \mathbb{E}\left[\sum_{i=1}^{n} \left(\frac{\partial m}{\partial y_i}\right) \mathcal{N}_i(0, \Sigma)\right]\left[\sum_{j=1}^{n} \left(\frac{\partial b}{\partial y_j}\right) \mathcal{N}_j(0, \Sigma)\right]$$

$$\text{cov}(m, b) = \sum_{i=1}^{n}\sum_{j=1}^{n} \left(\frac{\partial m}{\partial y_i}\right)\left(\frac{\partial b}{\partial y_j}\right) \text{cov}(y_i, y_j).$$

In combination,

$$\begin{pmatrix} \sigma_m^2 & \text{cov}(m, b) \\ \text{cov}(m, b) & \sigma_b^2 \end{pmatrix} = \begin{pmatrix} \frac{\partial m}{\partial y_1} & \cdots & \frac{\partial m}{\partial y_n} \\ \frac{\partial b}{\partial y_1} & \cdots & \frac{\partial b}{\partial y_n} \end{pmatrix} \Sigma \begin{pmatrix} \frac{\partial m}{\partial y_1} & \frac{\partial b}{\partial y_1} \\ \vdots & \vdots \\ \frac{\partial m}{\partial y_n} & \frac{\partial b}{\partial y_n} \end{pmatrix}.$$

To find the partial derivatives,

$$\begin{pmatrix} m \\ b \end{pmatrix} = (\Sigma^{-1/2}\mathbf{A})^\dagger \Sigma^{-1/2} \begin{pmatrix} y_1 \\ \vdots \\ y_n \end{pmatrix},$$

giving

$$\begin{pmatrix} \sigma_m^2 & \text{cov}(m, b) \\ \text{cov}(m, b) & \sigma_b^2 \end{pmatrix} = (\Sigma^{-1/2}\mathbf{A})^\dagger (\Sigma^{-1/2}\mathbf{A})^{\dagger T}. \tag{8.17}$$

Example 8.1 Fit the data in the following table to a line with the assumption that y errors do not covary. Determine the variance and covariance of the fit parameters.

x	y	σ
1	15.6	6.0
3	36.6	6.3
5	58.2	7.4
7	64.2	7.4
9	98.8	10.2

Solution
First, we write the given problem as a matrix, with each row weighted by $1/\sigma$:

$$\begin{pmatrix} \frac{1}{6.0} & \frac{1}{6.0} \\ \frac{3}{6.3} & \frac{1}{6.3} \\ \frac{5}{7.4} & \frac{1}{7.4} \\ \frac{7}{7.4} & \frac{1}{7.4} \\ \frac{9}{10.2} & \frac{1}{10.2} \end{pmatrix} \begin{pmatrix} m \\ b \end{pmatrix} = \begin{pmatrix} \frac{15.6}{6.0} \\ \frac{36.6}{6.3} \\ \frac{58.2}{7.4} \\ \frac{64.2}{7.4} \\ \frac{98.8}{10.2} \end{pmatrix},$$

and the matrix $\mathbf{A}' = \mathbf{\Sigma}^{-1/2}\mathbf{A}$ on the left-hand side has a QR decomposition

$$\mathbf{A}' = \mathbf{Q}\mathbf{R}$$

$$= \begin{pmatrix} -0.1079 & -0.8115 & -0.3525 & -0.3632 & -0.2713 \\ -0.3084 & -0.4528 & 0.1093 & 0.5070 & 0.6564 \\ -0.4376 & -0.1130 & 0.8257 & -0.2463 & -0.2309 \\ -0.6126 & 0.1594 & -0.2977 & 0.4978 & -0.5127 \\ -0.5714 & 0.3133 & -0.3055 & -0.5501 & 0.4235 \end{pmatrix} \begin{pmatrix} -1.544 & -0.2649 \\ 0 & -0.1701 \\ \hline 0 & 0 \\ 0 & 0 \\ 0 & 0 \end{pmatrix}.$$

With this QR decomposition we can solve $\mathbf{R}\mathbf{x} = \mathbf{Q}^{\mathrm{T}}\mathbf{b}$ for $\mathbf{x}$, giving the slope and inter-cept. We can also solve $\mathbf{R}\mathbf{A}'^{\dagger} = \mathbf{Q}^{\mathrm{T}}$ for $\mathbf{A}'^{\dagger}$:

$$\mathbf{A}'^{\dagger} = \begin{pmatrix} -0.7482 & -0.2568 & 0.1694 & 0.5574 & 0.6859 \\ 4.770 & 2.661 & 0.6644 & -0.9370 & -1.842 \end{pmatrix}.$$

The solution, with the standard deviations and covariance from (8.17), is

$$m = 9.375$$
$$b = 7.122$$
$$\sigma_m = 1.198$$
$$\sigma_b = 5.877$$
$$\sigma_{mb} = -5.925.$$

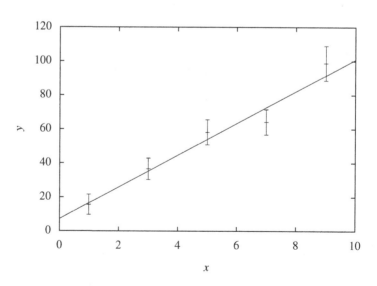

Figure 8.1 Data fitting with y error; Example 8.1.

The significance of the covariance is this. Error in b and error in m are related. How-ever, there are some linear combinations of m and b for which the variances are de-coupled. We write the matrix of variances, and diagonalize (with the QR method for eigenvalues)

$$\begin{pmatrix} 1.436 & -5.925 \\ -5.925 & 34.54 \end{pmatrix} \begin{pmatrix} -0.1710 & 0.9853 \\ 0.9853 & 0.1710 \end{pmatrix} = \begin{pmatrix} -0.1710 & 0.9853 \\ 0.9853 & 0.1710 \end{pmatrix} \begin{pmatrix} 35.57 & \\ & 0.4073 \end{pmatrix}$$

and the standard deviations are the squares root of the eigenvalues: 5.964 and 0.6382. When m, b covary in such a way that

$$m := m - 0.1710\delta$$
$$b := b + 0.9853\delta,$$

for some perturbation δ, the standard deviation is 5.964. When they covary in proportion to the other eigenvector, the standard deviation is 0.6382. The eigenvectors define the axes of an ellipse, and the eigenvalues specify the lengths of the semimajor and semiminor axes. Figure 8.2 shows the error bars, and the ellipse described by the eigenvalues and eigenvectors of the covariance matrix. Note that the ellipse implies that in increasing b it is most likely that m will decrease – which is what one would intuitively conclude from examination of Figure 8.1.

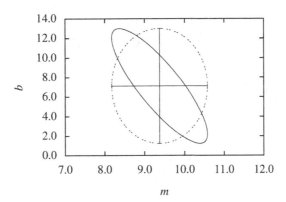

Figure 8.2 The covariance of slope and intercept error for data fitting of Example 8.1. The horizontal and vertical line segments are the 1σ error bars for m and b, respectively. The solid ellipse comes from the covariance matrix; the dashed ellipse is 1σ error envelope ignoring the covariance σ_{mb}, for reference.

8.4   Error in x and y

In most real problems there is error in both the x and the y data, and correlations may exist between all quantities. In this case, we might imagine that there is a true set of abscissas $\mathbf{X}$, and a true set of ordinates $\mathbf{Y}$, which obey $Y = mX + b$ pointwise. The measured $\mathbf{x}$ are approximations to $\mathbf{X}$, and the measured $\mathbf{y}$ are approximations to $\mathbf{Y}$, and we wish to minimize

$$D_M^2 = \left(\frac{\mathbf{x} - \mathbf{X}}{\mathbf{y} - m\mathbf{X} - b\mathbf{1}}\right)^{\mathrm{T}} \bar{\boldsymbol{\Sigma}}^{-1} \left(\frac{\mathbf{x} - \mathbf{X}}{\mathbf{y} - m\mathbf{X} - b\mathbf{1}}\right)$$

with

$$\bar{\boldsymbol{\Sigma}} = \left(\begin{array}{c|c} \mathrm{cov}(\mathbf{x}, \mathbf{x}) & \mathrm{cov}(\mathbf{x}, \mathbf{y}) \\ \hline \mathrm{cov}(\mathbf{x}, \mathbf{y}) & \mathrm{cov}(\mathbf{y}, \mathbf{y}) \end{array}\right).$$

Here D_M^2 is minimized with respect to all model parameters: $\mathbf{X}, m$, and b. As a quadratic form, we can equate that minimum with the solution to the nonlinear equation

$$\bar{\boldsymbol{\Sigma}}^{-1/2} \left(\begin{array}{c|c|c} \mathbf{I} & & \\ \hline \beta m\mathbf{I} & (1 - \beta)\mathbf{X} & \mathbf{1} \end{array}\right) \left(\begin{array}{c} \mathbf{X} \\ m \\ b \end{array}\right) = \bar{\boldsymbol{\Sigma}}^{-1/2} \left(\frac{\mathbf{x}}{\mathbf{y}}\right), \qquad (8.18)$$

where β is an arbitrary constant to split the nonlinear term. One could write

$$\mathbf{A} = \bar{\boldsymbol{\Sigma}}^{-1/2} \left(\begin{array}{c|c|c} \mathbf{I} & & \\ \hline \beta m\mathbf{I} & (1 - \beta)\mathbf{X} & \mathbf{1} \end{array}\right)$$

$$\left(\begin{array}{c} \mathbf{X} \\ m \\ b \end{array}\right) = \mathbf{A}^{\dagger} \bar{\boldsymbol{\Sigma}}^{-1/2} \left(\frac{\mathbf{x}}{\mathbf{y}}\right),$$

but this is not a solution since the left-hand side and $\mathbf{A}$ both depend on m and $\mathbf{X}$. However, one could interpret this as an iterative method:

$$\left(\begin{array}{c} \mathbf{X} \\ m \\ b \end{array}\right)^{(k+1)} = \mathbf{A}^{\dagger}(m^{(k)}, \mathbf{X}^{(k)}) \bar{\boldsymbol{\Sigma}}^{-1/2} \left(\frac{\mathbf{x}}{\mathbf{y}}\right), \qquad (8.19)$$

with

$$\mathbf{X}^{(0)} = \mathbf{x}$$
$$m^{(0)} = \text{some constant,}$$

and this can be made to converge to the desired solution when the errors are not too large. When $\mathbf{X}$ and m are far from their converged values, this iterative scheme can oscillate and diverge. One way to stabilize it is with *damping*: at the end of each iteration, replace the computed value with a weighed average of prior results. For example,

$$\left(\begin{array}{c} \mathbf{X} \\ m \end{array}\right)^{(k+1)} := (1 - \zeta) \left(\begin{array}{c} \mathbf{X} \\ m \end{array}\right)^{(k+1)} + \zeta \left(\begin{array}{c} \mathbf{X} \\ m \end{array}\right)^{(k)},$$

with $0 \le \zeta < 1$.

It is important that $\beta \ne 0$, because when $\beta = 0$ the top and bottom partitions of system (8.18) decouple. $\mathbf{X}$ becomes $\mathbf{x}$, and the bottom block reduces to the problem where there is error in $\mathbf{y}$ only. Also, if $\beta = 1$ then the slope m is not updated by the iteration; the matrix $\mathbf{A}$ is singular.

To determine the covariance of the fit parameters, differentiate (8.18) and rearrange to give

$$\bar{\Sigma}^{-1/2} \left(\begin{array}{c|c|c} \mathbf{I} & & \\ \hline m\mathbf{I} & \mathbf{X} & 1 \end{array} \right) \left(\begin{array}{c} d\mathbf{X} \\ \overline{dm} \\ \overline{db} \end{array} \right) = \bar{\Sigma}^{-1/2} \left(\frac{d\mathbf{x}}{d\mathbf{y}} \right).$$

The partial derivatives of the fit parameters with respect to the data are given by

$$\left(\begin{array}{c} d\mathbf{X} \\ \overline{dm} \\ \overline{db} \end{array} \right) = \mathbf{B}^{\dagger} \bar{\Sigma}^{-1/2} \left(\frac{d\mathbf{x}}{d\mathbf{y}} \right),$$

with

$$\mathbf{B} = \bar{\Sigma}^{-1/2} \left(\begin{array}{c|c|c} \mathbf{I} & & \\ \hline m\mathbf{I} & \mathbf{X} & 1 \end{array} \right).$$

Finally, $\mathbf{B}^{\dagger}\mathbf{B}^{\dagger T}$ is the $(n+2) \times (n+2)$ symmetric positive definite covariance matrix for the variables $\mathbf{X}, m, b$, and the m, b covariance matrix is the lower right 2×2 partition:

$$\left(\begin{array}{c|cc} \mathrm{cov}(\mathbf{X}, \mathbf{X}) & \mathrm{cov}(\mathbf{X}, m) & \mathrm{cov}(\mathbf{X}, b) \\ \hline \mathrm{cov}(\mathbf{X}^{\mathrm{T}}, m) & \sigma_m^2 & \mathrm{cov}(m, b) \\ \mathrm{cov}(\mathbf{X}^{\mathrm{T}}, b) & \mathrm{cov}(m, b) & \sigma_b^2 \end{array} \right) = \mathbf{B}^{\dagger}\mathbf{B}^{\dagger T}.$$

A noniterative approach to this problem was proposed by York [246] for the case that $\bar{\Sigma}$ is diagonal (i.e., error in all variables, but with no correlations). A recent review of approaches to this problem is given by Cantrell [32]. The iterative approach described here is new.

Example 8.2 As water drains from a tank, the height of the water column varies with time according to the following experimental measurements [205].

t	0	6.1	11.5	17.8	23.5	29.2	35.8	42.8	50.5	59.2	69.8	84.0
h	12	11	10	9	8	7	6	5	4	3	2	1

Assume that all measurement errors are independent and normally distributed, with $\sigma_h^2 = 0.0625$ and $\sigma_t^2 = 0.01$.

This data is to be fit to the model

$$-\frac{dh}{dt} = Ch^n$$

to find the exponent n.

Solution
Taking the log we obtain a line

$$\ln(-dh/dt) = n \ln(h) + \ln(C),$$

and we will estimate the derivative with centered differences. Therefore,

$$x_i = \ln(h_i)$$

$$y_i = \ln\left(\frac{h_{i-1} - h_{i+1}}{t_{i+1} - t_{i-1}}\right),$$

and

$$dx_i = \frac{1}{h_i} dh_i$$

$$dy_i = \frac{dh_{i+1} - dh_{i-1}}{h_{i+1} - h_{i-1}} + \frac{dt_{i-1} - dt_{i+1}}{t_{i+1} - t_{i-1}},$$

so

$$\sigma_{x_i}^2 = \left(\frac{1}{h_i}\right)^2 \sigma_h^2$$

$$\sigma_{y_i}^2 = \left(\frac{1}{h_{i+1} - h_{i-1}}\right)^2 2\sigma_h^2 + \left(\frac{1}{t_{i+1} - t_{i-1}}\right)^2 2\sigma_t^2$$

$$\mathrm{cov}(x_i, y_{i-1}) = \left(\frac{1}{h_i(h_i - h_{i-2})}\right) \sigma_h^2$$

$$\mathrm{cov}(x_i, y_{i+1}) = -\left(\frac{1}{h_i(h_{i+2} - h_i)}\right) \sigma_h^2$$

$$\mathrm{cov}(y_i, y_{i-2}) = -\left(\frac{1}{(h_{i+1} - h_{i-1})(h_{i-1} - h_{i-3})}\right) \sigma_h^2$$
$$\quad - \left(\frac{1}{(t_{i+1} - t_{i-1})(t_{i-1} - t_{i-3})}\right) \sigma_t^2,$$

with other covariances zero. Although the raw t, h data assumes uncorrelated error, the transformed x, y data has correlated errors with $\bar{\Sigma}$ a banded matrix:

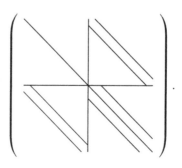

The derived quantities are given in Table 8.1. The resulting fit parameters are

slope ($m = n$)	4.483×10^{-1}
intercept ($b = \ln(C)$)	-2.741
σ_m^2	6.720×10^{-3}
σ_b^2	2.265×10^{-2}
$\text{cov}(m, b)$	-1.218×10^{-2},

using the iterative approach (8.19) with $\beta = \zeta = 0.5$. Convergence was judged by the change in parameters in an iteration prior to relaxation, and iteration was stopped when that variation was less than 10^{-6}. By this criterion, 23 iterations were required.

Table 8.1 Computed covariances for transformed data; Example 8.2.

i	x	y	$\sigma_{x_i}^2$	$\sigma_{y_i}^2$
1	2.398e+00	-1.749e+00	5.165e-04	3.140e-02
2	2.303e+00	-1.766e+00	6.250e-04	3.140e-02
3	2.197e+00	-1.792e+00	7.716e-04	3.139e-02
4	2.079e+00	-1.740e+00	9.766e-04	3.140e-02
5	1.946e+00	-1.816e+00	1.276e-03	3.138e-02
6	1.792e+00	-1.917e+00	1.736e-03	3.136e-02
7	1.609e+00	-1.995e+00	2.500e-03	3.134e-02
8	1.386e+00	-2.104e+00	3.906e-03	3.132e-02
8	1.099e+00	-2.267e+00	6.944e-03	3.130e-02
10	6.931e-01	-2.518e+00	1.562e-02	3.128e-02

i	cov (x_{i-1}, y_i)	cov (x_{i+1}, y_i)	cov (y_i, y_{i-2})
1		-3.125e-03	
2	2.841e-03	-3.472e-03	
3	3.125e-03	-3.906e-03	-1.570e-02
4	3.472e-03	-4.464e-03	-1.570e-02
5	3.906e-03	-5.208e-03	-1.569e-02
6	4.464e-03	-6.250e-03	-1.569e-02
7	5.208e-03	-7.812e-03	-1.568e-02
8	6.250e-03	-1.042e-02	-1.567e-02
9	7.812e-03	-1.562e-02	-1.566e-02
10	1.042e-02		-1.565e-02

8.5   Nonlinear least squares

Suppose we wish to fit data to a non-polynomial model, for example the three-parameter equation

$$y = b \exp(-ax) + c.$$

Our goal again will be to minimize

$$E^2 = \sum_{i=0}^{n} (y_i - y(x_i))^2,$$

but now the model does not lend itself to a system of linear equations. An approach to this problem is to construct an iterative method around a linearization of the model equations. Let $a^{(k)}$ be an iterative guess at parameter a, and let Δa be a perturbation to that guess so that $a^{(k+1)} = a^{(k)} + \Delta a$ is a better guess. Now, minimization of

$$E^2 = \sum_{i=0}^{n} \left(y_i - y^{(k)}(x_i) - \frac{\partial y^{(k)}(x_i)}{\partial a^{(k)}} \Delta a - \frac{\partial y^{(k)}(x_i)}{\partial b^{(k)}} \Delta b - \frac{\partial y^{(k)}(x_i)}{\partial c^{(k)}} \Delta c \right)^2$$

gives linear equations for Δa, Δb, and Δc just as (8.1) gave linear equations for m and b.

As with the linear example, we can write a matrix representation of the new problem

$$\begin{pmatrix} \frac{\partial y^{(k)}(x_1)}{\partial a^{(k)}} & \frac{\partial y^{(k)}(x_1)}{\partial b^{(k)}} & \frac{\partial y^{(k)}(x_1)}{\partial c^{(k)}} \\ \vdots & \vdots & \vdots \\ \frac{\partial y^{(k)}(x_n)}{\partial a^{(k)}} & \frac{\partial y^{(k)}(x_n)}{\partial b^{(k)}} & \frac{\partial y^{(k)}(x_n)}{\partial c^{(k)}} \end{pmatrix} \begin{pmatrix} \Delta a \\ \Delta b \\ \Delta c \end{pmatrix} = \begin{pmatrix} y_0 - y^{(k)}(x_0) \\ \vdots \\ y_n - y^{(k)}(x_n) \end{pmatrix}.$$

And, as with the linear example, we could solve this by the normal equations using $\mathbf{A}^T \mathbf{A}$ (but don't do this!), using Householder reduction, or SVD, or conjugate gradients.

After solving for the perturbations Δa, etc., one (i) adjusts a via $a^{(k+1)} = a^{(k)} + \Delta a$, etc., then (ii) re-computes the error vector $y_i - y^{(k+1)}(x_i)$, (iii) re-computes the matrix of partial derivatives, then (iv) solves for new perturbations. The cycle is repeated until convergence is achieved.

8.6   Fits in other norms

The quantity E appearing in (8.1) and (8.5) is the Euclidean or L^2 norm of the residual $\mathbf{r}$

$$\mathbf{r} = \mathbf{b} - \mathbf{A}\mathbf{x}$$

$$E = \|\mathbf{r}\|_2.$$

There are of course other vector norms, and sometimes it is appropriate to fit data to a model in such as way as to minimize the 1 norm, or the max norm (∞ norm):

$$L^1 = \sum |y_i - y(x_i)|$$
$$L^\infty = \max_i |y_i - y(x_i)|.$$

In these norms, the solution comes from linear programming.

First, consider the 1 norm. To solve this problem we have the following linear programming problem in standard form:

$$y_1 - mx_1 - b + \rho_1 - \sigma_1 = 0$$

$$\vdots$$

$$y_n - mx_n - b + \rho_n - \sigma_n = 0$$
$$\sigma_1 + \cdots + \sigma_n + \rho_1 + \cdots + \rho_n + f = 0$$
$$\text{maximize} \quad f$$

with restricted variables

$$I = \{\rho_1, \ldots, \rho_n, \sigma_1, \ldots, \sigma_n\}.$$

The artificial restricted variables ρ_i and σ_i are introduced so that $\rho_i + \sigma_i \geq |y_i - mx_i - b|$, and $-f$ is the sum of this quantity over data points. For this problem, the best solution will have, for each i, either $\sigma_i = 0$ and $\rho_i \geq 0$, or $\rho_i = 0$ and $\sigma_i \geq 0$. Both σ_i and ρ_i are allowed to be nonzero, but not when f is a maximum. At the optimum, $\rho_i + \sigma_i = |y_i - mx_i - b|$, and $-f = L^1$, so maximizing f minimizes the L^1 error.

The problem for L^∞ is similar. There will be one index, say q, such that

$$L^\infty = \max_i |y_i - y(x_i)| = |y_q - mx_q - b|,$$

and if we knew what that index was in advance, we could find m and b with the following linear programming problem in standard form:

$$y_1 - mx_1 - b + \rho_1 - \sigma_1 = 0$$

$$\vdots$$

$$y_n - mx_n - b + \rho_n - \sigma_n = 0$$
$$\sigma_1 + \rho_1 - \sigma_q - \rho_q = 0$$

$$\vdots$$

$$\sigma_{q-1} + \rho_{q-1} - \sigma_q - \rho_q = 0 \qquad (8.20)$$
$$\sigma_{q+1} + \rho_{q+1} - \sigma_q - \rho_q = 0$$

$$\vdots$$

$$\sigma_n + \rho_n - \sigma_q - \rho_q = 0$$
$$\sigma_q + \rho_q + f = 0$$
$$\text{maximize} \quad f,$$

with restricted variables

$$I = \{\rho_1, \ldots, \rho_n, \sigma_1, \ldots, \sigma_n\}.$$

Here our goal is to minimize the error of $|y_q - y(x_q)|$ while making the error of other terms less than or equal to this quantity.

Now, the equality $\sigma_p + \rho_p - \sigma_q - \rho_q = 0$ will be obeyed in this linear programming problem. So, we can add or subtract this equation wherever we like without changing the problem statement or the optimal solution. An exactly equivalent problem to (8.20) is the following:

$$y_1 - mx_1 - b + \rho_1 - \sigma_1 = 0$$

$$\vdots$$

$$y_n - mx_n - b + \rho_n - \sigma_n = 0$$
$$\sigma_1 + \rho_1 - \sigma_p - \rho_p = 0$$

$$\vdots$$

$$\sigma_{p-1} + \rho_{p-1} - \sigma_p - \rho_p = 0 \qquad\qquad (8.21)$$
$$\sigma_{p+1} + \rho_{p+1} - \sigma_p - \rho_p = 0$$

$$\vdots$$

$$\sigma_n + \rho_n - \sigma_p - \rho_p = 0$$
$$\sigma_p + \rho_p + f = 0$$
$$\text{maximize} \quad f.$$

This gives the L^∞ solution on the assumption that point p gives the maximum error, and it is equal to the solution that assumed that q gives the maximum error, so evidently any index in $[1, n]$ may be used.

Example 8.3   Fit the data in the following table to a line, using the 1, 2, and ∞ norms.

x	y
1	1
2	1
3	3
4	5
5	5

Solution

For the least squares L^2 fit, first set the problem up as an overdetermined matrix equation:

$$\begin{pmatrix} 1 & 1 \\ 2 & 1 \\ 3 & 1 \\ 4 & 1 \\ 5 & 1 \end{pmatrix} \begin{pmatrix} m \\ b \end{pmatrix} = \begin{pmatrix} 1 \\ 1 \\ 3 \\ 5 \\ 5 \end{pmatrix}.$$

Next, solution by Householder QR reduction gives the intermediate result:

$$\underbrace{\begin{pmatrix} -7.42 & -2.02 \\ 0.00 & -0.953 \\ 0.00 & 0.00 \\ 0.00 & 0.00 \\ 0.00 & 0.00 \end{pmatrix}}_{\mathbf{R}} \begin{pmatrix} m \\ b \end{pmatrix}$$

$$= \underbrace{\begin{pmatrix} -0.135 & -0.270 & -0.405 & -0.539 & -0.674 \\ -0.763 & -0.477 & -0.191 & 0.095 & 0.381 \\ -0.374 & -0.008 & 0.862 & -0.206 & -0.274 \\ -0.365 & 0.371 & -0.159 & 0.665 & -0.512 \\ -0.357 & 0.750 & -0.179 & -0.464 & 0.251 \end{pmatrix}}_{\mathbf{Q}^{\mathrm{T}}} \underbrace{\begin{pmatrix} 1.00 \\ 1.00 \\ 3.00 \\ 5.00 \\ 5.00 \end{pmatrix}}_{\mathbf{b}} = \begin{pmatrix} -7.69 \\ 0.572 \\ -0.199 \\ 0.294 \\ -1.21 \end{pmatrix}.$$

The result obtained by back substitution is presented in Table 8.2 and plotted in Figure 8.3.

Table 8.2 Result of fit in multiple norms; Example 8.3.

fit	m	b	r_1	r_2	r_∞
L^1	1.000	0.000	2.000	1.414	1.000
L^2	1.200	−0.600	2.400	1.265	0.800
L^∞	1.333	−1.000	3.667	1.333	0.667

To fit the data in the L^1 norm we express the linear programming problem in standard form as follows:

$$\text{maximize} \quad x_{13}$$
$$x_3 + \cdots + x_{12} + x_{13} = 0$$
$$1x_1 + x_2 + x_3 - x_8 = 1$$
$$2x_1 + x_2 + x_4 - x_9 = 1$$
$$3x_1 + x_2 + x_5 - x_{10} = 3$$
$$4x_1 + x_2 + x_6 - x_{11} = 5$$
$$5x_1 + x_2 + x_7 - x_{12} = 5.$$

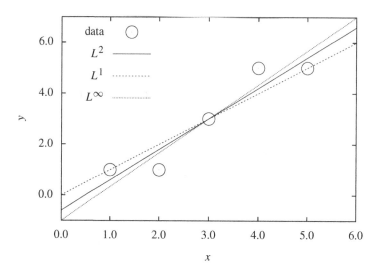

Figure 8.3 Data and fits in multiple norms; Example 8.3.

Here variable x_1 is the slope, x_2 is the intercept. The parameters σ are $x_3, \ldots, x_7$ and the parameters ρ are $x_8, \ldots, x_{12}$. The σ and ρ variables are restricted. We find a feasible basis using phase I, then optimize the objective x_{13} with phase II. At the end of the calculation we have the result:

$$
\begin{pmatrix} x_3 \\ x_1 \\ x_9 \\ x_6 \\ x_2 \\ x_{13} \end{pmatrix} = \begin{pmatrix} 0 \\ 1.00 \\ 1.00 \\ 1.00 \\ 0 \\ -2.00 \end{pmatrix}
$$

$$
+ \begin{pmatrix} 0 & -1.00 & 1.00 & 2.00 & -2.00 & 0 & 1.00 \\ 0 & -0.500 & 0 & 0.500 & -0.500 & 0 & 0.500 \\ 1.00 & 0.500 & 0 & -1.50 & 1.50 & 0 & -0.500 \\ 0 & 0.500 & 0 & 0.500 & -0.500 & 1.00 & -0.500 \\ 0 & 1.50 & 0 & -2.50 & 2.50 & 0 & -1.50 \\ -2.00 & -1.00 & -2.00 & -2.00 & 0 & -2.00 & -1.00 \end{pmatrix} \begin{pmatrix} x_4 \\ x_7 \\ x_8 \\ x_5 \\ x_{10} \\ x_{11} \\ x_{12} \end{pmatrix} ,
$$

which indicates successful termination.

To solve the problem in L^∞ we write the standard form linear programming problem:

$$\text{maximize} \quad x_{13}$$

$$x_3 + x_8 + x_{13} = 0$$

$$1x_1 + x_2 + x_3 - x_8 = 1$$

$$2x_1 + x_2 + x_4 - x_9 = 1$$

$$3x_1 + x_2 + x_5 - x_{10} = 3$$
$$4x_1 + x_2 + x_6 - x_{11} = 5$$
$$5x_1 + x_2 + x_7 - x_{12} = 5$$
$$x_4 + x_9 - x_3 - x_8 = 0$$
$$x_5 + x_{10} - x_3 - x_8 = 0$$
$$x_6 + x_{11} - x_3 - x_8 = 0$$
$$x_7 + x_{12} - x_3 - x_8 = 0.$$

The σ and ρ variables are labeled here as in the previous calculation. Again, they are all restricted variables. The initial feasible basis is determined by phase I of the simplex method. Then, phase II finds the optimum value of the objective x_{13}. The final calculation of the method is:

$$
\begin{pmatrix} x_6 \\ x_1 \\ x_{12} \\ x_{10} \\ x_2 \\ x_7 \\ x_9 \\ x_5 \\ x_3 \\ x_{13} \end{pmatrix}
=
\begin{pmatrix} 0.667 \\ 1.33 \\ 0.667 \\ 0.333 \\ -1.00 \\ 0 \\ 0.667 \\ 0.333 \\ 0.667 \\ -0.667 \end{pmatrix}
+
\begin{pmatrix} 1.00 & -0.667 & 0.667 \\ 0 & 0.667 & -0.667 \\ 0 & 1.33 & -0.333 \\ 0 & 0.667 & 0.333 \\ -1.00 & -1.00 & 2.00 \\ 1.00 & -1.00 & 1.00 \\ 0 & 0.333 & 0.667 \\ 1.00 & -0.333 & 0.333 \\ 1.00 & 0.333 & -0.333 \\ -1.00 & -0.333 & -0.667 \end{pmatrix}
\begin{pmatrix} x_4 \\ x_{11} \\ x_8 \end{pmatrix}.
$$

This indicates successful termination. The results are given in Table 8.2 and Figure 8.3.

8.7   Splines

One problem with polynomial data fitting is that degree n polynomials have n extrema. When the data is noisy, or derived from a non-polynomial function, a polynomial fit can have larger oscillations than the original data. In such cases, it would be good if smoothness could be imposed. One way to achieve smoothness is to use a smooth, possibly nonlinear, model, say one based on Gaussian forms. Another approach uses linear functions with compact support. B-splines are general piece-wise polynomial functions with wide applicability.

We will model the data with the following series,

$$y(x) = \sum_{i=0}^{M-1} P_i B_{i,d}(x). \tag{8.22}$$

Here, coefficients P are amplitudes, and $B_{i,d}(x)$ are B-spline basis functions of degree d (B is for "basic"). (If the degree is d, then the order is $d+1$. Some authors label splines

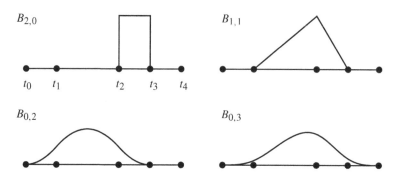

Figure 8.4 Some B-splines on a non-uniform knot partition.

by their order.) Some representative B-splines are shown in Figure 8.4. To evaluate these basis functions, one needs also a set of support abscissas, *knots* in the spline jargon.

One way to construct B-splines is with a simple recursion [53]:

$$B_{i,0}(x) = \begin{cases} 1 & t_i \le x < t_{i+1} \\ 0 & \text{otherwise,} \end{cases} \qquad (8.23)$$

and

$$B_{i,j}(x) = \frac{x - t_i}{t_{i+j} - t_i} B_{i,j-1}(x) + \frac{t_{i+j+1} - x}{t_{i+j+1} - t_{i+1}} B_{i+1,j-1}(x) \qquad (8.24)$$

for $j = 1, ..., d$. The number of knots needed to support calculation (8.22) is $M + d + 1$.

The elementary bases $B_{i,0}(x)$ clearly have compact support; they are zero outside the interval $[t_i, t_{i+1}]$. They are lines (degree zero polynomials) inside the support region.

Through the recursion (8.24), it can be seen that $B_{i,j}(x)$ is zero outside $[t_i, t_{i+j+1}]$ and that it will be composed of $j + 1$ degree j polynomials. Spline $B_{i,j}(x)$ has C^{j-1} continuity.

B-splines are a *partition of unity*:

$$1 = \sum_{i=i_a}^{i_b} B_{i,d}(x)$$

if, for $t_k \le x < t_{k+1}$, $i_a \le k - d + 1$ and $i_b \ge k + 1$. This is easily shown by induction. First, from the definition of $B_{i,0}$ (8.23), it is clear that this relation works for $d = 0$. Now assume the partition of unity property holds for all degrees $0, ..., d - 1$ and apply the recursion (8.24):

$$\sum_{i=k-d+1}^{k+1} B_{i,d}(x) = \sum_{i=k-d+1}^{k+1} \left[\frac{x - t_i}{t_{i+d} - t_i} B_{i,d-1}(x) + \frac{t_{i+d+1} - x}{t_{i+d+1} - t_{i+1}} B_{i+1,d-1}(x) \right]$$

$$= -\frac{x - t_{k-d}}{t_k - t_{k-d}} B_{k-d,d-1}(x) + \sum_{i=k-d}^{k+1} \left[\frac{x - t_i}{t_{i+d} - t_i} B_{i,d-1}(x) + \frac{t_{i+d} - x}{t_{i+d} - t_i} B_{i,d-1}(x) \right]$$

$$+ \frac{t_{k+d+2} - x}{t_{k+d+2} - t_{k+2}} B_{k+2,d-1}(x).$$

The first and third terms on the right-hand side are zero, and the second term simplifies to give

$$\sum_{i=k-d+1}^{k+1} B_{i,d}(x) = \sum_{i=k-d}^{k+1} B_{i,d-1}(x) = 1.$$

To fit data to a spline basis, one first chooses the degree d and knot set (*partition*), then solves a set of linear equations to determine the amplitudes. If the data to be fit are N (x, y) points,

$$\begin{pmatrix} B_{0,d}(x_1) & B_{1,d}(x_1) & \cdots & B_{M-1,d}(x_1) \\ B_{0,d}(x_2) & B_{1,d}(x_2) & \cdots & B_{M-1,d}(x_2) \\ \vdots & & & \vdots \\ B_{0,d}(x_N) & B_{1,d}(x_N) & \cdots & B_{M-1,d}(x_N) \end{pmatrix} \begin{pmatrix} P_0 \\ P_1 \\ \vdots \\ P_{M-1} \end{pmatrix} = \begin{pmatrix} y_1 \\ y_2 \\ \vdots \\ y_N \end{pmatrix}. \tag{8.25}$$

Although this matrix appears full, when the argument x_k lies outside the support $[t_i, t_{i+d+1}]$ the term is zero. Consequently when the abscissas are ordered $x_1 < x_2 < \dots < x_N$ the matrix is banded. All matrix elements are nonnegative, and because of the partition of unity property the sum of matrix elements across a row is one. If there is some support interval that contains no data point x, then the corresponding amplitude P_i cannot be determined. In this case matrix $\mathbf{A}$ is singular, and cannot be solved by Householder (for example), but may be solved by careful application of SVD.

Even if system (8.25) is solvable, it may not be the system you really want to solve. Consider the case $M = N$, with data chosen to coincide with consecutive knots. In this case, the interval $[x_1, x_N]$ is divided into $N - 1$ segments, each of which is modeled by a degree d polynomial. There are $(N - 1)(d + 1)$ model parameters. Degree-d splines possess C^{d-1} continuity, and this places $(N - 2)d$ constraints on the model parameters. We also ask that the model fit the data, which provides another N constraints. All together, this leaves $d - 1$ free parameters (Figure 8.5).

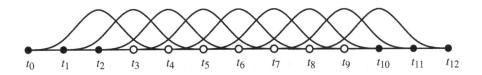

Figure 8.5 Nine cubic splines on a uniform partition. Interpolation at each of the interior knots $t_3, \dots, t_9$ is given by a linear combination of three splines. At knots t_2 and t_{10} only two splines contribute, and at t_1 and t_{11} only one contributes. When the data is collocated with knots, one generally wants each data point to receive the same number of degrees of freedom. In this case, place the data at the open circles (seven interior knots), while determining nine cubic spline amplitudes $P_0, \dots, P_8$. Two extra degrees of freedom remain.

By fitting N splines to N points, one is giving up $d - 1$ degrees of freedom to chance. Instead, use those degrees of freedom to achieve a better fit.

Cubic splines play a special role in data fitting because they are the smoothest, in a sense that will be made specific. With cubic splines, there are four common ways to solve the boundary condition problem:

- Let the second derivative of the solution be zero at the end points. This is the so-called *natural* cubic spline.
- Specify derivatives at the end points.
- If the data itself is periodic, enforce periodicity of slopes and second derivatives.
- Choose boundary values to make the third derivative continuous at the first interior knot: the *not-a-knot* condition [54, 55].

There are many other closures. Behforooz and Papamichael [11] consider a family of closures obtained by extrapolating interior knot values to the end points, and thereby achieving optimal accuracy [10].

Natural cubic splines minimize over all absolutely C^2 functions the quantity

$$\int_a^b |f''(x)|^2 dx. \tag{8.26}$$

In the theory of small deformation elasticity, the elastic energy of a bent rod is proportional to

$$\int_a^b \frac{|y''(x)|^2}{[1 + |y'(x)|^2]^{3/2}} dx \tag{8.27}$$

where the rod's center follows the curve $y(x)$. Thus, minimization of (8.26) approximates minimization of (8.27). This elastic energy is minimized when a draftsman uses a mechanical spline (typically a thin strip of wood) to construct a smooth curve through select points. This correspondence of the mathematical spline with the mechanical one was made by Schoenberg in his seminal work [212, p. 67] and is the origin of the name.

To specify a natural cubic spline, the second derivative at the knot must be specified. In the special case that $t_i - t_{i-1} = h$, the second derivative of (8.22) at knot t_3 (Figure 8.5) is

$$f''(t_3) = \frac{1}{h^2}(P_0 - 2P_1 + P_2).$$

When the knots are not regularly spaced, the numerical coefficients of P_0, P_1, and P_2 can be determined by differentiating the recursion (8.24).

Suppose we have $M + 4$ knots and N data points in the interval $[t_3, t_M]$ (Figure 8.5). To determine coefficients P_0 through P_{M-1} we include the constraint

$$f''(t_3) = L_0 P_0 + L_1 P_1 + L_2 P_2 = 0$$

for the left boundary, and the constraint

$$f''(t_M) = R_{M-3} P_{M-3} + R_{M-2} P_{M-2} + R_{M-1} P_{M-1} = 0$$

for the right boundary. Combine these two equations with N equations (8.25) to give a system of $N + 2$ equations in M unknowns.

Example 8.4   Fit a natural cubic spline to 20 regularly spaced samples of arctan(x) on $[-10, 10]$.

Solution

This this function is smooth and C^∞ with no extrema. Yet, polynomial fits to this function are quite poor and exhibit oscillations. As the polynomial degree increases, the magnitude of these oscillations increases, so with increasing accuracy comes diminished stability (Figure 8.6). Fitting these 20 data points to 22 cubic splines, and using

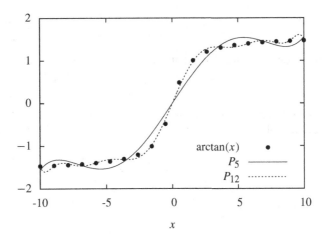

Figure 8.6 Some polynomial fits to arctan(x); Example 8.4.

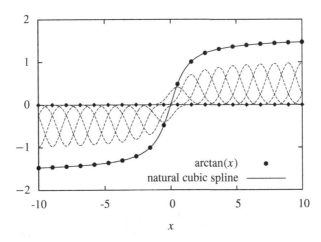

Figure 8.7 A natural cubic spline fit to arctan(x); Example 8.4. The component B-splines are shown by dashed curves. Their sum is the solid curve which passes through the support data.

the natural boundary conditions at the end points ± 10 gives a banded 22×22 matrix equation for the spline amplitudes. The component splines and their sum are shown in Figure 8.7. The fit exhibits no spurious oscillations.

The recursion for differentiation is

$$B'_{i,0}(x) = \delta(x - t_i)$$

$$B'_{i,j}(x) = \frac{(x - t_i)B'_{i,j-1}(x) + B_{i,j-1}(x)}{t_{i+j} - t_i}$$

$$+ \frac{(t_{i+j+1} - x)B'_{i_1,j-1}(x) - B_{i+1,j-1}(x)}{t_{i+j+1} - t_{i+1}},$$

where $\delta(x - t_i)$ is the Dirac delta. It is infinite when the argument is zero, zero otherwise, and its integral over all x is one. B'_{i0} is clearly unbounded at t_i and t_{i+1}, and B'_{i1} is also unbounded at its knots. However, $B_{i,2}$ is C^1 so its derivative is definite.

A simpler result emerges from the definition of B-splines in terms of divided differences (Problem 8.4). Let f_x^d be the function f_x of t, with parameter x,

$$f_x(t) = (t - x)_+ = \begin{cases} (t - x) & t > x \\ 0 & t \le x, \end{cases}$$

raised to the power d. Then,

$$B_{i,d}(x) = (t_{i+d+1} - t_i)f_x^d[t_i, ..., t_{i+d+1}] \tag{8.28}$$

where $f_x^d[\cdots]$ is a divided difference in the sense of Newton's interpolation tableau. Differentiating with respect to the parameter,

$$\frac{d}{dx}f_x^d = -df_x^{d-1}$$

$$\frac{d}{dx}f_x^d[t_i, ..., t_{i+d+1}] = -df_x^{d-1}[t_i, ..., t_{i+d+1}]$$

$$= -\frac{d}{t_{i+d+1} - t_i}(f_x^{d-1}[t_{i+1}, ..., t_{i+d+1}] - f_x^{d-1}[t_i, ..., t_{i+d}])$$

$$\frac{d}{dx}B_{i,d}(x) = \frac{d}{t_{i+d} - t_i}B_{i,d-1}(x) - \frac{d}{t_{i+d+1} - t_{i+1}}B_{i+1,d-1}(x). \tag{8.29}$$

And, differentiating again,

$$\frac{d^2}{dx^2}f_x^d[t_i, ..., t_{i+d+1}] = d(d-1)f_x^{d-2}[t_i, ..., t_{i+d+1}]$$

$$= \frac{d(d-1)}{t_{i+d+1} - t_i}\left[\frac{f_x^{d-2}[t_{i+2}, ..., t_{i+d+1}] - f_x^{d-2}[t_{i+1}, ..., t_{i+d}]}{t_{i+d+1} - t_{i+1}}\right.$$

$$\left. - \frac{f_x^{d-2}[t_{i+1}, ..., t_{i+d}] - f_x^{d-2}[t_i, ..., t_{i+d-1}]}{t_{i+d} - t_i}\right]$$

$$\frac{d^2}{dx^2}B_{i,d}(x) = \frac{d(d-1)}{(t_{i+d+1}-t_{i+1})(t_{i+d+1}-t_{i+2})}B_{i+2,d-2}(x)$$
$$-\frac{d(d-1)}{(t_{i+d+1}-t_{i+1})(t_{i+d}-t_{i+1})}B_{i+1,d-2}(x)$$
$$-\frac{d(d-1)}{(t_{i+d}-t_i)(t_{i+d}-t_{i+1})}B_{i+1,d-2}(x)$$
$$+\frac{d(d-1)}{(t_{i+d}-t_i)(t_{i+d-1}-t_i)}B_{i,d-2}(x).$$

It is sometimes necessary to integrate B-splines, and a simple relation is available for that purpose also [16]. Using (8.29), the derivative of a linear combination of basis splines is

$$\frac{d}{dx}\sum_{i=-\infty}^{+\infty}c_i B_{i,d+1} = (d+1)\sum_{i=-\infty}^{+\infty}\frac{c_i-c_{i-1}}{t_{i+d+1}-t_i}B_{i,d}.$$

Now let $c_i = c_{i+1}...$ be 1, and let $c_{i-1} = c_{i-2}...$ be zero,

$$\frac{d}{dx}\sum_{j=i}^{+\infty}B_{j,d+1} = \frac{d+1}{t_{i+d+1}-t_i}B_{i,d},$$

and integrate

$$\int_{-\infty}^{x}B_{i,d}(s)ds = \frac{t_{i+d+1}-t_i}{d+1}\sum_{j=i}^{\infty}B_{j,d+1}(x).$$

Recall that $B_{j,d+1}(x)$ is zero when $x \notin [t_j, t_{j+d+2}]$ so, if $t_k \le x < t_{k+1}$, then

$$\int_{-\infty}^{x}B_{i,d}(s)ds = \frac{t_{i+d+1}-t_i}{d+1}\begin{cases}\sum_{j=\max(k-d-1,i)}^{k}B_{j,d+1}(x) & k \ge i \\ 0 & \text{otherwise.}\end{cases}$$

Problems

8.1 To appreciate the meaning of "best fit" in the Lebesgue norms, plot in 2D the "circles" $L^1(\mathbf{x}) = 1$, $L^2(\mathbf{x}) = 1$, and $L^\infty(\mathbf{x}) = 1$.

8.2 [142] Solve

$$\begin{pmatrix}1 & 1 & 1 & 1 & 1 \\ \mu & & & & \\ & \mu & & & \\ & & \mu & & \\ & & & \mu & \\ & & & & \mu\end{pmatrix}\begin{pmatrix}x_1 \\ x_2 \\ x_3 \\ x_4 \\ x_5\end{pmatrix} = \begin{pmatrix}\mu \\ 0 \\ -5 \\ 5 \\ -5 \\ 0\end{pmatrix}$$

with $\mu = 10^{-9}$.

8.3 Find the least squares fit of the following data

x	y	f
1	1	1.985
4	2	5.990
1	6	7.970
3	4	6.982

to the model

$$f(x, y) = a + bx + cy$$
$$a \geq 0$$
$$b \geq 0$$
$$c \geq 0$$
$$b \geq c.$$

This is most easily done with the Goldfarb–Idnani algorithm.

8.4 [179, 246] Find the slope and intercept, and the variances and covariance of the fit parameters given the following data. Assume that all x and y measurements are independent and normally distributed.

x	y	σ_x^2	σ_y^2
0	5.9	1.000e-03	1.000e+00
0.9	5.4	1.000e-03	5.556e-01
1.8	4.4	2.000e-03	2.500e-01
2.6	4.6	1.250e-03	1.250e-01
3.3	3.5	5.000e-03	5.000e-02
4.4	3.7	1.250e-02	5.000e-02
5.2	2.8	1.667e-02	1.429e-02
6.1	2.8	5.000e-02	1.429e-02
6.5	2.5	5.556e-01	1.000e-02
7.4	1.5	1.000e+00	2.000e-03

8.5 Show that B-spline $B_{i,d}(x)$ can be given by the divided difference (8.28).

9 Integration

Integration is the solution of the problem

$$I = \int_a^b f(x)dx. \tag{9.1}$$

The closely related problem of finding $y(b)$ given

$$\frac{dy}{dx} = f(x, y)$$

$$y(a) = y_a$$

is an example of an ordinary differential equation, to be described in Chapter 10. The difference is that for ODEs $f = y'$ depends on y as well as on x. The mathematical approaches are similar, but there are also important differences.

9.1 Newton–Cotes

The simplest approach to solving a problem like (9.1) is to construct a polynomial interpolation function

$$f(x) \approx P_n(x)$$

using $n + 1$ support points $a, a + h, ..., a + (n - 1)h, b$, with

$$h = \frac{b - a}{n}.$$

One then integrates the polynomial to get the desired result.

Example 9.1 Use a degree two polynomial to integrate (9.1).

Solution
The polynomial can be constructed by the Newton tableau interpolation method:

a	f_0		
		$\frac{f_1 - f_0}{h}$	
$a + h$	f_1		$\frac{f_2 - 2f_1 + f_0}{2h^2}$
		$\frac{f_2 - f_1}{h}$	
$a + 2h$	f_2		

$$P_2(x) = f_0 + \frac{f_1 - f_0}{h}(x - a) + \frac{f_2 - 2f_1 + f_0}{2h^2}(x - a)(x - a - h)$$

$$I \approx \int_a^{a+2h} P_2(x)dx = \frac{h}{3}(f_0 + 4f_1 + f_2).$$

The steps of this example need be carried out only once, and the coefficients tabulated. The results of this exercise are given in Table 9.1. Boole's rule is attributed to G. Boole [19, p. 47], and Weddle's rule to T. Weddle [237], though both of these and the $n = 5$ case were described by T. Simpson over 100 years earlier [217, p. 117], and Simpson's rule was first written by Newton (as acknowledged by Simpson [217, p. vii]). Weddle actually proposed an approximation to the $n = 6$ Newton–Cotes formula,

$$I = \int_0^{6h} f(x)dx \approx \frac{3h}{10}(f_0 + f_2 + f_3 + f_4 + f_6 + 5(f_1 + f_3 + f_5)).$$

R. Cotes is credited with popularizing Newton's work through his posthumous contribution [43].

The $n = 8$ integration rule is not numerically stable because its coefficients do not have uniform sign. This allows for the possibility that the mean function value $\bar{f}$,

$$\bar{f} = \frac{\left[989(f_0 + f_8) + 5888(f_1 + f_7) - 928(f_2 + f_6) + 10496(f_3 + f_5) - 4540f_4\right]}{28350}$$

$$\approx \frac{1}{8h}\int_0^{8h} f(x)dx,$$

is not inside the interval spanned by the nine function evaluations. For example, with $f_0, f_1, f_2, f_3, f_4, f_5, f_6, f_7, f_8 = 1, 1, -1, 1, -1, 1, -1, 1, 1$, $\bar{f} = 6875/4725 \approx 1.45 \notin [-1, 1]$.

We know how to estimate the error of a polynomial interpolation function, and we can leverage this to estimate the error of the integral. This will turn out to give pessimistic error bounds, and a slightly more sophisticated analysis will give more realistic error bounds.

The error of polynomial interpolation is

$$P_n(x) - f(x) = -\frac{f^{(n+1)}(\xi)}{(n+1)!}\underbrace{(x - a)(x - a - h)\cdots(x - a - nh)(x - b)}_{\omega(x)}.$$

Therefore, the approximation to (9.1)

$$\tilde{I} = \int_a^b P_n(x)dx,$$

will have an error given by

$$\tilde{I} - I = -\int_a^b \frac{f^{(n+1)}(\xi(x))}{(n+1)!}\omega(x)dx,$$

Table 9.1 Coefficients of Newton–Cotes integrals for several degrees n.

name	n	factor	coefficients							
trapezoidal sum rule	1	$\frac{h}{2}$	1	1						
Simpson's rule	2	$\frac{h}{3}$	1	4	1					
Simpson's 3/8 rule	3	$\frac{3h}{8}$	1	3	3	1				
Boole's rule	4	$\frac{2h}{45}$	7	32	12	32	7			
	5	$\frac{5h}{288}$	19	75	50	50	75	19		
Weddle's rule	6	$\frac{h}{140}$	41	216	27	272	27	216	41	
	7	$\frac{7h}{17280}$	751	3577	1323	2989	2989	1323	3577	751
	8	$\frac{4h}{14175}$	989	5888	−928	10496	−4540	10496	−928	5888	989

with a bound

$$|\tilde{I} - I| \le \int_a^b \left| \frac{f^{(n+1)}(\xi(x))}{(n+1)!} \omega(x) \right| dx \le \frac{|f^{(n+1)}(\bar{\xi})|}{(n+1)!} \int_a^b |\omega(x)| dx, \qquad (9.2)$$

where $\bar{\xi}$ is some point in (a, b). Because of the absolute value sign in the integral, some care is required to carry out the integration. We know that the polynomial integrand is zero at $0, h, 2h$, etc., and between these special points the integrand will have constant sign. Therefore,

$$\int_a^b |\omega(x)| dx = \sum_{i=0}^{n-1} \left| \int_{a+ih}^{a+(i+1)h} \omega(x) dx \right|,$$

which is tedious, but not difficult to evaluate. Some error estimates by this method are given in Table 9.2.

These error bounds are correct, but they are not very useful. For example, consider the behavior of a method $n = 2$ operating on a function $f(x)$ that happens to be polynomial of degree 3, i.e., $f^{(3)} = c$, a constant. Then,

$$\int_a^b \frac{f^{(3)}(\xi(x))}{3!} \omega(x) dx = \frac{c}{3!} \int_a^{a+2h} (x-a)(x-a-h)(x-a-2h) dx = 0.$$

For this example, the Newton–Cotes formula with $n = 2$ is exact, not of *order of accuracy* h^3.

There are several methods of deriving tighter, hence more useful, bounds, and they are all a bit lengthy though not terribly complicated. Here is perhaps the easiest [98]. We define

$$f(x) = P_n(x) + A_m(x)g(x),$$

Table 9.2 An estimate of the Newton–Cotes error by (9.2). Do not use this.

poly. degree n	one polynomial: $h = (b-a)/n$	composite rule: N polynomials, $h = (b-a)/(Nn)$
1	$\frac{1}{12}|f^{(2)}(\xi)|h^3$	$\frac{1}{12}(b-a)|f^{(2)}(\xi)|h^2$
2	$\frac{1}{12}|f^{(3)}(\xi)|h^4$	$\frac{1}{24}(b-a)|f^{(3)}(\xi)|h^3$
3	$\frac{49}{720}|f^{(4)}(\xi)|h^5$	$\frac{49}{2160}(b-a)|f^{(4)}(\xi)|h^4$
4	$\frac{19}{360}|f^{(5)}(\xi)|h^6$	$\frac{19}{1440}(b-a)|f^{(5)}(\xi)|h^5$
5	$\frac{2459}{60480}|f^{(6)}(\xi)|h^7$	$\frac{2459}{302400}(b-a)|f^{(6)}(\xi)|h^6$
6	$\frac{71}{2240}|f^{(7)}(\xi)|h^8$	$\frac{71}{13440}(b-a)|f^{(7)}(\xi)|h^7$
7	$\frac{91463}{3628800}|f^{(8)}(\xi)|h^9$	$\frac{91463}{25401600}(b-a)|f^{(8)}(\xi)|h^8$
8	$\frac{18593}{907200}|f^{(9)}(\xi)|h^{10}$	$\frac{18593}{7257600}(b-a)|f^{(9)}(\xi)|h^9$

where $f(x)$ is the function we wish to integrate, $P_n(x)$ is the polynomial of degree n that we will actually integrate, and $A_m g$ is the difference. $A_m(x)$ resembles $\omega(x)$ above, but the polynomial degree may differ:

$$A_m(x) = (x - x_0) \cdots (x - x_m)$$

$$m = \begin{cases} n & \text{if } n \text{ is even} \\ n-1 & \text{if } n \text{ is odd.} \end{cases}$$

With A_m defined as above, g must be

$$g(x) = \frac{f(x) - P_n(x)}{A_m(x)}.$$

Here g is everywhere bounded: when $x = x_i$, $i = 0, ..., m$, $g(x_i) = (f'(x_i) - P_n'(x_i))/A_m'(x_i)$ by L'Hôpital's rule.

For the integration error we have then

$$\int_a^b P_n(x)dx - \int_a^b f(x)dx = -\int_a^b A_m(x)g(x)dx$$

$$= -\left(g(x) \int_a^x A_m(x')dx' \right)\Big|_a^b + \int_a^b g'(x) \int_a^x A_m(x')dx'dx, \quad (9.3)$$

with integration by parts.

The first term on the right-hand side is zero. $\int_a^x A_m(x')dx'$ is zero for $x = a$, trivially. When $x = b$ and $n = m$ the integral is zero because A_m is an odd function on the

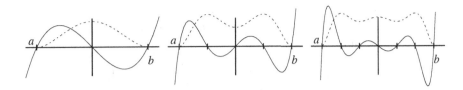

Figure 9.1 $A_m(x)$ vs. x for $m = 2, 4, 6$. The scaled amplitude of $A_m(x)$ is drawn as a solid curve, and the integral $\int_a^x A_m(x')dx'$ is shown dashed (the vertical scales are arbitrary).

interval $[a, b]$. When $x = b$ with $m = n - 1$, the integral will not be zero, and $A_m(b)$ is not zero, but $g(b)$ is zero since $b = x_n$ is a support abscissa of P_n.

We also have $\int_a^x A_m(x')dx' \geq 0$ for all $x \in [a, b]$. This can be appreciated by noting first that

$$\int_a^{(a+b)/2+|y|} A_m(x')dx' = \int_a^{(a+b)/2-|y|} A_m(x')dx' + \int_{(a+b)/2-|y|}^{(a+b)/2+|y|} A_m(x')dx',$$

where the last integral is zero because of the symmetry. Therefore, to assess the sign of $\int_a^x A_m(x')dx'$ for arbitrary x we need only consider values of the integral for $x \leq (a+b)/2 - h = a + (m-2h)/2$.

The argument that the integral is positive in this range begins by considering the interval $a \leq x \leq a + h$. By construction, $A_m(x) \geq 0$ in this interval, and therefore $\int_a^x A_m dx' \geq 0$ for any $x \in [a, a+h]$. Also by construction, $A_m(x) \leq 0$ for $a + h \leq x \leq a + 2h$, etc. If the negative values in the interval $[a + h, a + 2h]$ are each smaller in magnitude than the positive values in the interval $[a, a + h]$, then the integral $\int_a^x A_m dx' \geq 0$ for any $x \in [a + h, a + 2h]$ as well. Extending the integral to $[a + 2h, a + 3h]$ retains the positive sign, and extending again to $[a + 3h, a + 4h]$ we will again have positivity if $|A_m(x)| > A_m(x + h)|$. So, the integral is positive if it can be proved $|A_m(x)| \geq |A_m(x + h)|$ for all $x \leq a + (m-2)h/2$.

Let us assume this inequality is true and see if any contradictions arise:

$$|A_m(x)| \geq |A_m(x + h)|$$
$$|(x - a)(x - a - h)\cdots(x + h - a - mh)(x - a - mh)| \geq$$
$$|(x + h - a)(x + h - a - h)\cdots(x + 2h - a - mh)(x + h - a - mh)|$$
$$|x - a - mh| \geq |x - a + h|$$
$$mh - (x - a) \geq (x - a) + h$$
$$\frac{(m-1)h}{2} \geq (x - a).$$

For all $0 \leq x - a \leq (m-2)h/2$ it is true that $|A_m(x)| \geq |A_m(x+h)|$, which completes the logic showing that $\int_a^x A_m(x')dx'$ is positive.

The positivity of this integral allows us to use the mean value theorem to simplify (9.3):

$$\int_a^b P_n(x)dx - \int_a^b f(x)dx = g'(\xi) \int_a^b \int_a^x A_m(x')dx', \qquad (9.4)$$

for some $\xi \in [a, b]$.

Finally, we want to express $g'(\xi)$ in terms of the function f. To do this we analyze the function

$$F(t) = f(t) - P_n(t) - A_m(t)[g(x) + (t - x)g'(x)]$$
$$F(x) = 0 \tag{9.5}$$
$$F(x_i) = 0 \quad \text{for } i \in [0, ..., m],$$

using derivatives of $f(t) - P_n(t) = g(t)A_m(t)$ to simplify somewhat:

$$F'(t) = f'(t) - P_n'(t) - A_m'(t)[g(x) + (t{-}x)g'(x)] - A_m(t)g'(x)$$
$$= A_m'(t)[g(t){-}g(x)] + A_m(t)[g'(t){-}g'(x)] - A_m'(t)(t{-}x)g'(x) \tag{9.6}$$
$$F'(x) = 0,$$

and

$$F''(t) = A_m(t)g''(t) - A_m''(t)[g(x){-}g(t)] - 2A_m'(t)[g'(x){-}g'(t)]$$
$$- A_m''(t)(t{-}x)g'(x) \tag{9.7}$$
$$F''(x) = 0 \quad \text{if } x = x_i \text{ for some } i \in [0, ..., m].$$

If $x \neq x_i$, $i \in [0, ..., m]$, then $F(t)$ has $m + 2$ distinct zeros. The zero x is a zero of multiplicity 2 because of (9.5) and (9.6). By application of Rolle's theorem, accounting for the multiplicity (see page 355), $F^{(m+2)}(t)$ must have at least one zero in $[a, b]$.

If $x = x_i$, $i \in [0, ..., m]$, then $F(f)$ has $m + 1$ distinct zeros by (9.5). According to (9.6) and (9.7) the zero x_i has multiplicity 3. Again, application of Rolle's theorem, accounting for the multiplicity, tells us that $F^{(m+2)}(t)$ must have at least one zero in $[a, b]$.

Differentiate (9.5) $m + 2$ times, and evaluate at the zero $\xi \in [a, b]$:

$$F^{(m+2)}(\xi) = 0 = f^{(m+2)}(\xi) - P_n^{(m+2)}(\xi) - (m + 2)!g'(x)$$
$$g'(x) = \frac{f^{(m+2)}(\xi)}{(m + 2)!} \tag{9.8}$$

(P_n is a polynomial of degree $n \leq m + 1$, so it has no $m + 2$ derivative).

All together, (9.4) and (9.8) give

$$\int_a^b P_n(x)dx - \int_a^b f(x)dx = \frac{f^{(m+2)}(\xi)}{(m + 2)!} \int_a^b dx \int_a^x A_m(x')dx'. \tag{9.9}$$

The right-hand side is tabulated in Table 9.3.

In Tables 9.2 and 9.3 the errors are expressed in two forms, one proportional to some power of h alone, and one proportional to $(b - a)$ times a lower power of h. These expressions are completely equivalent when one writes $b - a = nh$. The reason for writing the second form is that when one wishes to compare methods for a given integral $\int_a^b f(x)dx$, the range $b - a$ is fixed, although the subdivisions h may be allowed to vary.

Table 9.3 Better estimates of Newton–Cotes error using (9.9).

poly. degree n	one polynomial: $h = (b-a)/n$	composite rule: N polynomials, $h = (b-a)/(Nn)$
1	$\frac{1}{12}f^{(2)}(\xi)h^3$	$\frac{1}{12}(b-a)f^{(2)}(\xi)h^2$
2	$\frac{1}{90}f^{(4)}(\xi)h^5$	$\frac{1}{180}(b-a)f^{(4)}(\xi)h^4$
3	$\frac{3}{80}f^{(4)}(\xi)h^5$	$\frac{1}{80}(b-a)f^{(4)}(\xi)h^4$
4	$\frac{8}{945}f^{(6)}(\xi)h^7$	$\frac{2}{945}(b-a)f^{(6)}(\xi)h^6$
5	$\frac{275}{12096}f^{(6)}(\xi)h^7$	$\frac{55}{12096}(b-a)f^{(6)}(\xi)h^6$
6	$\frac{9}{1400}f^{(8)}(\xi)h^9$	$\frac{3}{2800}(b-a)f^{(8)}(\xi)h^8$
7	$\frac{8183}{518400}f^{(8)}(\xi)h^9$	$\frac{1169}{518400}(b-a)f^{(8)}(\xi)h^8$
8	$\frac{2368}{467775}f^{(10)}(\xi)h^{11}$	$\frac{296}{467775}(b-a)f^{(10)}(\xi)h^{10}$

A *composite integration rule* comes about from repeating a simple integration rule across an interval. For example,

$$\int_a^b f(x)dx = \sum_{i=0}^{N-1} \int_{a_i}^{b_i=a_{i+1}} f(x)dx$$

$$a_i = a + (b-a)\frac{i}{N},$$

where each integral on the right-hand side is further subdivided into segments of length h according to the chosen integration rule. If the trapezoidal sum rule is used,

$$\int_a^b f(x)dx \approx \frac{h}{2}\left[f(a_0) + f(a_1)\right] + \frac{h}{2}\left[f(a_1) + f(a_2)\right] + \cdots + \frac{h}{2}\left[f(a_{N-1}) + f(a_N)\right]$$

$$= \frac{h}{2}f(a_0) + h\left[f(a_1) + \cdots + f(a_{N-1})\right] + \frac{h}{2}f(a_N),$$

with $h = (b-a)/N$, and error $\frac{1}{12}(b-a)f^{(2)}(\xi)h^2$.

One can develop integration methods for any choice of interpolation method using the techniques above. An interesting example incorporates derivative information using the Hermite tableau (cf. (7.1)):

$$a \qquad f_0$$

$$f_0'$$

$$a \qquad f_0 \qquad\qquad \frac{f_1 - f_0 - h f_0'}{h^2}$$

$$\frac{f_1 - f_0}{h} \qquad\qquad \frac{h f_1' + 2 f_0 - 2 f_1 + h f_0'}{h^3}$$

$$a + h \quad f_1 \qquad\qquad \frac{h f_1' + f_0 - f_1}{h^2}$$

$$f_1'$$

$$a + h \quad f_1$$

$$P_3(x) = f_0 + (x - a) f_0' + (x - a)^2 \frac{f_1 - f_0 - h f_0'}{h^2} +$$
$$(x - a)^2 (x - a - h) \frac{h f_1' + 2 f_0 - 2 f_1 + h f_0'}{h^3}$$

$$I \approx \frac{h^2}{12} f_0' + \frac{h}{2} (f_0 + f_1) - \frac{h^2}{12} f_1'. \tag{9.10}$$

The error associated with this method is

$$\int_a^{a+h} P_3(x) dx - \int_a^{b=a+h} f(x) = -\frac{h^4}{720} (b - a) f^{(4)}(\xi), \tag{9.11}$$

for some $\xi \in [a, a + h]$. This is $\mathcal{O}(h^2)$ better than the trapezoidal sum rule. The composite rule for this method is

$$\frac{h^2}{12} f'(a_0) + \frac{h}{2} f(a_0) + h \left[f(a_1) + \cdots + f(a_{N-1}) \right] + \underbrace{\frac{h}{2} f(a_N)}_{} - \frac{h^2}{12} f'(a_N), \tag{9.12}$$

$$\underbrace{\hphantom{\frac{h^2}{12} f'(a_0) + \frac{h}{2} f(a_0) + h \left[f(a_1) + \cdots + f(a_{N-1}) \right] + \frac{h}{2} f(a_N)}}_{\text{composite trapezoidal sum rule}}$$

and in this case the extra two orders of accuracy come at the cost of needing to know only 2 numbers – the derivatives of the function at the endpoints of the integration interval. Error formula (9.11) is given by the following example, which illustrates an alternative technique for finding the error of the Newton–Cotes formulas.

Example 9.2   Use the *Peano kernel* to determine the integration error of (9.10).

Solution

We consider here the linear operator $\mathcal{L}$

$$\mathcal{L} f = \int_a^b P(x | \{x_i, f(x_i)\}) dx - \int_a^b f(x) dx,$$

where the first term on the right is the integral of an interpolation polynomial $P(x)$ fit to a set of support data $\{x_i, f(x_i)\}$. We expand f in a Taylor series,

$$f(x) = f(a) + \cdots + \frac{(x - a)^3}{6} f^{(3)}(a) + \int_a^x \frac{(x - t)^3}{6} f^{(4)}(t) dt,$$

with the series on the right-hand side being a polynomial of degree 3. Since $\mathcal{L} P_3(x) = 0$ with the method (9.10), $\mathcal{L} f$ will simplify to

$$\mathcal{L} f = \mathcal{L} \int_a^x \frac{(x-t)^3}{6} f^{(4)}(t) dt = \int_a^b \mathcal{L} \frac{(x-t)_+^3}{6} f^{(4)}(t) dt,$$

and (see page 127 for details)

$$K = \mathcal{L}(x-t)_+^3 = \frac{3h^2}{12}(a-t)_+^2 + \frac{h}{2}(a-t)_+^3 + \frac{h}{2}(b-t)_+^3 - \frac{3h^2}{12}(b-t)_+^2$$

$$- \int_a^b (x-t)_+^3 dx$$

$$= \frac{h}{2}(b-t)^3 - \frac{3h^2}{12}(b-t)^2 - \int_0^{b-t} u^3 du$$

$$K(t) = -\frac{1}{4}(b-t)^2(a-t)^2.$$

The sign of the kernel is uniform in $[a, b]$, so the mean value theorem may be applied to give

$$\tilde{I} - I = -f^{(4)}(\xi) \int_a^b \frac{1}{24}(b-t)^2(a-t)^2 = -\frac{h^4}{720}(b-a) f^{(4)}(\xi).$$

Example 9.3 Numerically integrate

$$I = \int_0^2 \frac{dx}{1+x^2} \tag{9.13}$$

with composite Newton–Cotes methods.

Solution
The exact solution to this problem is arctan(2), which facilitates a comparison of the calculated numerical error with the theoretical results. Figure 9.2 displays the error $|I - \arctan(2)|$ for a variety of composite integration schemes and a range of step sizes h. Lines of slope h^n are drawn for reference. The integration of a polynomial approximation of degree 1 (i.e., trapezoidal sum rule, p_1) has a theoretical integration error proportional to h^2 for a composite method. This trend is observed in the numerical results. As the order of the method increases, the maximum step size for which the expected accuracy is seen decreases. At the same time, as $h \to 0$, numerical errors begin to dominate the solution – note the errors never improve beyond $\approx 10^{-16} \approx \epsilon \arctan(2)$. These competing trends act to diminish the range of h over which the expected order of accuracy is observed.

Order h^{10} behavior for p_8 is never observed because of the trend noted above, and because this method is unstable.

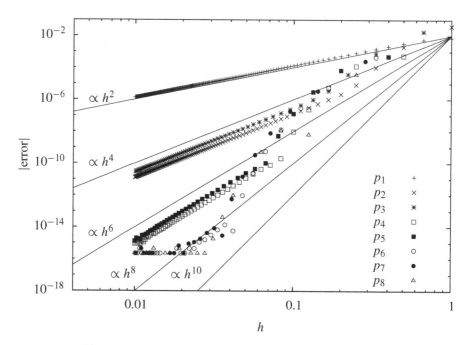

Figure 9.2: Newton–Cotes integration error; Example 9.3.

9.2   Extrapolation

We have seen that Newton–Cotes formulas carry errors proportional to powers of the step size h, so this approximation error should go to zero as h approaches zero. Of course this is not possible, but we can always compute solutions with h, $2h$, $4h$, etc., and extrapolate these to $0h$ by standard methods.

The underlying idea was first proposed by L. F. Richardson in 1911 [191, p. 310] in the context of finite differences, and more fully developed in 1927 [192]. The basic idea is that if a function $f(x)$ has a series

$$f(h) = f_0 + h^p f_p + \cdots, \tag{9.14}$$

then two measurements $f(h_1)$ and $f(h_2)$ can be used to estimate f_0. In the most common case when $h_2 = 2h_1 = 2h$,

$$f_0 \approx f(h) - \frac{f(2h) - f(h)}{2^p - 1}. \tag{9.15}$$

This is called *Richardson extrapolation*. In the context of integration, we consider $f(h)$ to be the desired integral taken with a composite Newton–Cotes method using step size h, where p is the order of accuracy of the chosen Newton–Cotes formula.

The method of *Romberg integration* [196] combines this extrapolation idea with the simple trapezoidal sum rule, and it applies the extrapolation step repeatedly. There is a special feature of this integration rule that makes it especially attractive. To see this we need a bit of a mathematical digression.

The *Bernoulli polynomials* $B_n(x)$ are functions of x with the following properties:

$$B_0(x) = 1 \tag{9.16a}$$

$$B_n'(x) = n B_{n-1}(x) \tag{9.16b}$$

$$\int_0^1 B_n(x)dx = 0, \quad n \geq 1 \tag{9.16c}$$

$$B_n(0) = B_n(1) = 0 \quad n = 3, 5, 7, \dots \tag{9.16d}$$

Equation (9.16c) supplies the constant of integration needed to evaluate (9.16b). Some examples of Bernoulli polynomials are

$$B_1(x) = x - \frac{1}{2}$$

$$B_2(x) = x^2 - x + \frac{1}{6}$$

$$B_3(x) = x^3 - \frac{3}{2}x^2 + \frac{1}{2}x \tag{9.17}$$

$$B_4(x) = x^4 - 2x^3 + x^2 - \frac{1}{30}.$$

The numbers $B_n(1)$ or B_n are called Bernoulli numbers.

Integrating property (9.16b) gives

$$B_{n+1}(x) = (n+1) \int_0^x B_n(x')dx' + B_{n+1}(0), \tag{9.18}$$

and using a change of variables one also obtains

$$B_{n+1}(1-x) = (n+1) \int_0^1 B_n(x')dx' - (n+1) \int_{1-x}^1 B_n(x')dx' + B_{n+1}(0)$$

$$= -(n+1) \int_{1-x}^1 B_n(x')dx' + B_{n+1}(0) \quad n \geq 1. \tag{9.19}$$

If B_n is odd on the interval (0, 1) then the integrals in (9.18) and (9.19) are the negative of each other. Taking account of the sign preceding the second integral, we conclude that B_{n+1} is even on (0, 1).

If B_n is even on the interval (0, 1) then integral in (9.18) is equal to the integral in (9.19). Accounting for the sign of the second integral, B_{n+1} will be an odd function if $B_{n+1}(0) = 0$ and $n \geq 1$. This is indeed the case. To show it, integrate (9.18), then switch the order of integration on the double integral:

$$\int_0^1 dx\, B_{n+1}(x) = 0 = (n+1) \int_0^1 (1-x) B_n(x)dx + B_{n+1}(0) \quad n \geq 1. \tag{9.20}$$

When B_n is even, the integrand is odd so the integral is zero. This forces $B_{n+1}(0)$ to be zero.

Finally, evaluation of (9.18) at $x = 1$, using property (9.16c), gives $B_{n+1}(1) = B_{n+1}(0)$ for all n. This, with the result $B_{2n+1}(0) = 0$ for $n \geq 1$, gives the important result (9.16d)

Using the Bernoulli functions we can now derive an important result to help analyze Romberg integration. Begin with an integral and use property (9.16a). With $h = b - a$,

$$\int_a^b f(x)dx = \int_a^b f(x) \underbrace{B_0 \left(\frac{x-a}{h} \right)}_{1 \text{ per } (9.16a)} dx$$

$$= \int_a^b hf(x) \frac{d}{dx} B_1 \left(\frac{x-a}{h} \right) dx,$$

with property (9.16b),

$$= \left(hf(x) B_1 \left(\frac{x-a}{h} \right) \right) \Big|_a^b - h \int_a^b f'(x) B_1 \left(\frac{x-a}{h} \right) dx,$$

integrating by parts,

$$= \underbrace{\frac{h}{2}[f(a) + f(b)]}_{\text{TSR: } T(h)} - h \int_a^b f'(x) B_1 \left(\frac{x-a}{h} \right) dx$$

$$= T(h) - \frac{h^2}{2} \int_a^b f'(x) \frac{d}{dx} B_2 \left(\frac{x-a}{h} \right) dx,$$

with property (9.16b),

$$= T(h) - \frac{h^2}{2} \left(f'(x) B_2 \left(\frac{x-a}{h} \right) \right) \Big|_a^b + \frac{h^2}{2} \int_a^b f''(x) B_2 \left(\frac{x-a}{h} \right) dx,$$

by parts,

$$= T(h) - \frac{h^2 B_2(0)}{2} (f'(b) - f'(a)) + \frac{h^3}{3!} \int_a^b f'''(x) \frac{d}{dx} B_3 \left(\frac{x-a}{h} \right) dx$$

$$= T(h) - \frac{h^2 B_2(0)}{2}(f'(b) - f'(a)) + \underbrace{\frac{h^3}{3!}\left(f'''(x)B_3\left(\frac{x-a}{h}\right)\right)\Big|_a^b}_{0 \text{ per } (9.16d)}$$

$$- \frac{h^3}{3!}\int_a^b f^{(4)}(x)B_3\left(\frac{x-a}{h}\right)dx$$

$$\vdots$$

$$= TSR - \sum_{p=1}^{\infty}\frac{h^{2p}B_{2p}(0)}{(2p)!}\left[f^{(2p-1)}(b) - f^{(2p-1)}(a)\right].$$

This result is called the Euler–Maclaurin sum formula, first written by Euler in 1738 [59]. Because the terms $f^{(2p-1)}(a)$ and $f^{(2p-1)}(b)$ have different signs, when applied to the composite trapezoidal sum rule the result is unchanged:

$$T(h) = \int_a^b f(x)dx + \sum_{p=1}^{\infty}\frac{h^{2p}B_{2p}(0)}{(2p)!}[f^{(2p-1)}(b) - f^{(2p-1)}(a)]. \qquad (9.21)$$

Without this analysis, we knew only (Table 9.3) that the first term in the error series is $\mathcal{O}(h^2)$. Now we know that all terms are even, which suggests that extrapolation should be based on an interpolation polynomial $P(h^2)$, not $P(h)$. The exact integral is the coefficient of h^0. Since we only care about the numerical value of the interpolation function at $h = 0$, the most suitable extrapolation method is Neville's.

Example 9.4   Integrate (9.13) using Romberg integration.

Solution
In the special case that the h sequence changes by factors of 2, i.e., $h_0, h_0/2, h_0/4$, etc., and we are interested only in interpolation to $h = 0$, the Neville interpolation formulae take a particularly convenient form. The first column of interpolants P_i are obtained by direct integration. The calculation for the second column of interpolants (5.10) becomes

$$P_{i,i+1} = \frac{4P_{i+1} - P_i}{3},$$

and the third column (5.11) becomes

$$P_{i,i+2} = \frac{16P_{i+2} - P_i}{15},$$

and so on.

In the case that $h_0 = 1/8$, (i.e., the interval $[0, 2]$ is subdivided into 16 steps of the composite trapezoidal sum rule), the following Neville tableau can be calculated:

1.106 316 631 718 353

        1.107 148 406 151 106

1.106 940 462 542 918   1.107 148 717 751 239

        1.107 148 698 276 231

1.107 096 639 342 903

Numbers in the first column of interpolants, having come from the trapezoidal sum rule, converge at rate h^2. The numbers in the second column of interpolants, being a degree one fit to the Euler–Maclaurin formula, converge at rate h^4, etc.

Figure 9.3 displays the error $|I - \arctan(2)|$ for the Romberg integration of (9.13) using extrapolation with two resolutions (p_1), three resolutions as shown above (p_2), and four resolutions (p_3). These should be expected to converge at rates h^4, h^6, and h^8, respectively. The h^4 and h^6 trends are seen in the figure. The h^8 trend is not observed because numerical error dominates the approximation error in the asymptotic regime.

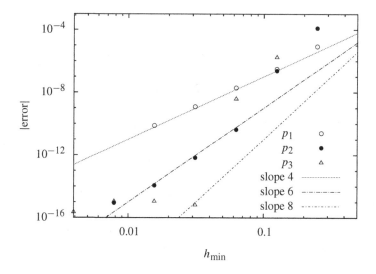

Figure 9.3 Error for (9.13) by Romberg integration; Example 9.4.

To continue the idea of Richardson extrapolation, consider the question "how can one extrapolate when the order p of the method is not known?" Using (9.14) one can combine three measurements as follows:

$$f(h) = f_0 + (h)^p f_p + \cdots$$
$$f(2h) = f_0 + (2h)^p f_p + \cdots$$
$$f(4h) = f_0 + (4h)^p f_p + \cdots$$

$$f(2h) - f(h) = (2^p - 1)(h)^p f_p + \cdots$$
$$f(4h) - f(2h) = 2^p(2^p - 1)(h)^p f_p + \cdots$$

$$2^p \approx \frac{f(4h) - f(2h)}{f(2h) - f(h)} \tag{9.22a}$$

$$p \approx \log_2 \frac{f(4h) - f(2h)}{f(2h) - f(h)} = \frac{\log \frac{f(4h)-f(2h)}{f(2h)-f(h)}}{\log 2}. \tag{9.22b}$$

This is an important application of Richardson extrapolation to the measurement of the order of accuracy.

One can now combine (9.22a) with (9.15) to extrapolate to $h = 0$ without knowing the order p in advance:

$$f(0) = f(h) - \frac{(f(2h) - f(h))^2}{f(h) - 2f(2h) + f(4h)}.$$

This is another application of Aitken's Δ^2 method (cf. (6.29)).

9.3 Adaptivity

For any given Newton–Cotes formula, approximation error may be reduced by subdividing the interval $[a, b]$ into a number of finer intervals, giving the associated composite Newton–Cotes formula. This process decreases h, which decreases the approximation error, but it increases the number of function evaluations, which may be expensive. As the number of quadrature points increases, one might expect numerical error to grow and ultimately dominate the numerical solution.

To control this cost and prevent excessive numerical error, it is advantageous to adaptively partition the domain $[a, b]$ to use small steps h only where necessary to achieve the desired accuracy.

An approach to this problem based on Simpson's rule was first proposed by McKeeman and analyzed and enhanced by many others [77, 150, 156, 157]. The goal is to find a partition, using successive trisections, such that the integral on $[a, b]$ has a error less than some tolerance ϵ_t.

For any coarse interval $[x_0, x_6]$, with associated error target ϵ, we can associate three finer partitions of equal size, $[x_0, x_2]$, $[x_2, x_4]$, $[x_4, x_6]$, with associated error targets $\epsilon/3$. In the interval $[x_0, x_6]$ there are seven quadrature points, hence seven function evaluations. If this process is invoked recursively, all seven of these function evaluations are reused.

If $S[a, b]$ is the Simpson's rule integral on domain $[a, b]$, then

$$I_c = S[x_0, x_6]$$

and

$$I_f = S[x_0, x_2] + S[x_2, x_4] + S[x_4, x_6]$$

are two estimates of the same integral, and

$$e_c = \frac{80}{81}|I_c - I_f| \approx |I_c - I_f|$$

is an estimate of the error associated with I_c. The estimate $\frac{80}{81}|I_c - I_f|$ comes from Richardson extrapolation, but the simplification is good enough for current purposes. McKeeman's [156] original algorithm compared this error with ϵ to assess whether or not additional refinement is indicated.

Gander and Gautschi [77] developed different criteria that include safety checks that safeguard against infinite recursion. They reason that if I_e is an estimate of the integral, and if the goal ϵ is of order machine precision, then $|I_c - I_f|$ should be negligible compared to $|I_e|$, i.e.,

$$\text{round}(I_e + I_c - I_f) = \text{round}(I_e). \tag{9.23}$$

If there are insufficient machine representable numbers between x_0 and x_6 then

$$\text{round}(x_1) \leq \text{round}(x_0) \quad \text{or} \quad \text{round}(x_5) \geq \text{round}(x_6). \tag{9.24}$$

If either criteria (9.23) or (9.24) are satisfied, then the recursion is terminated. Otherwise, each interval is recursively refined. To deal with error goals ϵ other than machine precision, the estimate I_e may be scaled by the ratio of ϵ to machine precision.

The criterion of McKeeman and those of Gander and Gautschi are based on the assumption that overall error is dominated by approximation error, i.e., that refinement of any interval will give a more accurate result. This assumption is not always valid: recall Figure 9.2. If the integration error goal ϵ is too small, the computed solution will carry a numerical error that exceeds its approximation error.

Example 9.5   Estimate

$$I = \int_{1/50}^{100} \cos(1/x)\,dx$$

using adaptive Simpson's rule, with error tolerance of 10^{-6}.

Solution
This function has a lot of structure where x is small, and very little when x is large, so it is a great candidate for adaptive integration. The function is displayed in Figure 9.4 along with the recursion depth d, which is related to the step size h by

$$h = \frac{(b - a)}{2 \times 3^d}.$$

On 7/9 of the domain, $d = 2$ or 3 – relatively coarse partitioning equivalent to the division of $[a, b]$ into only 18 or 54 equal parts. Roughly 0.01% of the domain has $d = 12$ – the finest partition equivalent to division into one half million parts.

The computed result and the "exact" result differ by 7×10^{-7}, so the algorithm achieved the error target within an order of magnitude.

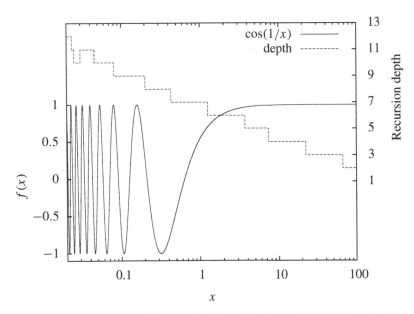

Figure 9.4: Adaptive Simpson's rule applied to $\cos(1/x)$; Example 9.5.

9.4 Gaussian quadrature

The Newton–Cotes formulas estimate a function using a polynomial constructed from uniformly spaced support abscissas $x_i = a + ih$. In this section we look at a class of methods called Gaussian quadrature that effectively optimize the placement of the x_i in order to achieve the highest accuracy. The cost of this improved accuracy is (i) a loss of flexibility in the end points a, b, and (ii) a very high cost for the calculation of quadrature weights and quadrature abscissas. However, in common with Newton–Cotes these quantities have already been worked out for most cases of interest, and can therefore be found in tables.

The method described here was first described by Gauss [78], and the mathematical development advanced by Christoffel [38] and Stieltjes [224]. The methods described here for the construction of new Gaussian quadrature formulas come principally from Golub and Welsch [94] and Gautschi [80, 81].

We will be concerned with integrals of the form

$$I = \int_a^b \omega(x) f(x) dx,$$

and we will seek approximations of the form

$$I \approx \sum_{i=1}^n w_i f(x_i),$$

where w_i are the quadrature weights and x_i are the quadrature abscissas, both to be determined.

The function $\omega(x)$ is required to be nonnegative, and all moments

$$\int_a^b x^k \omega(x)dx, \quad k \geq 0$$

exist and are finite, and for nonnegative functions f,

$$\int_a^b \omega(x)f(x)dx = 0 \quad \Longrightarrow \quad f(x) = 0, \quad x \in [a, b]. \tag{9.25}$$

The method of construction will be concerned with real polynomials $x^n + a_{n-1}x^{n-1} + \cdots + a_0$ (note the coefficient a_n of the highest order term is taken to be 1). We define the scalar product (f, g) to be

$$(f, g) = \int_a^b \omega(x)f(x)g(x)dx,$$

and we say polynomial $f(x)$ is orthogonal to polynomial $g(x)$ if $(f, g) = 0$.

There is a systematic way to construct orthogonal polynomials for a given weight function $\omega(x)$ and interval $[a, b]$. It proceeds as follows. Let

$$\begin{aligned}
p_0(x) &= 1 \\
p_1(x) &= (x - \delta_1)p_0(x) \\
p_i(x) &= (x - \delta_i)p_{i-1}(x) - \gamma_i^2 p_{i-2}(x), \quad i \geq 2,
\end{aligned} \tag{9.26}$$

where

$$\delta_i = (xp_{i-1}, p_{i-1})/(p_{i-1}, p_{i-1}) \tag{9.27}$$

$$\gamma_i^2 = (p_{i-1}, p_{i-1})/(p_{i-2}, p_{i-2}). \tag{9.28}$$

That is, by evaluating scalar products (xp, p) and (p, p), we define the coefficients δ, γ that allow high degree orthogonal polynomials to be constructed from lower degree ones. The recursive evaluation of $p_i(x)$ using equations (9.26) is referred to as Stieltjes' algorithm. It is numerically stable.

The reason this works is as follows. We can write a generic polynomial of degree $i + 1$ as:

$$p_{i+1} = (x - \delta_{i+1})p_i + c_{i-1}p_{i-1} + \cdots + c_0 p_0.$$

We assume that polynomials $p_0, \ldots, p_i$ are orthogonal, and use $(p_{i+1}, p_k) = 0, k = 0, \ldots, i$, to determine the coefficients of the model.

First, with $k = i$, we get

$$(p_{i+1}, p_i) = (xp_i, p_i) - \delta_{i+1}(p_i, p_i) = 0,$$

which requires that δ_{i+1} be determined by (9.27).

Then, with $k < i$ we get

$$(p_{i+1}, p_k) = (xp_i, p_k) + c_k(p_k, p_k) = 0$$

$$c_k = -\frac{(xp_i, p_k)}{(p_k, p_k)} = -\frac{(p_i, xp_k)}{(p_k, p_k)}.$$

To analyze this, we write

$$p_{k+1} = (x - \delta_{k+1})p_k - \gamma_{k+1}p_{k-1}$$

$$xp_k = p_{k+1} - \delta_{k+1}p_k + \gamma_{k+1}p_{k-1}$$

$$(xp_k, p_i) = \begin{cases} (p_i, p_i) & k = i - 1 \\ 0 & k < i - 1, \end{cases}$$

and therefore

$$c_{i-1} = -\frac{(p_i, p_i)}{(p_{i-1}, p_{i-1})}$$

$$c_k = 0, \quad k < i - 1.$$

Finally, since $\omega(x)$ is nonnegative and p_i^2 and p_{i-1}^2 are nonnegative, the coefficient c_{i-1} is negative, which permits us to write $c_{i-1} = -\gamma_{i+1}^2$, which establishes (9.28).

Having defined the orthogonal polynomials, we now define the quadrature weights and abscissas. The n abscissas of a method are given by the zeros of the polynomial p_n, and the n weights are defined by the system of linear equations

$$\sum_{i=1}^{n} w_i p_k(x_i) = \begin{cases} (p_0, p_0) & k = 0 \\ 0 & \text{otherwise} \end{cases} \tag{9.29}$$

(the weights are defined this way, but not calculated this way – do not use this).

Let's first show that $p_n(x)$ has n real simple (multiplicity 1) roots, and that they lie in $[a, b]$. Let $y_1, ..., y_m$ be the collection of real roots that lie in $[a, b]$, and that are of odd multiplicity (i.e., so p_n changes sign at such a root). We make the polynomial

$$q_m(x) = (x - y_1) \cdots (x - y_m)$$

and consider

$$(p_n, q_m) = \int_a^b \omega(x) p_n(x) q_m(x) dx,$$

since q_m changes sign wherever p_n does, the product $p_n q_m$ is nonnegative and $(p_n, q_m) \neq 0$ by (9.25). However, p_n is orthogonal to all polynomials of degree lower than n, and therefore $m = n$: there are n real roots in the interval $[a, b]$, and they must be simple.

Second, we show that the system of equations (9.29) is solvable. We can write this system in matrix form

$$\begin{pmatrix} p_0(x_1) & \cdots & p_0(x_n) \\ \vdots & & \vdots \\ p_{n-1}(x_1) & \cdots & p_{n-1}(x_n) \end{pmatrix} \begin{pmatrix} w_1 \\ \vdots \\ w_n \end{pmatrix} = \begin{pmatrix} (p_0, p_0) \\ 0 \\ \vdots \end{pmatrix}. \tag{9.30}$$

If this matrix is singular, then there is a nonzero vector $(c_0, \ldots, c_{n-1})$ such that

$$\begin{pmatrix} c_0 & \cdots & c_{n-1} \end{pmatrix} \begin{pmatrix} p_0(x_1) & \cdots & p_0(x_n) \\ \vdots & & \vdots \\ p_{n-1}(x_1) & \cdots & p_{n-1}(x_n) \end{pmatrix} = \begin{pmatrix} 0 & \cdots & 0 \end{pmatrix},$$

or

$$g(x) = c_0 p_0(x) + \cdots + c_{n-1} p_{n-1}(x),$$

a degree $n-1$ polynomial, has n zeros $x_1, \ldots, x_n$. This implies that $g(x)$ is identically zero by the fundamental theorem of algebra. Since the coefficients are not all zero, the only way this can be true is if the polynomials are linearly dependent. This contradicts the fact that they are orthogonal. Since the assumption of singularity leads to this contradiction, the matrix is nonsingular and the quadrature weights may be calculated.

Third, we show that with the weights and quadratures defined above, the quadrature formula

$$\int_a^b \omega(x) p(x) = \sum_{i=1}^n w_i p(x_i) \tag{9.31}$$

is exact for all polynomials p of degree up to $2n-1$. An arbitrary polynomial of degree $2n-1$ can be written

$$p = p_n(x) \left(\sum_{i=0}^{n-1} c_i p_i(x) \right) + \left(\sum_{j=0}^{n-1} d_i p_i(x) \right).$$

Then, the left-hand side of (9.31) becomes

$$\sum_{i=0}^{n-1} c_i (p_i, p_n) + \sum_{j=0}^{n-1} d_i (p_0, p_i) = d_0 (p_0, p_0),$$

and the right-hand side of (9.31) becomes

$$\sum_{i=1}^n w_i \overbrace{p_n(x_i)}^{\text{zero}} \sum_{j=1}^{n-1} c_j p_j(x_i) + \sum_{i=1}^n w_i \sum_{j=0}^{n-1} d_j p_j(x_i)$$

$$= \sum_{j=0}^{n-1} d_j \underbrace{\sum_{i=1}^n w_i p_j(x_i)}_{\substack{(p_0, p_0) \text{ if } j = 0, \\ 0 \text{ if } j \neq 0}} = d_0 (p_0, p_0).$$

The equality of the left and right sides of (9.31) establishes its truth.

Fourth, it can be shown that the quadrature weights w_i are positive, which means that this quadrature rule (9.31) has good stability properties (cf. page 244). Positivity is established by considering specially engineered polynomials of degree $2n-2$:

$$f_i = (x - x_1)^2 \cdots (x - x_{i-1})^2 (x - x_{i+1})^2 \cdots (x - x_n)^2$$

$$\int_a^b \omega(x) f_i(x) dx = w_i \prod_{k=1, k \neq i}^{n} (x_i - x_k)^2.$$

Since f_i is positive and $\omega(x)$ is positive, w_i must be positive.

The error associated with Gaussian quadrature can be established by considering the degree $2n - 1$ polynomial P_{2n-1} constructed via Hermite interpolation with

$$P_{2n-1}(x_i) = f(x_i)$$
$$P'_{2n-1}(x_i) = f'(x_i), \tag{9.32}$$

for $i = 1, ..., n$. Since P has degree less than $2n$, it is integrated exactly by the quadrature method. The error associated with this interpolation polynomial is

$$f(x) - P_{2n-1}(x) = \frac{f^{(2n)}(\xi)}{(2n)!}(x - x_1)^2 \cdots (x - x_n)^2.$$

The polynomial on the right-hand side of this equation is p_n^2 since it has degree $2n$, the same n multiplicity-2 roots, and the same normalization:

$$f(x) - P_{2n-1}(x) = \frac{f^{(2n)}(\xi)}{(2n)!} p_n(x)^2.$$

So,

$$\int_a^b \omega(x)(f(x) - P_{2n-1}(x)) dx = \frac{f^{(2n)}(\xi)}{(2n)!}(p_n, p_n).$$

On the other hand, we can write

$$\int_a^b \omega(x)(f(x) - P_{2n-1}(x)) dx = \int_a^b \omega(x) f(x) dx - \int_a^b \omega(x) P_{2n-1}(x) dx$$
$$= \int_a^b \omega(x) f(x) dx - \sum_{i=1}^{n} w_i P_{2n-1}(x_i)$$
$$= \int_a^b \omega(x) f(x) dx - \sum_{i=1}^{n} w_i f_{2n-1}(x_i),$$

where in the last step we used (9.32) that define the polynomial P_{2n-1}. In combination, we have the error formula

$$\sum_{i=1}^{n} w_i f(x_i) - \int_a^b \omega(x) f(x) dx = -\frac{f^{(2n)}(\xi)}{(2n)!}(p_n, p_n).$$

Some representative abscissa and weight pairs are given in Tables 9.4–9.5. For Gauss–Chebyshev integrals,

$$\int_{-1}^{+1} \frac{f(x)}{\sqrt{1-x^2}} dx,$$

the weights and quadrature points can be calculated from analytic formulas,

$$x_i = \cos\left(\frac{2i-1}{2n}\pi\right)$$

$$w_i = \frac{\pi}{n}.$$

A related Gauss–Chebyshev formula is

$$\int_{-1}^{+1} \sqrt{1-x^2}f(x)dx,$$

with

$$x_i = \cos\left(\frac{i}{n+1}\pi\right)$$

$$w_i = \frac{\pi}{n+1}\sin^2\left(\frac{i}{n+1}\pi\right).$$

Table 9.4 Some abscissas and weights for $\int_{-1}^{+1} f(x)dx$ (Legendre polynomials) [2].

n	$\pm x_i$	w_i
2	0.577 350 269 189 626	1.000 000 000 000 000
3	0.000 000 000 000 000	0.888 888 888 888 889
	0.774 596 669 241 483	0.555 555 555 555 556
4	0.339 981 043 584 856	0.652 145 154 862 546
	0.861 136 311 594 053	0.347 854 845 137 454
5	0.000 000 000 000 000	0.568 888 888 888 889
	0.538 469 310 105 683	0.478 628 670 499 366
	0.906 179 845 938 664	0.236 926 885 056 189

Table 9.5 Some abscissas and weights for $\int_{-\infty}^{\infty} e^{-x^2}f(x)dx$ (Hermite polynomials) [2].

n	$\pm x_i$	w_i
2	0.707 106 781 186 548	$8.862\,269\,254\,528\times10^{-1}$
3	0.000 000 000 000 000	1.181 635 900 604
	1.224 744 871 391 589	$2.954\,089\,751\,509\times10^{-1}$
4	0.524 647 623 275 290	$8.049\,140\,900\,055\times10^{-1}$
	1.650 680 123 885 785	$8.131\,283\,544\,725\times10^{-2}$
5	0.000 000 000 000 000	$9.453\,087\,204\,829\times10^{-1}$
	0.958 572 464 613 819	$3.936\,193\,231\,522\times10^{-1}$
	2.020 182 870 456 086	$1.995\,324\,205\,905\times10^{-2}$

Table 9.6 Some abscissas and weights for $\int_0^\infty e^{-x} f(x)dx$ (Laguerre polynomials) [2].

n	x_i	w_i
2	0.585 786 437 627	$8.535\,533\,905\,93 \times 10^{-1}$
	3.414 213 562 373	$1.464\,466\,094\,07 \times 10^{-1}$
3	0.415 774 556 783	$7.110\,930\,099\,29 \times 10^{-1}$
	2.294 280 360 279	$2.785\,177\,335\,69 \times 10^{-1}$
	6.289 945 082 937	$1.038\,925\,650\,16 \times 10^{-2}$
4	0.322 547 689 619	$6.031\,541\,043\,42 \times 10^{-1}$
	1.745 761 101 158	$3.574\,186\,924\,38 \times 10^{-1}$
	4.536 620 296 921	$3.888\,790\,851\,50 \times 10^{-2}$
	9.395 070 912 301	$5.392\,947\,055\,61 \times 10^{-4}$
5	0.263 560 319 718	$5.217\,556\,105\,83 \times 10^{-1}$
	1.413 403 059 107	$3.986\,668\,110\,83 \times 10^{-1}$
	3.596 425 771 041	$7.594\,244\,968\,17 \times 10^{-2}$
	7.085 810 005 859	$3.611\,758\,679\,92 \times 10^{-3}$
	12.640 800 844 276	$2.336\,997\,238\,58 \times 10^{-5}$

Example 9.6 Solve (9.13) with Gaussian quadrature.

Solution
The integral is not in a form that corresponds to any of the given Gaussian quadrature formulas, but with some simple manipulations it can be transformed into a form compatible with the Gauss–Legendre quadrature (Table 9.4):

$$I = \int_{-2}^{+2} \frac{dx}{2(1+x^2)} = \int_{-1}^{+1} \frac{du}{1+4u^2} = \arctan(2). \tag{9.33}$$

The integration results are given in Table 9.7.

Table 9.7: Gauss–Legendre evaluation of (9.13); Example 9.6.

n	I	$|I - \arctan(2)|$
2	8.5714285714285687e-01	2.500×10^{-1}
3	1.2156862745098045e+00	1.085×10^{-1}
4	1.0673234811165848e+00	3.983×10^{-2}
5	1.1226986849628355e+00	1.555×10^{-2}

For this particular example, the Gauss–Legendre quadrature does not converge very rapidly. That is because the integrand $f(x) = 1/(1+4x^2)$ does not have a convergent Taylor series in the integration region (there are poles at $\pm i/2$ that limit the radius of convergence). Consequently, $f^{(2n)}(x)/(2n)!$ also increases with n (Figure 9.5) so convergence is very slow.

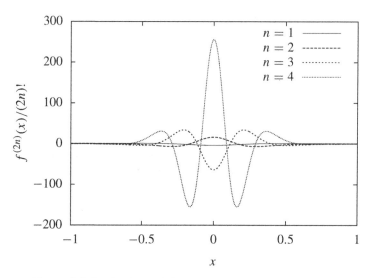

Figure 9.5: Derivatives of the integrand in (9.33); Example 9.6.

Integration of (9.13) by Newton–Cotes methods, including Romberg integration, converged in an expected manner (Examples 9.3 and 9.4). However, the convergence appears diminished in the context of Gauss–Legendre quadrature (Example 9.6). The reason for this observation concerns the significance of the number of quadrature points for the various methods.

With Newton–Cotes (including Romberg), as the number n of quadrature points increases, a factor weighted by the term $f^{(m)}(\xi)$ decreases as some power of $h = (b - a)/(n - 1)$ (e.g., for the trapezoidal sum rule, the error decreases with increasing n like $f^{(2)}(\xi)h^2$). The important point is that the derivative power m is unrelated to the number of quadrature points n. When n is large enough, ξ will not depend significantly on n (this defines the *asymptotic regime* in this context), and the error will be observed to decrease as a power of h. With Gaussian quadrature, however, as the number of quadrature points increases the derivative power m also changes: $m = 2n$. Therefore, the error cannot be expected to diminish with increasing n as a power of $(b - a)/(n - 1)$ if $f^{(2n)}(\xi)$ grows with n. In Example 9.6, the maximum magnitude of the derivative increases with n (Figure 9.5), which is what caused the improvement with increasing n to be somewhat disappointing. To show a more favorable example of the Gaussian quadrature method, at the same time testing the understanding behind the poor performance of the previous example, consider the following.

Example 9.7   Use Gauss–Legendre quadrature to numerically integrate

$$I = \int_{-1}^{+1} \frac{dx}{4 + x^2} = \arctan \frac{1}{2}. \tag{9.34}$$

Solution

This integrand has poles at $\pm 2i$, which makes possible a convergent Taylor series over the entire interval. The numerical integrals converge rapidly (Table 9.8). For comparison, some derivatives of this function are displayed in Figure 9.6.

Table 9.8 Gauss–Legendre evaluation of (9.34); Example 9.7.

n	I	$\lvert I - \arctan(1/2)\rvert$
2	4.6153846153846151e-01	2.109×10^{-3}
3	4.6376811594202921e-01	1.205×10^{-4}
4	4.6364080039043437e-01	6.809×10^{-6}
5	4.6364799181376293e-01	3.828×10^{-7}

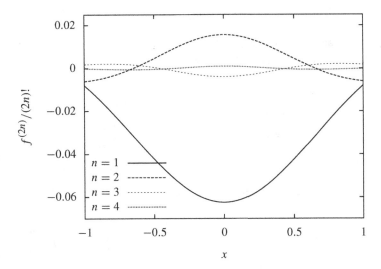

Figure 9.6 Derivatives of the integrand of (9.34); Example 9.7.

It was remarked above that system (9.30) is not the preferred way to calculate the weights w_i. The preferred way takes advantage of the fact that the characteristic equation of the following matrix is the orthogonal polynomial p_n:

$$
\mathbf{J}_n =
\begin{pmatrix}
\delta_1 & \gamma_2 & & \\
\gamma_2 & \delta_2 & \ddots & \\
& \ddots & \ddots & \gamma_n \\
& & \gamma_n & \delta_n
\end{pmatrix},
$$

which means that the quadrature abscissas are the eigenvalues of $\mathbf{J}_n$. Then, the vector

$$\mathbf{v}_i = \begin{pmatrix} p_0(x_i) \\ p_1(x_i)/\gamma_2 \\ \vdots \\ p_{n-1}(x_i)/\gamma_2 \cdots \gamma_n \end{pmatrix} \tag{9.35}$$

is the eigenvector corresponding to eigenvalue x_i. That the characteristic equation of $\mathbf{J}_n$ is p_n is easily shown by induction. That $\mathbf{v}_i$ is an eigenvector is shown as follows. The kth row of the equation $\mathbf{J}_n \mathbf{v}_i = x_i \mathbf{v}_i$ is the equation

$$\gamma_k \frac{p_{k-2}(x_i)}{\gamma_2 \cdots \gamma_{k-1}} + \delta_k \frac{p_{k-1}(x_i)}{\gamma_2 \cdots \gamma_k} + \gamma_{k+1} \frac{p_k(x_i)}{\gamma_2 \cdots \gamma_{k+1}} = x_i \frac{p_{k-1}(x_i)}{\gamma_2 \cdots \gamma_k}$$

$$\frac{\gamma_{k+1}}{\gamma_2 \cdots \gamma_{k+1}} \left[(x_i - \delta_k) p_{k-1}(x_i) - \gamma_k^2 p_{k-2}(x_i) \right] = \frac{\gamma_{k+1}}{\gamma_2 \cdots \gamma_{k+1}} p_k(x_i)$$

$$(x_i - \delta_k) p_{k-1}(x_i) - \gamma_k^2 p_{k-2}(x_i) = p_k(x_i),$$

which agrees with (9.26). With $\mathbf{v}_i$ an eigenvector, $\alpha \mathbf{v}_i$ is also an eigenvector: we are free to pick a normalization. If we arrange the eigenvectors columnwise in matrix form we get

$$(\mathbf{v}_1 ... \mathbf{v}_n) = \begin{pmatrix} 1 & & & \\ & \gamma_2^{-1} & & \\ & & \ddots & \\ & & & (\gamma_2 \cdots \gamma_n)^{-1} \end{pmatrix} \begin{pmatrix} p_0(x_1) & \cdots & p_0(x_n) \\ \vdots & & \vdots \\ p_{n-1}(x_1) & \cdots & p_n(x_n) \end{pmatrix},$$

then rearrange with (9.30) to give

$$\begin{pmatrix} 1 & & & \\ & \gamma_2 & & \\ & & \ddots & \\ & & & (\gamma_2 \cdots \gamma_n) \end{pmatrix} (\mathbf{v}_1 ... \mathbf{v}_n) \begin{pmatrix} w_1 \\ \vdots \\ w_n \end{pmatrix} = \begin{pmatrix} (p_0, p_0) \\ 0 \\ \vdots \end{pmatrix}.$$

The matrix $\mathbf{J}_n$ is symmetric, which means that its eigenvectors are orthogonal:

$$\begin{pmatrix} \mathbf{v}_1^T \\ \vdots \\ \mathbf{v}_n^T \end{pmatrix} (\mathbf{v}_1 \quad \cdots \quad \mathbf{v}_n) \begin{pmatrix} w_1 \\ \vdots \\ w_n \end{pmatrix} = \begin{pmatrix} \mathbf{v}_1^T \mathbf{v}_1 & & \\ & \ddots & \\ & & \mathbf{v}_n^T \mathbf{v}_n \end{pmatrix} \begin{pmatrix} w_1 \\ \vdots \\ w_n \end{pmatrix}$$

$$= (p_0, p_0) \begin{pmatrix} \mathbf{v}_1^T \mathbf{e}_1 \\ \vdots \\ \mathbf{v}_n^T \mathbf{e}_1 \end{pmatrix}$$

$$\begin{pmatrix} w_1 \\ \vdots \\ w_n \end{pmatrix} = (p_0, p_0) \begin{pmatrix} \mathbf{v}_1^T \mathbf{e}_1 / \mathbf{v}_1^T \mathbf{v}_1 \\ \vdots \\ \mathbf{v}_n^T \mathbf{e}_1 / \mathbf{v}_n^T \mathbf{v}_n \end{pmatrix}.$$

This result depends on the particular normalization for $\mathbf{v}$ given by (9.35). With that normalization, $\mathbf{v}_i^\mathrm{T}\mathbf{e}_1 = 1$, so one could write

$$w_i = (p_0, p_0)\frac{(\mathbf{v}_i^\mathrm{T}\mathbf{e}_1)^2}{\mathbf{v}_i^\mathrm{T}\mathbf{v}_i},$$

and this form is independent of the normalization of $\mathbf{v}$. In the special case that $\mathbf{v}_i^\mathrm{T}\mathbf{v}_i = (p_0, p_0)$ one obtains

$$w_i = (\mathbf{v}_i^\mathrm{T}\mathbf{e}_1)^2 \qquad \text{if } \mathbf{v}_i^\mathrm{T}\mathbf{v}_i = (p_0, p_0).$$

A significant source of numerical error in this procedure can be the construction and subsequent integration of the orthogonal polynomials. A procedure that mitigates these concerns is based on the evaluation of integrals of polynomials $q(x)$ that are not necessarily orthogonal, but that are assumed to obey a three-point recursion

$$q_0(x) = 1$$
$$q_1(x) = (x - d_1)q_0(x)$$
$$\vdots \tag{9.36}$$
$$q_i(x) = (x - d_i)q_{i-1}(x) + g_i q_{i-2} \quad i \geq 2,$$

with known constant coefficients d and g. Like the orthogonal polynomials p, we assume the coefficient of the highest power of x is unity. Without loss of generality, one can express these polynomials in terms of the orthogonal polynomials

$$q_i(x) = \ell_{i0}p_0 + \ell_{i1}p_1 + \cdots + \ell_{ii}p_i, \tag{9.37}$$

or

$$\mathbf{q} = \mathbf{L}\mathbf{p},$$

where $\mathbf{L}$ is a left triangular matrix. Since both p and q are normalized to have 1 as the leading coefficient, the diagonal elements ℓ_{ii} must be 1. Then,

$$\mathbf{q}\mathbf{q}^\mathrm{T} = \mathbf{L}\mathbf{p}\mathbf{p}^\mathrm{T}\mathbf{L}^\mathrm{T},$$

and, multiplying by $w(x)$ and integrating over the interval $[a, b]$,

$$\mathbf{M} = \tilde{\mathbf{L}}\tilde{\mathbf{L}}^\mathrm{T} = \mathbf{L}\mathrm{diag}((1, 1), ..., (p_i, p_i), ...)\mathbf{L}^\mathrm{T}, \qquad M_{ij} = \int_a^b w(x)q_i(x)q_j(x)dx.$$

So, one may perform elementary integration of products of the polynomials q to determine the symmetric positive definite matrix $\mathbf{M}$. Then, Cholesky decomposition of that matrix determines the transformation matrix $\tilde{\mathbf{L}} = \mathbf{L}\mathrm{diag}(\sqrt{(p_i, p_i)})$. Since the diagonal elements of $\mathbf{L}$ are 1, the diagonals of $\tilde{\mathbf{L}}$ give the quantities $\sqrt{(p_i, p_i)}$ which in turn determine the coefficients γ_i via (9.28), or

$$\gamma_i = \frac{\tilde{\ell}_{i-1,i-1}}{\tilde{\ell}_{i-2,i-2}}.$$

The transformation matrix $\mathbf{L} = \tilde{\mathbf{L}}\mathrm{diag}(\tilde{\ell}_{00}^{-1}, \tilde{\ell}_{11}^{-1}, ...)$ also associates the coefficients

δ of the orthogonal polynomials (9.26) with the coefficients d of the polynomials q (9.36). First, combine expansion (9.37) with the recursion (9.36):

$$p_0 = 1$$
$$\ell_{10} p_0 + p_1 = (x - d_1) p_0$$
$$\vdots$$
$$\sum_{k=0}^{j} \ell_{jk} p_k = (x - d_j) \sum_{k=0}^{j-1} \ell_{j-1,k} p_k + g_j \sum_{k=0}^{j-2} \ell_{j-2,k} p_k.$$

From the second of these equations we obtain

$$\delta_1 = d_1 + \ell_{10} = d_1 + \frac{\tilde{\ell}_{10}}{\tilde{\ell}_{00}}.$$

For the general case, multiply

$$p_j = \sum_{k=0}^{j-1} \left[(x - d_j)\ell_{j-1,k} - \ell_{jk} \right] p_k + g_j \sum_{k=0}^{j-2} \ell_{j-2,k} p_k$$

by $w(x) p_{j-1}(x)$ and integrate. After simplifying,

$$\delta_j = d_j + \ell_{j,j-1} - \ell_{j-1,j-2} = d_j + \frac{\tilde{\ell}_{j,j-1}}{\tilde{\ell}_{j-1,j-1}} - \frac{\tilde{\ell}_{j-1,j-2}}{\tilde{\ell}_{j-2,j-2}}.$$

A particularly simple class of polynomials $q_i(x)$ of degree i are the functions $q_i(x) = x^i$, for which $d_i = 0$. However, the computation of the coefficients δ, γ from these moments is not numerically stable [80, 81]. The resulting matrices are Vandermonde matrices. See Problems 2.3 and 3.6 for examples demonstrating the poor condition of Vandermonde systems.

Evaluation of integrals of orthogonal polynomials, e.g., (p_i, p_i) and (xp_i, p_i), are stable when the polynomials are evaluated numerically with Stieltjes' recursion. If the polynomials were expanded as a power series in x, then the resulting integrals can be associated with the moment integrals. An analogy can be drawn to Wilkinson's polynomial: when expanded as a power series, numerical errors are greatly exaggerated. Likewise, the expansion of orthogonal polynomial integrals in moments greatly amplifies the input (rounding) error.

Polynomials $q_i \neq x^i$ give rise to what are known as "modified moments" [81, 208]. A good choice for such a polynomial family is a classical orthogonal polynomial on $[a, b]$ with some weighting function $w_0(x)$, where ideally $w_0(x)/w(x)$ is slowly varying. Then, **M** will be diagonally dominant, approaching $\mathrm{diag}(..., (p_i, p_i), ...)$ in the limit $w_0/w \to 1$. If such a choice can be found, the modified moment method is a stable alternative to the Stieltjes construction.

9.5 Special cases

When faced with numerical integration, one always has recourse to the techniques of calculus to make a problem more tractable or a solution more accurate. A variety of special cases are described to emphasize this point.

Improper integrals are ones in which a limit of integration is not bounded, e.g.,

$$I = \int_0^\infty \frac{\sin^4 x}{x^4} dx. \tag{9.38}$$

It should be obvious that you do not want to discretize the interval $[a, b] = [0, \infty]$ into finite pieces for application with a Newton–Cotes formula. In this case, seek a change of variables that makes the integral proper. The variable $u = 1/x$ will transform the limit $x \to \infty$ to the limit $u = 0$, but this change creates troubles for the limit $x = 0$. However, we can always break the original integral into two pieces before doing the change of variables:

$$I = I_1 + I_2$$
$$I_1 = \int_0^1 \frac{\sin^4 x}{x^4} dx$$
$$I_2 = \int_1^\infty \frac{\sin^4 x}{x^4} dx = \int_0^1 u^2 \sin^4(u^{-1}) du.$$

These transformations give two proper integrals, for which subdivision of the intervals into finite segments is not a problem. However, we are still faced with two related difficulties.

In the evaluation of I_1, a Newton–Cotes formula will require that we evaluate the integrand $\sin^4 x / x^4$ at $x = 0$. Assisted by the Taylor series for $\sin(x)$, we know that

$$\lim_{x \to 0} \frac{\sin^4 x}{x^4} = 1.$$

So, for this particular support point we evaluate the integrand using this limit, versus numerically evaluating the function $\sin(0.0)/0.0$, which would generate an error. We are faced with a similar problem in the limit $u = 0$ of I_2. There, we use the fact that $|\sin(x)| \le 1$ for any x, so

$$\lim_{u \to 0} u^2 \sin^4(u^{-1}) = 0.$$

In summary, using the composite trapezoidal sum rule in N parts, we can evaluate improper integral (9.38) by writing

$$h_1 = \frac{1}{N_1}$$

$$I_1 = \frac{h_1}{2} + \sum_{i=1}^{N_1-1} h_1 \frac{\sin^4(ih_1)}{(ih_1)^4} + \frac{h_1}{2}\sin^4 1$$

$$h_2 = \frac{1}{N_2}$$

$$I_2 = \sum_{i=1}^{N_2-1} h_2(ih_2)^2 \sin^4\left(\frac{1}{ih_2}\right) + \frac{h_2}{2}\sin^4 1$$

$$I = I_1 + I_2.$$

The accuracy of Newton–Cotes methods relies on the differentiability of the integrand. We have implicitly assumed that the integrand $f(x)$ has as many derivatives as we like when we derived the error formulas. If the derivatives do not exist, the error formulas are inaccurate and greater error should be expected. Therefore, when faced with an integrand that does not possess derivatives at select points, it is advisable to restore differentiability if possible. One example is the piece-wise differentiable function,

$$f(x) = \begin{cases} g(x) & \text{for } x < x_1 \\ d(x) & \text{for } x > x_1, \end{cases}$$

where $a < x_1 < b$, and where $g(x)$ and $d(x)$ individually have enough derivatives for the method being used. Here, $f(x)$ is integrable on $[a, b]$ if $g(x)$ is integrable on $[a, x_1]$ and if $d(x)$ is integrable on $[x_1, b]$. However, $f(x)$ does not have a derivative at x_1, and this would spoil the accuracy of a Newton–Cotes method.

The fix in this case is to break up the integral at the point x_1 where the failure occurs:

$$\int_a^b f(x)dx = \int_a^{x_1} g(x)dx + \int_{x_1}^b d(x)dx.$$

The two separate integrals can be solved by standard methods while retaining the error properties of sufficiently differentiable functions.

A more subtle problem of this sort is an integral like

$$\int_0^1 \frac{e^x - 1}{\sqrt{x}}dx. \tag{9.39}$$

The problem in this case is that the integrand does not have a derivative at $x \to 0$. With the change of variables $x = u^2$, we get

$$\int_0^1 2(e^{u^2} - 1)du \tag{9.40}$$

and this integrand has arbitrarily many derivatives.

Finally, approximations may be beneficial in the case of numerically unstable integrands. Equations (9.39) and (9.40) are examples of this, because $e^x - 1$ has large cancellation errors for x small. As described in Chapter 1, one can replace $e^x - 1$ with $x + x^2/2$ to obtain a numerically stable approximation with acceptable accuracy in the range $[0, x_0]$, with $x_0 \approx 8.7 \times 10^{-6}$ in double precision floating point. One could use this approximation to improve the accuracy of (9.40) by writing

$$\int_0^{\sqrt{x_0}} 2\left(u^2 + \frac{1}{2}u^4\right) du + \int_{\sqrt{x_0}}^1 2(e^{u^2} - 1) du.$$

Problems

9.1 Use the Peano kernel to determine the error of the trapezoidal sum rule.

9.2 Calculate the numerical convergence of (9.39) and (9.40) with the composite trapezoidal sum rule to show that the possession of derivatives does matter.

9.3 Show that Romberg integration with two levels of resolution h_0 and $h_1 = h_0/2$ gives Simpson's rule, and with three levels of resolution ($h_2 = h_0/4$) gives Boole's rule.

9.4 The previous problem shows that Romberg integration can reproduce high-order Newton–Cotes formulas, but Newton–Cotes based on degree 8 and higher polynomials is unstable. Show that for $h_{i+1} = bh_i$, $0 < b < 1$, Romberg integration is stable when $1/b \geq 2$ is an integer.

9.5 Use the Euler–Maclaurin sum formula to show that the following modified composite trapezoidal sum rule is a method of $\mathcal{O}(h^3)$:

$$\hat{T} = h\left(\frac{5}{12}f(a) + \frac{13}{12}f(a+h) + f(a+2h) + \ldots + f(b-2h) + \frac{13}{12}f(b-h) + \frac{5}{12}f(b)\right)$$

(cf. 9.12). Construct a similar method of $\mathcal{O}(h^4)$.

9.6 [130] For integrals of the type

$$\int_0^1 \frac{\exp(-x/2)}{\sqrt{x}} f(x) dx$$

compute the first 5 Gaussian quadrature rules.

9.7 Compute erf(2) with approximation error less than 10^{-6} using a Newton–Cotes method. The function is given by the integral

$$\text{erf}(x) = \frac{2}{\sqrt{\pi}} \int_0^x e^{-t^2} dt.$$

Cramér's bounds on the derivative are given in Problem 5.7.

9.8 The fugacity coefficient ϕ is a measure of thermodynamic nonideality. It may be computed from experimentally determined compressibility factors Z through the equation

$$\ln \phi = \int_0^P \frac{dP}{P}(Z - 1).$$

$(Z = P\bar{V}/RT$, where P is pressure, $\bar{V}$ is molar volume, R is the gas constant, and T is temperature. $Z = 1$ for an ideal gas, but may differ from 1 for real gasses except when $P = 0$ where real gasses are ideal.) Determine the fugacity coefficient for CO_2 as a function of pressure at 100°F from the following experimental data [189].

P [psi]	Z	P [psi]	Z	P [psi]	Z	P [psi]	Z
0	1.000	1 250	0.315	3 500	0.465	8 000	0.926
200	0.941	1 500	0.255	4 000	0.519	9 000	1.023
400	0.875	1 750	0.272	4 500	0.572	10 000	1.116
600	0.798	2 000	0.299	5 000	0.624		
800	0.704	2 500	0.355	6 000	0.727		
1 000	0.580	3 000	0.411	7 000	0.828		

10 Ordinary differential equations

Ordinary differential equations (ODEs) have the form

$$\frac{dy}{dx} = f(x, y) \tag{10.1}$$

where y and f may be vectors or scalars. They differ from simple integrals by the dependence of f on y.

Since f does depend on y, we cannot solve a problem like (10.1) without knowing something about y. This gives rise to two classes of problem. Initial value problems (IVPs) are ODEs where the value of y is given somewhere, e.g.,

$$\frac{dy}{dx} = f(x, y)$$
$$y(x_0) = y_0.$$

Boundary value problems (BVPs) are ODEs where y is not completely specified anywhere, but instead some constraint is provided, e.g.,

$$\frac{dy}{dx} = f(x, y)$$
$$r(y(a), y(b)) = 0.$$

The plan of this chapter is to discuss first IVPs in various forms, and then to show how the solution to BVPs can be derived from IVP methods.

10.1 Initial value problems I: one-step methods

To begin the discussion, here is an IVP algorithm similar to the left Riemann sum for integration. It is called the *forward Euler method*:

$$y_{i+1} = y_i + hf(x_i, y_i)$$
$$x_{i+1} = x_i + h. \tag{10.2}$$

This has the form of a simple *one-step method*:

$$y_{i+1} = y_i + h\Phi(x_i, y_i)$$
$$x_{i+1} = x_i + h.$$

To analyze these methods we need some new concepts. The first is the *local discretization error*, τ, defined as

$$\tau(x, y; h) = \Delta(x, y; h) - \Phi(x, y; h), \tag{10.3}$$

where $\Phi(x, y; h)$ is the one-step method, and where

$$\Delta(x, y; h) = \begin{cases} \frac{y(x+h)-y(x)}{h} & \text{if } h \neq 0 \\ f(x, y) & \text{otherwise.} \end{cases} \tag{10.4}$$

By its construction, an algorithm like $y_{i+1} = y_i + h\Delta(x_i, y; h)$ would give the exact solution, ignoring numerical errors. τ is measure of error such that $h\tau$ is the error of a single integration step assuming the initial value is exact.

The way one applies (10.3) is to use Taylor series to express all y relative to some reference point. For the forward Euler method,

$$y_{i+1} = y_i + hf(x_i, y_i) + \frac{h^2}{2}\left[\frac{\partial f(x_i, y_i)}{\partial x} + \frac{\partial f(x_i, y_i)}{\partial y}f(x_i, y_i)\right] + \cdots$$

$$\Delta(x_i, y_i) = f_i + \frac{h}{2}(f_x + ff_y)_i + \text{h.o.t.}$$

$$\tau(x_i, y_i) = \frac{h}{2}(f_x + ff_y)_i + \text{h.o.t.}$$

We say that method Φ has *order of accuracy* p if $\tau = \mathcal{O}(h^p)$.

A method is *consistent* if $\lim_{h\to 0} \tau = 0$. A method is consistent if its order p satisfies $p > 0$.

The forward Euler method is therefore consistent and first order. Also, from the analysis one can see that the method

$$\Phi(x_i, y_i; h) = f(x_i, y_i) + \frac{h}{2}\left[\frac{\partial f(x_i, y_i)}{\partial x} + \frac{\partial f(x_i, y_i)}{\partial y}f(x_i, y_i)\right] \tag{10.5}$$

would be a consistent method of second order. We will find simpler ways to achieve a higher order of accuracy – in general we don't want to rely on derivatives of f since these can be difficult to calculate.

Before moving on to other methods, we introduce the next new concept: *global discretization error*. The global discretization error corresponds with our normal sense of error. It is the difference between the approximate value of $y(x)$ obtained with method Φ, and the exact value of $y(x)$. To distinguish these, write $\eta(x)$ for the approximate value. The global discretization error e is then

$$e(x; h) = \eta(x; h) - y(x). \tag{10.6}$$

Before deriving an expression for $e(x; h)$ we need a useful inequality. If numbers ξ_i satisfy

$$|\xi_{i+1}| \leq (1+\delta)|\xi_i| + B, \tag{10.7}$$

with $\delta > 0$ and $B \geq 0$, then

$$|\xi_n| \leq e^{n\delta}|\xi_0| + \frac{e^{n\delta} - 1}{\delta}B. \tag{10.8}$$

That this is true is shown as follows:

$$|\xi_1| \leq (1+\delta)|\xi_0| + B$$
$$|\xi_2| \leq (1+\delta)|\xi_1| + B = (1+\delta)^2|\xi_0| + (1+\delta)B + B$$

$$\vdots$$

$$|\xi_n| \leq (1+\delta)^n|\xi_0| + B(1 + (1+\delta) + \cdots + (1+\delta)^{n-1}) \tag{10.9}$$
$$= (1+\delta)^n|\xi_0| + B\frac{(1+\delta)^n - 1}{\delta}$$
$$\leq e^{\delta n}|\xi_0| + B\frac{e^{\delta n} - 1}{\delta}.$$

The global discretization error can now be determined. If the local discretization error satisfies

$$|\tau(x, y)| = |\Delta(x, y) - \Phi(x, y)| \leq N|h|^p,$$

and if the y-dependence of the method Φ obeys a Lipschitz continuity condition

$$|\Phi(x, y) - \Phi(x, \eta)| \leq M|y - \eta|,$$

then

$$|e_n| = |\eta(x_n) - y(x_n)| \leq \frac{e^{|x_n - x_0|M} - 1}{M}N|h|^p. \tag{10.10}$$

The derivation is simple given the definitions and the inequality (10.8):

$$e_n = \eta_n - y_n$$
$$= (\eta(x_{n-1}) + h\Phi(x_{n-1}, \eta_{n-1})) - (y(x_{n-1}) + h\Delta(x_{n-1}, y_{n-1}))$$
$$= e_{n-1} + h\left(\Phi(x_{n-1}, \eta_{n-1}) - \Phi(x_{n-1}, y_{n-1})\right) \tag{10.11}$$
$$+ h\left(\Phi(x_{n-1}, y_{n-1}) - \Delta(x_{n-1}, y_{n-1})\right)$$
$$|e_n| \leq |e_{n-1}| + |h|M|e_{n-1}| + N|h|^{p+1}.$$

This is a sequence of the form (10.7), and so with $e_0 = 0$,

$$|e_n| \leq \frac{e^{n|h|M} - 1}{hM}N|h|^{p+1} = \frac{e^{|x_n - x_0|M} - 1}{M}N|h|^p. \tag{10.12}$$

According to this result, if $f_y(x, y)$ is bounded, then a Lipschitz constant M exists and $e = \mathcal{O}(h)$ for the forward Euler method.

Some other simple one-step methods include

- Heun's method

$$\Phi(x, y; h) = \frac{1}{2}\left[f(x, y) + f(x + h, y + hf(x, y))\right];$$

- the modified Euler method

$$\Phi(x, y; h) = f\left(x + \frac{h}{2}, y + \frac{h}{2}f(x, y)\right);$$

• and the fourth-order Runge–Kutta method [137, 203]

$$\Phi(x, y; h) = \frac{1}{6}[k_1 + 2k_2 + 2k_3 + k_4]$$

$$k_1 = f(x, y)$$

$$k_2 = f\left(x + \frac{h}{2},\ y + \frac{h}{2}k_1\right)$$

$$k_3 = f\left(x + \frac{h}{2},\ y + \frac{h}{2}k_2\right)$$

$$k_4 = f(x + h,\ y + hk_3).$$

10.2   Initial value problems II: multistep methods

The ODE methods so far developed have taken the form

$$y_i = y_{i-1} + h\Phi(x_{i-1}, y_{i-1})$$
$$x_i = x_{i-1} + h. \tag{10.13}$$

We know from the integration methods that higher order accuracy can be achieved by integrating an interpolation polynomial fit to a number of support points. Here, we apply that idea to ODEs.

Before formally developing this idea, we begin with the generic model

$$y_{i+2} + a_1 y_{i+1} + a_0 y_i = h\left[b_2 f_{i+2} + b_1 f_{i+1} + b_0 f_i\right], \tag{10.14}$$

where $f_i = f(x_i, y_i)$, etc. We will pick the coefficients a and b to achieve as high an order of accuracy as possible. Accordingly, we will construct the local discretization error of the method, then choose the coefficients to make the order of accuracy of the local discretization error be as large as possible.

This model will lead to an unstable method: do not use this. This risk of instability was not encountered in one-step methods. Equation (10.13) reduces to the governing ODE as $h \to 0$, and so the numerical method will have the same stability properties as the equation being modeled. However, (10.14) is not the discrete analog of the governing ODE, and its stability may be unrelated to the stability of the ODE. These ideas were developed by Dahlquist [48].

The local discretization error of a multistep method is $1/h$ times the difference between the left-hand side of the discretization and the right-hand side, expanded in a Taylor series. τ_i for one-step methods can be described this way also. With the substitutions

$$y_{i+2} = y_i + 2h(y')_i + 2h^2(y'')_i + \frac{4}{3}h^3(y''')_i + \frac{2}{3}h^4(y'''')_i + \cdots$$

$$y_{i+1} = y_i + h(y')_i + \frac{1}{2}h^2(y'')_i + \frac{1}{6}h^3(y''')_i + \frac{1}{24}h^4(y'''')_i + \cdots$$

$$f_{i+2} = (y')_i + 2h(y'')_i + 2h^2(y''')_i + \frac{4}{3}h^3(y'''')_i + \cdots$$

$$f_{i+1} = (y')_i + h(y'')_i + \frac{1}{2}h^2(y''')_i + \frac{1}{6}h^3(y'''')_i + \cdots,$$

where $y' = f$, $y'' = f_x + ff_y$, etc., one has

$$
\begin{aligned}
\tau_i = {} & y_i(1 + a_0 + a_1)h^{-1} + \\
& + (y')_i(2 + a_1 - b_0 - b_1 - b_2) \\
& + (y'')_i(2 + \frac{1}{2}a_1 - b_1 - 2b_2)h \\
& + (y''')_i(\frac{4}{3} + \frac{1}{6}a_1 - \frac{1}{2}b_1 - 2b_2)h^2 \\
& + (y'''')_i(\frac{2}{3} + \frac{1}{24}a_1 - \frac{1}{6}b_1 - \frac{4}{3}b_2)h^3 + \cdots.
\end{aligned}
$$

The definitions of consistency and order carry over from the one-step discussion. Thus, for consistency we need

$$1 + a_0 + a_1 = 0$$

$$2 + a_1 - b_0 - b_1 - b_2 = 0.$$

To achieve second-order accuracy we need also

$$2 + \frac{1}{2}a_1 - b_1 - 2b_2 = 0;$$

to achieve third-order accuracy we need also

$$\frac{4}{3} + \frac{1}{5}a_1 - \frac{1}{2}b_1 - 2b_2 = 0;$$

and to achieve fourth-order accuracy we need also

$$\frac{2}{3} + \frac{1}{24}a_1 - \frac{1}{6}b_1 - \frac{4}{3}b_2 = 0.$$

The most accurate method, $\mathcal{O}(h^4)$, gives the coefficients $a_0 = 7$, $a_1 = -8$, $b_0 = -9/2$, $b_1 = -2$, and $b_2 = 1/2$, or

$$y_{j+2} - 8y_{j+1} + 7y_j = h\left[\frac{1}{2}f(x_{j+2}, y_{j+2}) - 2f(x_{j+1}, y_{j+1}) - \frac{9}{2}f(x_j, y_j)\right].$$

This scheme is *implicit* because y_{j+2} appears on both the left- and right-hand sides. It would be evaluated iteratively in practice.

We can develop an *explicit* scheme by requiring $b_2 = 0$, but this choice will limit our accuracy to $\mathcal{O}(h^3)$:

$$y_{j+2} + 4y_{j+1} - 5y_j = h\left[4f(x_{j+1}, y_{j+1}) + 2f(x_j, y_j)\right]. \tag{10.15}$$

Method (10.15) is illustrated with the simple ODE $y' = 0$, for which the exact solution is $y(x) = y(0)$. To employ the method we need two starting values. The results plotted in Figure 10.1 use $h = 0.1$, and $y(0) = 1$, $y(h) = 1 + 2^{-52}$. The value $y(h)$ is not exact, but differs by only one bit in the least significant digit from the exact value. The first few calculated results are clearly compatible with the exact solution, but it is also clear that the solution begins to oscillate wildly in a manner that is incompatible with the exact solution. The magnitude of the instability is better appreciated by looking at some tabulated numbers – see Table 10.1.

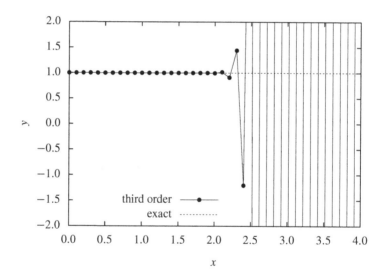

Figure 10.1 Unstable multistep method (10.15).

A generic multistep algorithm, generalizing on the example of (10.14) might look like

$$\underbrace{\begin{pmatrix} \eta_{j+q} \\ \vdots \\ \eta_{j+1} \end{pmatrix}}_{\mathbf{H}_j} = \underbrace{\begin{pmatrix} -a_{q-1} & -a_{q-2} & \cdots & -a_1 & -a_0 \\ 1 & & & & 0 \\ & 1 & & & 0 \\ & & \ddots & & \vdots \\ & & & 1 & 0 \end{pmatrix}}_{\mathbf{A}} \underbrace{\begin{pmatrix} \eta_{j+q-1} \\ \vdots \\ \eta_j \end{pmatrix}}_{\mathbf{H}_{j-1}} + \underbrace{\begin{pmatrix} 1 \\ 0 \\ \vdots \\ 0 \end{pmatrix}}_{\mathbf{b}} hF \tag{10.16a}$$

$$F = \sum_{s=0}^{q} b_s f(x_{j+s}, \eta_{j+s}). \tag{10.16b}$$

Table 10.1 Some data from unstable simulation

x_i	η_i
1.5	+1.0000011293772630e+00
1.6	+9.9999435311368501e-01
1.7	+1.0000282344315752e+00
1.8	+9.9985882784212432e-01
1.9	+1.0007058607893786e+00
2.0	+9.9647069605310712e-01
2.1	+1.0176465197344646e+00
2.2	+9.1176740132767709e-01
2.3	+1.4411629933616148e+00
2.4	-1.2058149668080738e+00
2.5	+1.2029074834040369e+01
2.6	-5.4145374170201848e+01
2.7	+2.7672687085100921e+02
2.8	-1.3776343542550460e+03
2.9	+6.8941717712752297e+03
3.0	-3.4464858856376151e+04
3.1	+1.7233029428188075e+05
3.2	-8.6164547140940372e+05
3.3	+4.3082333570470186e+06
3.4	-2.1541160785235092e+07
3.5	+1.0770580992617546e+08
3.6	-5.3852904363087726e+08
3.7	+2.6926452241543865e+09
3.8	-1.3463226114771933e+10
3.9	+6.7316130579859665e+10
4.0	-3.3658065289329834e+11

For the specific example resulting in the unstable behavior of Figure 10.1, $F = 0$, so the growth of η_j must be a consequence of $\mathbf{A}$ having eigenvalues λ_i with $|\lambda_i| > 1$.

Note that $\mathbf{A}$ in (10.16a) is just the transpose of the *Frobenius normal matrix* of (6.16) on page 169. Recall that the eigenvalues of matrix (6.16) are the roots of a polynomial. In the present case, the polynomial is

$$\phi(\mu) = \mu^q + a_{q-1}\mu^{q-1} + \cdots + a_1\mu + a_0, \qquad (10.17)$$

so we can conclude that instability occurred because the polynomial (10.17) corresponding to the left-hand side of method (10.15) has roots with magnitude greater than 1:

$$\mu^2 + 4\mu - 5 = (\mu + 5)(\mu - 1)$$

so there was indeed a root -5 with magnitude larger than 1. Also, this root is negative, and that explains the oscillatory behavior seen in Figure 10.1 and Table 10.1.

To emphasize the relation between ODE solution y_j and eigenvalue λ, note that $y_j = \lambda^j$ is a solution of the *linear difference equation*

$$y_{j+q} + a_{q-1}y_{j+q-1} + \cdots + a_0 y_j = 0. \qquad (10.18)$$

Show this by substituting $y_j = \lambda^j$ to rewrite (10.18) as

$$\lambda^j \left(\lambda^q + a_{q-1}\lambda^{q-1} + \cdots + a_0 \right) = 0. \tag{10.19}$$

Now, if λ is a root of multiplicity 2 of (10.17), then λ solves $\phi(\lambda) = 0$ and also $\phi'(\lambda) = 0$. This results in $j\lambda^j$ being a solution of the linear difference equation. This can be shown by substituting $y_j = j\lambda^j$ into (10.18),

$$\begin{aligned}
0 =& (j + q)\lambda^{j+q} + (j + q - 1)a_{q-1}\lambda^{j+q-1} + \cdots + (j)a_0\lambda^j \\
=& j\lambda^j \left(\lambda^q + a_{q-1}\lambda^{q-1} + \cdots + a_0 \right) \\
& + \lambda^j \left((q)\lambda^q + (q - 1)a_{q-1}\lambda^{q-1} + \cdots + a_1 \right).
\end{aligned} \tag{10.20}$$

The first term on the right-hand side of (10.20) is zero because λ is a root of (10.17). The second term on the right-hand side of (10.20) is zero because λ is a root of multiplicity 2: therefore, $\phi'(\lambda) = 0$.

This idea can be extended to show that if λ is a root of order m, then $y_j = P_{m-1}(j)\lambda^j$ is a solution of the linear difference equation, where P_{m-1} is a polynomial of degree $m - 1$.

The stability of a multistep method requires that:

- the polynomial (10.17) has no root λ with $|\lambda| > 1$; and
- if λ is a root of (10.17), and $|\lambda| = 1$, then λ must be a simple root.

If these properties are obeyed, then solutions of the linear difference equations will not grow unstably.

It should be noted that having roots λ_i with $|\lambda_i| < 1$ also leads to problems. The example problem described above, for which $y(x) = 1$ for all x, would show $|y|$ decreasing monotonically with x if the method were stable, but some $|\lambda_i| < 1$. Ideally, the roots will satisfy $|\lambda_i| = 1$, with all roots simple. A polynomial (10.17) with this character is called *conservative*.

Note that the algebraic growth of y_j if λ is a root of multiplicity $m \geq 2$ is reminiscent of a circumstance that occurred in Chapter 3 where we considered vector iteration with a Jordan matrix $\mathbf{C}$ (3.2). Powers k of $\mathbf{C}$, $\mathbf{C}^k$ (3.22), caused factors like $k\lambda^k$ to appear which in turn caused vector iteration to produce the desired eigenvector at an algebraic rate, versus the expected geometric rate.

The relation between the local discretization error and the global discretization error is similar to the result obtained for one-step methods, although the derivation takes a little extra explaining. The function F of (10.16b) is assumed to obey a Lipschitz condition

$$|F(x_j, ...; \eta_j, ...) - F(x_j, ...; y_j, ...)| \leq M \sum_{s=0}^{q} |\eta_{j+s} - y_{j+s}|.$$

The derivation begins with an expression like (10.16a) for the exact solution $\mathbf{y}$:

$$\mathbf{Y}_j = \mathbf{A}\mathbf{Y}_{j-1} + \mathbf{b}h(F(x_j, ..., x_{j+q}; y_j, ...y_{j+q}) + \tau_j).$$

The error vector $\mathbf{E} = \mathbf{H} - \mathbf{Y}$ then obeys the equation

$$\|\mathbf{E}_j\|_2 \le \|\mathbf{A}\|_2 \|\mathbf{E}_{j-1}\|_2 + \|\mathbf{b}\|_2 h(Nh^p + M \sum_{s=0}^{q} |\eta_{j+s} - y_{j+s}|), \qquad (10.21)$$

using the L^2 norm for convenience, and writing $|\tau| \le Nh^p$ as before. We simplify with $\|\mathbf{b}\|_2 = 1$, and $\|\mathbf{A}\|_2 \le 1$ if the method is stable. The $q+1$ terms of the sum $\sum |\eta - y|$ can be rearranged as follows:

$$\sum_{s=0}^{q} |\eta_{j+s} - y_{j+s}| \le \|\mathbf{E}_{j-1}\|_1 + \|\mathbf{E}_j\|_1 \le \sqrt{q}(\|\mathbf{E}_{j-1}\|_2 + \|\mathbf{E}_j\|_2),$$

using (2.17) to replace the L^1 norms with L^2 norms. Inequality (10.21) can then be written

$$(1 - hM\sqrt{q})\|\mathbf{E}_j\|_2 \le (1 + hM\sqrt{q})\|\mathbf{E}_{j-1}\|_2 + Nh^{p+1}.$$

We now restrict h to values such that $hM\sqrt{q} < \frac{1}{2}$. This restriction allows the substitutions:

$$\frac{1}{1 - hM\sqrt{q}} \le 2$$

$$\frac{1 + hM\sqrt{q}}{1 - hM\sqrt{q}} \le 1 + 4hM\sqrt{q}.$$

Then,

$$\|\mathbf{E}_j\|_2 \le (1 + 4hM\sqrt{q})\|\mathbf{E}_{j-1}\|_2 + 2Nh^{p+1}.$$

This inequality has the form (10.7), which permits use of the inequality (10.8):

$$\|\mathbf{E}_n\|_2 \le e^{4hM\sqrt{q}n}\|\mathbf{E}_0\|_2 + \frac{Nh^p}{2M\sqrt{q}}\left(e^{4hM\sqrt{q}n} - 1\right). \qquad (10.22)$$

With inequalities (2.16) and (2.17), this expression can be modified to provide bounds on $\|\mathbf{E}_n\|_1$ or $\|\mathbf{E}_n\|_\infty$. A stable multistep method with order p local discretization error has order p global discretization error, provided the initial error $\|\mathbf{E}_0\|_2$ also scales like h^p. We ignored this term for one-step methods, but for multistep methods this start-up error term is important.

The multistep methods so far discussed are unstable. A family of stable, high-order methods is systematically constructed by assuming :

- $a_k = 0$, except for a_{q-1} for which $a_{q-1} = -1$. With this assumption, the polynomial (10.17) has degree 1, and the single root $\lambda = 1$. The polynomial is stable and conservative;
- the coefficients b are determined by integrating an interpolation polynomial with support points $(x_j, f(x_j, y_j)), ..., (x_{j+q-1}, f(x_{j+q-1}, y_{j+q-1}))$ for explicit methods, or ..., $(x_{j+q}, f(x_{j+q}, y_{j+q}))$ for implicit methods.

To be specific,

$$y_{j+1} = y_j + h[b_{q-1}f_j + b_{q-2}f_{j-1} + \cdots + b_0 f_{j-q+1}] \tag{10.23}$$

gives a family of $\tau = \mathcal{O}(h^q)$ explicit Adams–Bashforth methods [6]. The coefficients b are determined by constructing a Lagrange interpolation polynomial, then integrating it:

$$y_{j+1} = y_j + \int_{x_j}^{x_{j+1}} dx \sum_{s=j+1-q}^{j} L_s(x) f_s$$

$$b_{s-(j+1-q)} = \frac{1}{h} \int_{x_j}^{x_{j+1}} dx \prod_{k=j+1-q, k \neq s}^{j} \frac{x - x_k}{x_s - x_k}$$

$$= \int_{j}^{j+1} du \prod_{k=j+1-q, k \neq s}^{j} \frac{u - k}{s - k},$$

with $b_q = 0$. The last form of the coefficients b assumes uniformly spaced abscissas, $x_{j+1} = x_j + h$. The results of this model for several orders q are displayed in Table 10.2.

The implicit version of this model, developed by F. R. Moulton [168] to understand the ballistics of army ordnance, are called Adams–Moulton methods. These have $\tau = \mathcal{O}(h^{q+1})$:

$$y_{j+1} = y_j + h[b_q f_{j+1} + b_{q-1}f_j + b_{q-2}f_{j-1} + \cdots + b_0 f_{j-q+1}] \tag{10.24}$$

$$y_{j+1} = y_j + \int_{x_j}^{x_{j+1}} dx \sum_{s=j+1-q}^{j+1} L_s(x) f_s$$

$$b_{s-(j+1-q)} = \frac{1}{h} \int_{x_j}^{x_{j+1}} dx \prod_{k=j+1-q, k \neq s}^{j+1} \frac{x - x_k}{x_s - x_k}$$

$$= \int_{j}^{j+1} du \prod_{k=j+1-q, k \neq s}^{j+1} \frac{u - k}{s - k}.$$

Some coefficients b for this model are displayed in Table 10.3. The simplest of these also goes by the name of the *backward Euler method*

$$y_{j+1} = y_j + hf(x_{j+1}, y_{j+1}).$$

Table 10.2 Coefficients for Adams–Bashforth (10.23) multistep methods ($b_q = 0$).

q	α	αb_5	αb_4	αb_3	αb_2	αb_1	αb_0
1	1						1
2	2					3	−1
3	12				23	−16	5
4	24			55	−59	37	−9
5	720		1901	−2774	2616	−1274	251
6	1440	4277	−7923	9982	−7298	2877	−475

Table 10.3 Coefficients for Adams–Moulton (10.24) multistep methods.

q	α	αb_6	αb_5	αb_4	αb_3	αb_2	αb_1	αb_0
0	1							1
1	2						1	1
2	12					5	8	−1
3	24				9	19	−5	1
4	720			251	646	−264	106	−19
5	1 440		475	1 427	−798	482	−173	27
6	60 480	19 087	65 112	−46 461	37 504	−20 211	6 312	−863

Example 10.1   Integrate the ODE

$$y' = y(1 - y)$$
$$y(0) = 2 \tag{10.25}$$

to find $y(1)$ using the fourth-order Adams–Moulton method.

Solution

The fourth-order ($q = 3$) method is

$$y_{i+1} = y_i + \frac{h}{24}(9f(y_{i+1}, x_{i+1}) + 19f_i - 5f_{i-1} + f_{i-2}), \tag{10.26}$$

so we are faced with two immediate problems: (i) the problem statement (10.25) provides a single initial value, but formula (10.26) requires three (y_i, y_{i-1}, and y_{i-2}); and (ii) the equation (10.26) is implicit with y_{i+1} appearing on both the left and right-hand sides.

For this problem (10.25) we have an exact solution

$$y(x) = 1 - \frac{1}{1 + \frac{y(0)\exp(x)}{1-y(0)}}, \tag{10.27}$$

so the first problem could be overcome using (10.27) to obtain $y(h)$, and $y(2h)$, then evaluate the $f(x, y) = y'$ using these exact values to give exact values of $f(h, y(h))$ and $f(2h, y(2h))$.

The second problem can be solved by iteration. First, we need a starting guess. That can come from the third-order ($q = 3$) Adams–Bashforth method:

$$y_{i+1}^{(0)} = y_i + \frac{h}{12}(23f_i - 16f_{i-1} + 5f_{i-2}).$$

Then, a simple iterative method is

$$y_{i+1}^{(k+1)} = y_i + \frac{h}{24}(9f(y_{i+1}^{(k)}, x_{i+1}) + 19f_i - 5f_{i-1} + f_{i-2}). \tag{10.28}$$

In practice, a single iteration is sufficient (for the present application) to achieve the desired accuracy. This is shown by computing $y(1)$ using three different step sizes, then estimating the order using Richardson extrapolation (9.22b):

h	$\eta(1)$
0.005	1.2253996763448902e+00
0.010	1.2253997176173284e+00
0.020	1.2254003631215991e+00
order	3.97

These results use exact starting information to get the multistep scheme going, but only use one iteration of the implicit step (10.28). If $y(h)$ and $y(2h)$ are calculated using the fourth-order Runge–Kutta scheme (10.13), then the following results are obtained:

h	$\eta(1)$
0.005	1.2253996763453268e+00
0.010	1.2253997176310287e+00
0.020	1.2254003635430712e+00
order	3.98

Again, the expected order of 4 is found. However, if the starting data $y(h)$ and $y(2h)$ are obtained from the first-order forward Euler scheme (10.2), then:

h	$\eta(1)$
0.005	1.2253788825834124e+00
0.010	1.2253162353148805e+00
0.020	1.2250640683346430e+00
order	2.01

and the order of accuracy is diminished by the inaccurate starting values.

Equation (10.22) shows that the overall order of convergence of a multistep method will depend on the order of the multistep scheme itself, and the order of accuracy of the multiple initial values. It is not surprising that using exact values, or fourth-order Runge–Kutta values, we observed an overall rate of $\approx h^4$.

Likewise, it is not surprising that using the low order Euler scheme to provide initial values caused the observed order to diminish. However, it may be surprising to see that the *first*-order Euler method allowed *second*-order overall convergence.

The reason for this apparent discrepancy lies in our analysis of the one-step methods. Consider (10.11) and the first few steps implied by this sequence:

$$
\begin{aligned}
&|e_0| = 0 \\
&|e_1| \leq N|h|^{p+1} = \mathcal{O}(h^{p+1}) \\
&|e_2| \leq 2N|h|^{p+1} + NM|h|^{p+2} = \mathcal{O}(h^{p+1}) \\
&|e_3| \leq 3N|h|^{p+1} + 3NM|h|^{p+2} + NM^2|h|^{p+3} = \mathcal{O}(h^{p+1}) \\
&\cdots \\
&|e_k| \leq kN|h|^{p+1} + \text{h.o.t.} = \mathcal{O}(h^{p+1}).
\end{aligned}
\tag{10.29}
$$

These first few steps give each carry an apparent order that is one greater than the error (10.12) from a large number of steps. It is when the number of steps increases at the same time that h decreases that the order is p. The case described by (10.12) is like (10.29) with $k \sim 1/h$, in which case the error is $\mathcal{O}(h^p)$. Likewise, for integration we see one order of accuracy difference between integration over a step (second column of Table 9.3), and integration over a fixed interval (last column of Table 9.3).

In the present circumstance of seeding a multistep method, the number of steps required of a one-step method is fixed. We conclude that to achieve an overall order of accuracy p using a multistep method of order p, one can seed the method using a one-step method of order $p - 1$ or higher.

10.3 Adaptivity

As with integration, one can construct adaptive strategies for one-step and multistep IVPs. The motivation is similar: the structure of the ODE may require increased resolution in only part of the domain. If the (absolute, not relative) error associated with one function evaluation and one addition is ϵ_1, i.e.,

$$
\eta_{i+1} = \eta_i + h\Phi(x_i, \eta_i) + \epsilon_i, \qquad |\epsilon_i| \leq \epsilon_1,
$$

then consideration of numerical error gives (cf. (10.10))

$$
|e_n| = |\eta(x_n) - y(x_n)| \leq \frac{e^{|x_n - x_0|M} - 1}{M}\left(N|h|^p + \frac{\epsilon_1}{|h|}\right),
$$

where the first term in parenthesis is related to the approximation error of the method, and the second is due to numerical error. This formula displays the trade-off that must be balanced: approximation error considerations call for h to be small, but numerical error considerations call for h to be large.

Approaches for one-step and multistep methods are driven by the same considerations, but the data requirements of multistep methods make their practical implementation significantly different.

10.3.1   One-step methods

For each step h beginning at x_i one computes both $\eta(x_i + h; h)$, the function at $x_i + h$ using step size h, and also $\eta(x_i + h; h/2)$, the function at $x_i + h$ with two steps of size $h/2$. Now if the local discretization error is $\mathcal{O}(h^p)$, then

$$\eta(x_i + h; h) = \eta(x_i + h) + Ch^{p+1}$$

$$\eta(x_i + h; h/2) = \eta(x_i + h) + 2C\frac{h^{p+1}}{2^{p+1}};$$

both representations of $h\tau$. The coefficient C is then

$$|C| = \frac{2^p}{h^{p+1}} \frac{|\eta(x_i + h; h) - \eta(x_i + h; h/2)|}{2^p - 1}.$$

If one had instead computed $\eta(x_i + \tilde{h}; \tilde{h}/2)$, etc., then the more accurate calculation would have an error bound e

$$|e| = \frac{\tilde{h}^{p+1}}{h^{p+1}} \frac{|\eta(x_i + h; h) - \eta(x_i + h; h/2)|}{2^p - 1}.$$

Now if we wished each step to have an approximation error ϵ_1, approximating the absolute numerical error, then the step size should be

$$\tilde{h} = h \sqrt[p+1]{\frac{\epsilon_1(2^p - 1)}{|\eta(x_i + h; h) - \eta(x_i + h; h/2)|}}.$$

If $\tilde{h} \ll h$, then the step should be recomputed with $h = \tilde{h}$. Otherwise, proceed to the next step using $h = \tilde{h}$, taking the more accurate $\eta_{i+1} = \eta(x_i + h; h/2)$.

Example 10.2   [225] Compute $y(0)$ given

$$y' = -200xy^2$$

$$y(-3) = \frac{1}{901}. \tag{10.30}$$

This system has an exact solution $y(x) = 1/(1 + x^2)$, so $y(0) = 1$.

Solution

We use an adaptive fourth-order Runge–Kutta integrator choosing an error per step of $2 \times 10^{-16}y$, i.e., assuming a relative error per step of double machine precision. The calculation begins with a step size $h = 0.1$, and recalculates each step for which $\tilde{h}/h < 0.5$.

The adaptive solver obtains a solution with error -1.930×10^{-11} using 3966 steps, for which the average step size is $3/3966 = 7.564 \times 10^{-4}$. The minimum step size used by the adaptive scheme is 2.185×10^{-4}, and a computation using that step size uniformly would take 13 729 steps. A comparison of the adaptive solution with solutions obtained holding h or N constant are shown in Table 10.4. Both of these alternative methods give less accurate solutions.

Table 10.4: Adaptive fourth-order Runge–Kutta, Example 10.2.

method	h_{min}	steps N	error
adaptive	2.185×10^{-4}	3966	-1.930×10^{-11}
h constant	2.185×10^{-4}	13729	-3.599×10^{-10}
N constant	7.564×10^{-4}	3966	-1.000×10^{-9}

10.3.2 Multistep methods

Multistep methods of the Adams–Bashforth and Adams–Moulton variety can be constructed by writing

$$\eta_{k+1} = \eta_k + \int_{x_k}^{x_{k+1}} P(x)dx, \tag{10.31}$$

where P is a polynomial that is fit to data

$$(x_{k-p+1}, f_{k-p+1}), \ (x_{k-p+2}, f_{k-p+2}), \ \ldots, \ (x_{k-1}, f_{k-1}), \ (x_k, f_k),$$

for an $\mathcal{O}(h^p)$ predictor method, or

$$(x_{k-p+2}, f_{k-p+2}), \ (x_{k-p+3}, f_{k-p+3}), \ \ldots, \ (x_k, f_k), \ (x_{k+1}, f_{k+1}),$$

for an order $\mathcal{O}(h^p)$ corrector method. The challenge now is to evaluate the polynomial and its integral in the case that $h_k = x_{k+1} - x_k$ is not uniform.

In the present application, it is convenient to use the Newton interpolation representation of the polynomial. For the predictor,

$$P = f[x_k] + (x - x_k)f[x_k, x_{k-1}] + \cdots + (x - x_k)\cdots(x - x_{k-p+2})f[x_k, \ldots, x_{k-p+1}]$$

$$\int_{x_k}^{x_{k+1}} P dx = f[x_k] \int_{x_k}^{x_{k+1}} dx + f[x_k, x_{k-1}] \int_{x_k}^{x_{k-1}} (x - x_k)dx + \cdots \tag{10.32}$$

$$+ f[x_k, \ldots, x_{k-p+1}] \int_{x_k}^{x_{k-1}} (x - x_k)\cdots(x - x_{k-p+2})dx.$$

And, for the corrector,

$$P = f[x_{k+1}] + (x - x_{k+1})f[x_{k+1}, x_k] + \cdots$$
$$+ (x - x_{k+1})\cdots(x - x_{k-p+3})f[x_{k+1}, \ldots, x_{k-p+2}]$$

$$\int_{x_k}^{x_{k+1}} P dx = f[x_{k+1}] \int_{x_k}^{x_{k+1}} dx + f[x_{k+1}, x_k] \int_{x_k}^{x_{k-1}} (x - x_{k+1})dx + \cdots \tag{10.33}$$

$$+ f[x_{k+1}, \ldots, x_{k-p+2}] \int_{x_k}^{x_{k-1}} (x - x_{k+1})\cdots(x - x_{k-p+3})dx.$$

The function value f_{k+1} is here assumed known: first from the predictor, and later (if required) from iterations of the corrector.

The integrals in both cases are easily constructed with a clever recursive procedure. Define

$$g_{ij} = \int_{x_k}^{x_{k+1}} (x - x_{k+1})^{j-1}(x - x_k) \cdots (x - x_{k-i+1})dx. \qquad (10.34)$$

The predictor uses $g_{11}, ..., g_{p-1,1}$, and the corrector uses $g_{12}, ..., g_{p-2,2}$, and also $\int (x - x_{k+1})dx = -\frac{1}{2}(x_k - x_{k+1})^2$. Both the predictor and corrector use the trivial $\int 1 dx = (x_{k+1} - x_k)$. For reasons to be described shortly, we also want $g_{p,2}$ and $g_{p+1,2}$.

The recursion,

$$g_{ij} = (x_{k+1} - x_{k+1-i})g_{i-1,j} + g_{i-1,j+1} \qquad (10.35)$$

is based on the definition of g_{ij} with the algebraic substitution

$$(x - x_{k-i+1}) = (x_{k+1} - x_{k-i+1}) + (x - x_{k+1}).$$

One begins the recursion with

$$g_{1j} = \frac{(x_k - x_{k+1})^{j+1}}{j(j+1)}, \qquad j = 1, ..., p+2, \qquad (10.36)$$

which follows the definition (10.34), and uses integration by parts.

With model (10.31), integral expansions (10.32) and (10.33), and recursion (10.36) and (10.35), one can solve an IVP with a multistep method with nonuniform step sizes $h_k = x_{k+1} - x_k$. The next objective is to derive an estimate of the error, and use it to adjust the step size.

Suppose one computes $\eta_{k+1}^{(0)}$, an estimate of y_{k+1}, with an $\mathcal{O}(h^p)$ predictor. With $f_{k+1} = f(x_{k+1}, \eta_{k+1}^{(0)})$, a single iteration of the corresponding $\mathcal{O}(h^p)$ corrector gives $\eta_{k+1}^{(1)}$.

Let $f_{k+1}^{(1)} = f(x_{k+1}, \eta_{k+1}^{(0)})$ be the function value computed using the result $\eta_{k+1}^{(0)}$ of the predictor. This is the function data used to construct the corrector:

$$(x_{k-p+2}, f_{k-p+2}), \ (x_{k-p+3}, f_{k-p+3}), \ ..., \ (x_k, f_k), \ (x_{k+1}, f_{k+1}^{(1)}). \qquad (10.37)$$

A higher-order method using available data might use

$$(x_{k-p+1}, f_{k-p+1}), \ (x_{k-p+2}, f_{k-p+2}), \ (x_{k-p+3}, f_{k-p+3}), \ ..., \ (x_k, f_k), \ (x_{k+1}, f_{k+1}^{(1)}).$$
$$(10.38)$$

The difference between the polynomial of degree $p - 1$ fitting (10.37) and the polynomial of degree p fitting (10.38) is

$$\Delta P = (x - x_{k+1}) \cdots (x - x_{k-p+2}) f^{(1)}[x_{k+1}, ..., x_{k-p+1}],$$

so

$$E_p = \left| g_{p2} f^{(1)}[x_{k+1}, ..., x_{k-p+1}] \right|$$

is a measure of error associated with the corrector. Since the corrector has local discretization error p, the error committed in a single step is $\mathcal{O}(h^{p+1})$. If the numerical error in the given step is estimated to be ϵ_1 (absolute, not relative), then

$$\tilde{h} = h \left(\frac{\epsilon_1}{E_p} \right)^{1/(p+1)} \tag{10.39}$$

is the step size indicated. To operate the multistep scheme fixing the order p, the procedure follows the one-step algorithm, except for the need to start the method with p initial values: if $\tilde{h}/h_k \ll 1$, then recompute the step with $h_k = \tilde{h}$; otherwise, proceed with $h_{k+1} = \tilde{h}$. This strategy is called *h-adaptive* because the step size h is adjusted to control the error. With multistep methods, one has an opportunity to achieve the same goal changing also the order of the method.

The error estimate, obtained by using a higher-order polynomial, can be adapted to ask what error an $\mathcal{O}(h^{p-1})$ scheme would have:

$$E_{p-1} = \left| g_{p-1,2} f^{(1)}[x_{k+1}, ..., x_{k-p+2}] \right|,$$

or, with a larger tableau, what error an $\mathcal{O}(h^{p+1})$ scheme would have:

$$E_{p+1} = \left| g_{p+1,2} f^{(1)}[x_{k+1}, ..., x_{k-p}] \right|.$$

Then,

$$r_- = \left(\frac{\epsilon_1}{E_{p-1}} \right)^{1/p}, \quad r_0 = \left(\frac{\epsilon_1}{E_p} \right)^{1/(p+1)}, \quad r_+ = \left(\frac{\epsilon_1}{E_{p+1}} \right)^{1/(p+2)} \tag{10.40}$$

are the relative step sizes indicated for orders $p - 1$, p, and $p + 1$ methods, respectively. One could pursue a so-called *hp-adaptive* strategy as follows: compute the three factors of (10.40). If $\tilde{h} \ll h$ (10.39), then redo the step with step size $\tilde{h} = r_0 h$. Otherwise, proceed with the next step. For this next step,

- use step size $h := r_- h$, and drop order, $p := p - 1$, if $r_- > \max(r_0, r_+)$; or
- use step size $h := r_0 h$, and keep order p, if $r_0 > \max(r_-, r_+)$; or
- use step size $h := r_+ h$, and increase order, $p := p + 1$, if $r_+ > \max(r_-, r_0)$.

Example 10.3   Solve (10.30) with the *hp*-adaptive multistep method.

Solution
Begin with order 1, i.e., forward and backward Euler methods, and allow the order to grow as large as 20. The error per step is estimated as $10^{-15} y$. When the order is 1, a step size of $h = 2.1200 \times 10^{-15}$ is required, but as the order climbs to 20 the step size diminishes to a maximum of 5.2856×10^{-2}. Only 298 steps are required, and the final error is 9.9003×10^{-12}. The relative error of $10\times$ machine precision is optimistic when the order is large, and so increasing the order ceiling does not improve the result unless the error target is adjusted upward appropriately. The example code is listed on page 430.

10.4   Boundary value problems

A boundary value problem differs from an initial value problem in two significant ways. First, there is insufficient information at either of the boundaries, taken individually, to solve the given problem as an initial value problem. Second, boundary value problems need not have solutions, or can have multiple solutions.

Our approach to boundary value problems will be to recast them as initial value problems by supplying guesses for those parameters that are missing relative to an initial value problem solution. We then integrate the problem by initial value problem methods until we reach the opposite boundary. Since the initial guesses will only coincidentally be correct, we expect that the information at the final boundary will not match the given conditions: some adjustment of the guessed data will be required. This adjustment occurs through iteration, e.g., by using a Newton–Raphson strategy to modify the guessed data. This combination of initial value problem techniques with Newton–Raphson is called the *simple shooting method*.

For example, assume that the boundary problem is expressed as follows for vector $\mathbf{y} \in \mathbb{R}^n$:

$$\frac{d\mathbf{y}}{dx} = \mathbf{f}(x, \mathbf{y})$$

$$y_i(a) = (\mathbf{y}_0)_i \quad i = 1, ..., m < n$$

$$y_i(b) = (\mathbf{y}_1)_i \quad i = 1, ..., n - m. \tag{10.41}$$

There are n total boundary conditions specified. To solve this as an initial value problem, values of $y_i(a)$ for $i = m + 1, ..., n$ must be supplied as guesses.

With standard techniques we can then compute a solution at $x = b$, $\mathbf{y}(b)$. We have a vector error $\mathbf{e}^{(k)}$ given by the mismatch of the boundary data (10.41):

$$\mathbf{e}_i^{(k)} = y_i^{(k)}(b) - (\mathbf{y}_1)_i, \quad i = 1, ..., n - m.$$

An improved set of initial guesses $\mathbf{y}^{(k)}(a)$ could be obtained by the Newton–Raphson method

$$y_i^{(k+1)}(a) = y_i^{(k)}(a) - \left(\frac{\partial \mathbf{y}^{(k)}(b)}{\partial \mathbf{y}^{(k)}(a)}\right)_{ij}^{-1} \mathbf{e}_j^{(k)}, \quad \begin{matrix} i = m + 1, ..., n \\ j - 1, ..., n - m. \end{matrix} \tag{10.42}$$

Here, $\partial \mathbf{y}(b)/\partial \mathbf{y}(a)$ is an $(n - m) \times (n - m)$ matrix expressing the partial derivatives of the far boundary values with respect to the initial guesses.

This matrix can be estimated by performing several IVP calculations with slightly varying $\mathbf{y}$, i.e., using finite differences to estimate the partial derivatives. An alternative strategy is to augment the system of equations with a set of equations that will determine those partial derivatives directly.

To develop this augmented strategy, we can write

$$\frac{\partial \mathbf{y}(x)}{\partial x} = \mathbf{f}(x, \mathbf{y})$$

$$\frac{\partial}{\partial \mathbf{y}(a)} \frac{\partial \mathbf{y}(x)}{\partial x} = \frac{\partial}{\partial \mathbf{y}(a)} \mathbf{f}(x, \mathbf{y}) = \frac{\partial \mathbf{f}(x, \mathbf{y})}{\partial \mathbf{y}(x)} \frac{\partial \mathbf{y}(x)}{\partial \mathbf{y}(a)}$$

$$\mathbf{W}(x) \equiv \frac{\partial \mathbf{y}(x)}{\partial \mathbf{y}(a)}$$

$$\frac{\partial}{\partial x} \mathbf{W}(x) = \frac{\partial \mathbf{f}(x, \mathbf{y})}{\partial \mathbf{y}} \mathbf{W}(x)$$

with initial condition

$$\mathbf{W}(a) = \mathbf{I}.$$

To solve this augmented system we write

$$\mathbf{Y} = \begin{pmatrix} \mathbf{y} \\ w_{1,1} \\ \vdots \\ w_{i,j} \\ \vdots \\ w_{n,n} \end{pmatrix} \qquad \mathbf{Y}(a) = \begin{pmatrix} \mathbf{y}_0^{(k)} \\ I_{1,1} \\ \vdots \\ I_{i,j} \\ \vdots \\ I_{n,n} \end{pmatrix} \qquad \mathbf{F} = \begin{pmatrix} \mathbf{f}(x, \mathbf{y}) \\ \sum_k \frac{\partial f_1(x,y)}{\partial y_k} W_{k1} \\ \vdots \\ \sum_k \frac{\partial f_i(x,y)}{\partial y_k} W_{kj} \\ \vdots \\ \sum_k \frac{\partial f_n(x,y)}{\partial y_k} W_{kn} \end{pmatrix},$$

which is a well-posed initial value problem for the ODE $d\mathbf{Y}/dx = \mathbf{F}$ in $\mathbb{R}^{n+n^2}$. At $x = b$, the $\mathbf{W}$ components of vector $\mathbf{Y}$ can be assembled to make the matrix $\partial \mathbf{y}(b)/\partial \mathbf{y}(a)$ needed in (10.42) to adjust the initial values. Not all components of $\mathbf{W}$ need be determined – only those required to support the final Newton–Raphson step.

Example 10.4   Solve the BVP:

$$y'' = -y(y')^2$$
$$y(0) = 1$$
$$y(1) = 2.$$

Solution

The first step is to express this as a system of first-order equations:

$$\frac{d\mathbf{Y}}{dx} = \frac{d}{dx} \begin{pmatrix} y \\ y' \end{pmatrix} = \mathbf{F} = \begin{pmatrix} Y_2 \\ -Y_1 Y_2^2 \end{pmatrix}.$$

Second, we find the derivative of the solution with respect to the starting solution:

$$\mathbf{W} = \begin{pmatrix} \partial Y_1(x)/\partial Y_1(0) & \partial Y_1(x)/\partial Y_2(0) \\ \partial Y_2(x)/\partial Y_1(0) & \partial Y_2(x)/\partial Y_2(0) \end{pmatrix}$$

$$\frac{\partial \mathbf{W}}{\partial x} = \frac{\partial \mathbf{F}}{\partial \mathbf{Y}}\mathbf{W} = \begin{pmatrix} 0 & 1 \\ -Y_2^2 & -2Y_1Y_2 \end{pmatrix}\mathbf{W} \tag{10.43}$$

$$= \begin{pmatrix} W_{21} & W_{22} \\ -Y_2^2 W_{11} - 2Y_1Y_2 W_{21} & -Y_2^2 W_{12} - 2Y_1Y_2 W_{22} \end{pmatrix}.$$

We don't need all of $\mathbf{W}$ – just enough to evaluate $\partial y(1)/\partial y'(0)$. The Newton–Raphson step requires only W_{12}, but the ODE for W_{12} depends on W_{22}, so the W_{22} equation is also included. The augmented system of equations is

$$\frac{d\mathbf{Y}}{dx} = \frac{d}{dx}\begin{pmatrix} y \\ y' \\ W_{12} \\ W_{22} \end{pmatrix} = \mathbf{F} = \begin{pmatrix} Y_2 \\ -Y_1 Y_2^2 \\ W_{22} \\ -Y_2^2 W_{12} - 2Y_1 Y_2 W_{22} \end{pmatrix}$$

$$\mathbf{Y}(0) = \begin{pmatrix} 1 \\ y'^{(k)}(0) \\ 0 \\ 1 \end{pmatrix}.$$

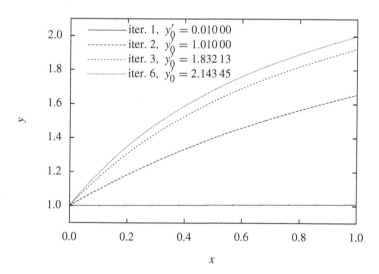

Figure 10.2 Simple shooting method example; Example 10.4.

Solve this to $x = 1$ using $h = 0.01$ and the fourth-order adaptive Runge–Kutta method. The initial step size is $h = 0.01$, and the error target per step is $2 \times 10^{-6}|y|$.

The initial guess used for $y'(0)$ is 0.01. The iteration to improve $y'(0)$ is

$$y'^{(k+1)}(0) = y'^{(k)}(0) - (y^{(k)}(1) - 2)/W_{12}^{(k)}(1).$$

The Newton–Raphson method converges in about six iterations and the resulting trajectory is shown in Figure 10.2. Note that although the Newton–Raphson method converged to a final value of $y'(0)$, the resulting number carries both numerical error and an error from the initial value problem integration.

With the simple shooting method, the error incurred by the IVP "shot" grows exponentially with the distance between boundaries (10.12). The circumstance can occur that an error in the initial conditions at $x = a$ on the order of machine precision can be amplified to such an extent that no meaningful matching of boundary conditions at $x = b$ is possible. In such a case, the simple shooting method is not effective.

A solution to this problem is the *multiple shooting method* – an idea first proposed by Morrison *et al.* [167], and made practical by Bulirsch [30]. In the multiple shooting method, the interval $b - a$ is subdivided into several increments, say K. Let us write

$$\frac{d\mathbf{y}}{dx} = \mathbf{f}$$
$$\mathbf{R}(\mathbf{y}(b), \mathbf{y}(a)) = 0$$

as the BVP specification.

At each of the interior boundaries x_i, one has an estimate of the state variables y_i. On each interval (x_{i-1}, x_i) one can conduct a simple shooting method. With y_{i-1} specified, one computes $y(x_i; y_{i-1})$, the value at x_i starting with y_{i-1} at x_{i-1}, and at convergence this would match y_i. The error at the end of this interval is

$$\mathbf{H}_i = \mathbf{y}(x_i; y_{i-1}^{(k)}) - y_i^{(k)}, \quad i = 1, ..., K$$

on iteration k, and across the domain one has

$$\mathbf{R}\left(y_K^{(k)}, y_0^{(k)}\right)$$

as the solution error.

Corrections $\Delta\mathbf{y}$ designed to zero these errors may be computed via

$$\mathbf{H}_i + \frac{\partial \mathbf{H}_i}{\partial \mathbf{y}_{i-1}} \Delta\mathbf{y}_{i-1} - \Delta\mathbf{y}_i = 0, \quad i = 1, ..., K$$

$$\mathbf{R} + \frac{\partial \mathbf{R}}{\partial \mathbf{y}_K} \Delta\mathbf{y}_K + \frac{\partial \mathbf{R}}{\partial \mathbf{y}_0} \Delta\mathbf{y}_0 = 0. \tag{10.44}$$

This is really just a Newton–Raphson step, as in the simple shooting method, and should be compared with Section 8.5. The simultaneous solution of these coupled error equations may be written

$$
\begin{pmatrix}
\frac{\partial \mathbf{H}_1}{\partial \mathbf{y}_0} & -\mathbf{I} & & & & \\
& \frac{\partial \mathbf{H}_2}{\partial \mathbf{y}_1} & -\mathbf{I} & & & \\
& & \ddots & \ddots & & \\
& & & \ddots & \ddots & \\
& & & & \frac{\partial \mathbf{H}_K}{\partial \mathbf{y}_{K-1}} & -\mathbf{I} \\
\frac{\partial \mathbf{R}}{\partial \mathbf{y}_0} & & \cdots & & & \frac{\partial \mathbf{R}}{\partial \mathbf{y}_K}
\end{pmatrix}
\begin{pmatrix}
\Delta \mathbf{y}_0 \\
\Delta \mathbf{y}_1 \\
\vdots \\
\vdots \\
\Delta \mathbf{y}_{K-1} \\
\Delta \mathbf{y}_K
\end{pmatrix}
=
\begin{pmatrix}
-\mathbf{H}_1 \\
-\mathbf{H}_2 \\
\vdots \\
\vdots \\
-\mathbf{H}_K \\
-\mathbf{R}
\end{pmatrix}.
$$

This is nearly block bidiagonal, which would be very easy to solve, but that form is spoiled by the block $\partial \mathbf{R}/\partial \mathbf{y}_0$. The bidiagonal form can be established by expressing $\Delta \mathbf{y}_K$ in terms of $\Delta \mathbf{y}_1$ as follows:

$$
\Delta \mathbf{y}_1 = \mathbf{H}_1 + \frac{\partial \mathbf{H}_1}{\partial \mathbf{y}_0} \Delta \mathbf{y}_0
$$

$$
\Delta \mathbf{y}_2 = \mathbf{H}_2 + \frac{\partial \mathbf{H}_2}{\partial \mathbf{y}_1} \Delta \mathbf{y}_1
$$

$$
= \mathbf{H}_2 + \frac{\partial \mathbf{H}_2}{\partial \mathbf{y}_1} \mathbf{H}_1 + \frac{\partial \mathbf{H}_2}{\partial \mathbf{y}_1} \frac{\partial \mathbf{H}_1}{\partial \mathbf{y}_0} \Delta \mathbf{y}_0
$$

$$
\Delta \mathbf{y}_3 = \mathbf{H}_3 + \frac{\partial \mathbf{H}_3}{\partial \mathbf{y}_2} \Delta \mathbf{y}_2
$$

$$
= \mathbf{H}_3 + \frac{\partial \mathbf{H}_3}{\partial \mathbf{y}_2} \mathbf{H}_2 + \frac{\partial \mathbf{H}_3}{\partial \mathbf{y}_2} \frac{\partial \mathbf{H}_2}{\partial \mathbf{y}_1} \mathbf{H}_1 + \frac{\partial \mathbf{H}_3}{\partial \mathbf{y}_2} \frac{\partial \mathbf{H}_2}{\partial \mathbf{y}_1} \frac{\partial \mathbf{H}_1}{\partial \mathbf{y}_0} \Delta \mathbf{y}_0
$$

$$
\vdots
$$

$$
\Delta \mathbf{y}_K = \sum_{i=1}^{K} \left(\prod_{j=i+1}^{K} \frac{\partial \mathbf{H}_j}{\partial \mathbf{y}_{j-1}} \right) \mathbf{H}_i + \left(\prod_{j=1}^{K} \frac{\partial \mathbf{H}_j}{\partial \mathbf{y}_{j-1}} \right) \Delta \mathbf{y}_0.
$$

Then, with (10.44),

$$
\underbrace{\left(\frac{\partial \mathbf{R}}{\mathbf{y}_0} + \frac{\partial \mathbf{R}}{\partial \mathbf{y}_K} \left(\prod_{j=1}^{K} \frac{\partial \mathbf{H}_j}{\partial \mathbf{y}_{j-1}} \right) \right)}_{\mathbf{A}} \Delta \mathbf{y}_0 = \underbrace{-\mathbf{R} - \frac{\partial \mathbf{R}}{\partial \mathbf{y}_K} \sum_{i=1}^{K} \left(\prod_{j=i+1}^{K} \frac{\partial \mathbf{H}_j}{\partial \mathbf{y}_{j-1}} \right) \mathbf{H}_i}_{\mathbf{b}}. \quad (10.45)
$$

In matrix form,

$$
\begin{pmatrix}
\mathbf{A} & & & & \\
\frac{\partial \mathbf{H}_1}{\partial \mathbf{y}_0} & -\mathbf{I} & & & \\
& \frac{\partial \mathbf{H}_2}{\partial \mathbf{y}_1} & -\mathbf{I} & & \\
& & \ddots & \ddots & \\
& & & \ddots & \ddots \\
& & & & \frac{\partial \mathbf{H}_K}{\partial \mathbf{y}_{K-1}} & -\mathbf{I}
\end{pmatrix}
\begin{pmatrix}
\Delta \mathbf{y}_0 \\
\Delta \mathbf{y}_1 \\
\vdots \\
\vdots \\
\Delta \mathbf{y}_{K-1} \\
\Delta \mathbf{y}_K
\end{pmatrix}
=
\begin{pmatrix}
\mathbf{b} \\
-\mathbf{H}_1 \\
-\mathbf{H}_2 \\
\vdots \\
\vdots \\
-\mathbf{H}_K
\end{pmatrix},
$$

and this is easily solved by block back-substitution. The matrix $\mathbf{A}$ and vector $\mathbf{b}$ look a little daunting in (10.45), but are easily constructed in practice:

$$A = I$$
$$b = 0,$$

for $i = 1$ to K,

$$b := H_i + \frac{\partial H_i}{\partial y_{i-1}} b$$

$$A := \frac{\partial H_i}{\partial y_{i-1}} A,$$

then, finally,

$$A := \frac{\partial R}{\partial y_0} + \frac{\partial R}{\partial y_K} A$$

$$b := -R - \frac{\partial R}{\partial y_K} b.$$

Note that the matrices $\partial H_i / \partial y_{i-1}$ are the matrices W (10.43) on the interval (x_{i-1}, x_i).

Example 10.5 [225] Solve

$$y'' = 400y + 400 \cos^2 \pi x + 2\pi^2 \cos 2\pi x$$
$$y(0) = 0$$
$$y(1) = 0$$

with the multiple shooting method.

Solution
The interval $[0, 1]$ is subdivided into 20 equal segments, and for each segment we assume $y = 0$ and $y' = 0$ as initial conditions. The 20 corresponding trajectories are shown in Figure 10.3. The analytical solution has modes like $\exp(\pm 20x)$, which make the numerical solution $y(1)$ very sensitive to the initial condition $y'(0)$. However, the ODE is linear, so $y(1)$ is linearly proportional to $y'(0)$, which leads (with exact math) to convergence of the multiple shooting method in one cycle. The simple shooting method would also converge in one cycle with exact math, but the math is considerably less exact because of the mode $\exp(20x)$. Roughly, the error in approximation $y'(0)$ is amplified by e^{20} for the simple shooting method, but by e^1 for the multiple shooting method with 20 intervals – a difference of approximately 2×10^8.

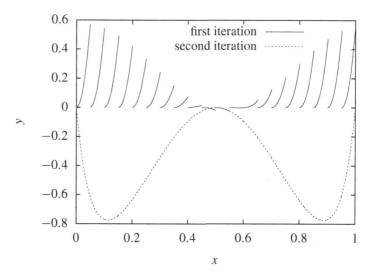

Figure 10.3: Multiple shooting method; Example 10.5.

10.5 Stiff systems

Stiffness is a characteristic of a problem with multiple and very different time scales. Such a problem is stiff if the shortest time scale must be respected for stability reasons. The best way to appreciate this is through the system of equations

$$\frac{\partial y_1}{\partial t} = -\frac{\lambda_1 + \lambda_2}{2} y_1 - \frac{\lambda_1 - \lambda_2}{2} y_2$$

$$\frac{\partial y_2}{\partial t} = -\frac{\lambda_1 - \lambda_2}{2} y_1 - \frac{\lambda_1 + \lambda_2}{2} y_2 \qquad (10.46)$$

$$y_1(0), y_2(0), \lambda_1, \lambda_2 > 0$$

$$\lambda_1 \gg \lambda_2.$$

For this system, the exact solution is

$$y_1(t) = \frac{y_1(0) + y_2(0)}{2} e^{-\lambda_1 t} + \frac{y_1(0) - y_2(0)}{2} e^{-\lambda_2 t}$$

$$y_2(t) = \frac{y_1(0) + y_2(0)}{2} e^{-\lambda_1 t} - \frac{y_1(0) - y_2(0)}{2} e^{-\lambda_2 t},$$

and the solution has the property that

$$\lim_{t \to \infty} y_1(t) = \lim_{t \to \infty} y_2(t) = 0.$$

Since $\lambda_1 \gg \lambda_2$, the terms proportional to $e^{-\lambda_1 t}$ will decay quickly, and on a time scale $t \gg 1/\lambda_1$ the solution will look like

$$y_1(t) \approx \frac{y_1(0) - y_2(0)}{2} e^{-\lambda_2 t}$$

$$y_2(t) \approx \frac{y_2(0) - y_1(0)}{2} e^{-\lambda_2 t},$$

i.e., smoothly decaying exponentials with time scales $1/\lambda_2$.

If we were to estimate the numerical solution of (10.46) by the forward Euler method, we have the algorithm

$$y_1^{n+1} = y_1^n - h\frac{\lambda_1 + \lambda_2}{2} y_1^n - h\frac{\lambda_1 - \lambda_2}{2} y_2^n$$

$$y_2^{n+1} = y_2^n - h\frac{\lambda_1 - \lambda_2}{2} y_1^n - h\frac{\lambda_1 + \lambda_2}{2} y_2^n$$

or

$$\mathbf{y}^{n+1} = \begin{pmatrix} 1 - h\frac{\lambda_1+\lambda_2}{2} & -h\frac{\lambda_1-\lambda_2}{2} \\ -h\frac{\lambda_1-\lambda_2}{2} & 1 - h\frac{\lambda_1+\lambda_2}{2} \end{pmatrix} \mathbf{y}^n,$$

and the matrix has eigenvalues $1 - h\lambda_1$ and $1 - h\lambda_2$. For the numerical solution to decay to zero like the exact one does, we need the eigenvalues to have magnitudes less than 1. With the constants λ positive, and h positive, this requires

$$h < 2/\lambda_1. \tag{10.47}$$

The time step is controlled by the smallest time scale of the system, regardless of the time scale over which we want to obtain a solution. If $h > 2/\lambda_1$, the solution will grow unphysically. In fact, it will oscillate wildly and unstably.

This example is typical of many chemical reaction systems. In fact, the concept of stiffness was first described by Curtiss and Hirschfelder [47] who were active in studying the reaction chemistry of rocket fuels.

The standard way to "fix" stiff systems is to use implicit methods. If we apply the backward Euler scheme to (10.46) we obtain the discretization

$$y_1^{n+1} = y_1^n - h\frac{\lambda_1 + \lambda_2}{2} y_1^{n+1} - h\frac{\lambda_1 - \lambda_2}{2} y_2^{n+1}$$

$$y_2^{n+1} = y_2^n - h\frac{\lambda_1 - \lambda_2}{2} y_1^{n+1} - h\frac{\lambda_1 + \lambda_2}{2} y_2^{n+1}$$

or

$$\begin{pmatrix} 1 + h\frac{\lambda_1+\lambda_2}{2} & +h\frac{\lambda_1-\lambda_2}{2} \\ +h\frac{\lambda_1-\lambda_2}{2} & 1 + h\frac{\lambda_1+\lambda_2}{2} \end{pmatrix} \mathbf{y}^{n+1} = \mathbf{y}^n$$

$$\mathbf{y}^{n+1} = \begin{pmatrix} \frac{2+h(\lambda_1+\lambda_2)}{2(1+h\lambda_1)(1+h\lambda_2)} & \frac{h(\lambda_2-\lambda_1)}{2(1+h\lambda_1)(1+h\lambda_2)} \\ \frac{h(\lambda_2-\lambda_1)}{2(1+h\lambda_1)(1+h\lambda_2)} & \frac{2+h(\lambda_1+\lambda_2)}{2(1+h\lambda_1)(1+h\lambda_2)} \end{pmatrix} \mathbf{y}^n.$$

The eigenvalues of this matrix are $1/(1 + h\lambda_1)$ and $1/(1 + h\lambda_2)$, and these are both positive and smaller than 1 for any positive h. By selecting an implicit method, we obtain a stable solution that decays like the exact solution, and we are free to pick any time step we like.

Problems

10.1 Solve Problem 3.3 as a boundary value problem.

10.2 Assess the stability of the Chebyshev polynomial recurrence formula – a linear difference equation:

$$T_{k+1}(x) = 2x T_k(x) - T_{k-1}(x),$$

for $x \in [-1, 1]$.

10.3 Two recurrence relations for the associated Legendre polynomials are

$$P_\ell^{m+1} + \frac{2mx}{\sqrt{1-x^2}} P_\ell^m + [\ell(\ell+1) - m(m-1)]P_\ell^{m-1} = 0,$$

for ascending m, and

$$(\ell - m + 1)P_{\ell+1}^m - (2\ell+1)x P_\ell^m + (\ell+m)P_{\ell-1}^m = 0,$$

for ascending ℓ, with $0 \le |m| \le \ell$, and $-1 \le x \le 1$. Analyze these as linear difference equations to assess their stability. Hint: consider the limit $\ell \gg m$ with $P_\ell^m = (m\lambda)^m$ for the m-recurrence, and $P_\ell^m = (\lambda^\ell/\ell)$ for the ℓ-recurrence.

10.4 The harmonic oscillator

$$\frac{d^2 x}{dt^2} = -kx \qquad (k > 0)$$

$$x(0) = x_0$$

$$\dot{x}(0) = v_0,$$

has exact solution

$$x(t) = x_0 \cos(\sqrt{k}t) + \frac{v_0}{\sqrt{k}} \sin(\sqrt{k}t).$$

What solutions $x(t)$ come from the forward Euler method, the backward Euler method, and the modified Euler method?

10.5 Evaluate the order of accuracy of the Verlet algorithm [233] (a.k.a. Störmer's method [226]),

$$x^{n+1} - 2x^n + x^{n-1} = h^2 f(x^n),$$

for the second-order ODE

$$\frac{d^2 x}{dt^2} = f.$$

How does the Verlet algorithm solve the harmonic oscillator problem?

10.6 [169] This problem will apply ODEs to a problem of infectious disease modeling. The population to be modeled will consist of susceptible individuals S, zombies Z, and resurrectable corpses R.

Nominally, these populations are governed in a statistical sense by simple differential equations. We assume a constant birth rate B, death rate δS, and infection rate βSZ, so the susceptible population is governed by the differential equation

$$\frac{dS}{dt} = B - \beta SZ - \delta S.$$

The zombie population grows when susceptible people are infected, or when resurrectable corpses become infected (rate ζR), and it diminished when a zombie is decapitated (rate αSZ):

$$\frac{dZ}{dt} = \beta SZ + \zeta R - \alpha SZ.$$

The population of resurrectable corpses grows when susceptibles die and diminishes when corpses are infected:

$$\frac{dR}{dt} = \delta S - \zeta R.$$

An effective strategy to maintaining the susceptible population is to periodically launch a campaign to decapitate some fraction k of the zombie population,

$$\Delta Z = -kZ.$$

Assuming these impulsive eradication events occur every 5 days, find the kill factor k so that on the 365th day, immediately following the 73rd event, one has $S/Z = 100$.

Assume initial conditions $S = 100$, $R = 40$, $Z = 0$ (i.e., a small village with a very small cemetery), and rate parameters $B = 10^{-4}$, $\delta = 10^{-4}$, $\beta = 5.5 \times 10^{-3}$, $\zeta = 9 \times 10^{-2}$, and $\alpha = 7.5 \times 10^{-3}$.

11 Introduction to stochastic ODEs

Viewed at certain time scales, processes occurring at much shorter time scales can appear random and chaotic. Perhaps the best known example is the random walk of pollen grains in water, or Brownian motion [27], described by the Langevin equations [141]:

$$\frac{d}{dt}x = v$$

$$\frac{d}{dt}v = -\gamma v + \sigma \xi,$$

(11.1)

where x is the position, v the velocity, and t time; ξ is a particular random function; $-\gamma v$ represents Stokes drag – the tendency of the pollen grain to move with the surrounding fluid (assumed stationary); and $\sigma \xi$ represents the change of velocity due to collision of water molecules with the pollen grain – a process that appears random when viewed at macroscopic time scales. Another very well-known example is the modeling of stock option prices S,

$$\frac{dS}{dt} = \mu S + \sigma S \xi,$$

(11.2)

here given by the Black–Scholes equations [18]. Here μS is the drift in stock prices, e.g., the tendency for the market experience a smooth long-term change. $\sigma S\xi$ represents the volatility of the stock – a fluctuation that appears random and chaotic when viewed at a time scale of days. The physics and mathematics behind such models are of considerable current interest. Both of these examples contributed to Nobel prizes: J.-B. Perrin for Physics in 1926 for his work in Brownian motion, and R. Merton and M. Scholes for Economics in 1997. Perrin's experimental work confirmed earlier theory developed by Nobel Laureate A. Einstein.

If σ is constant, then (11.1) is said to have "additive noise." In contrast, (11.2) has "multiplicative noise" because the coefficient of ξ depends on the solution S.

In these equations, ξ is a so-called "Gaussian-distributed white noise" – a random time-dependent variable with unbounded magnitude and motion at all time scales. If ξ were some smooth function – something possessing a Taylor series – then normal methods of ODE integration could solve these equations. But, the random and unbounded nature of ξ make the meaning of (11.1) and (11.2) different, necessitating unique mathematical and numerical approaches. Here, a brief introduction to this exciting new field will be presented.

The following discussions will draw analogies where possible to deterministic ODEs, to emphasize both similarities and differences. This is facilitated by using concepts like the method Φ and local discretization error τ, that are not ideally suited to the SDE (stochastic differential equation) application. Be aware that the notation used here is somewhat nonstandard in the SDE literature.

This introductory presentation will focus on explicit numerical methods derived from the so-called Itô–Taylor series, analogous to ODE IVP one-step methods derived from a Taylor expansion (e.g., (10.5)). The reader should be aware that there exist numerous other approaches, including implicit, multistep, and Runge–Kutta ones. See the monographs of Kloeden and Platen [131] and of Milstein and Tretyakov [164] for numerous examples.

Frequent reference will be made to the generic SDE

$$dy = f(y)dt + g(y)dW. \tag{11.3}$$

11.1 White noise and the Wiener process

We define $R(t, s)$ to be the covariance of the distribution,

$$R(t, s) = \mathbb{E}\xi(t)\xi(s),$$

where $\mathbb{E}$ is the expectation operator. When $R(t, s)$ has the form $c(|t - s|)$, one says that ξ is "wide sense stationary." In this case, the spectrum of the covariance is its Fourier transform

$$S(\omega) = \frac{1}{\sqrt{2\pi}} \int_{-\infty}^{+\infty} e^{i\omega t} c(t)dt,$$

called the Wiener–Khinchin theorem after [128, 239] (note: there are a variety of conventions for writing the Fourier transform, differing by signs and factors of 2 and π; this definition also varies). If the distribution is completely uncorrelated one expects

$$c(t - s) = S_0\delta(t - s),$$

for some constant S_0, which leads to

$$S(\omega) = \frac{1}{\sqrt{2\pi}} S_0,$$

and the power of the spectrum, $|S(\omega)|^2$ is clearly independent of ω. This is the definition of a white noise: a random signal whose power $|S(\omega)|^2$ is independent of ω. Random signals with different frequency dependencies are called colored noises.

Let $\xi(t)$ be a white noise,

$$\mathbb{E}\xi(t) = 0$$
$$\mathbb{E}\xi(t)\xi(s) = S_0\delta(t - s). \tag{11.4}$$

This signal is not bounded (the amplitude of the Dirac delta is infinite), not continuous, and not differentiable. It is therefore not amenable to many normal mathematical or computational treatments.

However, it is integrable. Let us assume $S_0 = 1$, and evaluate the covariance of the integrals

$$W(t) = \int_0^t \xi(\tau)d\tau. \tag{11.5}$$

The covariance $R(t, s)$ of the W functions is

$$R(t, s) = \mathbb{E}W(s)W(t) = \mathbb{E}\int_0^s d\sigma \int_0^t d\tau \xi(\tau)\xi(\sigma)$$

$$= \int_0^s d\sigma \int_0^t d\tau \delta(\tau - \sigma)$$

$$= \int_0^s d\sigma \int_{-\sigma}^{t-\sigma} dp \delta(p).$$

The second integral is 1 if $t - \sigma$ is positive, so

$$R(t, s) = \int_0^t d\sigma = t \qquad s > t.$$

But $R(t, s)$ is symmetric in its arguments, i.e., $R(t, s) = R(s, t)$, so

$$R(t, s) = \min(t, s).$$

It is not wide sense stationary.

When ξ is Gaussian-distributed, i.e., according to (8.10), W is called the Wiener process. It takes its name from N. Wiener who developed the integral calculus of Brownian motion and other stochastic processes (e.g., [240]). It is characterized by

$$\mathbb{E}W(t) = 0 \tag{11.6a}$$

$$\mathbb{E}W(t)W(s) = \min(t, s), \tag{11.6b}$$

and

$$W(0) = 0.$$

The first two properties follow from (11.4), while the last follows from (11.5). From (11.6b) the standard deviation of the Wiener process $W(t)$ is $\sqrt{t}$, and its mean is zero (11.6a). Therefore, being Gaussian distributed, $W(t) = \sqrt{t}\mathcal{N}(0, 1)$ (8.11).

A particular Wiener process, or path, $W(t)$ for $t \in [0, T]$, can be constructed by discretizing the time interval T into N intervals of width $h = T/N$, then drawing a particular sequence of Gaussian normal pseudorandom numbers:

$$W(0) = 0$$
$$W(h) = \sqrt{h}\,\mathcal{N}(0, 1)$$
$$W(2h) = W(h) + \sqrt{h}\,\mathcal{N}(0, 1) \tag{11.7}$$
$$\vdots$$
$$W(T) = W(T - h) + \sqrt{h}\,\mathcal{N}(0, 1).$$

This is a particular realization of the random events consistent with an underlying white noise correlation and Gaussian normal distribution.

Computationally, the need will arise to estimate random numbers whose distribution is Gaussian, $\mathcal{N}(0, 1)$. Unfortunately, the simpler problem addressed by most compilers is the generation of a uniformly distributed random number, say on the interval $[0, 1]$. The interconversion of uniform deviates (i.e., random numbers uniformly distributed on $[0, 1]$) to Gaussian deviates (random numbers distributed according to $\mathcal{N}(0, 1)$) uses a clever geometric analysis due to Box and Muller [22].

Their idea is to connect a uniform distribution on the unit square to the Gaussian distribution in $\mathbb{R}^2$. This can be accomplished with a single change of variables, which we will break down into four simple steps:

$$1 = \int_0^1 du_1 \int_0^1 du_2; \tag{11.8a}$$

let $r^2 = u_1$ and $\theta = 2\pi u_2$ to map the square to the unit circle

$$1 = \int_0^1 dr \int_0^{2\pi} r d\theta \frac{1}{\pi}; \tag{11.8b}$$

let $\rho = -\ln r$ to map the unit circle to $\mathbb{R}^2$,

$$1 = \int_0^\infty d\rho \int_0^{2\pi} \rho d\theta \frac{e^{-2\rho}}{\pi\rho}; \tag{11.8c}$$

then let $R = 2\sqrt{\rho}$,

$$1 = \int_0^\infty dR \int_0^{2\pi} R d\theta \frac{e^{-R^2/2}}{2\pi}; \tag{11.8d}$$

and finally move to Cartesian coordinates,

$$1 = \int_{-\infty}^{+\infty} dx \int_{-\infty}^{+\infty} dy \frac{e^{-(x^2+y^2)/2}}{2\pi} = \left(\int_{-\infty}^{+\infty} dx \frac{e^{-x^2/2}}{\sqrt{2\pi}} \right)^2. \tag{11.8e}$$

The uniform distribution on a unit square has become two Gaussian normal distributions with unit variance, $P(x)$ and $P(y)$ (cf. (8.10)).

One begins the process by drawing two uniformly distributed numbers u_1, u_2 on $[0, 1]$. Then, following the sequence of variable changes outlined above, one finds

$$x = \sqrt{-2 \ln u_1} \cos 2\pi u_2$$
$$y = \sqrt{-2 \ln u_1} \sin 2\pi u_2,$$

as a pair of Gaussian normal deviates drawn from $\mathcal{N}(0, 1)$.

A variation on this Box–Muller transform due to Marsaglia and Bray [154] chooses two uniform deviates u_1 and u_2 subject to the constraint

$$u_1^2 + u_2^2 < 1. \tag{11.9}$$

Then

$$1 = \int_{-1}^{+1} du_1 \int_{-\sqrt{1-u_1^2}}^{+\sqrt{1-u_1^2}} du_2 \frac{1}{\pi},$$

and changing variables $u_1 = r \cos\theta, u_2 = r \sin\theta$ yields (11.8b). The mapping $(u_1, u_2) \mapsto (x, y)$ uses

$$R = \sqrt{-2 \ln(u_1^2 + u_2^2)}$$
$$\cos\theta = \frac{u_1}{\sqrt{u_1^2 + u_2^2}}$$
$$\sin\theta = \frac{u_2}{\sqrt{u_1^2 + u_2^2}},$$

so finally

$$x = u_1 \sqrt{\frac{-2 \ln(u_1^2 + u_2^2)}{u_1^2 + u_2^2}}$$
$$y = u_2 \sqrt{\frac{-2 \ln(u_1^2 + u_2^2)}{u_1^2 + u_2^2}}.$$

This form avoids calculation of the sine and cosine, but at the expense of sometimes drawing and discarding uniform deviates that violate (11.9).

11.2   Itô and Stratonovich calculus

Stochastic processes obey different calculus systems than do usual deterministic systems. The two common varieties are "Itô" and "Stratonovich." They yield different results – a slightly unsettling affair. But, given the setting and time scale of the physics modeled there is often cause to choose one calculus over the other with confidence.

The Itô calculus derives from an integral approach after the forward Euler method:

$$\int_0^T h(t) dW(t) = \lim_{N \to \infty} \sum_{i=0}^{N-1} h(t_i)(W_{i+1} - W_i). \tag{11.10}$$

For example, by Itô's calculus,

$$
\begin{aligned}
I_I = \int_0^T W(t)\,dW(t) &= \sum_i W_i(W_{i+1} - W_i) \\
&= \sum_i \frac{1}{2}\left[-(W_{i+1} - W_i)^2 - W_i^2 + W_{i+1}^2 \right] \\
&= \frac{1}{2}W_N^2 - \frac{1}{2}\sum_i (W_{i+1} - W_i)^2 - \frac{1}{2}W_0^2 \\
&= \frac{1}{2}W(T)^2 - \frac{T}{2}.
\end{aligned}
\tag{11.11}
$$

The Stratonovich calculus derives from the trapezoidal sum rule:

$$
\int_0^T h(t)\,dW(t) = \lim_{N \to \infty} \sum_{i=0}^{N-1} \frac{h(t_{i+1}) + h(t_i)}{2}(W_{i+1} - W_i).
\tag{11.12}
$$

With this calculus,

$$
\begin{aligned}
I_S = \int_0^T W(t)\,dW(t) &= \frac{1}{2}\sum_i (W_{i+1} + W_i)(W_{i+1} - W_i) \\
&= \frac{1}{2}\sum_i (W_{i+1}^2 - W_i^2) \\
&= \frac{1}{2}W_N^2 - \frac{1}{2}W_0^2 \\
&= \frac{1}{2}W(T)^2.
\end{aligned}
$$

Clearly these results differ. In expectation, $\mathbb{E}I_I = 0$, while $\mathbb{E}I_S = T/2$.

Qualitatively, one can see that (11.10) is *nonanticipatory*: a step of the integral does not assume knowledge of the integrand h other than as an initial condition. In contrast, the Stratonovich method (11.12) is *anticipatory*: in a given step, knowledge of the integrand h is assumed as both the initial and final conditions. So, in a sense the Itô method respects causality more so than does the Stratonovich method.

This qualitative judgment is made more quantitative by Kupferman, Pavliotis, and Stuart [136] who study the stochastic dynamics of physical processes for which the noise spectrum is not strictly white, but rather colored. That is, the correlation is not strictly a delta function, but a steep distribution in frequency space with width given by the inverse of the correlation time. They show that for physical systems with dissipative timescales greater than the correlation time, Itô's calculus is appropriate. And, when the correlation time exceeds the physical dissipative time scale, Stratonovich's calculus is most appropriate.

In the model problem (11.1), the relaxation time scale is $1/\gamma$. For pollen in water the relaxation time is of the order 10^{-4}s, while the autocorrelation time of hydrodynamic fluctuations is of the order 10^{-13}s: Itô's calculus is suggested. In model problem (11.2), the relaxation time is $1/\mu$ – years, and the autocorrelation time is around a day (see

Problem 5.8), but with a significant ca. 100 day tail by some analyses. (The correlation is significant when the market crashes, and in such catastrophic cases cannot be treated like a white noise. The faulty assumption of white noise behavior may have contributed to the recession of 2008 [64]. In 1998 the investment company Long-Term Capital Management lost billions of dollars because of overreliance on Black–Scholes, necessitating what was then the largest government bailout of financial institutions. Nobel Laureates Merton and Scholes were on the board of directors.) Again Itô's calculus seems most appropriate. Accordingly, the remainder of this chapter will emphasize the Itô calculus.

11.3   Itô's formula

From the properties (11.6) of a Wiener process, it follows that

$$\mathbb{E}(W(t) - W(s))^2 = t - s,$$

when $t \geq s$. Then, taking the limit $t \to s + dt$ one infers

$$\mathbb{E}dW^2 = dt.$$

It is not true that $dW^2 = dt$, but it is true that

$$\int_s^{s+h} f dW^2 = \int_s^{s+h} f dt,$$

almost surely (a.s.), for any h. To show this, break the interval into pieces in which f has uniform sign, and in each such piece apply the mean value theorem:

$$\int_s^{s+h} f dW^2 = \sum_i \int_{s_i}^{s_{i+1}} f dW^2 = \sum_i f(\xi_i) \int_{s_i}^{s_{i+1}} dW^2$$

$$= \sum_i f(\xi_i) \lim_{n\to\infty} \sum_{j=1}^n (W(s_i + jh + h) - W(s_i + jh))^2,$$

with $h = (s_{i+1} - s_i)/n$,

$$= \sum_i f(\xi_i) \lim_{n\to\infty} n \frac{s_{i+1} - s_i}{n}$$

$$= \sum_i f(\xi_i) \int_{s_i}^{s_{i+1}} dt = \int_s^{s+h} f(x)dt.$$

The trick here, as in (11.11), is the replacement of an integral with an infinite sum, and the average summand is the expectation. So, with the understanding that any differential expression will be integrated, we may treat

$$dW^2 = dt$$

as an identity.

Itô's formula [117] comes from the derivative of a general function $x(y)$, where y is governed by the generic stochastic process (11.3):

$$dx = x(y + dy) - x(y)$$

$$= x(y) + \frac{dx}{dy}dy + \frac{1}{2}\frac{d^2x}{dy^2}dy^2 + \ldots - x(y)$$

$$= \frac{dx}{dy}(f dt + g dW) + \frac{1}{2}\frac{d^2x}{dy^2}(f dt + g dW)^2 + \text{h.o.t.} \qquad (11.13)$$

$$dx = \left(x'f + \frac{1}{2}x''g^2\right)dt + x'g dW \quad \text{a.s.}$$

using $dW^2 = dt$, and retaining terms to order dt.

For example, consider $y = W$, $dy = dW$ ($f = 0$, $g = 1$), and $x = y^2$. Application of Itô's formula (11.13) gives

$$x(t) - x(0) = \int_0^t dx = \int_0^t ds + \int_0^t 2W dW,$$

where $\frac{1}{2}x''g^2 = 1$ in the dt integral and $x'g = 2W$ in the dW integral. With $W(0) = 0$ one has then

$$W(t)^2 = t + 2\int_0^t W dW,$$

which is the result previously obtained, (11.11).

11.4 The Itô–Taylor series

Platen and Wagner [182, 235] developed a systematic technique to construct stochastic Taylor series of any order, based on the Itô calculus. As with ODEs, it is useful as a generator of numerical methods, and is required to assess concepts like truncation error.

The idea is best explained by example. We begin with the generic stochastic equation, and integrate it to generate a solution:

$$dy = f(y)dt + g(y)dW$$

$$y(t) = y(t_0) + \int_{t_0}^t ds_0 f(y) + \int_{t_0}^t dW_{s_0} g(y). \qquad (11.14)$$

Next, we note that $f(y)$ is a function in its own right, with its own SDE based on Itô's formula. The same is true of $g(y)$. To simplify notation, we will write $f(t)$ for $f(y(t))$, etc., and f_0 for $f(y(t_0))$, etc.:

$$df = \left(f_y f + \tfrac{1}{2} g^2 f_{yy} \right) dt + f_y g \, dW$$

$$f(s_0) = f_0 + \int_{t_0}^{s_0} ds_1 \left(f_y f + \tfrac{1}{2} g^2 f_{yy} \right) + \int_{t_0}^{s_0} dW_{s_1} f_y g$$

$$g(s_0) = g_0 + \int_{t_0}^{s_0} ds_1 \left(g_y f + \tfrac{1}{2} g^2 g_{yy} \right) + \int_{t_0}^{s_0} dW_{s_1} g_y g.$$

Now, substitute these expansions into (11.14) to give

$$y(t) = y_0 + g_0 \Delta W + f_0 \Delta t + \int_{t_0}^{t} ds_0 \int_{t_0}^{s_0} ds_1 \left(f_y f + \tfrac{1}{2} g^2 f_{yy} \right)$$

$$+ \int_{t_0}^{t} ds_0 \int_{t_0}^{s_0} dW_{s_1} f_y g$$

$$+ \int_{t_0}^{t} dW_{s_0} \int_{t_0}^{s_0} ds_1 \left(g_y f + \tfrac{1}{2} g^2 g_{yy} \right) \tag{11.15}$$

$$+ \int_{t_0}^{t} dW_{s_0} \int_{t_0}^{s_0} dW_{s_1} g_y g.$$

Each of these new integrands can in turn be expanded to give successively higher-order integrals terms. At some point, the process is terminated. The integrands can be evaluated at t_0, and the error committed by this approximation can be estimated. For example, if the expansion stopped at (11.15), we would use

$$\int_{t_0}^{t} ds_0 \int_{0}^{s_0} ds_1 \left(f_y f + \tfrac{1}{2} g^2 f_{yy} \right) = \left(f_y f + \tfrac{1}{2} g^2 f_{yy} \right)_0 \frac{t^2}{2} + \mathcal{O} \Delta t^{5/2}$$

$$\int_{t_0}^{t} ds_0 \int_{t_0}^{s_0} dW_{s_1} f_y g = (f_y g)_0 \left(\int_{t_0}^{t} ds_0 \int_{t_0}^{s_0} dW_{s_1} \right) + \mathcal{O} \Delta t^2 \tag{11.16}$$

$$\int_{t_0}^{t} dW_{s_0} \int_{t_0}^{s_0} ds_1 \left(g_y f + \tfrac{1}{2} g^2 g_{yy} \right) = \left(g_y f + \tfrac{1}{2} g^2 g_{yy} \right)_0 \left(\int_{t_0}^{t} dW_{s_0} \int_{0}^{s_0} ds_1 \right) + \mathcal{O} \Delta t^2 \tag{11.17}$$

$$\int_{t_0}^{t} dW_{s_0} \int_{t_0}^{s_0} dW_{s_1} g_y g = (g_y g)_0 \left(\int_{t_0}^{t} dW_{s_0} \int_{t_0}^{s_0} dW_{s_1} \right) + \mathcal{O} \Delta t^{3/2} \tag{11.18}$$

and conclude

$$y(t) = y_0 + f_0 \Delta t + g_0 \Delta W + (g_y g)_0 \left(\int_{t_0}^{t} dW_{s_0} \int_{t_0}^{s_0} dW_{s_1} \right) + \mathcal{O} \Delta t^{3/2},$$

by omitting all terms of order $\Delta t^{3/2}$ and higher. We recognize the term in parenthesis as being the example (11.11), so

$$y(t) = y_0 + g_0 \Delta W + f_0 \Delta t + (g_y g)_0 \frac{1}{2} (\Delta W^2 - \Delta t) + \mathcal{O} \Delta t^{3/2}. \tag{11.19}$$

To obtain an expansion with error $\mathcal{O}\Delta t^2$ it is necessary to further expand (11.18):

$$(g_y g)(s_1) = (g_y g)_0 + \int_{t_0}^{s_1} ds_2 \left((g_y g)_y f + \tfrac{1}{2}(g_y g)_{yy} g^2\right) + \int_{t_0}^{s_1} dW_{s_2}(g_y g)_y g$$

$$\int_{t_0}^{t} dW_{s_0} \int_{t_0}^{s_0} dW_{s_1} g_y g = (g_y g)_0 \tfrac{1}{2}(\Delta W^2 - \Delta t)$$

$$+ ((g_y g)_y g)_0 \left(\int_{t_0}^{t} dW_{s_0} \int_{t_0}^{s_0} dW_{s_1} \int_{t_0}^{s_1} dW_{s_2}\right) + \mathcal{O}\Delta t^2.$$

We are now faced with a new stochastic integral,

$$\frac{1}{2} \int_{t_0}^{t} dW_{s_0} ((W(s_0) - W_0)^2 - (t_0 - s_0)). \tag{11.20}$$

To evaluate this, apply Itô's formula to $(W(t) - W_0)^3$:

$$\frac{1}{3} d(W - W_0)^3 = (W - W_0)dt + (W - W_0)^2 dW$$

$$\int_{t_0}^{t} (W - W_0)^2 dW = \frac{\Delta W^3}{3} - \int_{t_0}^{t} (W - W_0)dt,$$

where $\int W dt$ requires evaluation both here and in (11.16). This integral is not reducible to a polynomial in t and W, so we give it a new symbol:

$$\Delta Z = \int_{t_0}^{t} (W - W_0)dt.$$

Remaining to be evaluated from (11.20), and also appearing in (11.17), is $\int t \, dW$:

$$d(t - t_0)(W - W_0) = (W - W_0)dt + (t - t_0)dW$$

$$\int_{t_0}^{t} (t - t_0)dW = \Delta t \Delta W - \int_{t_0}^{t} (W - W_0)dt = \Delta t \Delta W - \Delta Z.$$

Integral (11.20) is then

$$\frac{1}{2} \int_{t_0}^{t} dW_{s_0} ((W(s_0) - W_0)^2 - (t_0 - s_0)) = \frac{\Delta W^3}{6} - \frac{\Delta t \Delta W}{2},$$

and the next higher-order Itô–Taylor series is

$$y(t) = y_0 + g_0 \Delta W + f_0 \Delta t + (g_y g)_0 \tfrac{1}{2}(\Delta W^2 - \Delta t)$$
$$+ (f_y g)_0 \Delta Z + \tfrac{1}{2}((g_y g)_y g)_0 (\tfrac{1}{3} \Delta W^3 - \Delta t \Delta W)$$
$$+ (g_y f + \tfrac{1}{2} g^2 g_{yy})_0 (\Delta t \Delta W - \Delta Z) + \mathcal{O}\Delta t^2.$$

11.5 Orders of accuracy

The order of accuracy for stochastic methods must be defined in terms of expectations. After all, any solution depends on an assumed underlying Wiener process, and some processes may exhibit greater error than others. In expectation, this variability is eliminated. Let $y(t)$ be the exact solution of an SDE (subject to an assumed Wiener path), and let $\eta(t; h)$ be a computed solution evaluated with time discretization h. η depends on Wiener path assumptions, but in the present case may use different paths than $y(t)$. We can then define the *weak order of accuracy* p_w

$$|\mathbb{E}(g(\eta(t; h))) - \mathbb{E}(g(y(t)))| \le Ch^{p_w},$$

where g is a sufficiently continuous test function, and C is a constant independent of h. Here, expectations are compared to expectations, which is why these quantities may be calculated with different Wiener paths. The weak measure of error is the error of expectations.

When the exact solution is not known, one may estimate p_w with Richardson extrapolation, e.g.,

$$r_w^{(2h/h)} = |\mathbb{E}(g(\eta(t; 2h))) - \mathbb{E}(g(\eta(t; h)))|$$
$$r_w^{(4h/2h)} = |\mathbb{E}(g(\eta(t; 4h))) - \mathbb{E}(g(\eta(t; 2h)))|,$$

and

$$p_w = \log_2 \left(\frac{r_w^{(4h/2h)}}{r_w^{(2h/h)}} \right).$$

When integrating over many paths the results may exhibit significant variation. For example, in Figure 11.1 eight calculations of the Black–Scholes equations vary by as much as a factor of 20. To evaluate p_w numerically, the expectations need to be determined more accurately than the differences between expectations, and this requires that the solution be computed along a colossal number of paths (the *sampling error* in the computed expectation scales with the number of paths n like $\sigma/\sqrt{n}$, where σ is the standard deviation of the results, e.g., (1.11)). A computational strategy to cope with enormous sampling error is *variance reduction*, e.g., [37].

An entirely different measure of error occurs when one computes the expectation of error:

$$\sqrt{\mathbb{E}(|\eta(t; h)) - y(t)|^2)} \le Ch^{p_s}.$$

Here p_s is the *strong order of accuracy*. It is the expectation of error (vs. error of expectation). Since the error is computed first, η and y are each computed with the same Wiener paths.

Like the weak measure, one can estimate the strong order using Richardson extrapolation:

$$r_s^{(2h/h)} = \sqrt{\mathbb{E}((\eta(t; 2h) - \eta(t; h))^2)}$$
$$r_s^{(4h/2h)} = \sqrt{\mathbb{E}((\eta(t; 4h) - \eta(t; 2h))^2)},$$

and

$$p_s = \log_2 \left(\frac{r_s^{(4h/2h)}}{r_s^{(2h/h)}} \right).$$

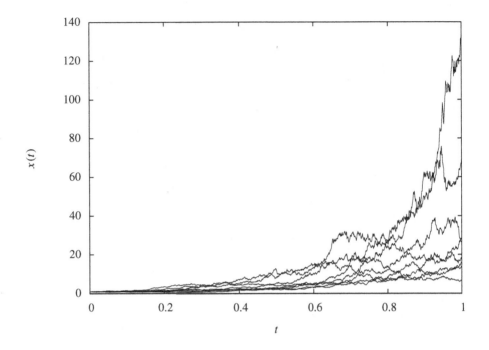

Figure 11.1 Idea of weak convergence. Eight different solutions of the same Black–Scholes equation are displayed, each assuming a different Wiener path.

In this case, for each path one calculates $\eta(t; h)$, $\eta(t; 2h)$ and $\eta(t; 4h)$. Numerically this is more easily computed than in the weak case. This is because the standard deviation of the pathwise error is much smaller than the standard deviation of the result over paths. In Figure 11.2 a particular path of the Black–Scholes problem is computed with two step sizes differing by a factor of 32, and the effect of step size is markedly less than the effect of choosing different paths, as in Figure 11.1.

For deterministic methods, convergence implies consistency and stability. This is not necessarily true in the case of stochastic methods when measured in the weak sense. To illustrate this point, consider a fictitious method such that

$$\eta(t; h) = y(t) + W(t),$$

and let test function $g(y) = y$. For this method

$$\mathbb{E}(\eta) - \mathbb{E}(y) = 0$$

so the method appears exact by the weak measure, even though it is not consistent. But, in the strong sense

$$\sqrt{\mathbb{E}(\eta - y)^2} = \sqrt{t},$$

this fictitious method has strong order zero: it is not convergent, and this implies its

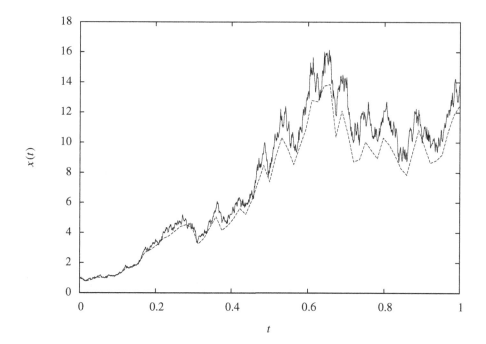

Figure 11.2 Idea of strong convergence. The Black–Scholes equation is solved with two discretizations of the same Wiener path.

inconsistency. For this reason, weak consistency is determined differently from convergence, and one should consider in principle all test functions.

11.6   Strong convergence

The criteria for a numerical method

$$\eta(t^{n+1}; h) = \eta(t^n; h) + h\Phi(t^n, \eta(t^n, h))$$

to achieve a given strong order of accuracy is given by a theorem of Milstein [163]. Define Δ as in (10.4). Let y be the exact solution (for a given stochastic path), and let η be the numerical approximation.

The starting point of the analysis is the error

$$y(t^{n+1}; t^n, y^n) - \eta(t^{n+1}; h, t^n, \eta^n) = y(t^n; t^{n-1}, y^{n-1}) - \eta(t^n; h, t^{n-1}, \eta^{n-1})$$
$$+ h\left[\Delta(t^n, y^n) - \Phi(t^n, y^n)\right] + h\left[\Phi(t^n, y^n) - \Phi(t^n, \eta^n)\right]$$

(cf. (10.11)). We then square this to give

$$\left(y(t^{n+1}; t^n, y^n) - \eta(t^{n+1}; t^n, \eta^n)\right)^2 \tag{11.21}$$

$$= \left(y(t^n; t^{n-1}, y^{n-1}) - \eta(t^n; t^{n-1}, \eta^{n-1})\right)^2$$
$$+ h^2 \left(\Delta(t^n, y^n) - \Phi(t^n, y^n)\right)^2 + h^2 \left(\Phi(t^n, y^n) - \Phi(t^n, \eta^n)\right)^2$$
$$+ 2h \left(y(t^n; t^n, y^n) - \eta(t^n; t^n, \eta^n)\right) \left(\Delta(t^n, y^n) - \Phi(t^n, y^n)\right)$$
$$+ 2h \left(y(t^n; t^{n-1}, y^{n-1}) - \eta(t^n; t^{n-1}, \eta^{n-1})\right) \left(\Phi(t^n, y^n) - \Phi(t^n, \eta^n)\right)$$
$$+ 2h^2 \left(\Delta(t^n, y^n) - \Phi(t^n, y^n)\right) \left(\Phi(t^n, y^n) - \Phi(t^n, \eta^n)\right).$$

One now proceeds to bound the expectations of the various terms. It will be convenient to define

$$\mathbb{E}_2(y) = \sqrt{\mathbb{E}(y^2)},$$

and $\mathbb{E}|y| < \mathbb{E}_2|y|$. The first two terms of (11.21) are easily connected with the definitions of $\mathbb{E}_2$ and of the global discretization error e (10.6):

$$\mathbb{E}\left(y(t^{n+1}; t^n, y^n) - \eta(t^{n+1}; t^n, \eta^n)\right)^2 = \left[\mathbb{E}_2|e^{n+1}|\right]^2$$
$$\mathbb{E}\left(y(t^n; t^{n-1}, y^{n-1}) - \eta(t^n; t^{n-1}, \eta^{n-1})\right)^2 = \left[\mathbb{E}_2|e^n|\right]^2.$$

For third term in (11.21) we define p_2 by

$$\mathbb{E}|\tau|^2 \leq N_2^2 |h|^{2p_2}, \tag{11.22}$$

and so

$$\mathbb{E}h^2 \left(\Delta(t^n, y^n) - \Phi(t^n, y^n)\right)^2 \leq N_2^2 |h|^{2p_2+2}.$$

Here p_2 is the strong order of the local discretization error, when defined analogously to an ODE.

For the fourth term of (11.21) we assume a Lipschitz property on the SDE coefficients

$$|f(x) - f(y)| + |g(x) - g(y)| \leq M|x - y|.$$

Then,

$$h\Phi(t^n, y^n) = \int_{t^n}^{t^n+h} g(y, t)dW + \text{h.o.t}$$

$$h\Phi(t^n, \eta^n) = \int_{t^n}^{t^n+h} g(\eta, t)dW + \text{h.o.t}$$

$$h(\Phi(t^n, y^n) - \Phi(t^n, \eta^n)) \leq M|e^n||\Delta W$$

$$h^2 \mathbb{E}|\Phi(t^n, y^n) - \Phi(t^n, \eta^n)|^2 \leq M^2 \left[\mathbb{E}_2|e^n|\right]^2 h$$

bounds the fourth term.

For the fifth term of (11.21) we define p_1 by

$$|\mathbb{E}\tau| \le N_1 |h|^{p_1}. \tag{11.23}$$

Here p_1 is the weak order of the method when interpreted as an ODE. Then,

$$\mathbb{E}\left|2h\left(y(t^n; t^{n-1}, y^{n-1}) - \eta(t^n; t^{n-1}, \eta^{n-1})\right)\left(\Delta(t^n, y^n) - \Phi(t^n, y^n)\right)\right|$$

$$\le 2N_1\mathbb{E}_2|e^n||h|^{p_1+1}$$

$$\le \left[\mathbb{E}_2|e^n|\right]^2 N_1^2 h + h^{2p_1+1}.$$

The last step uses the inequality

$$\sqrt{\alpha\beta} \le \tfrac{1}{2}[\alpha + \beta], \tag{11.24}$$

which relates geometric to arithmetic means.

For the sixth term of (11.21),

$$\mathbb{E}\left|2h\left(y(t^n; t^{n-1}, y^{n-1}) - \eta(t^n; t^{n-1}, \eta^{n-1})\right)\left(\Phi(t^n, y^n) - \Phi(t^n, \eta^n)\right)\right|$$

$$\le 2M\left[\mathbb{E}_2|e^n|\right]^2 h.$$

And finally the seventh term,

$$\mathbb{E}\left|2h^2\left(\Delta(t^n, y^n) - \Phi(t^n, y^n)\right)\left(\Phi(t^n, y^n) - \Phi(t^n, \eta^n)\right)\right|$$

$$\le 2MN_2\mathbb{E}_2|e^n||h|^{p_2+3/2}$$

$$\le M^2 N_2^2 \left[\mathbb{E}_2|e^n|\right]^2 h + h^{2p_2+2}.$$

Here we combine the first term of $\Delta - \Phi$ containing stochastic terms, order h^{p_2} with the ΔW of the second term, to give the leading order $h^{p_2+3/2}$. In the last step we used (11.24).

In combination,

$$\left[\mathbb{E}_2|e^{n+1}|\right]^2 \le \left[\mathbb{E}_2|e^n|\right]^2 + N_2^2|h|^{2p_2+2} + M_2\left[\mathbb{E}_2|e^n|\right]^2 h + \left[\mathbb{E}_2|e^n|\right]^2 N_1^2 h$$

$$+ h^{2p_1+1} + 2M\left[\mathbb{E}_2|e^n|\right]^2 h + M^2 N_2^2\left[\mathbb{E}_2|e^n|\right]^2 h + h^{2p_2+2}$$

or

$$\left[\mathbb{E}_2|e^{n+1}|\right]^2 \le \left[\mathbb{E}_2|e^n|\right]^2 (1 + K_1 h) + K_2|h|^{2p+1},$$

where $K_1 = M_2 + N_1^2 + 2M + 2M^2 N_2^2$, $K_2 = 2 + N_2^2$, and $p = \min(p_2 + \tfrac{1}{2}, p_1)$. Higher-order terms have been dropped.

Now, by (10.9) the order of accuracy of $[\mathbb{E}_2|e|]^2$ is $2p$, so the order of accuracy of $\mathbb{E}_2|e|$ will be p:

$$p_s = \min(p_2 + \tfrac{1}{2}, p_1). \tag{11.25}$$

Example 11.1   Find the strong order of the Euler–Maruyama method

$$y_{n+1} = y_n + f_n \Delta t_n + g_n \Delta W_n,$$ (11.26)

where $t_{n+1} = t_n + \Delta t_n$, $f_n = f(y(t_n), t_n)$, etc.

Solution

To determine the theoretical strong order of convergence, we need p_1 (11.23) and p_2 (11.22). With the ODE notation $y_{n+1} = y_n + h\Phi$, and $h = \Delta t$, we deduce

$$h\Phi = f_n \Delta t_n + g_n \Delta W_n.$$ (11.27)

For the exact difference $h\Delta$, we can refer to the Itô–Taylor series (11.19)

$$h\Delta = g_n \Delta W_n + f_n \Delta t_n + (g_y g)_n \tfrac{1}{2}(\Delta W_n^2 - \Delta t_n) + \mathcal{O}\Delta t^{3/2}.$$ (11.28)

Then,

$$h\tau = h\Delta - h\Phi = (g_y g)_n \tfrac{1}{2} \underbrace{(\Delta W_n^2 - \Delta t_n)}_{\mathcal{O}\Delta t} + \mathcal{O}\Delta t^{3/2}$$

$$\mathbb{E}h\tau = h\Delta - h\Phi = (g_y g)_n \tfrac{1}{2} \underbrace{\mathbb{E}(\Delta W_n^2 - \Delta t_n)}_{\text{zero}} + \mathcal{O}\Delta t^2$$

$$\mathbb{E}\tau = \mathcal{O}h^1, \qquad p_1 = 1.$$

Next, we find p_2,

$$\mathbb{E}\tau^2 = h^{-2}(g_y g)_n^2 \frac{1}{4} \underbrace{\mathbb{E}(\Delta W_n^2 - \Delta t_n)^2}_{2h^2} = \mathcal{O}h^0.$$

So, $p_2 = 0$, and $p_s = 1/2$ (11.25). Note that $\mathbb{E}h\tau$ has integer order because all terms of odd half integer order have zero expectation. The Euler–Maruyama method has strong order of convergence $1/2$.

Here we used the general result of Gaussian integrals,

$$\mathbb{E}\Delta W^n = \frac{1}{\sqrt{2\pi \Delta t}} \int_{-\infty}^{+\infty} dx\, x^n e^{-x^2/(2\Delta t)} = 2^{(n-1)/2} \frac{1 + (-1)^n}{\sqrt{2\pi}} \Gamma\frac{n+1}{2}$$

$$= 1, 0, 1, 0, 3, 0, 15, \ldots \times \Delta t^{n/2} = \Delta t^{n/2} \begin{cases} 0 & n \text{ odd} \\ (n-1)!! & n \text{ even.} \end{cases}$$ (11.29)

Based on this analysis, it is easy to decide which terms in an Itô–Taylor series need be retained to achieve a given strong order of accuracy. Let p_s be the desired strong order of accuracy. Define

$$p_i = \lceil p_s \rceil$$
$$p_W = p_s.$$

In the Itô–Taylor series, group terms by stochastic character (depending on any integral containing a dW), and deterministic character (depending only on integrals containing no dW). Keep all deterministic terms to order p_t, and keep all stochastic terms to order p_W. For example,

$p_s = 1/2$ (Euler–Maruyama), with $p_t = 1$ and $p_W = 1/2$:

$$y_{n+1} = y_n + g_n \Delta W + f_n \Delta t;$$

$p_s = 1$ (Milstein), with $p_t = 1$ and $p_W = 1$:

$$= \cdots + \frac{1}{2}(gg_y)_n(\Delta W^2 - \Delta t);$$

$p_s = 3/2$, with $p_t = 2$ and $p_W = 3/2$:

$$= \cdots + (f_y g)_n \Delta Z + (fg_y + \frac{1}{2}g^2 g_{yy})_n(\Delta W \Delta t - \Delta Z) \tag{11.30}$$

$$+ \frac{1}{2}(g_y g)_y g)_n(\frac{1}{3}\Delta W^3 - \Delta t \Delta W) + \frac{1}{2}(ff_y + \frac{1}{2}g^2 f_{yy})_n \Delta t^2.$$

11.7   Weak convergence

We anticipate here the case of vector SDEs, with $\mathbf{y}$ a d-dimensional stochastic vector function. To accommodate the arbitrary scalar test function g, we use a multidimensional Taylor series analysis, after Talay and Tubaro [229]. The d-dimensional multivariate Taylor series give

$$g(\boldsymbol{\eta}^{n+1}) = g(\mathbf{y}^n + h\boldsymbol{\Phi}) = \sum_{P=0} \sum_{|\mathbf{p}|=P} \frac{1}{\mathbf{p}!} \frac{\partial^{\mathbf{p}} g}{\partial \mathbf{y}^{\mathbf{p}}} h^P \boldsymbol{\Phi}^{\mathbf{p}}$$

$$g(\mathbf{y}^{n+1}) = g(\mathbf{y}^n + h\boldsymbol{\Delta}) = \sum_{P=0} \sum_{|\mathbf{p}|=P} \frac{1}{\mathbf{p}!} \frac{\partial^{\mathbf{p}} g}{\partial \mathbf{y}^{\mathbf{p}}} h^P \boldsymbol{\Delta}^{\mathbf{p}}$$

$$\mathbb{E}(g(\boldsymbol{\eta}^{n+1})) - \mathbb{E}(g(\mathbf{y}^{n+1})) = \sum_{P=1} \sum_{|\mathbf{p}|=P} \frac{1}{\mathbf{p}!} \mathbb{E}\frac{\partial^{\mathbf{p}} g}{\partial \mathbf{y}^{\mathbf{p}}} h^P (\boldsymbol{\Phi}^{\mathbf{p}} - \boldsymbol{\Delta}^{\mathbf{p}}). \tag{11.31}$$

Here P is a scalar, denoting the degree of the Taylor polynomial and $\mathbf{p}$ is a vector of integers in the range $[0, P]$ with sum P: $|\mathbf{p}| = \sum_{i=1}^{d} p_i = P$. The multiindex power is

$$\mathbf{y}^{\mathbf{p}} = \prod_{i=1}^{d} (y_i)^{p_i}$$

and the multiindex factorial is

$$\mathbf{p}! = \prod_{i=1}^{d} (p_i)!.$$

Now, drawing on the analogy to ODEs, if the one-step expansion (11.31) is of order

$p_w + 1$, then we would expect the global measure $\mathbb{E}(g(\eta(T))) - \mathbb{E}(g(\mathbf{y}(T)))$ to have order p_w. Considering $g(\mathbf{y}^n)$ to be a particular fixed quantity,

$$h^{|\mathbf{p}|}(\mathbb{E}\Phi^{\mathbf{p}} - \mathbb{E}\Delta^{\mathbf{p}}) \leq Ch^{p_w+1} \quad \forall \quad |\mathbf{p}| < 2p_w + 2. \tag{11.32}$$

Only monomial degrees up to and including $2p_w + 1$ need be considered for a method of order p_w. The reason is that if the coefficients f and g of the stochastic differential equation obey a Lipschitz condition, and if $h\Phi \to 0$ as $h \to 0$, then Φ will have the form $C\Delta W$ plus higher-order terms, and Φ^p will be $\mathcal{O}h^{p/2}$ plus higher-order terms. If the monomial degree p is $2p_w + 2$, then the smallest term in $(h\Phi)^p$ will be order h^{p_w+1}, which will satisfy (11.32) with equality for any C. That is, Φ could be completely wrong and satisfy (11.32) for $p \geq 2p_w + 2$, provided $\Phi \to 0$ as $h \to 0$ and the SDE satisfy a Lipschitz condition.

Example 11.2 Determine the weak order of convergence of the Euler–Maruyama method (11.26).

Solution
We will want to compare expectations of moments of $h\Delta$ (11.28) with expectations of moments of $h\Phi$ (11.27).

First, the 1 moment. Note that $\mathbb{E}\Delta W = 0$, $\mathbb{E}\Delta Z = 0$, etc. All stochastic terms of odd half integer order have zero expectation:

$$\mathbb{E}(h\Delta)^1 = f_n\Delta t + \mathcal{O}\Delta t^2$$
$$\mathbb{E}(h\Phi)^1 = f_n\Delta t$$
$$h^1(\mathbb{E}\Phi^1 - \mathbb{E}\Delta^1) \leq Ch^2.$$

So far, $p_w \leq 1$.
Now, the 2 moment:

$$\mathbb{E}h^2\Delta^2 = g_n^2\Delta t + f_n^2\Delta t^2 + \mathcal{O}\Delta t^3$$
$$\mathbb{E}h^2\Phi^2 = g_n^2\Delta t + f_n^2\Delta t^2$$
$$h^2(\mathbb{E}\Phi^2 - \mathbb{E}\Delta^2) \leq Ch^3.$$

This permits $p_w \leq 2$, but from the 1 moment we have only $p_w \leq 1$.
Next, the 3 moment:

$$\mathbb{E}h^3\Delta^3 = f_n^3\Delta t^3 + \mathcal{O}\Delta t^4$$
$$\mathbb{E}h^3\Phi^3 = f_n^3\Delta t^3$$
$$h^3(\mathbb{E}\Phi^3 - \mathbb{E}\Delta^3) \leq Ch^4.$$

We can see that the 1 moment provides the tightest bound, and $p_w = 1$. The Euler–Maruyama method has weak order 1.

Based on this analysis, it is easy to decide which terms in an Itô–Taylor series need be retained to achieve a given weak order of accuracy. Let p_w be the desired weak order of accuracy: one includes all terms of the Itô–Taylor series with multiple integrals of p_w order. That is, if $p_w = 1$, include all simple integrals $\int dt$ and $\int dW$. If $p_w = 2$, include also all double integrals: $\int dt \int dt$, $\int dt \int dW$, $\int dW \int dt$, and $\int dW \int dW$. For $p_w = 3$, all triple integrals are also included, etc.

For example, the Euler–Maruyama method

$$y_{n+1} = y_n + g_n \Delta W + f_n \Delta t$$

has all simple integrals: $p_w = 1$; the method is first-order accurate in the weak sense. The method

$$y_{n+1} = y_n + g_n \Delta W + f_n \Delta t + \frac{1}{2}(gg_y)_n(\Delta W^2 - \Delta t) \tag{11.33}$$

$$+ (f_y g)_n \Delta Z + (fg_y + \frac{1}{2}g^2 g_{yy})_n(\Delta W \Delta t - \Delta Z)$$

$$+ \frac{1}{2}(ff_y + \frac{1}{2}g^2 f_{yy})_n \Delta t^2 \tag{11.34}$$

of Milstein and Talay includes all single and double integrals: $p_w = 2$; the method is second-order accurate in the weak sense (cf. (11.30)).

The reason for this selection is simple. First, the expectation of a stochastic integral multiplied with a second stochastic integral is zero unless they have the same number of dWs (Problem 11.4). Now consider an integer order p_w, and a stochastic integral $\int dW \int dW \cdots \int dW$ ($p_w + 1$ times). This integral has magnitude $\Delta t^{(p_w+1)/2}$, but does not affect the first moment $\mathbb{E}(\Delta^1) - \mathbb{E}(\Phi^1)$, because as a stochastic integral its expectation is zero. In the second moment, the stochastic integral under consideration will have nonzero expectation only when multiplied with itself, or with certain other higher-order integrals. The expectations of such products will have magnitude Δt^{p_w+1} or higher, and such terms need not be retained to achieve an order p_w method. Likewise, the given stochastic term is unimportant in higher-degree moments.

It should be immediately apparent that weak methods are considerably simpler than strong methods for a given order of accuracy (compare the weak method of order 2 (11.33) with the strong method of order 3/2 (11.30)).

11.8   Modeling

To illustrate some of the steps involved in modeling stochastic systems, and analyzing stochastic models, we will develop here a example of a two-state system after Berglund and Gentz [14]:

$$dx = f\,dt + g\,dW$$

$$f = \frac{x - x^3 + A\cos 2\pi t}{\epsilon}$$

$$g = \frac{\sigma}{\sqrt{\epsilon}},$$

with $x_0 = 1.15303$, $\epsilon = 0.0025$, $\sigma = 0.065$, and $A = 0.3799$. If one ignores the stochastic term $g\,dW$, the forcing f has three zeros $x(t)$. These correspond to two stable $x(t)$ paths, the minima of a time-varying "W"-shaped potential, and one unstable path, a critical point, corresponding to the bounded maximum of the "W." When the stochastic term is large enough the solution x can hop the barrier to transition from one stable path to another.

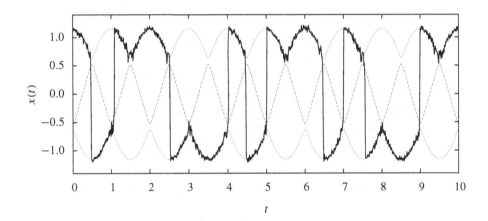

Figure 11.3 Two-state SDE.

A simulation of this system is shown in Figure 11.3. The dotted lines show the two stable states, and the dashed line shows the critical point. When the critical state and stable state are close, the barrier is low and hopping is facilitated.

A numerical method for this model can be made from an Itô–Taylor series. Since f has an explicit time dependence, the series here is slightly different than in Section 11.4.

$$x_{n+1} = x_n + \int_{t_n}^{t_{n+1}} \frac{x - x^3 + A\cos 2\pi s_0}{\epsilon}\,ds_0 + \int_{t_n}^{t_{n+1}} \frac{\sigma}{\sqrt{\epsilon}}\,dW_{s_0}$$

$$f(s_0) = f_n + \int_{t_n}^{s_0} \frac{\sigma(1 - 3x^2)}{\epsilon^{3/2}}\,dW_{s_1}$$

$$+ \int_{t_n}^{s_0} \frac{x(1 - 3\sigma^2) - 4x^3 + 3x^5 + A(1 - 3x^2)\cos 2\pi s_1 - 2A\epsilon\pi \sin 2\pi s_1}{\epsilon^2}\,ds_1$$

$$x_{n+1} = x_n + \left(\frac{x - x^3 + A \cos 2\pi t}{\epsilon}\right)_{t_n} \Delta t + \frac{\sigma}{\sqrt{\epsilon}} \Delta W$$

$$+ \int_{t_n}^{t_{n+1}} ds_0 \int_{t_n}^{s_0} dW_{s_1} \frac{\sigma(1 - 3x^2)}{\epsilon^{3/2}}$$

$$+ \int_{t_n}^{t_{n+1}} ds_0 \int_{t_n}^{s_0} ds_1 \epsilon^{-2} \Big[x(1 - 3\sigma^2) - 4x^3 + 3x^5 + A(1 - 3x^2) \cos 2\pi s_1$$

$$- 2A\epsilon\pi \sin 2\pi s_1 \Big]$$

(11.35)

$$x_{n+1} = x_n + \left(\frac{x - x^3 + A \cos 2\pi t}{\epsilon}\right)_{t_n} \Delta t + \frac{\sigma}{\sqrt{\epsilon}} \Delta W + \left(\frac{\sigma(1 - 3x^2)}{\epsilon^{3/2}}\right)_n \Delta Z$$

$$+ \frac{\Delta t^2}{2} \left(\frac{x(1 - 3\sigma^2) - 4x^3 + 3x^5 + A(1 - 3x^2) \cos 2\pi t_n - 2A\epsilon\pi \sin 2\pi t_n}{\epsilon^2}\right)_n .$$

To simulate this system we need to be able to construct a sequence ΔW_n^h (n labels the step, and h labels the step size). Since ΔW is a Gaussian-distributed quantity with mean 0 and variance h, one could construct ΔW by scaling uniform normal deviates as indicated in (11.7). However, we also need ΔZ_n^h which is constructed by a different integral of the same dW process sampled by ΔW. This consistency needs to be respected. To characterize the relation between these quantities, compute their covariance, and the variance of the Z process:

$$\mathbb{E}(\Delta Z_n^h)^2 = \mathbb{E} \int_{nh}^{(n+1)h} ds_0 \int_{nh}^{s_0} dW_{s_1} \int_{nh}^{(n+1)h} ds_2 \int_{nh}^{s_2} dW_{s_3}$$

$$= \int_{nh}^{(n+1)h} ds_0 \int_{nh}^{(n+1)h} ds_2 \int_{nh}^{\min(s_0, s_2)} ds_1$$

$$= \int_{nh}^{(n+1)h} ds_0 \int_{nh}^{s_0} ds_2 \int_{nh}^{s_2} ds_1 + \int_{nh}^{(n+1)h} ds_0 \int_{s_0}^{(n+1)h} ds_2 \int_{nh}^{s_0} ds_1$$

$$= \frac{h^3}{3}$$

$$\mathbb{E}\Delta Z_n^h \Delta W_n^h = \mathbb{E} \int_{nh}^{(n+1)h} ds_0 \int_{nh}^{s_0} dW_{s_1} \int_{nh}^{(n+1)h} dW_{s_2}$$

$$= \int_{nh}^{(n+1)h} ds_0 \int_{nh}^{s_0} ds_1 = \frac{h^2}{2}.$$

With this variance knowledge, and the Gaussian distribution property, we can construct a model of ΔW_n^h and ΔZ_n^h that is consistent with these being determined by the same underlying noise process. Let G_1 and G_2 be two independent Gaussian normal deviates:

$$G_1 = \mathcal{N}(0, 1)$$
$$G_2 = \mathcal{N}(0, 1)$$
$$\Delta W_n^h = \sqrt{h} G_1$$
$$\Delta Z_n^h = \frac{h^{3/2}}{2} G_1 + \frac{h^{3/2}}{2\sqrt{3}} G_2.$$

This model works because ΔW is defined to have a Gaussian distribution, and it may be shown that the distribution of ΔZ, conditional upon prior knowledge of ΔW, is also Gaussian. However, not all stochastic integrals have Gaussian distributions (see, e.g., [147] for a famous example), and such integrals require more elaborate models. A robust approach to modeling the general stochastic integral uses a Fourier expansion of each Wiener process on each time interval (e.g., [164]).

To measure the strong error, one requires stochastic integrals ΔW_n^{2h}, etc., that are consistent with those paths that were constructed using $\Delta t = h$. This consistency is enforced by manipulating the integral definitions [124]:

$$\Delta W_n^{2h} = \int_{2nh}^{2(n+1)h} dW = \int_{2nh}^{2nh+h} dW + \int_{2nh+h}^{2(n+1)h} dW$$

$$\Delta W_n^{2h} = \Delta W_{2n}^h + \Delta W_{2n+1}^n$$

$$\Delta Z_n^{2h} = \int_{2nh}^{2(n+1)h} ds_0 \int_{2nh}^{s_0} dW_{s_1}$$

$$= \int_{2nh}^{2nh+1} ds_0 \int_{2nh}^{s_0} dW_{s_1} + \int_{2nh+1}^{2(n+1)h} ds_0 \int_{2nh}^{s_0} dW_{s_1}$$

$$= \int_{2nh}^{2nh+1} ds_0 \int_{2nh}^{s_0} dW_{s_1} + \int_{2nh+1}^{2(n+1)h} ds_0 \int_{2nh}^{2nh+1} dW_{s_1}$$

$$+ \int_{2nh+1}^{2(n+1)h} ds_0 \int_{2nh+1}^{s_0} dW_{s_1}$$

$$\Delta Z_n^{2h} = \Delta Z_{2n}^h + \Delta t \Delta W_{2n}^h + \Delta Z_{2n+1}^h.$$

A convergence test of stochastic method (11.35), integrating to $t = 0.001$ and averaging over six million Wiener paths, is displayed for weak and strong error in Figures 11.4 and 11.5 respectively. The strong error displays a slope of 1.5, and the weak moments display slopes of 2.0.

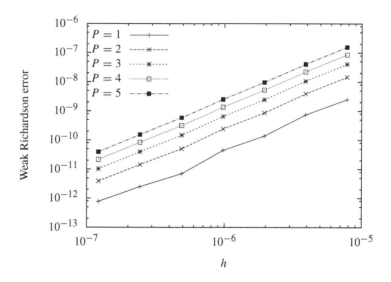

Figure 11.4 Two-state SDE weak convergence

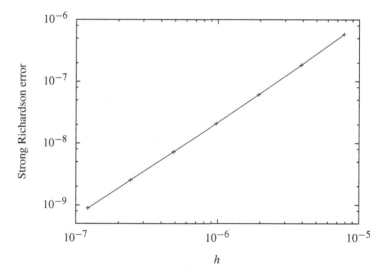

Figure 11.5 Two-state SDE strong convergence

Problems

11.1 Verify that the velocity equation of the Langevin equations (11.1) has exact solution

$$v(t) = v_0 e^{-\gamma t} + \sigma \int_0^t dW_s e^{-\gamma(t-s)}.$$

11.2 Verify that

$$S(t) = S(0)\exp\left[(\mu - \sigma^2/2)t + \sigma W(t)\right]$$

is the analytic solution of the Black–Scholes equation (11.2).

11.3 Show that

$$\Delta W_h = \frac{W(t+h) - W(t)}{h}$$

is wide sense stationary, and compare its power spectrum of with that of the white noise ξ.

11.4 Show that the following expectations are zero.

(i)

$$\left(\int_0^t ds_0 f(s_0)\int_0^{s_0} ds_1 g(s_1)\right)\left(\int_0^t ds_2 p(s_2)\int_0^{s_2} dW_3 q(s_3)\right)$$

(ii)

$$\left(\int_0^t dW_0 f(s_0)\int_0^{s_0} dW_1 g(s_1)\right)\left(\int_0^t ds_2 p(s_2)\int_0^{s_2} dW_3 q(s_3)\right)$$

(iii)

$$\left(\int_0^t dW_0 f(s_0)\right)\left(\int_0^t dW_1 p(s_1)\int_0^{s_1} q(s_2)dW_2\right),$$

and that the following expectation is not necessarily zero:

(iv)

$$\left(\int_0^t dW_0 f(s_0)\int_0^{s_0} ds_1 g(s_1)\right)\left(\int_0^t ds_2 p(s_2)\int_0^{s_2} dW_3 q(s_3)\right).$$

11.5 What is the expectation and variance of the Black–Scholes solution? Is Figure 11.1 compatible with your estimates (here, $S(0) = 1$, $\mu = 4$, and $\sigma = 1$)? How many paths need to be sampled to verify the theoretical expectation at $t = 1$ with 1% accuracy?

12 A big integrative example

The *ab initio* determination of molecular structure is an example that will use:

- singular value decomposition (Section 3.4) in particular the Löwdin decomposition (8.13) encountered in data fitting;
- the QR method for eigenvalues and eigenvectors (Section 3.3);
- interpolation (Chapter 5);
- fixed point iteration (Chapter 6) with stabilization by damping (Section 8.4);
- data fitting (Chapter 8), which involves solutions of linear systems (Chapter 2);
- integration: generally (Chapter 9), and Gaussian integration in particular (Section 9.4, Problem 9.6, which requires numerical root finding (Section 6.6) to set up), and the use of recursions, e.g., (10.35); and
- optimization with the variable metric method (Section 7.4).

In addition, concerns about numerical error (Chapter 1) are omnipresent. We will encounter the error function (Problems 5.7 and 9.7), and lots of Gaussian integrals (equations (8.10), (11.8), and (11.29)). We will use a paradigm called the variational principle that is essentially what motivated the conjugate gradient algorithm (Section 4.1).

The chemical physics problem is described in Section 12.1, and the Hartree–Fock–Roothaan equations are derived. These determine the energy of a particular molecular configuration. The HFR equations rely on a number of integrals of Gaussian functions that are introduced in their simplest form in Section 12.2. With these simple formulas we determine the energy of the H_2 molecule for prescribed geometry, demonstrating in Section 12.3 reasonable results compared to far more elaborate calculations. In Section 12.4 the more complex Gaussian integrals are addressed. Many of these are solved by way of Rys quadrature (Section 12.5). Finally, these components are assembled to find the optimum configuration of the water molecule in Section 12.6.

12.1 The Schrödinger equation

In 1900 M. Planck (Nobel Prize in Physics, 1918) presented a hypothesis that light waves are quantized, i.e., that they have a discrete particle-like character. This led L. de Broglie (Nobel Prize in Physics, 1929) to form the complementary hypothesis in his 1924 PhD thesis that all matter can be treated like a wave. E. Schrödinger (Nobel Prize in Physics 1933), in a series of papers in 1926 [213], developed the celebrated equation

$$H\Psi = E\Psi, \tag{12.1}$$

the wave equation for the wave description of matter. Here H is the Hamiltonian operator, E is the energy, and Ψ is the wave function. In the event that only electrons are to be described by waves, the Hamiltonian is

$$H = -\frac{1}{2}\sum_{i}^{N_e}\nabla_i^2 - \sum_{i}^{N_e}\sum_{s}^{N_n}\frac{Z_s}{r_{is}} + \sum_{i,j<i}^{N_e}\frac{1}{r_{ij}}, \tag{12.2}$$

where N_e is the number of electrons, N_n is the number of nuclei, and Z are the nuclear charges. The Laplace operator ∇^2 operates in $\mathbb{R}^{3N_e}$: (12.1) is a very high dimensional partial differential equation. Without loss of generality, one can take the wave function to be normalized

$$\langle \Psi \mid \Psi \rangle = \int \cdots \int d\mathbf{x}^{N_e} dm^{N_e} \Psi^* \Psi = 1, \tag{12.3}$$

where the integral on the right-hand side is over all degrees of freedom of the wave function ($3N_e$ spatial dimensions, and N_e spin dimensions, about which more will be said below). Equation (12.2) uses *atomic units*:

dimension	unit	formula	SI value
length	bohr	$\frac{\epsilon_0 h^2}{\pi m_e e^2}$	5.292×10^{-11} m
energy	hartree	$\frac{\pi m_e e^4}{\epsilon_0 h^2}$	4.360×10^{-18} J

where h is Planck's constant, ϵ_0 is the permittivity of vacuum, m_e is the mass of an electron, and e is the charge of an electron.

Early solutions of this equation were pioneered by D. Hartree, who also specialized in the construction of mechanical analog computers for the solution of differential equations. J. C. Slater and V. Fock [67] improved on Hartree's solutions by recognizing that (12.1) is subject to constraints. In particular:

1. The electron wave function is antisymmetric with respect to the exchange of any two particles (a quantum-mechanical property of fermions – the class of spin-1/2 particles that includes electrons).
2. No two electrons may share the same quantum state: the Pauli exclusion principle [177] (W. Pauli won the Nobel Prize in Physics in 1945).

A definitive approach to solving equation (12.1) with these constraints was proposed by C. Roothaan [197], giving what is known as the "Hartree–Fock–Roothaan" method. This is the method we will solve. Today a variety of very powerful commercial software packages are available to solve the Hartree–Fock–Roothaan equations, including the well-known "Gaussian" program developed by J. Pople (Nobel Prize in Chemistry 1998).

We will make three additional assumptions. First, that the motion of nuclei can be neglected in determining the wave function of electrons. This is known as the Born–Oppenheimer approximation [21] (M. Born won the Nobel Prize in Physics in 1954, R. Oppenheimer was the director of the Manhattan Project – the father of the atomic bomb). This is justified by the enormous mass difference between electrons and nuclei: the mass of a single proton is 1836 times that of a single electron. So, to a first approximation electrons circulate in the field of fixed nuclei.

Our second assumption will be that electrons are paired, the so-called *closed-shell* case. This is not at all true in general (e.g., a single hydrogen atom has but one electron and is therefore unpaired). However, it is true for a large number of important molecules (e.g., H_2 and H_2O). This assumption is made for convenience only – the Hartree–Fock–Roothaan equations are more difficult when there exist unpaired electrons.

Finally, we will restrict ourselves to single-determinant wave functions. Slater [218] proposed making wave function Ψ as a determinant of single-electron wave functions ψ, as follows:

$$\Psi = \frac{1}{\sqrt{N_e!}} \begin{vmatrix} \psi_1(1)\alpha(1) & \psi_1(2)\alpha(2) & \dots & \psi_1(N_e)\alpha(N_e) \\ \psi_1(1)\beta(1) & \psi_1(2)\beta(2) & \dots & \psi_1(N_e)\beta(N_e) \\ \psi_2(1)\alpha(1) & \psi_2(2)\alpha(2) & \dots & \psi_2(N_e)\alpha(N_e) \\ \psi_2(1)\beta(1) & \psi_2(2)\beta(2) & \dots & \psi_2(N_e)\beta(N_e) \\ \vdots & & & \\ \psi_{N_e/2}(1)\beta(1) & \psi_{N_e/2}(2)\beta(2) & \dots & \psi_{N_e/2}(N_e)\beta(N_e) \end{vmatrix}.$$

This particular Slater determinant consists of $N_e/2$ spatial orbitals ψ, coupled with spin wave functions α, β, for a total of N_e single-electron states. The equal number of α and β states is consistent with closed-shell assumption. These wave functions obey:

$$\langle \alpha(i) \mid \beta(i) \rangle = 0 \tag{12.4a}$$

$$\langle \alpha(i) \mid \alpha(i) \rangle = \langle \beta(i) \mid \beta(i) \rangle = 1 \tag{12.4b}$$

$$\langle \psi_i(k) \mid \psi_j(k) \rangle = \delta_{ij} \qquad \text{if } m_i = m_j, \tag{12.4c}$$

where m_i is the spin quantum number of orbital i. The integrals over spin coordinates in (12.3) amount to (12.4a) and (12.4b). Here $\alpha(i)$ denotes that electron i is spin $+1/2$, while $\beta(i)$ denotes that electron i has spin $-1/2$. The spin variables are abstract concepts, but do enter integrals like (12.3) through rules (12.4a) and (12.4b). The factor $1/\sqrt{N_e!}$ maintains consistency between normalization (12.3) and (12.4). The Slater determinant makes Ψ change sign when two electrons are interchanged, since any determinant changes sign when any two rows or columns are interchanged. It also enforces the Pauli principle, since if any two electrons had the same wave function, including spin, their columns in the determinant would be equal: the matrix would then be singular and the wave function would be zero. Therefore, any nonzero Slater determinant wave function obeys the Pauli exclusion principle. Of course Ψ could be made of the sum of several different Slater determinants. This is absolutely necessary to describe certain molecular wave functions, like that of H_2 when the atoms are widely separated (e.g., [159]).

It is convenient to write wave function Ψ in the Leibniz form (A.14)

$$\Psi = \frac{1}{\sqrt{N_e!}} \sum_{p}^{N_e!} \sigma(p) \chi_{p_1}(1) \chi_{p_2}(2)..., \tag{12.5}$$

where $p_1, p_2, ..., p_n$ is a permutation of $1, 2, ..., n$, $\chi = \psi\alpha$ or $\psi\beta$ is a *spin orbital*, and $\sigma(p) = \pm 1$ is the determinant of the permutation matrix. After Slater and Condon [41, 218], we will develop the Hartree–Fock–Roothaan equation by substituting expansion (12.5) into the integral

$$\langle \Psi \mid H \mid \Psi \rangle = E \langle \Psi \mid \Psi \rangle.$$

The right-hand side is just E with Ψ normalized. The left-hand side is simplified by expanding the determinants, and using the orthogonality property of the single-electron orbitals ψ:

$$\langle \Psi \mid H \mid \Psi \rangle = \frac{1}{N_e!} \sum_{i}^{N_e} \sum_{p,q}^{N_e!} \sigma(p)\sigma(q) \langle \chi_{p_i} \mid -\frac{1}{2}\nabla_i \mid \chi_{q_i} \rangle \prod_{j \neq i}^{N_e} \langle \chi_{p_j} \mid \chi_{q_j} \rangle \tag{12.6}$$

$$- \frac{1}{N_e!} \sum_{i}^{N_e} \sum_{s}^{N_n} \sum_{p,q}^{N_e!} \sigma(p)\sigma(q) \langle \chi_{p_i} \mid \frac{Z_s}{r_{is}} \mid \chi_{q_i} \rangle \prod_{j \neq i}^{N_e} \langle \chi_{p_j} \mid \chi_{q_j} \rangle$$

$$+ \frac{1}{N_e!} \sum_{i<j}^{N_e} \sum_{p,q}^{N_e!} \sigma(p)\sigma(q) \langle \chi_{p_i} \chi_{p_j} \mid \frac{1}{r_{ij}} \mid \chi_{q_i} \chi_{q_j} \rangle \prod_{k \neq i,j}^{N_e} \langle \chi_{p_k} \mid \chi_{q_k} \rangle.$$

In the first line, if $p \neq q$ the term is zero. There are $N_e!$ permutations in total, and $(N_e - 1)!$ permutations that use a particular spin orbital χ_p for electron i. So, the first line of (12.6) simplifies to

$$\frac{(N_e - 1)!}{N_e!} \sum_{i}^{N_e} \sum_{p}^{N_e} \langle \chi_{p_i}(i) \mid -\frac{1}{2}\nabla_i \mid \chi_{p_i}(i) \rangle = \sum_{i}^{N_e} \langle \chi_i \mid -\frac{1}{2}\nabla \mid \chi_i \rangle$$

$$= \sum_{i}^{N_e} \langle \psi_i \mid -\frac{1}{2}\nabla \mid \psi_i \rangle.$$

By the same logic, the second line becomes

$$- \sum_{i}^{N_e} \sum_{s}^{N_n} \langle \psi_i \mid \frac{Z_s}{r_{is}} \mid \psi_i \rangle.$$

For the third term, $p_k = q_k$ for all $k \neq i, j$. Thus, $q = p$ or q differs from p by a single row exchange

$$\frac{1}{N_e!} \sum_{i<j}^{N_e} \sum_{p}^{N_e!} \left(\langle \chi_{p_i}(i)\chi_{p_j}(j) \mid \frac{1}{r_{ij}} \mid \chi_{p_i}(i)\chi_{p_j}(j) \rangle - \langle \chi_{p_i}(i)\chi_{p_j}(j) \mid \frac{1}{r_{ij}} \mid \chi_{p_j}(i)\chi_{p_i}(j) \rangle \right)$$

$$= \sum_{i<j}^{N_e} \left(\langle \chi_i(i)\chi_j(j) \mid \frac{1}{r_{ij}} \mid \chi_i(i)\chi_j(j) \rangle - \langle \chi_i(i)\chi_j(j) \mid \frac{1}{r_{ij}} \mid \chi_j(i)\chi_i(j) \rangle \right)$$

$$= \sum_{i<j}^{N_e} \left(\langle \psi_i(i)\psi_j(j) \mid \frac{1}{r_{ij}} \mid \psi_i(i)\psi_j(j) \rangle - \langle \psi_i(i)\psi_j(j) \mid \frac{1}{r_{ij}} \mid \psi_j(i)\psi_i(j) \rangle \delta_{m_i m_j} \right).$$

Note that when $i = j$ the corresponding spin-orbital term is zero, so the third line of (12.6) reduces to

$$\frac{1}{2} \sum_{i,j}^{N_e} \left(\langle \psi_i(i)\psi_j(j) \mid \frac{1}{r_{ij}} \mid \psi_i(i)\psi_j(j) \rangle - \langle \psi_i(i)\psi_j(j) \mid \frac{1}{r_{ij}} \mid \psi_j(i)\psi_i(j) \rangle \delta_{m_i m_j} \right).$$

All together,

$$\langle \Psi \mid \Psi \rangle E = \langle \Psi \mid H \mid \Psi \rangle = \sum_i^{N_e} \langle \psi_i \mid -\frac{1}{2}\nabla \mid \psi_i \rangle - \sum_i^{N_e} \sum_s^{N_n} \langle \psi_i \mid \frac{Z_s}{r_{is}} \mid \psi_i \rangle \qquad (12.7)$$

$$+ \frac{1}{2} \sum_{i,j}^{N_e} \left(\langle \psi_i(i)\psi_j(j) \mid \frac{1}{r_{ij}} \mid \psi_i(i)\psi_j(j) \rangle - \langle \psi_i(i)\psi_j(j) \mid \frac{1}{r_{ij}} \mid \psi_j(i)\psi_i(j) \rangle \delta_{m_i, m_j} \right),$$

and the normalization condition (12.3) becomes

$$\langle \Psi \mid \Psi \rangle = \frac{1}{N_e!} \sum_{p,q}^{N_e!} \sigma(p)\sigma(q) \prod_j^{N_e} \langle \chi_{p_j} \mid \chi_{q_j} \rangle = \prod_j^{N_e} \langle \psi_j \mid \psi_j \rangle \qquad (12.8)$$

and this is 1 because of (12.4).

For a given set of nuclei and nuclear geometry, the molecular orbitals χ are not known in advance. One may suppose that they are made of a linear combination of atomic orbitals, or atomic basis functions g. This gives rise to what is called the LCAO approximation (linear combination of atomic orbitals). Given N_b basis functions g

$$\psi_i = \sum_j^{N_b} c_{ij} g_j,$$

and (12.7) becomes

$$E = \sum_{i}^{N_e} \sum_{jk}^{N_b} c_{ij}^* \langle g_j \mid -\frac{1}{2}\nabla \mid g_k \rangle c_{ik} - \sum_{s}^{N_n} \sum_{i}^{N_e} \sum_{jk}^{N_b} c_{ij}^* \langle g_j \mid \frac{Z_s}{r} \mid g_k \rangle c_{ik} \qquad (12.9)$$

$$+ \frac{1}{2} \sum_{i,j}^{N_e} \sum_{k\ell pq}^{N_b} \Big[c_{ik}^* c_{j\ell}^* \langle g_k g_\ell \mid r^{-1} \mid g_p g_q \rangle c_{ip} c_{jq}$$

$$- c_{ik}^* c_{j\ell}^* \langle g_k g_\ell \mid r^{-1} \mid g_p g_q \rangle c_{jp} c_{iq} \delta_{m_i m_j} \Big],$$

and (12.8) becomes

$$\langle \Psi \mid \Psi \rangle = c_{ij}^* \langle g_j \mid g_k \rangle c_{ik} = 1, \qquad (12.10)$$

or

$$\sum_{jk}^{N_b} c_{ij}^* \langle g_j \mid g_k \rangle c_{\ell k} = \delta_{i\ell}, \qquad (12.11)$$

to express the orthogonality of the individual molecular orbitals (12.4c).

We now have an equation (12.9) subject to the constraint (12.10). The variables are the total energy E and the LCAO expansion coefficients c_{ij}.

The solution is to treat this as a Rayleigh–Ritz variational problem [188, 193]. The ground state of the system is the lowest energy state; an extremum. Therefore, when the system is in the ground state, E (12.9) will be an extremum, subject to the constraints (12.10). As in the case of quadratic programming, we formulate this constrained optimization problem with Lagrange multipliers. The molecular orbitals ψ are individually orthogonal, so from (12.11) we have N_e^2 individual constraints and N_e^2 Lagrange multipliers $\mathbf{E}_{ij}$.

Simplifying

$$0 = \frac{\partial E}{\partial c_{ij}} - E_{ik} \frac{\partial}{\partial c_{ij}} \langle \Psi \mid \Psi \rangle,$$

one arrives at

$$(\mathbf{H} + \mathbf{J} - \mathbf{K})\mathbf{C} = \mathbf{SCE},$$

where

$$\mathbf{S}_{jk} = \langle g_j \mid g_k \rangle \qquad (12.12a)$$

$$\mathbf{H}_{jk} = \langle g_j \mid -\frac{1}{2}\nabla - \sum_{s}^{N_n} \frac{Z_s}{r} \mid g_k \rangle \qquad (12.12b)$$

$$\mathbf{J}_{jk} = \langle g_j g_p \mid r^{-1} \mid g_k g_q \rangle P_{pq} \qquad (12.12c)$$

$$\mathbf{K}_{jk} = \langle g_j g_p \mid r^{-1} \mid g_q g_k \rangle Q_{pq} \qquad (12.12d)$$

$$\mathbf{C}_{kl} = c_{lk}, \qquad (12.12e)$$

and where

$$\mathbf{P}_{\ell p} = \sum_{s}^{N_e} c_{s\ell}^* c_{sp}$$

and

$$\mathbf{Q}_{\ell p} = \sum_{s:m_s=m_i}^{N_e} c_{s\ell}^* c_{sp}.$$

In $\mathbf{Q}$, the sum is only over those spin orbitals whose spin is equal to the spin of χ_i. $\mathbf{S}$ is called the overlap matrix, $\mathbf{J}$ is the Coulomb matrix, and $\mathbf{K}$ is called the exchange matrix.

At this point we really introduce the closed-shell assumption. Then, with all spatial orbitals containing two electrons, we can redefine $\mathbf{P}$ and sort out the consequence:

$$\mathbf{P}_{kp} = \sum_{s}^{N_p} c_{sk}^* c_{sp},$$

and

$$\mathbf{J}_{jk} = \sum_{pq}^{N_b} \langle g_j g_p \mid r^{-1} \mid g_q g_k \rangle P_{pq}$$

$$\mathbf{K}_{jk} = \sum_{pq}^{N_b} \langle g_j g_p \mid r^{-1} \mid g_k g_q \rangle P_{pq}$$

$$\underbrace{(\mathbf{H} + 2\mathbf{J} - \mathbf{K})}_{\mathbf{F}} \mathbf{C} = \mathbf{SCE},$$

where $\mathbf{F}$ is called the Fock matrix.

Now, after Roothaan [197], we argue that $\mathbf{E}$ is Hermitian and diagonal. First, the orthogonality of the molecular orbitals ψ gives $\mathbf{C}^H \mathbf{SC} = \mathbf{I}$, so

$$\mathbf{C}^H \mathbf{FC} = \mathbf{E}.$$

Here $\mathbf{H}$, $\mathbf{J}$, and $\mathbf{K}$ are Hermitian, so $\mathbf{F}$ and $\mathbf{E}$ are Hermitian. $\mathbf{E}$ may therefore be diagonalized, and it has real eigenvalues.

Let $\mathbf{E} = \mathbf{U}\boldsymbol{\Lambda}\mathbf{U}^H$ be the eigenvalue decomposition of $\mathbf{E}$, with $\mathbf{U}$ unitary and $\boldsymbol{\Lambda}$ diagonal. Then,

$$\mathbf{F}(\mathbf{CU}) = \mathbf{S}(\mathbf{CU})\boldsymbol{\Lambda}.$$

Now $\boldsymbol{\Psi} = \mathbf{C}^T \mathbf{G}$, and $\mathbf{U}^{*H}\mathbf{C}^T\mathbf{G}$ is just an orthogonal transformation, so without loss of generality $\mathbf{C}' = \mathbf{CU}$ then drop the prime

$$\mathbf{FC} = \mathbf{SC}\boldsymbol{\Lambda}. \tag{12.13}$$

This is the Hartree–Fock–Roothaan equation in its simplest form. Matrix $\mathbf{E}$ has effectively been replaced by a diagonal matrix $\boldsymbol{\Lambda}$.

Equation (12.13) is a generalized eigenvalue problem. Since $\mathbf{S}$ is symmetric positive definite we could use its Cholesky decomposition $\mathbf{S} = \mathbf{LL}^H$ to solve (12.13) as

$$\underbrace{\mathbf{L}^{-1}\mathbf{F}\mathbf{L}^{-H}}_{\mathbf{A}} \overbrace{\mathbf{L}^H\mathbf{C}}^{\mathbf{X}} = \overbrace{\mathbf{L}^H\mathbf{C}}^{\mathbf{X}} \Lambda$$

$$\mathbf{AX} = \mathbf{X}\Lambda,$$

which is an eigenvalue problem. The nice thing about using Cholesky here is that with $\mathbf{F}$ symmetric, $\mathbf{A}$ is symmetric, the eigenvectors will be real, and only real math is required.

A more stable result comes from the Löwdin transform [149] (8.13)

$$\mathbf{S} = \mathbf{UD}^2\mathbf{U}^T$$

where $\mathbf{U}$ is unitary. With $\mathbf{S}$ symmetric, this is obtained from the SVD with $\mathbf{D}^2 = \Sigma$. So,

$$\underbrace{\mathbf{D}^{-1}\mathbf{U}^T\mathbf{F}\mathbf{U}\mathbf{D}^{-1}}_{\mathbf{A}} \overbrace{\mathbf{D}\mathbf{U}^T\mathbf{C}}^{\mathbf{X}} = \overbrace{\mathbf{D}\mathbf{U}^T\mathbf{C}}^{\mathbf{X}} \Lambda, \tag{12.14}$$

which is a simple eigenvalue problem. This approach is preferred since the Cholesky decomposition can fail numerically if $\mathbf{S}$ is poorly conditioned. Again, only real math is required.

We have reduced the Schrödinger equation to what appears to be a real symmetric eigenvalue problem – easily solved by QR. However, it is really a third-order nonlinear equation since the coefficients $\mathbf{C}$, obtained as eigenvectors, determine $\mathbf{P}$ hence $\mathbf{J}$ and $\mathbf{K}$. One approach is to make this into an iterative method:

$$\mathbf{P}_{ij}^{(k)} = \sum_s (c_{si}c_{sj})^{(k)}$$

$$\mathbf{J}_{ij}^{(k)} = \langle g_i g_p \mid r^{-1} \mid g_q g_j \rangle P_{pq}^{(k)}$$

$$\mathbf{K}_{ij}^{(k)} = \langle g_i g_p \mid r^{-1} \mid g_j g_q \rangle P_{pq}^{(k)}$$

$$\mathbf{F}^{(k)} = \mathbf{H} + 2\mathbf{J}^{(k)} - \mathbf{K}^{(k)}$$

$$\mathbf{F}^{(k)}\mathbf{C}^{(k+1)} = \mathbf{S}\mathbf{C}^{(k+1)}\Lambda.$$

At this point we have a strategy for solving the Hartree–Fock–Roothaan equations:

1. Determine a basis g.
2. Perform all necessary integrals with basis functions. Determine $\mathbf{H}$ and $\mathbf{S}$.
3. Assume initial value of $\mathbf{C}^{(0)}$, and set iteration index $k = 0$.
4. Iterate:
    (i) Create $\mathbf{J}$, $\mathbf{K}$, and $\mathbf{F} = \mathbf{H} + 2\mathbf{J} - \mathbf{K}$.
    (ii) Löwdin (SVD): $\mathbf{C} = \mathbf{UD}^2\mathbf{U}^T$.
    (iii) $\mathbf{A} = \mathbf{D}^{-1}\mathbf{U}^T\mathbf{F}\mathbf{U}\mathbf{D}^{-1}$.
    (iv) QR: $\mathbf{AX} = \mathbf{X}\Lambda$.
    (v) $\mathbf{C}^{(k+1)} = \mathbf{UD}^{-1}\mathbf{X}$.

Note that the matrix $\mathbf{A}$ is $N_b \times N_b$, where N_b is the number of basis functions, and $\mathbf{C}$ is $N_b \times N_p$, where $N_p \leq N_b$ is the number of electron pairs. Matrix $\mathbf{A}$ has N_b eigenvectors, so some criterion is needed to select only N_p of them to make $\mathbf{C}$. The ones selected are the lowest energy states (smallest eigenvalue).

Once the ground state orbital coefficients $\mathbf{C}$ are determined, the ground state energy E is given by

$$E = 2\sum_{i=1}^{N_p} \lambda_i - 2\sum_{p,q}^{N_b} P_{pq} J_{pq} + \sum_{pq}^{N_b} P_{pq} K_{pq} + \sum_{i,j<i}^{N_n} \frac{Z_i Z_j}{|r_{ij}|}, \tag{12.15}$$

assuming the basis set $\{g\}$ is complete. In the Hamiltonian (12.9) the kinetic and nuclear attraction terms scale like c^2, while the Coulomb and exchange terms scale like c^4. When (12.9) is differentiated to make the Fock matrix, the Coulomb and exchange terms are therefore weighted more strongly (by $2\times$) than in the Hamiltonian. Equation (12.15) corrects the Fock energy to account for this over-weighting of the Coulomb and exchange terms, and also adds the internuclear repulsion, which was not included in (12.2).

The ground state energy, by definition the smallest solution of (12.1), can be written

$$E = \min_{\langle \Psi | \Psi \rangle = 1} \langle \Psi \mid H \mid \Psi \rangle, \tag{12.16}$$

and therefore Ψ other than the solution to that in (12.16) will give an energy greater than or equal to the ground state energy. Any energy computed with a finite basis is an upper bound to the true ground state energy. This is essentially the *variational principle*. It can even be shown [7, 151] that if $\mathcal{G}_n$ is a finite basis set with n members, and

$$\mathcal{G}_{n+1} = \mathcal{G}_n \cup \{g_{n+1}\}$$

is the basis set constructed by adding one new function g_{n+1} to $\mathcal{G}_n$ then the n eigenvalues resulting from $\mathcal{G}_n$ lie between the $n+1$ eigenvalues of $\mathcal{G}_{n+1}$.

Similarly, the ground state wave function cannot be expected to be given by a single Slater determinant. The ground state of (12.16) is therefore approached as the number of basis sets increases and as the number of determinants increases. The sample calculations presented below make no attempt approach this ground state by optimizing the basis set.

As an iterative method, the algorithm sketched above converges at first-order, as demonstrated in Figure 12.1, if it converges at all. Without loss of generality we can think of our iterative method as

$$\mathbf{X}^{(k+1)} = \Phi(\mathbf{X}^{(k)}),$$

where $\mathbf{X}$ of (12.14) and $\mathbf{C}$ are simply related, and Φ is implicitly defined through

$$\mathbf{A}(\mathbf{X}^{(k)})\mathbf{X}^{(k+1)} = \mathbf{X}^{(k+1)}\mathbf{\Lambda}^{(k+1)}. \tag{12.17}$$

The convergence of the method is tied to the properties of Φ (Chapter 6). In particular, the method has first-order convergence if $0 < |\Phi(\Xi)'| < 1$, where Ξ is the converged value of $\mathbf{X}$.

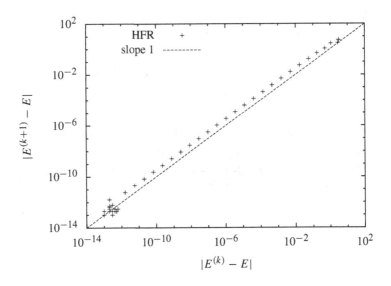

Figure 12.1 Hartree–Fock–Roothaan convergence.

To characterize Φ', one needs first-order perturbation theory. Let us write (12.17) component-wise,

$$\mathbf{A}\mathbf{x}_i = \mathbf{x}_i \lambda_i$$

$$\mathbf{A}'\mathbf{x}_i + \mathbf{A}\mathbf{x}_i' = \mathbf{x}_i'\lambda_i + \mathbf{x}_i \lambda_i' \tag{12.18a}$$

$$\mathbf{x}_i^H \mathbf{A}' \mathbf{x}_i + \underbrace{\mathbf{x}_i^H \mathbf{A}\, \mathbf{x}_i'}_{\lambda_i \mathbf{x}_i^H} = \mathbf{x}_i^H \mathbf{x}_i' \lambda_i + \underbrace{\mathbf{x}_i^H \mathbf{x}_i}_{1}\, \lambda_i'$$

$$\lambda_i' = \mathbf{x}_i^H \mathbf{A}' \mathbf{x}_i. \tag{12.18b}$$

The eigenvectors are orthogonal, so

$$\mathbf{x}_i^H \mathbf{x}_i = 1$$

$$\mathbf{x}_i'^{\,H} \mathbf{x}_i + \mathbf{x}_i^H \mathbf{x}_i' = 0,$$

and with $\mathbf{x}_i$ real, $\mathbf{x}_i^H \mathbf{x}_i' = \mathbf{x}_i^H \mathbf{x}_i'$, so

$$\mathbf{x}_i^H \mathbf{x}_i' = 0.$$

To find $\mathbf{x}'$, begin with a rearrangement of (12.18a)

$$(\mathbf{A} - \lambda_i \mathbf{I})\mathbf{x}_i' = \mathbf{x}_i \lambda_i' - \mathbf{A}' \mathbf{x}_i$$

$$\underbrace{\mathbf{x}_j^H (\mathbf{A} - \lambda_i \mathbf{I})}_{\mathbf{x}_j^H (\lambda_j - \lambda_i)}\, \mathbf{x}_i' = \underbrace{\mathbf{x}_j^H \mathbf{x}_i}_{0}\, \lambda_i' - \mathbf{x}_j^H \mathbf{A}' \mathbf{x}_i$$

$$\mathbf{x}_i' = \sum_{j \neq i} \frac{\mathbf{x}_j^H \mathbf{A}' \mathbf{x}_i}{\lambda_i - \lambda_j} \mathbf{x}_j,$$

assuming all eigenvalues are distinct. With this assumption, the iterative method $\mathbf{x}^{(k)} = \Phi(\mathbf{x}^{(k-1)})$ at the root $\mathbf{x} = \xi$ has derivative

$$\Phi_i'(\Xi) = \sum_{j \neq i} \frac{\xi_j^H \mathbf{A}' \xi_i}{\lambda_i - \lambda_j},$$

which is not a priori zero: the method is first order if it converges at all. For a first-order method to converge at all, one needs also $|\Phi'(\Xi)| < 1$, and that also is not a priori assured. To extend to degenerate cases (where one or more eigenvectors share the same eigenvalue), the result is modified as

$$\Phi_i'(\Xi) = \sum_{j:\lambda_j \neq \lambda_i} \frac{\xi_j^H \mathbf{A}' \xi_i}{\lambda_i - \lambda_j}.$$

Failure to converge can occur because of the nonlinearity of the process. Sometimes this failure is associated with a bad starting guess, and sometimes it is inherent to the problem. A number of tricks have been employed to assure convergence, and to accelerate it [198]. In unstable situations, *damping* can help stabilize the calculation (see also Section 8.4). In effect, one uses a linear combination of the newly calculated orbital coefficient $\mathbf{C}$ with the previously calculated ones,

$$\mathbf{C}^{(k)} := (1 - r)\mathbf{C}^{(k)} + r\mathbf{C}^{(k-1)}, \tag{12.19}$$

with $0 \leq r < 1$. This suppresses large changes in the $\mathbf{J}$ and $\mathbf{K}$ matrices from iteration to iteration. In the absence of convergence issues, $r \neq 0$ slows convergence, but the rate is still first order. It affects Φ by

$$\Phi_i(\mathbf{X}^{(k)}; r) = (1 - r)\Phi_i(\mathbf{X}^{(k)}) + r\mathbf{X}^{(k)}$$

$$\Phi_i'(\Xi; r) = r + (1 - r) \sum_{j:\lambda_j \neq \lambda_i} \frac{\xi_j^H \mathbf{A}' \xi_i}{\lambda_i - \lambda_j}.$$

By letting $r \to 1^-$ (i.e., letting r approach 1 from below) one can assure first-order convergence, but $|\Phi'(\Xi)| \to 1^-$ so convergence would be very slow. Also, if $\mathbf{F}$ has degenerate eigenvalues, the eigenvectors are not uniquely defined. Mixing them via (12.19) can be tricky in that case. Further, orbital orthogonality means $\mathbf{C}$ is $\mathbf{S}$-orthogonal ($\mathbf{C}^H\mathbf{S}\mathbf{C} = \mathbf{I}$), and this property is lost by mixing, though it can be reestablished through Gram–Schmidt orthogonalization (3.12) of the column vectors $\mathbf{D}\mathbf{U}^T\mathbf{C}$ (using the Löwdin decomposition of $\mathbf{S}$, $\mathbf{U}\mathbf{D}^2\mathbf{U}^T$).

To accelerate convergence, tricks like Aitken's method (Section 6.9) can be employed [198, 209].

12.2   Gaussian basis functions

Slater [219] recognized that electron orbitals on individual atoms are very well represented with simple functions of the form

$$g_i = \|\mathbf{x} - \mathbf{x}_i\|^{n_i - 1} e^{-\zeta_i \|\mathbf{x} - \mathbf{x}_i\|}, \quad n_i = 1, 2, \ldots,$$

for so-called "S" orbitals with principal quantum number n_i centered at $\mathbf{x}_i$. Orbitals with higher azimuthal quantum numbers ℓ are obtained by multiplying the S orbitals by an appropriate spherical harmonic function.

While these Slater functions are fantastic for modeling atomic orbitals, they are very difficult to evaluate in expressions like

$$\langle \mu\nu \mid \frac{1}{r} \mid \lambda\sigma \rangle = \iiint\limits_{-\infty}^{+\infty} d\mathbf{x}_1 \iiint\limits_{-\infty}^{+\infty} d\mathbf{x}_2 e^{-\zeta_\mu \|\mathbf{x}_1 - \mathbf{x}_\mu\| - \zeta_\nu \|\mathbf{x}_2 - \mathbf{x}_\nu\| - \zeta_\lambda \|\mathbf{x}_1 - \mathbf{x}_\lambda\| - \zeta_\sigma \|\mathbf{x}_2 - \mathbf{x}_\sigma\|}$$

since the various x, y, z components are not separable.

Boys [23] noted that if one used instead functions based on

$$g_i = e^{-\alpha_i \|\mathbf{x} - \mathbf{x}_i\|^2}$$

(an S-like Gaussian basis function), then the six-dimensional integrals of the type $\langle \mu\nu \mid \frac{1}{r} \mid \lambda\sigma \rangle$ are separable into three two-dimensional integrals which can be analytically evaluated, at least for S orbitals. Hehre, Stewart, and Pople [99] showed how Slater functions can be approximated with a linear combination of Gaussian functions, giving rise to such basis functions as STO-3G (Slater Type Orbitals modeled by three Gaussian functions). The inaccuracy of individual Gaussians to model atomic orbitals is a drawback that is amply compensated by the ability to determine six-dimensional integrals rapidly and accurately.

The relevant integrals in the case of S atomic orbitals are worked out in the following sections, after Boys [23] and Shavitt [216], which equips us to solve the example of H_2. Since we will be concerned only with real Gaussian basis functions from here out, the complex conjugate $*$ will be omitted in the $\langle | \rangle$ integrals.

12.2.1   Overlap

The overlap integrals are the components of the overlap matrix $\mathbf{S}$ (12.12a):

$$\langle \mu \mid \lambda \rangle = \iiint_{-\infty}^{+\infty} d^3\mathbf{x}_1 g_\mu(\mathbf{x}_1) g_\lambda(\mathbf{x}_1)$$

$$g_\mu(\mathbf{x}_1) = \exp(-\alpha_\mu \|\mathbf{x}_1 - \mathbf{x}_\mu\|^2)$$

$$g_\lambda(\mathbf{x}_1) = \exp(-\alpha_\lambda \|\mathbf{x}_1 - \mathbf{x}_\lambda\|^2).$$

(12.20)

By expanding the product $g_\mu g_\lambda$, and completing the square, one sees immediately an advantage of the Gaussian form:

$$g_\mu(\mathbf{x}_1) g_\lambda(\mathbf{x}_1) = \exp(-\alpha_\mu \mathbf{x}_1^2 - \alpha_\lambda \mathbf{x}_1^2 - 2\alpha_\mu \mathbf{x}_1 \cdot \mathbf{x}_\mu - 2\alpha_\lambda \mathbf{x}_1 \cdot \mathbf{x}_\lambda - \alpha_\mu \mathbf{x}_\mu^2 - \alpha_\lambda \mathbf{x}_\lambda^2)$$

$$= \exp\left(-\frac{\alpha_\mu \alpha_\lambda}{\alpha_\mu + \alpha_\lambda} \|\mathbf{x}_\mu - \mathbf{x}_\lambda\|^2\right) \exp\left(-(\alpha_\mu + \alpha_\lambda)\left(\mathbf{x}_1 - \frac{\alpha_\mu \mathbf{x}_\mu + \alpha_\lambda \mathbf{x}_\lambda}{\alpha_\mu + \alpha_\lambda}\right)^2\right).$$

The product of two Gaussians is a single Gaussian, and all 3D Gaussian functions are separable as a product of 1D Gaussians. The terms that appear here will appear again several times, so it is useful at this point to introduce some notation:

$$\alpha \equiv \alpha_\mu + \alpha_\lambda$$

$$\mathbf{x}_a \equiv \frac{\alpha_\mu \mathbf{x}_\mu + \alpha_\lambda \mathbf{x}_\lambda}{\alpha_\mu + \alpha_\lambda}$$

$$C_a \equiv \exp\left(-\frac{\alpha_\mu \alpha_\lambda}{\alpha_\mu + \alpha_\lambda} \|\mathbf{x}_\mu - \mathbf{x}_\lambda\|^2\right),$$

and

$$\beta \equiv \alpha_\nu + \alpha_\sigma$$

$$\mathbf{x}_b \equiv \frac{\alpha_\nu \mathbf{x}_\nu + \alpha_\sigma \mathbf{x}_\sigma}{\alpha_\nu + \alpha_\sigma}$$

$$C_b \equiv \exp\left(-\frac{\alpha_\nu \alpha_\sigma}{\alpha_\nu + \alpha_\sigma} \|\mathbf{x}_\nu - \mathbf{x}_\sigma\|^2\right).$$

Then, $g_\mu g_\lambda = C_a \exp(-\alpha \|\mathbf{x} - \mathbf{x}_a\|^2)$ and $g_\nu g_\sigma = C_b \exp(-\beta \|\mathbf{x} - \mathbf{x}_b\|^2)$.

Integral (12.20) is determined by steps analogous to those of the Box–Muller method (Section 11.1). In one dimension,

$$\int_{-\infty}^{+\infty} dx\, e^{-\alpha x^2} = \sqrt{\frac{\pi}{\alpha}},$$

so all together

$$\langle \mu \mid \lambda \rangle = C_a \left(\frac{\pi}{\alpha}\right)^{3/2}.$$

12.2.2 Kinetic energy

The kinetic energy integrals, which are part of $\mathbf{H}$ (12.12b), are easily derived from the overlap integrals. One uses differentiation by the scale factor α to relate the two integrals (cf. (8.12))

$$\langle \mu \mid -\frac{1}{2}\nabla \mid \lambda \rangle = \int\!\!\!\int\!\!\!\int_{-\infty}^{+\infty} d^3x e^{-\alpha_\mu \|\mathbf{x}-\mathbf{x}_\mu\|^2} \left(3\alpha_\lambda - 2\alpha_\lambda^2 \|\mathbf{x} - \mathbf{x}_\lambda\|^2\right) e^{-\alpha_\lambda \|\mathbf{x}-\mathbf{x}_\lambda\|^2}$$

$$= 3\alpha_\lambda \langle \mu \mid \lambda \rangle + 2\alpha_\lambda^2 \frac{\partial}{\partial \alpha_\lambda} \langle \mu \mid \lambda \rangle$$

$$= \langle \mu \mid \lambda \rangle \left(3\alpha_\lambda + 2\alpha_\lambda^2 \left[-\frac{\alpha_\mu^2 \|\mathbf{x}_\mu - \mathbf{x}_\lambda\|^2}{(\alpha_\mu + \alpha_\lambda)^2} - \frac{3}{2(\alpha_\mu + \alpha_\lambda)}\right]\right)$$

$$\langle \mu \mid -\frac{1}{2}\nabla \mid \lambda \rangle = \langle \mu \mid \lambda \rangle \frac{\alpha_\mu \alpha_\lambda}{\alpha} \left(3 - 2\frac{\alpha_\mu \alpha_\lambda}{\alpha} \|\mathbf{x}_\mu - \mathbf{x}_\lambda\|^2\right).$$

12.2.3 Nuclear attraction

The key to the nuclear attraction integral is the equation

$$\frac{1}{|r|} = \frac{2}{\sqrt{\pi}} \int_0^\infty du e^{-u^2 r^2},$$

which is a simple rearrangement of the Gaussian integral. Using this trick, and the completing-the-square trick to merge Gaussian functions,

$$\langle \mu \mid r_{1i}^{-1} \mid \lambda \rangle = \int\!\!\!\int\!\!\!\int_{-\infty}^{+\infty} d\mathbf{x}_1 g_\mu(\mathbf{x}_1) g_\lambda(\mathbf{x}_1) \frac{1}{|\mathbf{x}_i - \mathbf{x}_1|}$$

$$= \frac{2}{\sqrt{\pi}} C_a \int_0^\infty du \int\!\!\!\int\!\!\!\int_{-\infty}^{+\infty} d\mathbf{x}_1 e^{-\alpha \|\mathbf{x}_1 - \mathbf{x}_a\|^2 - u^2 \|\mathbf{x}_1 - \mathbf{x}_i\|^2}$$

$$= \frac{2}{\sqrt{\pi}} C_a \int_0^\infty du e^{-\frac{\alpha u^2}{\alpha+u^2} \|\mathbf{x}_a - \mathbf{x}_i\|^2} \int\!\!\!\int\!\!\!\int_{-\infty}^{+\infty} d\mathbf{x}_1 e^{-(\alpha+u^2)\left(\mathbf{x}_1 - \frac{\alpha \mathbf{x}_a + u^2 \mathbf{x}_i}{\alpha+u^2}\right)}$$

$$= \frac{2}{\sqrt{\pi}} C_a \int_0^\infty du \frac{\pi^{3/2}}{(\alpha + u^2)^{3/2}} e^{-\frac{\alpha u^2}{\alpha+u^2} \|\mathbf{x}_a - \mathbf{x}_i\|^2}.$$

Now, change variables

$$t^2 = \frac{u^2}{\alpha + u^2}$$

$$u^2 = \frac{\alpha t^2}{1 - t^2} \tag{12.21}$$

$$dt = \frac{\alpha du}{(\alpha + u^2)^{3/2}},$$

to give

$$\langle \mu \mid r_{1i}^{-1} \mid \lambda \rangle = \frac{2\pi}{\alpha} C_a \int_0^1 dt e^{-t^2 \|\mathbf{x}_a - \mathbf{x}_i\|^2} = \left(\frac{\pi}{\alpha}\right)^{3/2} C_a \frac{\text{erf}(\sqrt{\alpha}\|\mathbf{x}_a - \mathbf{x}_i\|)}{\|\mathbf{x}_a - \mathbf{x}_i\|}.$$

12.2.4   Electron repulsion and exchange

The electron repulsion and exchange integrals use the same steps as the nuclear attraction integrals:

$$
\langle \mu\nu \mid r_{12}^{-1} \mid \lambda\sigma \rangle = \iiint_{-\infty}^{+\infty} d\mathbf{x}_1 \iiint_{-\infty}^{+\infty} d\mathbf{x}_2 g_\mu(\mathbf{x}_1) g_\lambda(\mathbf{x}_1) g_\nu(\mathbf{x}_2) g_\sigma(\mathbf{x}_2) \frac{1}{|\mathbf{x}_1 - \mathbf{x}_2|}
$$

$$
= C_a C_b \frac{2}{\sqrt{\pi}} \int_0^\infty du \iiint_{-\infty}^{+\infty} d\mathbf{x}_1 \iiint_{-\infty}^{+\infty} d\mathbf{x}_2 e^{-\alpha\|\mathbf{x}_1-\mathbf{x}_a\|^2 - \beta\|\mathbf{x}_1-\mathbf{x}_b\|^2 - u^2\|\mathbf{x}_1-\mathbf{x}_2\|^2}
$$

$$
= \frac{2\pi^{5/2}}{(\alpha\beta)\sqrt{\alpha+\beta}} C_a C_b \int_0^1 dt \exp\left[-t^2 \frac{\alpha\beta\|\mathbf{x}_a - \mathbf{x}_b\|^2}{\alpha+\beta}\right]
$$

$$
= \frac{\pi^3}{(\alpha\beta)^{3/2}} C_a C_b \frac{\mathrm{erf}\left(\|\mathbf{x}_a - \mathbf{x}_b\|\sqrt{\frac{\alpha\beta}{\alpha+\beta}}\right)}{\|\mathbf{x}_a - \mathbf{x}_b\|},
$$

where u and t are related by the change of variables

$$
t^2 = \frac{u^2}{\frac{\alpha\beta}{\alpha+\beta} + u^2}. \tag{12.22}
$$

A real issue with these two-electron integrals is the large number of them. If there are N_b basis functions, then there are order N_b^4 such integrals. The number may be decreased somewhat using symmetry. First, there is 1–2 symmetry (the label of the electron does not matter)

$$
\langle \mu\nu \mid \frac{1}{r_{12}} \mid \lambda\sigma \rangle = \langle \nu\mu \mid \frac{1}{r_{12}} \mid \sigma\lambda \rangle.
$$

Second, there is left–right symmetry

$$
\langle \mu\nu \mid \frac{1}{r_{12}} \mid \lambda\sigma \rangle = \langle \mu\sigma^* \mid \frac{1}{r_{12}} \mid \lambda\nu^* \rangle
$$

and

$$
\langle \mu\nu \mid \frac{1}{r_{12}} \mid \lambda\sigma \rangle = \langle \lambda\sigma \mid \frac{1}{r_{12}} \mid \mu\nu \rangle^*,
$$

but complex conjugation is not really an issue with real basis functions. So, all together,

$$
\langle \mu\nu \mid r_{12}^{-1} \mid \lambda\sigma \rangle = \langle \mu\sigma \mid r_{12}^{-1} \mid \lambda\nu \rangle = \langle \lambda\nu \mid r_{12}^{-1} \mid \mu\sigma \rangle = \langle \lambda\sigma \mid r_{12}^{-1} \mid \mu\nu \rangle
$$
$$
= \langle \nu\mu \mid r_{12}^{-1} \mid \sigma\lambda \rangle = \langle \nu\lambda \mid r_{12}^{-1} \mid \sigma\mu \rangle = \langle \sigma\mu \mid r_{12}^{-1} \mid \nu\lambda \rangle = \langle \sigma\lambda \mid r_{12}^{-1} \mid \nu\mu \rangle.
$$

This reduces the number of two-electron integrals by almost a factor of eight to

$$
\frac{1}{2}\left(\frac{N_b(N_b+1)}{2}+1\right)\frac{N_b(N_b+1)}{2} = \frac{N_b(N_b+1)(N_b^2+N_b+2)}{8}.
$$

12.3   Results I: H$_2$

In 1968 Kołos and Wolniewicz undertook a very careful study of the hydrogen molecule, using as many as 100 basis functions [133]. This study is famous for its accuracy, especially given the state of the art in computing at the time. Kołos and Wolniewicz determined the dissociation energy of the hydrogen molecule from their computations, obtaining a value that differed from experiment by 1 part in 10 000. This was very troubling at the time, but it turned out that the error was in the experiments, not the computation (experimentalist G. Herzberg [102] won the Nobel Prize in Chemistry in 1971).

For our example, we will use three Gaussians per atom with exponents α taken from the STO-3G fit of Hehre, Stewart, and Pople [99]. They intended for these 3 Gaussians to be used in a fixed relative proportion, a so-called *contracted Gaussian basis function*. By allowing the coefficients to vary, a somewhat improved result is obtained by the variational principle.

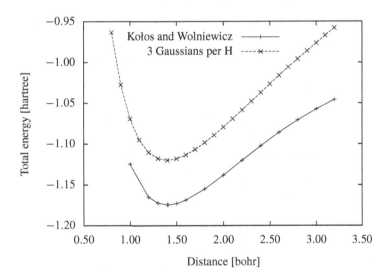

Figure 12.2 The hydrogen molecule.

As shown in Figure 12.2, the very simple three-Gaussian model does a remarkable job of predicting the interatomic spacing and ground state energy. Consistent with the variational principle, the three-Gaussian model has a higher total energy than does the 100-basis calculation of Kołos and Wolniewicz (though our basis is not included in their basis, so this analysis is a bit specious).

12.4   Angular momentum

For orbitals with angular momentum the appropriate Gaussian basis is like

$$(x - x_\mu)^{n_x}(y - y_\mu)^{n_y}(z - z_\mu)^{n_z}\exp(-\alpha\|\mathbf{x} - \mathbf{x}_\mu\|^2),$$

where $n_x, n_y, n_z \geq 0$ and $L = n_x + n_y + n_z$ is the orbital angular momentum quantum number; $L = 0$ for S orbitals, 1 for P orbitals, 2 for D orbitals, etc.

These orbitals can be obtained from S orbitals by manipulation of derivatives with respect to the nuclear coordinate $\mathbf{x}_\mu$. This trick, similar to that employed in solving for the kinetic energy integrals, has been worked out in detail by Taketa, Huzinaga, and O-Ohata [228]. The resulting expressions are very awkward to program, however, and computationally expensive to evaluate.

To simplify the problem, two general approaches have evolved. McMurchie and Davidson [158] developed a family of recursion relations which reduce complicated differential expressions to simple algebraic ones. Rys, King, and Dupuis [130] favored an approach based on Gaussian quadrature, for which the polynomial integrand can be evaluated by simple recursion relations [207]. Both of these methods remain in widespread use today. Both will be explored below.

12.4.1   Overlap

Consider the integral

$$\int_{-\infty}^{+\infty} dx_1 (x_1 - x_\mu)^n e^{-\alpha(x_1 - x_a)^2},$$

which represents part of the overlap integral with some angular momentum on center μ. Now integrate by parts,

$$= -\int_{-\infty}^{+\infty} dx_1 \frac{(x_1 - x_\mu)^{n+1}}{n + 1}(-2\alpha(x_1 - x_a))e^{-\alpha(x_1 - x_a)^2}$$

$$= \int_{-\infty}^{+\infty} dx_1 \left[\frac{2\alpha}{n + 1}(x_1 - x_\mu)^{n+2} + \frac{2\alpha}{n + 1}(x_\mu - x_a)(x_1 - x_\mu)^{n+1}\right]e^{-\alpha(x_1 - x_a)^2}.$$

Writing $d_{n,m}^x$ to represent $(x_1 - x_\mu)^n (x_1 - x_\lambda)^m$, we obtain the recursion

$$d_{n+2,0}^x = (x_a - x_\mu)d_{n+1,0}^x + \frac{n + 1}{2\alpha}d_{n,0}^x,$$

with initial value $d_{0,0}^x = 1$.

To transfer momentum from center μ to center λ,

$$(x_1 - x_\lambda) = (x_1 - x_\mu) + (x_\mu - x_\lambda)$$

so

$$d_{n,m+1}^x = d_{n+1,m}^x + (x_\mu - x_\lambda)d_{n,m}^x.$$

Since the factors d are numbers independent of x_1, they can be taken out of the integral, and

$$\langle \mu^{\mathbf{n}} \mid \lambda^{\mathbf{m}} \rangle = d_{\mathbf{n},\mathbf{m}} \langle \mu^{\mathbf{0}} \mid \lambda^{\mathbf{0}} \rangle,$$

where

$$d_{\mathbf{n},\mathbf{m}} = d^x_{n_x,m_x} d^y_{n_y,m_y} d^z_{n_z,m_z}$$

is the product of d-factors for the three directions.

12.4.2 Kinetic energy

First, note that

$$\langle \mu \mid -\frac{1}{2}\Delta \mid \lambda \rangle = \frac{1}{2}\langle \nabla\mu \cdot \nabla\lambda \rangle,$$

using integration by parts. The relevant recursion uses

$$\psi_n = (x - x_0)^n \exp(-\alpha \|\mathbf{x} - \mathbf{x}_0\|^2)$$

$$\frac{\partial}{\partial x}\psi_n = n\psi_{n-1} - 2\alpha\psi_{n+1}.$$

Using this, the kinetic energy integral can be expressed as a weighted sum of overlap integrals:

$$
\begin{aligned}
\frac{1}{2}\langle \nabla\mu \cdot \nabla\lambda \rangle = {} & \frac{n_x m_x}{2}\langle \mu^{\mathbf{n}_\mu - \mathbf{e}_x} \mid \lambda^{\mathbf{n}_\lambda - \mathbf{e}_x} \rangle - n_x \alpha_\lambda \langle \mu^{\mathbf{n}_\mu - \mathbf{e}_x} \mid \lambda^{\mathbf{n}_\lambda + \mathbf{e}_x} \rangle \\
& - m_x \alpha_\mu \langle \mu^{\mathbf{n}_\mu + \mathbf{e}_x} \mid \lambda^{\mathbf{n}_\lambda - \mathbf{e}_x} \rangle + 2\alpha_\mu \alpha_\lambda \langle \mu^{\mathbf{n}_\mu + \mathbf{e}_x} \mid \lambda^{\mathbf{n}_\lambda + \mathbf{e}_x} \rangle \\
& + \frac{n_y m_y}{2}\langle \mu^{\mathbf{n}_\mu - \mathbf{e}_y} \mid \lambda^{\mathbf{n}_\lambda - \mathbf{e}_y} \rangle - n_y \alpha_\lambda \langle \mu^{\mathbf{n}_\mu - \mathbf{e}_y} \mid \lambda^{\mathbf{n}_\lambda + \mathbf{e}_y} \rangle \\
& - m_y \alpha_\mu \langle \mu^{\mathbf{n}_\mu + \mathbf{e}_y} \mid \lambda^{\mathbf{n}_\lambda - \mathbf{e}_y} \rangle + 2\alpha_\mu \alpha_\lambda \langle \mu^{\mathbf{n}_\mu + \mathbf{e}_y} \mid \lambda^{\mathbf{n}_\lambda + \mathbf{e}_y} \rangle \\
& + \frac{n_z m_z}{2}\langle \mu^{\mathbf{n}_\mu - \mathbf{e}_z} \mid \lambda^{\mathbf{n}_\lambda - \mathbf{e}_z} \rangle - n_z \alpha_\lambda \langle \mu^{\mathbf{n}_\mu - \mathbf{e}_z} \mid \lambda^{\mathbf{n}_\lambda + \mathbf{e}_z} \rangle \\
& - m_z \alpha_\mu \langle \mu^{\mathbf{n}_\mu + \mathbf{e}_z} \mid \lambda^{\mathbf{n}_\lambda - \mathbf{e}_z} \rangle + 2\alpha_\mu \alpha_\lambda \langle \mu^{\mathbf{n}_\mu + \mathbf{e}_z} \mid \lambda^{\mathbf{n}_\lambda + \mathbf{e}_z} \rangle.
\end{aligned}
$$

12.4.3 Nuclear attraction

Recursion formulas of the sort used for overlap can be developed for the nuclear attraction term. However, here a slightly different approach will be developed whose real benefit is the repulsion/exchange integrals. Consider the nuclear attraction integral for Gaussian function with angular momentum n in the x direction of center μ. The relevant integral to be solved is

$$\int_0^\infty du \int_{-\infty}^{+\infty} dx_1 (x_1 - x_\mu)^n e^{-\alpha(x_1 - x_a)^2 - u^2(x_1 - x_i)^2}.$$

Now use integration by parts to give

$$\frac{2}{n+1}\int_0^\infty du \int_{-\infty}^{+\infty} dx_1 (x_1 - x_\mu)^{n+1} \left[\alpha(x_1 - x_a) + u^2(x_1 - x_i)\right] e^{-\alpha(x_1 - x_a)^2 - u^2(x_1 - x_i)^2}.$$

Let A_{n0}^x stand for the angular momentum function $(x_1 - x_\mu)^n$. Comparing the first and second expressions, the recursion

$$A_{n0}^x = \frac{2(\alpha + u^2)}{n+1} A_{n+2,0}^x + \frac{2\alpha}{n+1}(x_\mu - x_a)A_{n+1,0}^x + \frac{2u^2}{n+1}(x_\mu - x_i)A_{n+1,0}^x$$

appears. Rearrangement with the change of variables from u to t with (12.21) gives

$$A_{n+2,0}^x = t^2(x_i - x_a)A_{n+1,0}^x + (x_a - x_\mu)A_{n+1,0}^x + \frac{(n+1)(1-t^2)}{2\alpha}A_{n,0}^x.$$

To generate solutions with angular momentum on center λ, use

$$(x - x_\lambda) = (x - x_\mu) + (x_\mu - x_\lambda),$$

which gives the recursion

$$A_{n,m+1}^x = A_{n+1,m}^x + (x_\mu - x_\lambda)A_{n,m}^x$$

to transfer momentum from center μ to center λ, and

$$\langle \mu^n | r_{i1}^{-1} | \lambda^m \rangle = \frac{2\pi}{\alpha} C_a \int_0^1 dt\, A_{nm}(t^2)e^{-\alpha \|\mathbf{x}_c - \mathbf{x}_i\|^2 t^2},$$

where $A_{\mathbf{nm}} = A_{n_x m_x}^x A_{n_y m_y}^y A_{n_z m_z}^z$. Since $A_{\mathbf{nm}}(t^2)$ is a polynomial, and $\exp(-\alpha \|\mathbf{x}_c - \mathbf{x}_i\| t^2)$ is a positive weight factor, the integral can be evaluated exactly with Gaussian quadrature. That is,

$$\int_0^1 dt\, e^{-\alpha \|\mathbf{x}_c - \mathbf{x}_i\|^2 t^2} A_{\mathbf{n,m}}(t^2) = \sum_i W_i A_{\mathbf{n,m}}(t_i^2),$$

where W_i are the Gaussian quadrature weights, and t_i are the Gaussian quadrature support abscissas associated with the weight function $\exp(-\alpha \|\mathbf{x}_c - \mathbf{x}_i\|^2 t^2)$ on the interval $[0, 1]$.

12.4.4   Electron repulsion and exchange

In a manner analogous to the nuclear attraction case, we begin with the integral

$$\int_{-\infty}^{\infty} du \int_{-\infty}^{\infty} dx_1 \int_{-\infty}^{\infty} dx_2 e^{-\alpha(x_1 - x_a)^2 - \beta(x_2 - x_b)^2 - u^2(x_1 - x_2)^2}.$$

Into this integral we can insert

$$G_{n,m}^x = (x_1 - x_\mu)^n (x_2 - x_\nu)^m,$$

then perform integration by parts to deduce the recurrence relations.

Using integration by parts with the x_1 variable,

$$2\alpha G_{n+1,m}^x = -2\alpha(x_\mu - x_a)G_{n,m}^x + nG_{n-1,m}^x - 2u^2(x_1 - x_2)G_{n,m}^x$$

$$2(\alpha + u^2)G_{n+1,m}^x = -2\left[\alpha(x_\mu - x_a) - u^2(x_\nu - x_\mu)\right]G_{n,m}^x + nG_{n-1,m}^x$$

$$+ 2u^2 G_{n,m+1}^x,$$

and, using instead the x_2 variable,

$$2\beta G^x_{n,m+1} = -2\beta(x_\nu - x_b)G^x_{n,m} + mG^x_{n,m-1} + 2u^2(x_1 - x_2)G^x_{n,m}$$

$$2(\beta + u^2)G^x_{n,m+1} = -2\left[\beta(x_\nu - x_b) + u^2(x_\nu - x_\mu)\right]G^x_{n,m} + mG^x_{n,m-1}$$
$$+ nG^x_{n-1,m} + 2u^2G^x_{n+1,m}.$$

Eliminating $G^x_{n+1,m}$ from these two equations gives

$$G^x_{n,m+1} = \frac{m}{2\beta}\left(1 - \frac{\alpha t^2}{\alpha + \beta}\right)G^x_{n,m-1} + \frac{nt^2}{2(\alpha + \beta)}G^x_{n-1,m}$$
$$+ \left((x_b - x_\nu) + \frac{\alpha(x_a - x_b)t^2}{\alpha + \beta}\right)G^x_{n,m},$$

and eliminating $G^x_{n,m+1}$ gives

$$G^x_{n+1,m} = \frac{n}{2\alpha}\left(1 - \frac{\beta t^2}{\alpha + \beta}\right)G^x_{n-1,m} + \frac{mt^2}{2(\alpha + \beta)}G^x_{n,m-1}$$
$$+ \left((x_a - x_\mu) + \frac{\beta(x_b - x_a)t^2}{\alpha + \beta}\right)G^x_{n,m},$$

where t^2 and u^2 are related by (12.22).

Finally, momentum can be transferred from μ to λ and from ν to σ as occurred in the nuclear attraction case,

$$G^x_{n,m,p,q} = G^x_{n-1,m,p+1,q} + (x_\lambda - x_\mu)G^x_{n-1,m,p,q}$$
$$G^x_{n,m,p,q} = G^x_{n,m-1,p,q+1} + (x_\sigma - x_\nu)G^x_{n,m-1,p,q},$$

where $G^x_{n,m,0,0} = G^x_{n,m}$. The overall integrals,

$$\langle \mu^n \nu^m \mid r_{12}^{-1} \mid \lambda^p \sigma^q \rangle = \frac{2\pi^{5/2}}{(\alpha\beta)\sqrt{\alpha + \beta}}C_aC_b \int_0^1 dt\, G_{\mathbf{nmpq}}(t^2)e^{-t^2\frac{\alpha\beta\|x_a - x_b\|^2}{\alpha + \beta}}$$

can be solved exactly with Gaussian integration, as in the kinetic energy case. $G_{\mathbf{nmpq}}$ is $G^x_{n_x m_x p_x q_x} G^y_{n_y m_y p_y q_y} G^z_{n_z m_z p_z q_z}$.

12.5 Rys polynomials

The nuclear attraction and electron repulsion and exchange integrals contain an integral easily solved by an appropriate Gaussian quadrature method,

$$\int_0^1 e^{-\gamma t^2}P(t^2)dt = \sum_i W_i P(t_i^2).$$

The orthogonal polynomials underlying this method are called Rys polynomials – they were created expressly for this quantum mechanics application. The rub is that the quadrature weights W_i and points t_i depend on the parameter γ, which varies depending on the atoms and basis functions being considered. So, to make this method practical one must first find the quadrature points and weights for a wide range of γ values, then fit them to sensible models so they can be estimated accurately for any γ. There are two principal concerns: accuracy, and speed of implementation. King and Dupuis [130] did this with considerable care. Here, the general outline of their strategy will be described without emphasizing operation count.

We will also limit our attention to quadrature with four support points. This will enable us to evaluate all relevant integrals for problems with D orbitals. For maximum efficiency, one would work out the following steps for $n = 1, n = 2$, etc., support points in order to optimize the efficiency of the overall method.

King and Dupuis noted that the scheme

$$\int_0^1 P(t^2)e^{-\gamma t^2}\,dt = \frac{1}{\sqrt{\gamma}}\int_0^{\sqrt{\gamma}} e^{-s^2}\,ds = \sum_{i=1}^n P(t_i^2)W_i \qquad (12.23)$$

can be compared with the Gauss–Hermite quadrature

$$\int_{-\infty}^{+\infty} P(s^2)e^{-s^2}\,ds$$

in the limit that γ is large. We will restrict our attention to Gauss–Hermite method with $2n$ quadrature points r_i and associated weights ρ_i. Since the integrand is even,

$$\int_0^\infty P(s^2)e^{-s^2}\,ds = \sum_{i=1}^n \rho_{2i} P(r_{2i}^2). \qquad (12.24)$$

Then, comparing (12.24) with (12.23) they deduce

$$\lim_{\gamma\to\infty} t_i \to \frac{r_{2i}}{\sqrt{\gamma}}$$

$$\lim_{\gamma\to\infty} W_i \to \frac{\rho_{2i}}{\sqrt{\gamma}}.$$

So, when γ is large, we might seek a fit of the Rys quadrature data to a function of the type

$$t_i^2(\gamma) = \frac{r_{2i}^2}{\gamma} + \sum_{k=k_{min}}^{k_{max}} q_k^t \gamma^k$$

$$W_i(\gamma) = \frac{\rho_{2i}}{\sqrt{\gamma}} + \sum_{k=k_{min}}^{k_{max}} q_k^W \gamma^k.$$

They deduced that when $\gamma > 50$ the first term in these fit expressions is adequate, and no polynomial is necessary, and when $\gamma < 20$, this large γ approximation is not very good. Consequently, quadrature data in the γ range $[20, 50]$ are needed to fit this model. King and Dupuis wished to approximate a fit in L^∞, and they did this by writing a nonlinear L^p error expression and fitting data to it with a nonlinear Newton-based approach. When $p = 4$, they found the fit adequate. We can use directly the simplex-based L^p-norm fitting strategy to accomplish this same goal. It turns out, however, that when seeking very high accuracy by using a very large number (say 150) of fitting points, the simplex constraints become numerically degenerate and the simplex procedure does poorly. Fortunately, a simple least squares fit does quite well, which suggests no need for an L^p model. The fit presented below used 150 evenly spaced data on [20, 50], and used the model parameters $k_{min} = -3$ and $k_{max} = 7$.

Away from the asymptotic limit, i.e., when $\gamma < 20$, King and Dupuis used polynomial fits over restricted intervals: [0, 1], [1, 5], [5, 10], [10, 15], and [15, 20]. Rather than use data fitting, or polynomial interpolation, they used something of a hybrid approach. They used a polynomial interpolation method in each interval, expressed in terms of a Chebyshev series instead of a power series. In this form, the series converges rapidly. So, while interpolated with say 48 points, it can be reconstructed with fewer, say 20, terms for the problem at hand. That is,

$$P(x) = \sum_{i=0}^{47} a_i x^i = \sum_{i=0}^{47} b_i T_i(x)$$

$$P(x) \approx \sum_{i=0}^{19} b_i T_i(x). \tag{12.25}$$

Rather than first finding coefficients a by a procedure like Newton's method, one can construct the coefficients b directly if the support points are roots of a Chebyshev polynomial. This is because the Chebyshev polynomials posses a discrete orthogonality property,

$$\sum_{k=0}^{n-1} T_i(x_k) T_j(x_k) = \begin{cases} 0 & i \neq j \\ n & i = j = 0 \\ n/2 & i = j \neq 0, \end{cases}$$

where points x_k are roots of $T_n(x)$. Then,

$$P(x_k) = \sum_{i=0}^{n-1} b_i T_i(x_k)$$

$$\sum_{k=0}^{n-1} P(x_k) T_j(x_k) = \sum_{i,k=0}^{n-1} b_i T_i(x_k) T_j(x_k) = \begin{cases} nb_j & j = 0 \\ nb_j/2 & j \neq 0 \end{cases}$$

$$b_0 = \frac{1}{n} \sum_{k=0}^{n-1} T_0(x_k) f(x_k)$$

$$b_j = \frac{2}{n} \sum_{k=0}^{n-1} T_j(x_k) f(x_k) \quad j = 1, ..., n-1.$$

The evaluation of series (12.25) is very efficient by a recursion discovered by Clenshaw [39]. For a series with m terms, he defines the sequence

$$y_{m+1} = 0$$

$$y_m = 0$$

$$y_k = 2x y_{k+1} - y_{k+2} + b_k, \quad k = m-1, ..., 0.$$

Then,

$$P = \sum_{i=0}^{m-1} b_i T_i = \sum_{i=0}^{m-1} (y_i - 2xy_{i+1} + y_{i+2})T_i$$

$$= \sum_{i=2}^{m-1} y_i \underbrace{(T_i - 2xT_{i-1} + T_{i-2})}_{0} + y_1 \underbrace{T_1}_{x} + \underbrace{(y_0 - 2xy_1)}_{b_0 - y_2} \underbrace{T_0}_{1}$$

$$P = b_0 - y_2 + xy_1.$$

The results of this fitting exercise are displayed in Figures 12.3 through 12.6. It can be seen that without too much effort a relative error of about 100ϵ has been achieved.

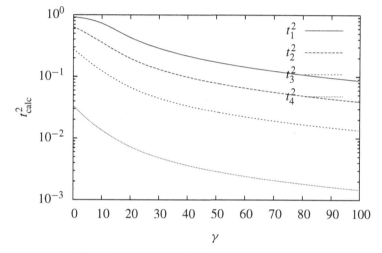

Figure 12.3: Calculated Rys quadrature points.

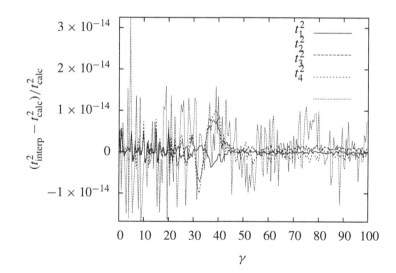

Figure 12.4: Calculated interpolation error for Rys quadrature points.

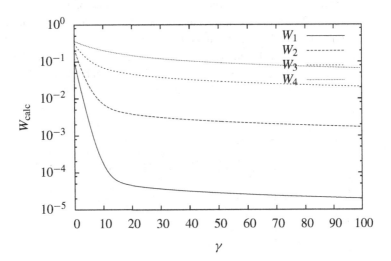

Figure 12.5: Calculated Rys quadrature weights.

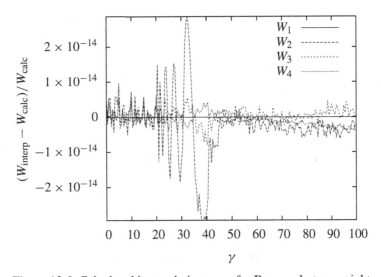

Figure 12.6: Calculated interpolation error for Rys quadrature weights.

12.6   Results II: H$_2$O

Now that we have the capacity to include P orbitals in our calculation, we can take on the water molecule. Water has 10 electrons in total, and involves the 2S and 2P orbitals of oxygen. The STO-3G basis of Hehre *et al.* assigns 6 primitive Gaussian functions to the 2S oxygen orbital in order to facilitate orthogonality with the 1S orbital. With 3 primitive Gaussians for each other orbital, a total of 24 Gaussians are suggested. As with the H_2 calculation, we do not use contracted bases – we will allow the 24 Gaussian primitives to take on whatever weight minimizes the total energy.

Figure 12.7 shows the results of our calculation holding fixed the O–H distance, and varying $\angle$HOH. The results are compared to a careful study by Rosenberg, Ermler, and Shavitt [199, 200]. It can be seen that our simple calculation does a respectable job of estimating the bond angle, but the energy is too high by about 0.88 hartree. The sign of the error is roughly consistent with the variational principle.

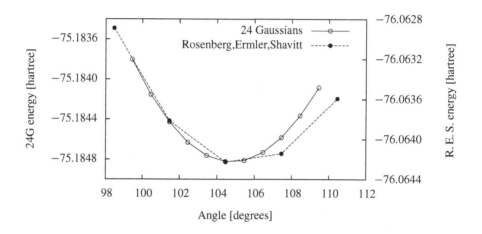

Figure 12.7 The water molecule.

This example shows that using the O–H distance found by Rosenberg, Ermler, and Shavitt, the optimum angle is close to the optimum angle they discovered. A more challenging problem is the discovery of the optimum geometry from scratch. This is an optimization problem in 2D (angle and distance, assuming both O-H distances are equal). The variable metric method provides a Hessian-free estimate, but requires gradient information. Gradients are easily estimated by finite differences. Direct calculation of the energy gradients is possible, but requires new integrals.

Table 12.1 *Ab initio* geometry optimization, H_2O.

	24 Gaussian	Rosenberg *et al.*
energy [hartree]	−75.1848	−76.0642
distance [bohr]	1.8200	1.8111
angle [degree]	105.01	104.45

Using the Davidon–Fletcher–Powell version of the VMM, one finds the results of Table 12.1. A very respectable estimate is obtained with a rudimentary basis and relatively little effort. To improve the convergence of geometry optimization, *relaxation* in the chemical literature, one could use analytical derivatives. Using (12.18b), one sees that the derivative of the total energy can be found by computing expectations of the gradient of the Hamiltonian, $\langle \Psi \mid \nabla H \mid \Psi \rangle$, holding fixed the wave function Ψ. Analytical Hessians can similarly be computed. These are useful not only for geometry optimization, but also for computation of vibrational frequencies (see the algorithms of Wilson, Decius, and Cross [245]).

Appendix A Mathematical background

A.1   Continuity

A function $f(x)$ is said to be continuous at x_0 if for any $\epsilon > 0$ there is a $\delta > 0$ such that

$$|x - x_0| < \delta \quad \Longrightarrow \quad |f(x) - f(x_0)| < \epsilon. \qquad (A.1)$$

In (A.1), x being within distance δ of x_0 implies that $f(x)$ is within distance ϵ of $f(x_0)$ (e.g., Cauchy [34]).

Taking the limit $\epsilon \to 0$, we conclude that $f(x)$ approaches $f(x_0)$ from the left and from the right. The graph of $f(x)$ is a single unbroken curve when f is continuous.

If continuity holds for all $x_0 \in [a, b]$, we say $f \in C[a, b]$, or $C^0[a, b]$. In this case we also say that f is *uniformly continuous* to make a distinction with pointwise continuity described at point x_0 in (A.1).

If $f(x)$ has a derivative and the derivative is continuous, we say $f \in C^1$. In general, $f \in C^n[a, b]$ means that the nth derivative of f, $f^{(n)}$, exists and is continuous on the interval $[a, b]$.

A function is Lipschitz continuous [148] if

$$|f(x) - f(y)| \leq K|x - y|,$$

for some real constant K, called the Lipschitz constant.

A function that is Lipschitz continuous is also uniformly convergent: consider the case $\delta = \epsilon/K$. However, the converse is not true. For example, $f = \sqrt[3]{x}$ is continuous on $[-1, 1]$ but with $x = 1/z^3$ and $y = -1/z^3$ for $|z| > 1$,

$$|f(x) - f(y)| = |2/z|,$$

whereas

$$|x - y| = |2/z^3|.$$

For this function to be Lipschitz continuous, K must satisfy

$$K \geq z^2,$$

but there is no finite K for which this will be true.

A.2   Triangle inequality

The triangle inequality states that if

$$y = x_1 + x_2$$

then

$$|y| = |x_1 + x_2| \le |x_1| + |x_2|.$$

This may be applied recursively to deduce that

$$y = \sum_{i=1}^{n} x_i$$

$$|y| \le \sum_{i=1}^{n} |x_i|.$$

Some unintuitive forms of the triangle inequality are important:

$$a = a - b + b$$
$$|a| \le |a - b| + |b|$$
$$|a - b| \ge |a| - |b|.$$

Also,

$$b = b - a + a$$
$$|b| \le |b - a| + |a|$$
$$|b - a| \ge |b| - |a|,$$

and in combination

$$|a - b| \ge \Big| |a| - |b| \Big|. \tag{A.2}$$

A.3   Rolle's theorem

Rolle's theorem [195] states that if a function $f(x)$ is continuous and continuously differentiable on the interval $[a, b]$, i.e., $f \in C^1[a, b]$, and if $f(a) = f(b)$, then there is some point c, $a \le c \le b$, where $f'(c) = 0$.

If f is constant, then $f' = 0$ everywhere, and the theorem is trivial. If f is not constant, but $f(a) = f(b)$, then there must exist an extremum; a point where $f' = 0$ if f is continuously differentiable. But, if f were not continuously differentiable, e.g., the triangle function $f(a) = f(a) + \min(|x - a|, |x - b|)$, then at the extremum $f = (a + b)/2$ the derivative would not be zero and Rolle's theorem would not apply.

If $f(x)$ is the derivative of some function $g(x)$, and $g(x)$ has continuous and differentiable derivatives, then Rolle's theorem holds that if $g'(a) = g'(b)$, then there is some point c, $a \le c \le b$, where $g''(c) = 0$.

The most general result is that if $f(x)$ is C^{n+1} on $[a, b]$ and $f^{(n)}(a) = f^{(n)}(b)$, then there is some point c, $a \leq c \leq b$ where $f^{(n+1)}(c) = 0$.

A.3.1 Rolle's theorem for zeros of multiplicity m

One says that a is a zero of multiplicity m of $f(x)$ if $f(x)$ can be written

$$f(x) = (x - a)^m g(x), \quad g(a) \neq 0,$$

and this affects Rolle's theorem as follows.

Suppose a is a zero of multiplicity m of $f(x)$, and b is a simple zero (i.e., of multiplicity 1). Rolle's theorem tells us that $f'(x)$ has a somewhere between a and b. In addition, because a is a root of multiplicity m then a is a root of $f'(x)$ with multiplicity $m - 1$. Therefore, all together f' is zero m times in $[a, b]$. Similarly, f'' is zero $m - 1$ times, ..., and $f^{(m)}$ has one zero in $[a, b]$.

Thus, if f has k zeros on $[a, b]$ when counting each zero by its multiplicity, then f' has $k - 1$ zeros $[a, b]$, f'' has $k - 2$ zeros, and $f^{(k-1)}$ has one zero on $[a, b]$.

A.4 Mean value theorem

If a function $f(x)$ is uniformly continuous on the closed interval $[a, b]$, then there exists a point $c \in [a, b]$ such that

$$\frac{f(b) - f(a)}{b - a} = f'(c). \tag{A.3}$$

We can show this using Rolle's theorem, applied to $F(x) = f(x) - kx$ for some special value of k. Choose k so that $F(a) = F(b)$, i.e.,

$$k = \frac{f(b) - f(a)}{b - a}. \tag{A.4}$$

Now because f is continuous, so too is F. And, with k given by (A.4) $F(a) = F(b)$ so Rolle's theorem says that there is a point $c \in [a, b]$ where $F'(c) = 0$. This gives

$$0 = F'(c) = f'(c) - k,$$

which reduces to (A.3).

The mean value theorem for integration states that if $f(t)$ does not change sign on $[a, b]$, and if $g(t)$ is $C^1[a, b]$, then

$$\int_a^b f(x)g(x)dx = g(\xi) \int_a^b f(x)dx,$$

for some $\xi \in [a, b]$.

Without loss of generality, assume $b \geq a$. Let m be the minimum value of $g(x)$ on $[a, b]$, and let M be the maximum:

$$m \leq g \leq M \quad \forall x \in [a, b],$$

and since g is C^1 it will take all values $[m, M]$. It is also true that

$$\phi(x) = \frac{f(x)}{\int_a^b f(x')dx'}$$

is a positive function that integrates to 1:

$$\int_a^b \phi(x)dx = 1$$

(ϕ is a *"convex partition of unity"*). It follows that

$$m \leq \int_a^b \phi(x)g(x)dx \leq M,$$

where $\int \phi g dx$ is a weighted average of the values of g on $[a, b]$ and the convex partition of unity property guarantees that the weighted average is with the range of g. Written differently,

$$m \leq \frac{\int_a^b f(x)g(x)dx}{\int_a^b f(x)dx} \leq M,$$

and therefore the ratio $\int fg dx / \int f dx$ has the same range as function g. Any value that this ratio can take on corresponds to some value $g(\xi)$, $\xi \in [a, b]$.

A.5   Geometric series

The geometric series for $|x| < 1$ is

$$\frac{1}{1+x} = 1 - x + x^2 - x^3 + \cdots.$$

Proof: First show the following identity is true:

$$\frac{1 - x^N}{1 - x} = 1 + x + x^2 + x^3 + \cdots + x^{N-1}. \tag{A.5}$$

Multiply both sides of the equation by $(1 - x)$:

$$1 - x^N = 1 + x + x^2 + x^3 + \cdots + x^{N-1}$$
$$- x - x^2 - x^3 - \cdots - x^{N-1} - x^N$$
$$= 1 - x^N.$$

Next, if $|x| < 1$, then in the limit $N \to \infty$ we have $x^N \to 0$:

$$\frac{1}{1 - x} = \sum_{n=0}^{\infty} x^n \quad \text{if } |x| < 1.$$

Finally, one could replace x by $-x$ to give

$$\frac{1}{1 + x} = \sum_{n=0}^{\infty} (-x)^n \quad \text{if } |x| < 1.$$

A.6　Taylor series

The basic idea, credited to B. Taylor [230] (but first discovered by Gregory, Newton, Leibniz, Bernoulli, and de Moivre [63]), is to represent a function $f(x)$ by a series

$$f(x) = a_0 + a_1 x + a_2 x^2 + \cdots + a_n x^n + \cdots$$

$$= \sum_{n=0}^{\infty} a_n x^n.$$

When $x = 0$ the left-hand side becomes $f(0)$ and the right-hand side becomes a_0, so

$$a_0 = f(0).$$

Now let's differentiate the series k times

$$f^{(k)} = \sum_{n=k}^{\infty} a_n \frac{n!}{(n-k)!} x^{n-k},$$

then set $x = 0$. On the left we get $f^{(k)}(0)$ and on the right $k! a_k$, so

$$a_k = \frac{f^{(k)}(0)}{k!}.$$

The series can therefore be written

$$f(x) = f(0) + f'(0)x + \frac{1}{2} f''(0)x^2 + \cdots + \frac{1}{n!} f^{(n)}(0)x^n + \cdots .$$

To be more general, write a series in powers of $x - x_0$, and determine the coefficients by considering the limits $x \to x_0$:

$$f(x) = f(x_0) + f'(x_0)(x - x_0) + \frac{1}{2} f''(x_0)(x - x_0)^2 + \cdots$$

$$+ \frac{1}{n!} f^{(n)}(x_0)(x - x_0)^n + \cdots .$$

In practice we will want to use finite series approximations of functions. In that case, equality cannot be expected and there is a remainder term:

$$f_N(x) = \sum_{n=0}^{N} \frac{1}{n!} f^{(n)}(x_0)(x - x_0)^n \tag{A.6}$$

$$f(x) = f_N(x) + R_N(x),$$

where $R_N(x)$ is the difference between the infinite and finite series, and can be written in the Lagrange form [139, 138] (cf. [35, pp. 257–261])

$$R_N(x) = \frac{f^{(N+1)}(\xi)}{(N+1)!} (x - x_0)^{N+1} \quad \text{for some } \xi \in [x, x_0]. \tag{A.7}$$

The Lagrange formula for the remainder can be derived by consideration of Rolle's theorem. From (A.6) it follows that

$$R_N(x_0) = 0$$
$$R'_N(x_0) = 0$$
$$R''_N(x_0) = 0$$
$$\vdots$$
$$R_N^{(N)}(x_0) = 0.$$

And, for the function $h(t)$,

$$h(t) = (f(t) - f_N(t)) - \frac{f(x) - f_N(x)}{(x - x_0)^{N+1}}(t - x_0)^{N+1},$$

it is true that

$$h(x_0) = 0 \tag{A.8a}$$
$$h'(x_0) = 0 \tag{A.8b}$$
$$h''(x_0) = 0$$
$$\vdots$$
$$h^{(N)}(x_0) = 0,$$

and

$$h(x) = 0. \tag{A.8c}$$

Let us assume that f, therefore h, is C^∞. Then, consideration of (A.8a) and (A.8c) in light of Rolle's theorem says that there is some constant c_0 between x_0 and x such that

$$h'(c_0) = 0. \tag{A.9}$$

Now, consideration of (A.8b) and (A.9) in view of Rolle's theorem says that there is some constant c_1 between c_0 and x_0 such that

$$h''(c_1) = 0.$$

Carrying on similarly, we conclude that there is some constant ξ between x and x_0 such that

$$h^{(N+1)}(\xi) = 0. \tag{A.10}$$

But

$$h^{(N+1)}(t) = f^{(N+1)}(t) - (N + 1)!\frac{f(x) - f_N(x)}{(x - x_0)^{N+1}}$$
$$= f^{(N+1)}(t) - (N + 1)!\frac{R_N(x)}{(x - x_0)^{N+1}}$$

so

$$R_N(x) = \frac{f^{(N+1)}(\xi)}{(N + 1)!}(x - x_0)^{N+1}, \tag{A.11}$$

for some ξ between x and x_0.

Note that the application of Rolle's theorem to deduce (A.10) required that $f(x)$ possess C^{N+1} continuity.

There are many other representations for the error in a Taylor series. Another important form is due to Cauchy [35, p. 212]. Begin with the fundamental theorem of calculus,

$$f(x) = f(x_0) + \int_{x_0}^{x} f'(t)dt$$
$$= f(x_0) + \int_{x_0}^{x} \frac{(x-t)^0}{0!} f'(t)dt,$$

then integrate by parts

$$f(x) = f(x_0) - \left((x-t)^1 f'(t) \Big|_{x_0}^{x} + \int_{x_0}^{x} \frac{(x-t)^1}{1!} f''(t)dt \right)$$
$$= f(x_0) + (x - x_0)f'(x_0) + \int_{x_0}^{x} \frac{(x-t)^1}{1!} f''(t)dt.$$

This integration by parts can be continued as many times as needed to give

$$f(x) = f(x_0) + f'(x_0)(x - x_0) + \dots + \frac{1}{n!} f^{(n)}(x - x_0)^n + \int_{x_0}^{x} \frac{(x-t)^n}{n!} f^{(n+1)}(t)dt.$$

Therefore, for the remainder we have

$$R_N(x) = \int_{x_0}^{x} \frac{(x-t)^N}{N!} f^{(N+1)}(t)dt. \tag{A.12}$$

In this expression, $f^{(N+1)}$ must exist, but need not be continuous. That is, this remainder form requires $f \in C^N$.

If f were in C^{N+1}, then the mean value theorem could be applied to (A.12) to give (A.11). Or, if $f^{(N+1)}$ were not continuous, but had a Lipschitz constant M,

$$|f^{(N+1)}(y) - f^{(N+1)}(y')| \le M|y - y'| \quad \forall y, y' \in [x_0, x],$$

then

$$R_N(x) - \frac{(x - x_0)^{N+1}}{(N+1)!} f^{(N+1)}(x) = \int_{x_0}^{x} \frac{(x-t)^N}{N!} (f^{(N+1)}(t) - f^{(N+1)}(x))dt$$

$$\left| R_N(x) - \frac{(x - x_0)^{N+1}}{(N+1)!} f^{(N+1)}(x) \right| \le M \int_{x_0}^{x} \frac{(x-t)^{N+1}}{N!} dt = \frac{M(x - x_0)^{N+2}}{(N+2)N!}.$$

Some common Taylor series are

$$e^x = 1 + x + \frac{1}{2!}x^2 + \frac{1}{3!}x^3 + \cdots$$

$$= \sum_{n=0}^{N} \frac{x^n}{n!} + \frac{e^\xi}{(N+1)!} x^{N+1}$$

$$\cos(x) = 1 - \frac{1}{2!}x^2 + \frac{1}{4!}x^4 + \cdots$$

$$= \sum_{n=0}^{N} \frac{(-1)^n x^{2n}}{(2n)!} + \frac{\cos(\xi)^{(2N+1)}}{(2N+1)!}x^{2N+1}$$

$$= \sum_{n=0}^{N} \frac{(-1)^n x^{2n}}{(2n)!} - \frac{(-1)^N \sin(\xi)}{(2N+1)!}x^{2N+1},$$

and

$$\sin(x) = x - \frac{1}{3!}x^3 + \frac{1}{5!}x^5 + \cdots$$

$$= \sum_{n=0}^{N} \frac{(-1)^n}{(2n+1)!}x^{2n+1} + \frac{\sin(\xi)^{(2(N+1))}}{(2(N+1))!}x^{2(N+1)}$$

$$= \sum_{n=0}^{N} \frac{(-1)^n}{(2n+1)!}x^{2n+1} - \frac{(-1)^N \sin(\xi)}{(2(N+1))!}x^{2(N+1)},$$

for some ξ between 0 and x.

To consider multivariate functions, e.g., $f(x_0, x_1, \ldots, x_n)$, it is convenient to use multiindex notation. Let $\mathbf{p}$ be a vector of nonnegative integers, and $\mathbf{x}$ the vector of variables. $\mathbf{x}^{\mathbf{p}}$ will then be shorthand for $x_0^{p_0} \cdots x_n^{p_n}$. We also have $\mathbf{p}! = p_0! \cdots p_n!$, and we will write $|\mathbf{p}| = p_0 + p_1 + \cdots + p_n$. A multivariate Taylor series might express $f(\mathbf{y})$ in terms of derivatives $f^{(\mathbf{p})}(\mathbf{x})$. To develop such a series, write

$$\mathbf{g}(t) = \mathbf{x} + t(\mathbf{y} - \mathbf{x}),$$

so $f(\mathbf{x}) = f(\mathbf{g}(0))$ and $f(\mathbf{y}) = f(\mathbf{g}(1))$. Now, the multivariate Taylor series can be obtained from the univariant function of t:

$$f(\mathbf{g}(1)) = \sum_{n=0}^{P-1} \frac{1}{n!}\frac{d^n}{dt^n}f(\mathbf{g}(0)) + \frac{1}{P!}\frac{d^P}{dt^P}f(\mathbf{g}(\tau)),$$

for some τ in $[0, 1]$. Here the Lagrange form of the remainder has been used. Now expanding the various total derivatives,

$$f(\mathbf{y}) = \sum_{|\mathbf{p}|=0}^{P-1} \frac{(\mathbf{y} - \mathbf{x})^{\mathbf{p}}}{\mathbf{p}!}f^{(\mathbf{p})}(\mathbf{x}) + \sum_{|\mathbf{p}|=P} \frac{(\mathbf{y} - \mathbf{x})^{\mathbf{p}}}{\mathbf{p}!}f^{(\mathbf{p})}(\xi).$$

The last term is Lagrange remainder, again, but now expanded into $\binom{n-1+P}{P}$ different partial derivatives of f. Each of these derivatives is evaluated at the same point ξ which lies somewhere on the line connecting $\mathbf{x}$ and $\mathbf{y}$.

A.7 Linear algebra

An $n \times 1$ column vector $\mathbf{x}$ is a set of n numbers x_i arranged as follows:

$$\mathbf{x} = \begin{pmatrix} x_1 \\ x_2 \\ \vdots \\ x_n \end{pmatrix}.$$

The product of a vector with a scalar is a vector:

$$c\mathbf{x} = \mathbf{x}c = \begin{pmatrix} cx_1 \\ cx_2 \\ \vdots \\ cx_n \end{pmatrix}.$$

The inner product of an $n \times 1$ vector $\mathbf{x}$ with another $n \times 1$ vector $\mathbf{y}$ is a scalar. This is also called the dot product:

$$\mathbf{x} \cdot \mathbf{y} = x_1 y_1 + x_2 y_2 + \cdots + x_n y_n = \sum_{i=1}^{n} x_i y_i.$$

The outer product of an $n \times 1$ vector $\mathbf{x}$ with an $m \times 1$ vector $\mathbf{y}$ is an $n \times m$ matrix. The outer product symbol is $\otimes$. Often it is necessary to examine the context to determine whether an inner product or outer product is indicated.

$$\mathbf{xy} = \mathbf{x} \otimes \mathbf{y} = \begin{pmatrix} x_1 y_1 & x_1 y_2 & \cdots & x_1 y_m \\ x_2 y_1 & x_2 y_2 & \cdots & x_2 y_m \\ \vdots & & & \vdots \\ x_n y_1 & x_n y_2 & \cdots & x_n y_m \end{pmatrix}.$$

As the previous example illustrated, an $n \times n$ matrix $\mathbf{A}$ is a set of n^2 numbers a_{ij} arranged as follows:

$$\mathbf{A} = \begin{pmatrix} a_{11} & a_{12} & \cdots & a_{1n} \\ a_{21} & a_{22} & \cdots & a_{2n} \\ \vdots & & & \vdots \\ a_{n1} & a_{n2} & \cdots & a_{nn} \end{pmatrix}. \tag{A.13}$$

The diagonal of matrix $\mathbf{A}$ refers to the matrix of components a_{ii}. The diagonal of matrix $\mathbf{A}$ (A.13) is

$$\text{diag}(a_{11}, a_{22}, \ldots, a_{nn}) = \begin{pmatrix} a_{11} & 0 & \cdots & 0 & 0 \\ 0 & a_{22} & \cdots & 0 & 0 \\ \vdots & & \ddots & & \vdots \\ 0 & 0 & \cdots & a_{n-1,n-1} & 0 \\ 0 & 0 & \cdots & 0 & a_{n,n} \end{pmatrix}.$$

A diagonal matrix is a matrix equal to its diagonal.

The $n \times n$ identity matrix $\mathbf{I}$ is the diagonal matrix with $I_{ii} = 1$:

$$
\mathbf{I} = \begin{pmatrix}
1 & 0 & \cdots & 0 & 0 \\
0 & 1 & \cdots & 0 & 0 \\
\vdots & & \ddots & & \vdots \\
0 & 0 & \cdots & 1 & 0 \\
0 & 0 & \cdots & 0 & 1
\end{pmatrix}.
$$

The transpose of matrix $\mathbf{A}$, denoted $\mathbf{A}^T$, is constructed by exchanging indices $(\mathbf{A})_{ij} = (\mathbf{A}^T)_{ji}$:

$$
\mathbf{A}^T = \begin{pmatrix}
a_{11} & a_{21} & \cdots & a_{n1} \\
a_{12} & a_{22} & \cdots & a_{n2} \\
\vdots & & & \vdots \\
a_{1n} & a_{2n} & \cdots & a_{nn}
\end{pmatrix}.
$$

A matrix of real numbers that is equal to its own transpose is called symmetric. If the matrix consists of complex numbers then its Hermitian transpose $(\mathbf{A}^H)$ is the transpose of its complex conjugate $\mathbf{A}^H = (\mathbf{A}^*)^T$ where $(x + iy)^* = (x - iy)$ assuming x and y are real. A matrix is called Hermitian if it is equal to its own Hermitian transpose. Such matrices have real diagonals.

A matrix is *skew Hermitian* if $\mathbf{A}^H = -\mathbf{A}$.

The product of a matrix with a scalar is a matrix:

$$
c\mathbf{A} = \begin{pmatrix}
ca_{11} & ca_{12} & \cdots & ca_{1n} \\
ca_{21} & ca_{22} & \cdots & ca_{2n} \\
\vdots & & & \vdots \\
ca_{n1} & ca_{n2} & \cdots & ca_{nn}
\end{pmatrix}.
$$

The product of a matrix with a vector is a new vector,

$$
\mathbf{A}\mathbf{x} = \mathbf{b}
$$

$$
\begin{pmatrix}
a_{11} & a_{12} & \cdots & a_{1n} \\
a_{21} & a_{22} & \cdots & a_{2n} \\
\vdots & & & \vdots \\
a_{n1} & a_{n2} & \cdots & a_{nn}
\end{pmatrix}
\begin{pmatrix}
x_1 \\
x_2 \\
\vdots \\
x_n
\end{pmatrix}
=
\begin{pmatrix}
a_{11}x_1 + a_{12}x_2 + \cdots + a_{1n}x_n \\
a_{21}x_1 + a_{22}x_2 + \cdots + a_{2n}x_n \\
\vdots \\
a_{n1}x_1 + a_{n2}x_2 + \cdots + a_{nn}x_n
\end{pmatrix}.
$$

The product of a matrix $\mathbf{A}$ with another matrix $\mathbf{B}$ is constructed similarly. Consider $\mathbf{B}$ to be a set of column vectors and apply the matrix-vector multiplication rule. In this example, let $\mathbf{A}$ be $n \times m$ and let $\mathbf{B}$ be $m \times p$

$$\mathbf{AB} = \mathbf{C}$$

$$\begin{pmatrix} a_{11} & a_{12} & \cdots & a_{1m} \\ a_{21} & a_{22} & \cdots & a_{2m} \\ & \vdots & & \vdots \\ a_{n1} & a_{n2} & \cdots & a_{nm} \end{pmatrix} \left(\begin{array}{c|c|c|c} b_{11} & b_{12} & \cdots & b_{1p} \\ b_{21} & b_{22} & \cdots & b_{2p} \\ \vdots & & & \vdots \\ b_{m1} & b_{m2} & \cdots & b_{mp} \end{array} \right)$$

$$= \left(\begin{array}{c|c|c|c} \sum_{k=1}^{m} a_{1k}b_{k1} & \sum_{k=1}^{m} a_{1k}b_{k2} & \cdots & \sum_{k=1}^{m} a_{1k}b_{kp} \\ \sum_{k=1}^{m} a_{2k}b_{k1} & \sum_{k=1}^{m} a_{2k}b_{k2} & \cdots & \sum_{k=1}^{m} a_{2k}b_{kp} \\ \vdots & & & \vdots \\ \sum_{k=1}^{m} a_{nk}b_{k1} & \sum_{k=1}^{m} a_{nk}b_{k2} & \cdots & \sum_{k=1}^{m} a_{nk}b_{kp} \end{array} \right).$$

A.7.1 The determinant

The determinant of a matrix is a particular scalar measure of the matrix best defined by its properties. All properties may be deduced from the following three defining properties [227]:

1. The determinant is a linear function of its first row

$$\det \begin{pmatrix} a_{11} + a'_{11} & a_{12} + a'_{12} \\ a_{21} & a_{22} \end{pmatrix} = \det \begin{pmatrix} a_{11} & a_{12} \\ a_{21} & a_{22} \end{pmatrix} \det \begin{pmatrix} a'_{11} & a'_{12} \\ a_{21} & a_{22} \end{pmatrix}$$

$$\det \begin{pmatrix} \alpha a_{11} & \alpha a_{12} \\ a_{21} & a_{22} \end{pmatrix} = \alpha \det \begin{pmatrix} a_{11} & a_{12} \\ a_{21} & a_{22} \end{pmatrix}.$$

2. The determinant changes sign when two rows are interchanged,

$$\det \begin{pmatrix} a_{11} & a_{12} & a_{13} \\ a_{21} & a_{22} & a_{23} \\ a_{31} & a_{32} & a_{33} \end{pmatrix} = -\det \begin{pmatrix} a_{21} & a_{22} & a_{23} \\ a_{11} & a_{12} & a_{13} \\ a_{31} & a_{32} & a_{33} \end{pmatrix} = +\det \begin{pmatrix} a_{21} & a_{22} & a_{23} \\ a_{31} & a_{32} & a_{33} \\ a_{11} & a_{12} & a_{13} \end{pmatrix}.$$

3. The determinant of the identity matrix is 1,

$$\det \mathbf{I} = 1.$$

A formula for the determinant due to Leibniz in 1693 [144] (from a letter to the Marquis de l'Hôpital, translated in [221, p. 267–269]) is

$$\det \mathbf{A} = \sum_{\sigma} a_{1\alpha} a_{2\beta} \cdots a_{n\nu} \det P_{\sigma}, \tag{A.14}$$

where σ denotes a permutation of the numbers $1, \ldots, n$ into the order $\alpha, \beta, \ldots, \nu$, and

$$P_{\upsilon} - \begin{pmatrix} \mathbf{e}_{\alpha}^{\mathrm{T}} \\ \mathbf{e}_{\beta}^{\mathrm{T}} \\ \vdots \\ \mathbf{e}_{\upsilon}^{\mathrm{T}} \end{pmatrix}$$

is the associated permutation matrix. The sum extends over all $n!$ possible permutations. This formula is consistent with the three defining characteristics, and its application gives rise to the well-known formulas for 2×2 and 3×3 matrices:

$$\det\mathbf{A} = |\mathbf{A}| = \begin{vmatrix} a_{11} & a_{12} \\ a_{21} & a_{22} \end{vmatrix}$$

$$= a_{11}a_{22} - a_{12}a_{21}.$$

$$\begin{vmatrix} a_{11} & a_{12} & a_{13} \\ a_{21} & a_{22} & a_{23} \\ a_{31} & a_{32} & a_{33} \end{vmatrix} = \begin{matrix} a_{11}a_{22}a_{33} + a_{12}a_{23}a_{31} + a_{13}a_{21}a_{32} \\ -a_{11}a_{23}a_{32} - a_{12}a_{12}a_{33} - a_{13}a_{12}a_{31}. \end{matrix} \tag{A.15}$$

These explicit formulas follow a particular pattern: there is a sum of products taken on top–left to bottom–right diagonals, minus the sum of products taken on the bottom–left to top–right diagonals.

Leibniz's formula for 4×4 and bigger square matrices do not follow this pattern. Instead, it is useful to interpret (A.14) as a cofactor expansion.

The $(n-1) \times (n-1)$ "minor" submatrix $\mathbf{M}_{ij}$ of $n \times n$ matrix $\mathbf{A}$ is the matrix that results by deleting the ith row and the jth column of $\mathbf{A}$. For example, $\mathbf{M}_{21}$ derived from the 3×3 example above is

$$\mathbf{M}_{21} = \begin{pmatrix} \cancel{a}_{11} & a_{12} & a_{13} \\ \cancel{a_{21}} & \cancel{a_{22}} & \cancel{a_{23}} \\ \cancel{a}_{31} & a_{32} & a_{33} \end{pmatrix} = \begin{pmatrix} a_{12} & a_{13} \\ a_{32} & a_{33} \end{pmatrix}.$$

The cofactor C_{ij} of element a_{ij} of matrix $\mathbf{A}$ is the determinant of minor $\mathbf{M}_{ij}$ with a sign factor:

$$C_{ij} = (-1)^{i+j} \det\mathbf{M}_{ij}.$$

With these definitions, the determinant of an arbitrary $n \times n$ matrix may be written

$$\det\mathbf{A} = \sum_{k=1}^{n} a_{ik} C_{ik}, \tag{A.16}$$

for any $1 \le i \le n$, or

$$\det\mathbf{A} = \sum_{i=1}^{n} a_{ik} C_{ik}, \tag{A.17}$$

for any $1 \le k \le n$. Equations (A.16) and (A.17) are just rearrangements of Leibniz's formula (A.14). For example, the determinant of a 5×5 matrix is given in terms of five 4×4 cofactors. These, in turn, are each determined by evaluation four 3×3 cofactors. These twenty 3×3 cofactors could be constructed from sixty 2×2 cofactors, or they could be evaluated following the example of (A.15). The evaluation of determinants by this recursive procedure takes a lot of algebra, and numerically this is not the best approach.

Some important properties of determinants include:

- The determinant of a matrix product is the product of matrix determinants:

$$\det(\mathbf{AB}) = \det(\mathbf{A})\det(\mathbf{B}).$$

It is easy to see that the determinant of the identity matrix is 1. Therefore, the determinant of the inverse of a matrix is the inverse of the determinant of the matrix

$$1 = \det\mathbf{I} = \det(\mathbf{A}^{-1}\mathbf{A}) = \det(\mathbf{A}^{-1})\det(\mathbf{A})$$
$$\det(\mathbf{A}^{-1}) = \frac{1}{\det(\mathbf{A})}.$$

- If a matrix is block partitioned in the form

$$\mathbf{A} = \left(\begin{array}{c|c} \mathbf{B} & \mathbf{C} \\ \hline \mathbf{0} & \mathbf{E} \end{array}\right), \quad \text{or } \mathbf{A}' = \left(\begin{array}{c|c} \mathbf{B} & \mathbf{0} \\ \hline \mathbf{D} & \mathbf{E} \end{array}\right), \tag{A.18}$$

with partitions $\mathbf{B}$ and $\mathbf{E}$ square, then

$$\det(\mathbf{A}) = \det(\mathbf{A}') = \det(\mathbf{B})\det(\mathbf{E}).$$

A matrix is *irreducible* if it cannot be converted to the block triangular form (A.18) through a similarity transformation.

- The determinant of a triangular matrix is the product of its diagonal entries.
- The determinant is the product of eigenvalues.

A.7.2 Eigenvalues and eigenvectors

Every $n \times n$ matrix $\mathbf{A}$ has n eigenvalues – numbers λ such that

$$\mathbf{Ax} = \lambda\mathbf{x}, \tag{A.19}$$

for some particular vector $\mathbf{x}$. That special vector is called the eigenvector corresponding to eigenvalue λ. Introducing the $n \times n$ identity matrix $\mathbf{I}$ we can write the eigenvalue equation (A.19) as

$$(\mathbf{A} - \lambda\mathbf{I})\mathbf{x} = 0. \tag{A.20}$$

A scalar equation $ax = 0$ would tell us that $a = 0$ or $x = 0$. In the matrix case, these are trivial solutions and we are interested in nontrivial results. What equation (A.20) signifies is that the matrix $(\mathbf{A} - \lambda\mathbf{I})$ is singular (does not possess an inverse). A matrix is singular if its determinant is zero.

From the previous review of determinants, it should be clear that $\det(\mathbf{A} - \lambda\mathbf{I})$ is an nth degree polynomial in λ. The polynomial itself is called the characteristic equation, and the eigenvalues of $\mathbf{A}$ are its roots.

For example, the $n \times n$ matrix *Frobenius normal matrix*

$$\mathbf{A} = \begin{pmatrix} 0 & 0 & 0 & \cdots & 0 & 0 & -a_0 \\ 1 & 0 & 0 & & & 0 & -a_1 \\ 0 & 1 & 0 & \ddots & & 0 & -a_2 \\ & \ddots & \ddots & \ddots & \ddots & \vdots & \vdots \\ & & \ddots & \ddots & 0 & 0 & -a_{n-3} \\ & & & \ddots & 1 & 0 & -a_{n-2} \\ & & & & 0 & 1 & -a_{n-1} \end{pmatrix} \tag{A.21}$$

has the degree n polynomial characteristic equation given by $\det(\mathbf{A} - \lambda\mathbf{I}) = 0$. We will find the determinant by expanding $(\mathbf{A} - \lambda\mathbf{I})$ in cofactors of the last column:

$$\det(\mathbf{A} - \lambda\mathbf{I}) = -a_0(-1)^{n+1}\det \begin{pmatrix} 1 & -\lambda & & & & \\ & 1 & -\lambda & & & \\ & & \ddots & \ddots & & \\ & & & \ddots & -\lambda & \\ & & & 1 & -\lambda \\ & & & & 1 \end{pmatrix}$$

$$ - a_1(-1)^{n+2}\det \left(\begin{array}{c|ccccc} -\lambda & & & & \\ \hline & 1 & -\lambda & & & \\ & & \ddots & \ddots & & \\ & & & \ddots & -\lambda & \\ & & & 1 & -\lambda \\ & & & & 1 \end{array} \right) - \cdots $$

$$ - a_j(-1)^{n+1+j}\det \left(\begin{array}{ccccc|ccc} -\lambda & & & & & & & \\ 1 & \ddots & & & & & & \\ & \ddots & \ddots & & & & \mathbf{0}_{j\times(n-j)} & \\ & & 1 & -\lambda & & & & \\ & & & 1 & -\lambda & & & \\ \hline & & & & 1 & -\lambda & & \\ & \mathbf{0}_{(n-j)\times j} & & & & \ddots & \ddots & \\ & & & & & & \ddots & -\lambda \\ & & & & & & & 1 \end{array} \right) $$

$$- \ldots - a_{n-2}(-1)^{2n-1}\det \left(\begin{array}{cccccc|c} -\lambda & & & & & & \\ 1 & -\lambda & & & & & \\ & 1 & -\lambda & & & & \\ & & & \ddots & & \ddots & \\ & & & & \ddots & -\lambda & \\ \hline & & & & & & 1 \end{array} \right)$$

$$- (a_{n-1} + \lambda)(-1)^{2n}\det \left(\begin{array}{ccccc} -\lambda & & & & \\ 1 & -\lambda & & & \\ & 1 & -\lambda & & \\ & & \ddots & \ddots & \\ & & & \ddots & -\lambda \\ & & & 1 & -\lambda \end{array} \right).$$

The cofactors in this expansion are determinants of block triangular form (A.18), with the diagonal blocks being triangular. Therefore the determinants are 1, $-\lambda$, ..., $(-\lambda)^j$, ..., $(-\lambda)^{n-2}$, $(-\lambda)^{n-1}$, and so

$$\det(\mathbf{A} - \lambda\mathbf{I}) = (-1)^n(a_0 + a_1\lambda + \cdots + a_{n-2}\lambda^{n-2} + a_{n-1}\lambda^{n-1} + \lambda^n). \quad \text{(A.22)}$$

This particular Frobenius matrix is known as the *companion matrix* because of its special relation to polynomial equations.

If $\mathbf{x}$ is an eigenvector corresponding to eigenvalue λ, then $c\mathbf{x}$, with c a constant, is also an eigenvector corresponding to eigenvalue λ. Eigenvalues may be uniquely specified, but eigenvectors are unique only up to multiplicative constants.

A *similarity transformation* is an operation that preserves the eigenvectors of a matrix. If $\mathbf{S}$ is an invertible square matrix, then matrix $\mathbf{B}$

$$\mathbf{B} = \mathbf{SAS}^{-1}$$

is similar to matrix $\mathbf{A}$, and $\mathbf{B}$ and $\mathbf{A}$ have the same eigenvalues:

$$\det(\mathbf{A} - \lambda\mathbf{I}) = \det\mathbf{S}(\mathbf{A} - \lambda\mathbf{I})\mathbf{S}^{-1} = \det(\mathbf{B} - \lambda\mathbf{I})$$

so $\mathbf{B}$ and $\mathbf{A}$ have the same characteristic equation and therefore the same eigenvalues.

The eigenvalues of a rotation matrix are $e^{i\theta}$ (once), $e^{-i\theta}$ (once), and 1 ($n - 2$ times), where θ is the angle of rotation.

The eigenvalues of a reflection matrix are -1 (once) and 1 ($n - 1$) times. The eigenvector corresponding to $\lambda = -1$ is parallel to the normal of the plane of the reflection.

A.7.3 Positive definite

A matrix is *positive definite* if for any $\mathbf{x} \neq 0$ one has

$$\mathbf{x}^T\mathbf{A}\mathbf{x} > 0.$$

In the most common case, $\mathbf{A}$ is symmetric and positive definite, i.e., *symmetric positive definite*. Such an $\mathbf{A}$ can be composed by multiplying two nonsingular matrices $\mathbf{B}$,

$$\mathbf{A} = \mathbf{B}^\mathsf{T}\mathbf{B},$$

and we know this is positive definite because

$$\mathbf{x}^\mathsf{T}\mathbf{A}\mathbf{x}^\mathsf{T} = (\mathbf{B}\mathbf{x})^\mathsf{T}(\mathbf{B}\mathbf{x}),$$

and $\mathbf{B}\mathbf{x}$ will not be zero when $\mathbf{x} \neq 0$ if $\mathbf{B}$ is not singular.

A matrix need not be symmetric to be positive definite. For example,

$$\begin{pmatrix} x_1 & x_2 \end{pmatrix} \begin{pmatrix} 1 & 1 \\ -1 & 1 \end{pmatrix} \begin{pmatrix} x_1 \\ x_2 \end{pmatrix} = x_1^2 + x_2^2.$$

A.7.4   Special matrices

The shape of a matrix is often important, and special shapes have special names. Here, "shape" refers to the pattern created by nonzero elements. Some special shapes include *right triangular*:

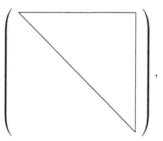

,

and *left triangular*:

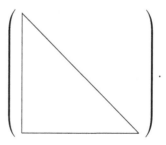

.

An *upper Hessenberg* matrix is a right triangular matrix, with the first subdiagonal also nonzero:

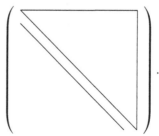

.

A *tridiagonal* matrix consists of the main diagonal, the first subdiagonal, and the first superdiagonal:

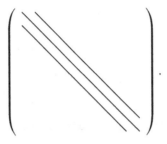

A matrix that is symmetric and upper Hessenberg (or symmetric and lower Hessenberg) is tridiagonal.

A *Frobenius matrix* is square, and differs from the identity matrix by at most one column,

$$
\begin{pmatrix}
1 & & & * & & \\
& \ddots & & \vdots & & \\
& & 1 & * & & \\
& & & * & & \\
& & * & 1 & & \\
& & \vdots & & \ddots & \\
& & * & & & 1
\end{pmatrix}.
$$

A *Frobenius normal matrix* (A.21) is not a *Frobenius matrix* by this definition.

A.8 Complex numbers

The methods and algorithms of this book are designed to work with real numbers, for the most part. However, real matrices can have complex eigenvalues and real polynomials can have complex roots, so some exposure to complex numbers is necessary.

We will write i for the pure imaginary number $\sqrt{-1}$. A complex number is a number that contains a real part and an imaginary part. If x and y are real, then

$$
z = x + iy
$$

is complex. The number

$$
z^* = x - iy
$$

is the *complex conjugate* of z.

It is sometimes easiest to think of a complex number in a polar coordinate system where instead of x and y we have a positive radius r and an angle θ, and these quantities are related through

$$
x = r \cos \theta
$$
$$
y = r \sin \theta.
$$

In this system our complex number can be written as

$$z = r(\cos\theta + i\sin\theta).$$

Euler's formula

$$e^{i\theta} = \cos\theta + i\sin\theta \qquad\qquad (A.23)$$

provides a more compact representation

$$z = re^{i\theta}.$$

Euler's formula can be proved by using the Taylor series

$$e^{i\theta} = \sum_{n=0}^{\infty} \frac{(i\theta)^n}{n!}$$

$$= 1 + i\theta + \frac{i^2\theta^2}{2!} + \frac{i^3\theta^3}{3!} + \frac{i^4\theta^4}{4!} + \frac{i^5\theta^5}{5!} + \frac{i^6\theta^6}{6!} + \frac{i^7\theta^7}{7!} + \cdots$$

$$= 1 + i\theta - \frac{\theta^2}{2!} - \frac{i\theta^3}{3!} + \frac{\theta^4}{4!} + \frac{i\theta^5}{5!} - \frac{\theta^6}{6!} - \frac{i\theta^7}{7!} + \cdots$$

$$= \left(1 - \frac{\theta^2}{2!} + \frac{\theta^4}{4!} - \frac{\theta^6}{6!} + \cdots\right) + i\left(\theta - \frac{\theta^3}{3!} + \frac{\theta^5}{5!} - \frac{\theta^7}{7!} + \cdots\right)$$

$$= \cos\theta + i\sin\theta.$$

Appendix B Sample codes

Some of the algorithms in this book (e.g., FFT) have an aspect of complexity due to data management issues. These can be explained through detailed and lengthy descriptions of the algorithm, or through examples, and this book favors the latter approach.

Accordingly, the codes contained in this appendix should be viewed as additional examples that emphasize the data management aspects of the numerical methods. The codes "work," but are not as robust or as efficient as one would want from "software." The emphasis is on clarity.

The algorithms selected for this appendix are those for which data management plays a significant role in practical implementations.

The codes are in C++. The main significance (and point of confusion) is that in C++ matrix and vector indices run from 0 to $n - 1$, vs. from 1 to n. Otherwise, the code examples closely follow the text.

For most algorithms, three files are listed

1. a header file $\cdots$. H that makes public the interface to the utility functions,
2. a file $\cdots$. cpp containing the routines that implement the given algorithm, and
3. a file $\cdots$ -test . cpp that demonstrates the use of the utility functions by solving some problem.

These codes are available online for download at http://www.cambridge.org/9781107021082.

B.1   Utility routines

Some algorithms are greatly simplified if one can dynamically resize objects like vectors and matrices. To facilitate such operations, a simple "reference counted pointer" class is helpful.

file: RCP.H

```
1   #ifndef __RCP_H__
2   #define __RCP_H__
3   // reference counted pointer class for objects of type T
4   // created via "new T[]", i.e, requiring "delete []" vs "delete"
5   // for deallocation.
6
```

```
 7    template <class T> class RCP
 8    {
 9      public:
10
11        inline RCP(T* ptr = 0)
12        {
13          m_ptr = ptr;
14          m_count = NULL;
15          if( m_ptr ) {
16            m_count = new int;
17            *m_count = 1;
18          }
19        }
20
21        inline RCP( const RCP<T> &a_rcp )
22        {
23          m_ptr   = a_rcp.m_ptr;
24          m_count = a_rcp.m_count;
25          ++(*m_count);
26        }
27
28        inline ~RCP()
29        {
30          if( m_count != NULL && --(*m_count) == 0 ) {
31            delete [] m_ptr;
32            delete m_count;
33            m_ptr   = NULL;
34            m_count = NULL;
35          }
36        }
37
38        inline const RCP<T>& operator =(const RCP<T>& a_rcp)
39        {
40          if( m_ptr != a_rcp.m_ptr ) {
41            if( m_count != 0 && --(*m_count) == 0 ) {
42              delete [] m_ptr;
43              delete m_count;
44            }
45            m_ptr = a_rcp.m_ptr;
46            m_count = a_rcp.m_count;
47            if( m_count != NULL ) ++(*m_count);
48          }
49          return *this;
50        }
51
52        inline T& operator [] (int n)
53        {
54          return m_ptr[n];
55        }
56
57        const inline T& operator [] (int n) const
58        {
59          return m_ptr[n];
60        }
61
62      protected:
63        T*    m_ptr;
64        int*  m_count;
65
66    };
67    #endif
```

We need classes to define vectors and matrices, sometimes of integers, and other times for floating point real numbers. This utility file `Matrix.H` below provides that functionality. It uses the RCP class to enable dynamic resizing.

`file: Matrix.H`

```
1    #ifndef __Matrix_H__
2    #define __Matrix_H__
3    #include <iostream>
4    #include <assert.h>
5    #include "RCP.H"
6
7    template <class T> class Vector
8    {
9
10     public:
11
12       Vector()
13       {
14         m_size = m_alloc_size = 0;
15       }
16
17       Vector(int a_size)
18       {
19         m_size = m_alloc_size = 0;
20         resize( a_size );
21       }
22
23       ~Vector()
24       {
25       }
26
27       void resize( int a_size )
28       {
29         assert( a_size >=0 );
30
31         if( a_size == m_size ) return;
32
33         if( a_size < m_alloc_size ) {
34           m_size = a_size;
35           return;
36         }
37
38         // The following few lines are why we need the RCP class.
39         // if "T" has allocated memory, then "spare[]=m_data[]"
40         // copies pointers, but the subsequent delete of m_data
41         // will remove allocated memory. "spare" will end up with
42         // (formerly) valid pointers that point to freed memory,
43         // not allocated memory.
44
45         T* spare = new T[a_size];
46         for( int i=0; i<m_size; i++ )
47           spare[i] = m_data[i];
48         // RCP<> operator= will handle the deletion of any memory
49         // allocated to m_data (that is not referenced elsewhere).
50         m_data = RCP<T>(spare);
51
52         m_alloc_size = m_size = a_size;
53       }
54
55       const T& operator[] (int n) const
56       {
57         assert( n>=0 );
58         assert( n<m_size );
59         return m_data[n];
60       }
61
62       T& operator[] (int n)
63       {
64         assert( n>=0 );
65         assert( n<m_size );
66         return m_data[n];
67       }
68
69       int size() const
70       {
```

```
71          return m_size;
72        }
73
74        friend std::ostream& operator<< (std::ostream& os, const Vector<T>&v )
75        {
76          for( int i=0; i<v.m_size; i++ ) {
77            os << " " << v[i] << std::endl;
78          }
79          return os;
80        }
81
82        friend std::istream& operator>> (std::istream& is, Vector<T>&v )
83        {
84          for( int i=0; i<v.m_size; i++ ) {
85            is >> v[i];
86          }
87          return is;
88        }
89
90      private:
91
92      RCP<T> m_data;
93      int  m_size;         // declared size
94      int  m_alloc_size;  // allocated size
95    };
96
97    template <class T> class Matrix
98    {
99      public:
100
101        Matrix()
102        {
103          m_rows = m_columns = 0;
104        }
105
106        Matrix(int a_rows, int a_columns)
107        {
108          m_rows = m_columns = 0;
109          resize( a_rows, a_columns );
110        }
111
112        ~Matrix()
113        {
114        }
115
116        void resize( int a_rows, int a_columns )
117        {
118          assert( a_rows >= 0 );
119          assert( a_columns >= 0 );
120
121          // first, rows
122          if( a_rows != m_rows ) {
123            m_data.resize( a_rows );
124            for( int i=m_rows; i<a_rows; i++ ) {
125              m_data[i].resize( m_columns );
126              for( int j=0; j<m_columns; j++ ) m_data[i][j] = 0;
127            }
128          }
129          m_rows = a_rows;
130
131          // next, columns
132          if( a_columns != m_columns ) {
133            for( int i=0; i<m_rows; i++ ) {
134              m_data[i].resize( a_columns );
135            }
136          }
137          m_columns = a_columns;
138        }
139
140        const Vector<T>& operator[] (int n) const
```

```
141        {
142          assert( n >= 0 );
143          assert( n < m_rows );
144          return m_data[n];
145        }
146
147        Vector<T>& operator[] (int n)
148        {
149          assert( n >= 0 );
150          assert( n < m_rows );
151          return m_data[n];
152        }
153
154        int rows() const
155        {
156          return m_rows;
157        }
158
159        int columns() const
160        {
161          return m_columns;
162        }
163
164        friend std::ostream& operator<< (std::ostream& os, const Matrix<T>&m )
165        {
166          for( int i=0; i<m.m_rows; i++ ) {
167            for( int j=0; j<m.m_columns; j++ ) {
168              os << " " << m[i][j];
169            }
170            os << std::endl;
171          }
172          return os;
173        }
174
175        friend std::istream& operator>> (std::istream& is, Matrix<T>&m )
176        {
177          for( int i=0; i<m.m_rows; i++ ) {
178            is >> m[i];
179          }
180          return is;
181        }
182
183      private:
184
185      Vector<Vector<T> > m_data;
186      int m_rows;
187      int m_columns;
188    };
189    #endif
```

B.2   Gaussian elimination

The examples here convert a square matrix into its LR decomposition using different pivot strategies.

file: GE.H

```
1    #ifndef __GE_H__
2    #define __GE_H__
3    enum GEPivotType { TRIVIAL = 0, PARTIAL = 1, IMPLICITSCALEDPP = 2,
4                       ROOK = 3, COMPLETE = 4 };
5    int ge( Matrix<double>&A, Vector<int>& Pl, Vector<int>&Pr, GEPivotType type );
```

```
6   void solveLR(   const Matrix<double>& A,
7                   const Vector<int>& Pl,
8                   const Vector<int>& Pr,
9                   Vector<double>& b );
10  void printLRP( const Matrix<double>&A,
11                   const Vector<int>& Pl,
12                   const Vector<int>& Pr );
13
14  #endif
```

file: GE.cpp

```
1   #include <iostream>
2   #include <iomanip>
3   #include <math.h>
4   #include "Matrix.H"
5   #include "GE.H"
6   using namespace std;
7
8   // Perform Gaussian Elimination with pivoting:
9   //  — pivot type is specified by variable "type".
10  //  — permutations are represented by vectors of indices "PermL" and "PermR".
11  //  — matrix "A" is replaced by "LR" in place.
12  // Return 0 on success, −1 of there's a problem.
13
14  int ge( Matrix<double>& A,
15          Vector<int>& PermL,
16          Vector<int>& PermR,
17          GEPivotType type )
18  {
19    int n = A.rows();
20    int m = A.columns();
21    assert( n == m );
22
23    // initialize the permutation list
24    for( int i=0; i<n; i++ ) PermL[i] = PermR[i] = i;
25
26    Vector<double> rowscale;
27    if( type == IMPLICITSCALEDPP ) {
28      rowscale.resize(n);
29      for( int r=0; r<n; r++ ) {
30        double s = 0;
31        for( int c=0; c<n; c++ ) s += fabs(A[r][c]);
32        rowscale[r] = s;
33      }
34    }
35
36    for( int d=0; d<n−1; d++ ) {
37
38      int pivot_c, pivot_r;
39      double pval;
40
41      // find pivot
42      if( type == TRIVIAL ) {
43        // Trival pivoting:  find first nonzero element in column
44        pivot_c = d;
45        for( int r=d; r<n; r++ ) {
46          if( A[r][d] != 0 ) {
47            pivot_r = r;
48            pval = fabs(A[r][d]);
49            break;
50          }
51        }
52      } else if( type == PARTIAL ) {
53        // Partial pivoting:  find largest element in column
54        pivot_c = d;
55        pivot_r = d;
```

```
56        pval = fabs(A[d][d]);
57        for( int r=d+1; r<n; r++ ) {
58          double t = fabs(A[r][d]);
59          if( t > pval ) {
60            pval = t;
61            pivot_r = r;
62          }
63        }
64      } else if( type == IMPLICITSCALEDPP ) {
65        pivot_c = d;
66        pivot_r = d;
67        pval = fabs(A[d][d])/rowscale[d];
68        for( int r=d+1; r<n; r++ ) {
69          double t = fabs(A[r][d])/rowscale[r];
70          if( t > pval ) {
71            pval = t;
72            pivot_r = r;
73          }
74        }
75      } else if( type == ROOK ) {
76        // find first that is largest in row and column
77        int num_pass = 0;        // number of searches where max passes
78        bool rowcol  = true;    // begin search in row.
79        pivot_c = pivot_r = d;   // starting guess
80        int r=pivot_r , c=pivot_c;
81        pval = fabs(A[r][c]);
82        do {
83          bool change = false;
84          c = pivot_c;
85          r = pivot_r;
86          for( int i=d; i<n; i++ ) {
87            if( rowcol ) {
88              r = i;
89            } else {
90              c = i;
91            }
92            double t = fabs(A[r][c]);
93            if( t > pval ) {
94              pval = t;
95              pivot_r = r;
96              pivot_c = c;
97              change = true;
98            }
99          }
100         if( !change ) {
101           num_pass ++;
102         } else {
103           num_pass = 1;
104         }
105         rowcol = !rowcol;
106       } while( num_pass != 2 );
107     } else if( type == COMPLETE ) {
108       // find largest in all rows and columns.
109       pval = 0;
110       for( int r=d; r<n; r++ ) {
111         for( int c=d; c<n; c++ ) {
112           double t = fabs(A[r][c]);
113           if( t > pval ) {
114             pval = t;
115             pivot_c = c;
116             pivot_r = r;
117           }
118         }
119       }
120     } else {
121       cout << "unknown pivot type" << endl;
122       return -1;
123     }
124
125     if( pval <= 0 ) {
```

```
126          cout << "bad pivot" << endl;
127          return −1;
128        }
129
130        if( pivot_r != d ) {
131          // a row permutation occurred.  Do it.
132          int k = PermL[pivot_r]; PermL[pivot_r] = PermL[d]; PermL[d] = k;
133          for( int i=0; i<n; i++ ) {
134            double t = A[d][i];  A[d][i] = A[pivot_r][i]; A[pivot_r][i] = t;
135          }
136
137          if( type == IMPLICITSCALEDPP ) {
138            double t = rowscale[d];
139            rowscale[d] = rowscale[pivot_r];  rowscale[pivot_r] = t;
140          }
141        }
142
143        if( pivot_c != d ) {
144          // a column permutation occurred.  Do it.
145          int k = PermR[pivot_c]; PermR[pivot_c] = PermR[d]; PermR[d] = k;
146          for( int i=0; i<n; i++ ) {
147            double t = A[i][d]; A[i][d] = A[i][pivot_c]; A[i][pivot_c] = t;
148          }
149        }
150
151        // do the elimination:  everything above this was pivoting overhead.
152        for( int r=d+1; r<n; r++ ) {
153          double f = A[r][d]/A[d][d];
154          A[r][d] = f;
155          for( int c=d+1; c<n; c++ ) {
156            A[r][c] −= f*A[d][c];
157          }
158        }
159      }
160
161    return 0;
162  }
163
164  // solve Ax=b after "A" has been replaced by "LR" in place, with
165  // associated permutations stored in index lists "PermL" and "PermR".
166  // The result "x" is returned in the vector "b".
167
168  void solveLR( const Matrix<double>& A,
169                const Vector<int>& PermL,
170                const Vector<int>& PermR,
171                Vector<double>& b )
172  {
173    assert( A.columns() == b.size() );
174    int n = A.rows();
175    Vector<double> y(b.size());
176
177    // let y = Pleft b
178    for( int i=0; i<n; i++ ) y[i] = b[PermL[i]];
179
180    // solve Lb = y for b
181    for( int i=0; i<n; i++ ) {
182      b[i] = y[i];
183      for( int j=0; j<i; j++ )
184        b[i] −= A[i][j]*b[j];
185    }
186
187    // solve Ry = b for y
188    for( int i=n−1; i>=0; i−− ) {
189      y[i] = b[i];
190      for( int j=i+1; j<n; j++ ) y[i] −= A[i][j]*y[j];
191      y[i] /= A[i][i];
192    }
193
194    // let b = Pright y
195    for( int i=0; i<n; i++ ) b[i] = y[PermR[i]];
```

```
196   }
197
198   void printLRP( const Matrix<double>&A,
199                 const Vector<int>& PermL,
200                 const Vector<int>& PermR )
201   {
202     ios_base::fmtflags ff = cout.flags();
203     streamsize prec = cout.precision();
204
205     cout << "L:" << endl << scientific << setprecision(16) << showpos;
206     for( int r=0; r<A.rows(); r++ ) {
207       for( int c=0; c<r;  c++ )
208         cout << " " << A[r][c];
209       cout << " " << 1.0;
210       for( int c=r+1; c<A.columns();  c++ )
211         cout << " " << 0.0;
212       cout << endl;
213     }
214
215     cout << "R:" << endl;
216     for( int r=0; r<A.rows(); r++ ) {
217       for( int c=0; c<r;  c++ )
218         cout << " " << 0.0;
219       for( int c=r; c<A.columns();  c++ )
220         cout << " " << A[r][c];
221       cout << endl;
222     }
223
224     cout << "PL:" << endl << noshowpos;
225     for( int r=0; r<A.rows(); r++ ) {
226       int rr = PermL[r];
227       for( int c=0; c<A.columns(); c++ )
228         cout << " " << ((c==rr)?1:0);
229       cout << endl;
230     }
231
232     cout << "PR:" << endl << noshowpos;
233     for( int r=0; r<A.rows(); r++ ) {
234       int rr = PermR[r];
235       for( int c=0; c<A.columns(); c++ )
236         cout << " " << ((c==rr)?1:0);
237       cout << endl;
238     }
239
240     cout.flags(ff);
241     cout << setprecision(prec);
242   }
```

---

file: GE-test.cpp

```
1    #include <iostream>
2    #include <iomanip>
3    #include <math.h>
4    #include "Matrix.H"
5    #include "GE.H"
6    using namespace std;
7
8    int main()
9    {
10     int code, n = 3;
11     Matrix<double> A(n,n);
12     Vector<double> b(n);
13     Vector<int>    pl(n), pr(n);
14
15     for( int k=0; k<5; k++ ) {
16
17       A[0][0] = 0.03;   A[0][1] = 300.;   A[0][2] = 8e7;
```

```
18        A[1][0] = −0.995; A[1][1] = −140000; A[1][2] = 1.6e7;
19        A[2][0] = 0.01;   A[2][1] = −19000;  A[2][2] = 1e6;
20
21        b[0] = A[0][0]+A[0][1]+A[0][2];
22        b[1] = A[1][0]+A[1][1]+A[1][2];
23        b[2] = A[2][0]+A[2][1]+A[2][2];
24
25        cout << "A" << endl << A << endl;
26        cout << "b" << endl << b << endl;
27
28        switch(k) {
29        case 0: cout << "trivial pivoting" << endl;
30                code = ge( A,pl,pr, TRIVIAL);
31                break;
32        case 1: cout << "partial pivoting" << endl;
33                code = ge( A,pl,pr, PARTIAL);
34                break;
35        case 2: cout << "implicit scaled pivoting" << endl;
36                code = ge( A,pl,pr, IMPLICITSCALEDPP);
37                break;
38        case 3: cout << "rook pivoting" << endl;
39                code = ge( A,pl,pr, ROOK );
40                break;
41        case 4: cout << "complete pivoting" << endl;
42                code = ge( A,pl,pr, COMPLETE );
43                break;
44        }
45        if( code != 0 ) {
46           cout << "error" << endl;
47        } else {
48           printLRP(A,pl,pr);
49           solveLR( A, pl, pr, b );
50           cout << "x= " << scientific  << setprecision(16) << endl << b << endl;
51        }
52     }
53  }
```

B.3   Householder reduction

These examples implement the Householder QR reduction of a matrix, possibly with
more rows than columns. Different pivoting strategies are implemented: nothing, Golub
[90], Powell-Reid [186], and Björck [17]. Björck's method sorts rows prior to decomposition. This is implemented by first creating a list of norms. This list is sorted in descending order, while simultaneously composing the permutation index, using a variation of
Hoare's Quicksort algorithm [109, 110]. The rows of **A** are then sorted according to the
permutation list with Boothroyd's algorithm [20].

file: Householder.H

```
1   #ifndef __Householder_H__
2   #define __Householder_H__
3
4   enum QRPivotType { NOTHING=0, GOLUB=1, POWELLREID=2, BJORCK=3 };
5
6   int Householder( Matrix<double>& A, Vector<int>& Pl, Vector<int>& Pr,
7                    Vector<double>& diag, QRPivotType type );
8   void QTb( const Matrix<double>&A, Vector<double>& b );
9   double solveQR( const Matrix<double>& A, const Vector<int>& Pl, const Vector<int>& Pr,
10                  const Vector<double>& diag, Vector<double>& b );
```

```
11    void printQR( const Matrix<double>&A , const Vector<int>& Pl, const Vector<int>& Pr,
12                  const Vector<double>& diag );
13    #endif
```

file: Householder.cpp

```
1     #include <iostream>
2     #include <iomanip>
3     #include <math.h>
4     #include "Matrix.H"
5     #include "Householder.H"
6     #include "HQsort.H"
7     using namespace std;
8
9     #ifndef MAX
10    #define MAX(A,B) (((A)>(B))?(A):(B))
11    #endif
12
13    // Replace A with composite of R (without diagonal)
14    // with columns w stored on and below diagonal.
15    // Diagonal of R goes into "diag".
16    // Return 0 on success, −1 if there's a problem (singular)
17
18    int Householder( Matrix<double>& A, Vector<int>& Pl, Vector<int>& Pr,
19                     Vector<double>& diag, QRPivotType type )
20    {
21      int n_eqn = A.rows();
22      int n_var = A.columns();
23      assert( n_eqn >= n_var );
24      assert( diag.size() >= n_var );
25
26      int    i, j, l, s=0;
27      double K, len, dot;
28
29      // initialize the permutation
30      for( int i=0; i<n_eqn; i++ ) Pl[i] = i;
31      for( int i=0; i<n_var; i++ ) Pr[i] = i;
32
33      // if Bjorck pivoting, first sort so that row inf−norm is ordered
34      if( type == BJORCK ) {
35        Vector<double> norm(n_eqn);
36        for( int i=0; i<n_eqn; i++ ) {
37          double m = fabs(A[i][0]);
38          for( int j=1; j<n_var; j++ ) m = MAX(m,fabs(A[i][j]));
39          norm[i] = m;
40        }
41
42        // sort with customized Quicksort.
43        hqsort( norm, Pl, 0, n_eqn−1 );
44
45        // rearrange rows
46        for( int i=n_eqn−1; i>=0; i−− ) {
47          int k = Pl[i];
48          while( k > i ) k = Pl[k];
49          if( k != i ) {
50            for( int j=0; j<n_var; j++ ) {
51              double d=A[k][j]; A[k][j]=A[i][j]; A[i][j]=d;
52            }
53          }
54        }
55      }
56
57      // for each column of A
58      for( j=0; j<n_var; j++ ) {
59
60        if( type != NOTHING ) {
61          // perform column pivoting
```

```
62          int bestcol   = j-1;
63          double bestval = -1;
64          for( int jj=j; jj<n_var; jj++ ) {
65            double val = 0;
66            for( int k=j; k<n_eqn; k++ ) val += A[k][jj]*A[k][jj];
67            if( val > bestval ) {
68              bestval = val;
69              bestcol = jj;
70            }
71          }
72          if( bestcol != j ) {
73            // column pivot
74            int p = Pr[j]; Pr[j]=Pr[bestcol]; Pr[bestcol]=p;
75            for( int kk=0; kk<n_eqn; kk++ ) {
76              double d = A[kk][j]; A[kk][j]=A[kk][bestcol]; A[kk][bestcol]=d;
77            }
78          }
79        }
80
81        if( type == POWELLREID ) {
82          // perform row pivoting
83          int bestrow = j-1;
84          double bestval = -1;
85          for( int jj=j; jj<n_eqn; jj++ ) {
86            double val = fabs(A[jj][j]);
87            if( val > bestval ) {
88              bestval = val;
89              bestrow = jj;
90            }
91          }
92          if( bestrow != j ) {
93            // row pivot
94            int p = Pl[j]; Pl[j]=Pl[bestrow]; Pl[bestrow]=p;
95            for( int kk=0; kk<n_var; kk++ ) {
96              double d = A[j][kk]; A[j][kk]=A[bestrow][kk]; A[bestrow][kk]=d;
97            }
98          }
99        }
100
101       // here begins the reduction: everything above this is pivoting overhead.
102       // find the length squared of the vector we want to transform into (K e_1)
103       // omit the first entry ...
104       len = 0.0;
105       for( i=j+1; i<n_eqn; i++ )
106         len += A[i][j]*A[i][j];
107
108       // make K --- sign for numerical stability
109       K = sqrt( len + A[j][j]*A[j][j] );
110       if( A[j][j] > 0 ) K = -K;
111
112       // K is the new diagonal entry
113       diag[j] = K;
114       if( !diag[j] ) s= -1; // singular
115
116       // replace the column of A with w = a - K e_1
117       // not yet normalized.
118       A[j][j] -= K;
119
120       // now update len to be length of w
121       len = sqrt( len + A[j][j]*A[j][j] );
122       if( len <= 0 ) {
123         // this means that the corresponding part of Q is part of I,
124         // w is zero.  diag is zero.
125         continue;
126       }
127
128       // normalize so |w|_2 =1
129       for( i=j; i<n_eqn; i++ )
130         A[i][j] /= len;
131
```

```
132        // now A[i][j], i=j to n_eqn is normalized w
133
134        // use w on subsequent columns of A
135        // a := a - 2 w (w.a)
136        for( l=j+1; l<n_var; l++ ) {
137
138          // make 2(w.a)
139          dot = 0.0;
140          for( i=j; i<n_eqn; i++ )
141            dot += A[i][l]*A[i][j];
142          dot *= 2;
143
144          // a := a - dot w
145          for( i=j; i<n_eqn; i++ )
146            A[i][l] -= dot*A[i][j];
147        }
148
149      }
150      return s;
151  }
152
153  // Replace b with Q^T b
154  // Everything needed to make Q is stored in w form in A.
155
156  void QTb( const Matrix<double>&A, Vector<double>& b )
157  {
158      int n_var = A.columns();
159      int n_eqn = A.rows();
160      int   i, j;
161      double dot;
162
163      // for each w
164      for( j=0; j<n_var; j++ ) {
165
166        // b := b - 2 w (w.b)
167        // first make 2*(w.b)
168        dot = 0;
169        for( i=j; i<n_eqn; i++ )
170          dot += A[i][j]*b[i];
171        dot *= 2;
172
173        // b := b - w*dot
174        for( i=j; i<n_eqn; i++ )
175          b[i] -= dot*A[i][j];
176      }
177  }
178
179  // replace b  with x, solving Ax=b after  A has been converted
180  // to QR, with diagonals of R in "diag".  Return L2 norm of residual
181  // (for overdetermined systems only).
182
183  double solveQR( const Matrix<double>& A, const Vector<int>& Pl, const Vector<int>& Pr,
184                  const Vector<double>& diag, Vector<double>& b )
185  {
186      int n_var = A.columns();
187      int n_eqn = A.rows();
188      int   i, j;
189
190      // first, make u = Q^T b
191      Vector<double> u(n_eqn);
192      for( j=0; j<n_eqn; j++ ) u[j] = b[Pl[j]];
193
194      // replace u with QT u
195      QTb( A, u );
196
197      double residual = 0;
198      for( int i=n_var; i<n_eqn; i++ )
199        residual += u[i]*u[i];
200      residual = sqrt(residual);
201
```

```
202    // now solve Rx=u (save x in b):  back substitution
203
204    for( i=n_var-1; i>=0; i-- ) {
205      b[i] = u[i];
206      for( j=i+1; j<n_var; j++ )
207        b[i] -= A[i][j]*b[j];
208      b[i] /= diag[i];
209    }
210
211    // and permute
212    for( i=n_var-1; i>=0; i-- ) {
213      int k = Pr[i];
214      while( k < i ) k = Pr[k];
215      double d = b[k]; b[k] = b[i]; b[i] = d;
216    }
217
218    return residual;
219  }
220
221  // print the QR form of the matrix in readable form.
222
223  void printQR( const Matrix<double>&A , const Vector<int>&Pl,
224                const Vector<int>&Pr, const Vector<double>& diag )
225  {
226    int n_eqn = A.rows();
227    ios_base::fmtflags ff = cout.flags();
228    streamsize prec = cout.precision();
229    cout << scientific << setprecision(16) << endl;
230
231    cout << "Pl: " << endl << noshowpos << Pl << showpos << endl;
232
233    cout << "Q:" << endl;
234    Vector<double> Qrow(n_eqn);
235    for( int i=0; i<n_eqn; i++ ) {
236      for( int j=0; j<n_eqn; j++ ) Qrow[j] = 0;
237      Qrow[i] = 1;
238      QTb(A,Qrow);
239      for( int j=0; j<n_eqn; j++ )
240        cout << " " << Qrow[j];
241      cout << endl;
242    }
243
244    cout << "R:" << endl;
245    for( int r=0; r<A.rows(); r++ ) {
246      int rmax = A.columns();
247      rmax = (rmax > r) ? r : rmax;
248      for( int c=0; c<rmax; c++ )
249        cout << " " << 0.0;
250      for( int c=rmax; c<A.columns(); c++ )
251        cout << " " << ((c==r) ? diag[c] : A[r][c]);
252      cout << endl;
253    }
254
255    cout << "Pr: " << endl << noshowpos << Pr << endl;
256
257    cout.flags(ff);
258    cout << setprecision(prec);
259  }
```

---

file: HQsort.H

```
1  #ifndef __HQ_SORT_H__
2  #define __HQ_SORT_H__
3
4  // norm,perm [0 to n-1] inclusive
5  // hqsort( norm, perm, 0, n-1 ) will place elements of norm
6  // in descending order.
```

```
 7
 8   void hqsort( Vector<double>& norm, Vector<int>& perm, int left, int right );
 9
10   // a simpler entry point with initialization
11
12   void hqsort0( Vector<double>& norm, Vector<int>& perm );
13
14   #endif
```

file: HQsort.cpp

```
 1   #include "Matrix.H"
 2
 3   void hqsort( Vector<double>& norm, Vector<int>& perm, int left, int right )
 4   {
 5     if( left < right ) {
 6       int    pivot = (left+right)/2;
 7       double pval  = norm[pivot];
 8       int lo=left;
 9       int hi=right;
10       while( lo <= hi ) {
11         while( norm[lo] > pval ) lo++;
12         while( norm[hi] < pval ) hi--;
13         if( lo <= hi ) {
14           double d=norm[lo]; norm[lo]=norm[hi]; norm[hi]=d;
15           int    p=perm[lo]; perm[lo]=perm[hi]; perm[hi]=p;
16           lo++;
17           hi--;
18         }
19       };
20       if( left < hi  ) hqsort( norm, perm, left, hi  );
21       if( lo < right ) hqsort( norm, perm, lo, right );
22     }
23   }
24
25   void hqsort0( Vector<double>& norm, Vector<int>& perm )
26   {
27     int sn = norm.size();
28     int sp = perm.size();
29     for( int i=0; i<sp; i++ ) perm[i] = i;
30     hqsort( norm, perm, 0, sn-1 );
31   }
```

file: Householder-test.cpp

```
 1   #include <iostream>
 2   #include <iomanip>
 3   #include <stdlib.h>
 4   #include <math.h>
 5   #include "Matrix.H"
 6   #include "Householder.H"
 7   using namespace std;
 8
 9   int main()
10   {
11     cout << scientific << setprecision(16) << showpos;
12
13     Matrix<double> A(3,3);
14     Vector<double> b(3);
15     Vector<double> diag(3);
16     Vector<int> Pl(3), Pr(3);
17     double d = 1.e-10;
18     A[0][0]=   2; A[0][1]=  -1; A[0][2]=   1;   b[0]=  2*(1.+d);
```

```
19    A[1][0]=   -1; A[1][1]=   d; A[1][2]=   d; b[1]=   -d;
20    A[2][0]=    1; A[2][1]=   d; A[2][2]=   d; b[2]=    d;
21
22    int ret = Householder( A, Pl, Pr, diag, NOTHING );
23    if( ret != 0 ) {
24      cout << "Householder failed" << endl;
25      exit(0);
26    }
27
28    printQR( A, Pl, Pr, diag );
29
30    for( int i=0; i<A.columns(); i++ ) {
31      cout << " column " << noshowpos << " k= " << showpos << diag[i] << endl;
32      for( int j=i; j<A.rows(); j++ ) {
33        cout << " w" << noshowpos << j << " " << showpos << A[j][i] << endl;
34      }
35    }
36    cout << "solve QRx = b for x:" << endl;
37    solveQR( A, Pl, Pr, diag, b );
38    for( int i=0; i<A.columns(); i++ )
39      cout << "x" << noshowpos << i << " " << showpos << b[i] << endl;
40  }
```

B.4   Cholesky reduction

These examples implement the Cholesky LL^T reduction of a real symmetric positive definite matrix.

file: Cholesky.H

```
1    #ifndef __Cholesky_H__
2    #define __Cholesky_H__
3    int Cholesky( Matrix<double>& A );
4    void solveLLH( const Matrix<double>& A, Vector<double>& b );
5    void printLLH( const Matrix<double>&A );
6    #endif
```

file: Cholesky.cpp

```
1    #include <iostream>
2    #include <iomanip>
3    #include <math.h>
4    #include "Matrix.H"
5    using namespace std;
6
7    // Replace Matrix "A" with L of L LH decomposition.
8
9    int Cholesky( Matrix<double>& A )
10   {
11     int n = A.rows();
12     int m = A.columns();
13     assert( n == m );
14
15     for( int i=0; i<n; i++ ) {
16       for( int j=i; j<n; j++ ) {
17         double p;
18         double x = A[i][j];
19         for( int k=i-1; k>=0; k-- )
```

```
20          x -= A[j][k]*A[i][k];
21        if( i==j ) {
22          if( x <= 0 ) return -1;
23          A[i][j] = sqrt(x);
24          p = 1./sqrt(x);
25        } else {
26          A[j][i] = x*p;
27        }
28      }
29    }
30    return 0;
31  }
32
33  // solve Ax=b after "A" has been replaced by "L" in place.
34  // The resut "x" is returned in the vector "b".
35
36  void solveLLH( const Matrix<double>& A, Vector<double>& b )
37  {
38    assert( A.columns() == b.size() );
39    int n = A.rows();
40    Vector<double> y(b.size());
41
42    // solve Ly = b for y
43    for( int i=0; i<n; i++ ) {
44      y[i] = b[i];
45      for( int j=0; j<i; j++ ) y[i] -= A[i][j]*y[j];
46      y[i] /= A[i][i];
47    }
48
49    // solve Rx = y for x -- store result in b
50    for( int i=n-1; i>=0; i-- ) {
51      b[i] = y[i];
52      for( int j=i+1; j<n; j++ ) b[i] -= A[j][i]*b[j];
53      b[i] /= A[i][i];
54    }
55  }
56
57  void printLLH( const Matrix<double>&A )
58  {
59    ios_base::fmtflags ff = cout.flags();
60    streamsize prec = cout.precision();
61
62    cout << "L:" << endl << scientific << setprecision(16) << showpos;
63    for( int r=0; r<A.rows(); r++ ) {
64      for( int c=0; c<=r; c++ )
65        cout << " " << A[r][c];
66      for( int c=r+1; c<A.columns(); c++ )
67        cout << " " << 0.0;
68      cout << endl;
69    }
70
71    cout << "LH:" << endl;
72    for( int r=0; r<A.rows(); r++ ) {
73      for( int c=0; c<r; c++ )
74        cout << " " << 0.0;
75      for( int c=r; c<A.columns(); c++ )
76        cout << " " << A[c][r];
77      cout << endl;
78    }
79
80    cout.flags(ff);
81    cout << setprecision(prec);
82  }
```

file: Cholesky-test.cpp

```
1  #include <iostream>
```

```
2   #include <iomanip>
3   #include <math.h>
4   #include "Matrix.H"
5   #include "Cholesky.H"
6   using namespace std;
7
8   int main()
9   {
10    cout << scientific << setprecision(4) << showpos;
11
12    Matrix<double> A(4,4);
13    Vector<double> b(4);
14    A[0][0] = 4;
15    A[0][1] = A[1][0] = -1;
16    A[0][2] = A[2][0] = 0;
17    A[0][3] = A[3][0] = 0;
18    A[1][1] = 4;
19    A[1][2] = A[2][1] = -1;
20    A[1][3] = A[3][1] = 0;
21    A[2][2] = 4;
22    A[2][3] = A[3][2] = -1;
23    A[3][3] = 4;
24    b[0] = 1;
25    b[1] = 2;
26    b[2] = 2;
27    b[3] = 1;
28
29    int ret = Cholesky( A );
30    if( ret ) {
31      cout << "Cholesky failed." << endl;
32      return -1;
33    }
34    printLLH(A);
35    solveLLH( A, b );
36    for( int i=0; i<b.size(); i++ )
37      cout << "x" << noshowpos << i << " " << showpos << b[i] << endl;
38
39    return 0;
40  }
```

B.5   The QR method with shifts for symmetric real matrices

file: Givens.H

```
1   #ifndef __Givens__H_
2   #define __Givens__H_
3   #include <math.h>
4
5   // find the coefficients of the Givens reflection matrix
6   // to eliminate y2:
7   // (c   s)(y1) = (*)
8   // (s  -c)(y2)   (0)
9
10  inline void Givens( const double& y1, const double& y2,
11                      double& c, double& s, double& v )
12  {
13    double mu = fabs(y1);
14    double tmp = fabs(y2);
15    if( tmp > mu ) mu = tmp;
16
17    if( mu == 0 ) {
18      c = 1;
```

```
19        s = 0;
20        v = 0;
21        return;
22    }
23
24    double s1 = y1/mu;
25    double s2 = y2/mu;
26    double p = mu*sqrt(s1*s1 + s2*s2);
27    if( y1 < 0 ) p = -p;
28    c = y1/p;
29    s = y2/p;
30    v = s/(1+c);
31  }
32  #endif
```

---

file: QR.H

```
1  #ifndef __QR_eig_H__
2  #define __QR_eig_H__
3  int QReig( Matrix<double>& A, Matrix<double>& P, Vector<double>& lambda );
4  #endif
```

---

file: QR.cpp

```
1   #include <math.h>
2   #include "Matrix.H"
3   #include "Givens.H"
4   #include <assert.h>
5   #include <iostream>
6
7   // input:  a symmetric real "A", and "P" is identity matrix.
8   // output: "A" is tridiagonal, and "P" is the similarity transform.
9
10  void makeTridiag(   Matrix<double>& A,
11                      Matrix<double>& P  )
12  {
13    int N = A.rows();
14
15    // first step — make tridiagonal.  Do N-2 Householder steps
16    Vector<double> w(N);
17    for( int c=0; c<N-2; c++ ) {
18
19      // a_i starts 1 below diagonal.
20
21      double len = 0;
22      for( int r=c+2; r<N; r++ ) len += A[r][c]*A[r][c];
23
24      if( len != 0 ) {
25        double K = sqrt( len + A[c+1][c]*A[c+1][c] );
26        if( A[c+1][c] > 0 ) K = -K;
27
28        // make W explicitly
29        for( int r=c+1; r<N; r++ ) w[r] = A[r][c];
30        w[c+1] -= K;
31        len = sqrt( len + w[c+1]*w[c+1] );
32
33        for( int r=c+1; r<N; r++ ) w[r] /= len;
34
35        // now process "A" — convert column "c" directly, and others numerically
36        A[c+1][c] = K;
37        tor( int r=c+2; r<N; r++ ) A[r][c] = 0;
38
39        for( int c2=c+1; c2<N; c2++ ) {
```

```
40            // d = w . a
41            double d=0;
42            for( int r=c+1; r<N; r++ ) d += w[r]*A[r][c2];
43            // a -= 2*d*w
44            d *= 2;
45            for( int r=c+1; r<N; r++ ) {
46              A[r][c2] -= d*w[r];
47            }
48          }
49
50          // Next operate from right — similarity transformation.
51          // same steps as above.
52
53          A[c][c+1] = K;
54          for( int r=c+2; r<N; r++ ) A[c][r] = 0;
55
56          for( int c2=c+1; c2<N; c2++ ) {
57            double d = 0;
58            for( int r=c+1; r<N; r++ ) d += w[r]*A[c2][r];
59            d *= 2;
60            for( int r=c+1; r<N; r++ ) {
61              A[c2][r] -= d*w[r];
62            }
63          }
64
65          // and change P.
66          // at start I A I, then introduce a Q
67          //           (I Q) Q A Q (Q I)
68          // so P := P Q from right
69
70          for( int c2=0; c2<N; c2++ ) {
71            double d = 0;
72            for( int r=c+1; r<N; r++ ) d += w[r]*P[c2][r];
73            for( int r=c+1; r<N; r++ ) P[c2][r] -= 2.*d*w[r];
74          }
75        }
76      }
77
78      // symmetrize
79      for( int i=0; i<N; i++ ) {
80        for( int j=0; j<i; j++ )
81          A[i][j] = A[j][i] = 0.5*( A[i][j] + A[j][i] );
82      }
83    }
84
85    // Perform one QR step assuming a symmetric tridiagonal matrix A.
86    // The "A" information is represented by a vector of diagonal
87    // values, and a vector of subdiagonal values.
88    // "P" collects the Q's and therefore becomes the eigenvectors.
89
90    // return "1" if last element of diag is an eigenvalue.
91
92    int oneQRstep( Vector<double>& diag,
93                   Vector<double>& subdiag,
94                   Matrix<double>& P,
95                   int             N )
96    {
97      int NP = P.rows();
98      double machine = 1e-16;
99
100     if( N == 1 ) return 1;  // A is a scalar
101
102     // check for reducibility: z will be the starting index
103     // for a reduced block.
104     int z = N-1;
105     do {
106       if( z == 0 ) break;
107       z--;
108       double sum = fabs(diag[z]) + fabs(diag[z+1]);
109       if( fabs(subdiag[z]) <= machine*sum ) {
```

```
110        // the matrix is reducible. set z to be the
111        // "zero" position of last partition.
112        ++z;
113        break;
114      }
115    } while(1);
116
117    // if partition contains only 1 diagonal element,
118    // no work needs to be done.
119    if( z >= N-1 ) return 1;
120
121    double shift = 0;
122
123    double qa = 1;
124    double qb = -(diag[N-1]+diag[N-2]);
125    double qc =  diag[N-1]*diag[N-2]-subdiag[N-2]*subdiag[N-2];
126    double qs = ((qb<0)?-1:1);
127
128    double mu1 = -qs*( fabs(qb)+sqrt(qb*qb-4*qa*qc) )/( 2 * qa );
129    double mu2 = -qs*( 2 * qc )/( fabs(qb)+sqrt(qb*qb-4*qa*qc) );
130    shift = mu2;
131    if( fabs(diag[N-1]-mu1) < fabs(diag[N-1]-mu2) ) shift = mu1;
132
133    // find first Givens factors to eliminate A[z+1][z] using A[z][z]
134    double c,s,v;
135    Givens( diag[z]-shift, subdiag[z], c, s, v );
136
137    // apply this Givens from left, AND right. the elements
138    // affected are   0,0   0,1   0,2
139    //                1,0   1,1   1,2
140    //                2,0   2,1
141    // or diag[z], subdiag[z], diag[z+1], subdiag[z+1], and
142    // "spare" = A[z+2][z] which lies outside tridiagonal form.
143
144    double di = diag[z];
145    double dip1 = diag[z+1];
146    double sdi = subdiag[z];
147    double sdip1 =0;
148    if( N-z > 2 ) sdip1 = subdiag[z+1];
149    double dip2;
150    double sdip2 =0;
151
152    diag[z  ] = c*(c*(di-dip1) + 2*s*sdi) + dip1;
153    diag[z+1] = c*(c*(dip1-di) - 2*s*sdi) + di;
154    subdiag[z] = c*(s*(di-dip1) - 2*c*sdi) + sdi;
155    if( N-z > 2 ) subdiag[z+1] = -c*sdip1;
156    double spare = s*sdip1;
157
158    // and apply the Givens reflection to the eigenvectors;
159    for( int j=0; j<NP; j++ ) {
160      double u1 = P[j][z  ];
161      double u2 = P[j][z+1];
162      P[j][z  ] = c*u1 + s*u2;
163      P[j][z+1] = v*(u1 + P[j][z]) - u2;
164    }
165
166    // select next Givens to eliminate this "spare" value,
167    // but operation on right makes a new spare ...
168
169    for( int i=z; i<N-2; i++ ) {
170
171      Givens( subdiag[i], spare, c, s, v );
172
173      dip1 = diag[i+1];
174      dip2 = diag[i+2];
175      sdi = subdiag[i];
176      sdip1 = subdiag[i+1];
177      if( i != N-3 ) sdip2 = subdiag[i+2];
178
179      diag[i+1] = c*(c*(dip1-dip2) + 2*s*sdip1) + dip2;
```

```
180        diag[i+2] = c*(c*(dip2-dip1) - 2*s*sdip1) + dip1;
181        subdiag[i   ] = s*spare + c*sdi;
182        subdiag[i+1] = c*(s*(dip1-dip2) - 2*c*sdip1) + sdip1;
183        if( i != N-3 ) {
184           subdiag[i+2] = -c*sdip2;
185           spare = s*sdip2;
186        }
187
188        // and apply the Givens reflection to the eigenvectors;
189        for( int j=0; j<NP; j++ ) {
190           double u1 = P[j][i+1];
191           double u2 = P[j][i+2];
192           P[j][i+1] = c*u1 + s*u2;
193           P[j][i+2] = v*(u1 + P[j][i+1]) - u2;
194        }
195     }
196
197     return 0;
198 }
199
200 // On input, A is a symmetric real matrix, and P is the identity matrix.
201 // On output, P is the matrix of eigenvectors and lambda is the vector
202 // of corresponding eigenvalues.  "A" is modified to tridiagonal form.
203 // return 0 on success
204
205 int QReig( Matrix<double>& A,
206            Matrix<double>& P,
207            Vector<double>& lambda )
208 {
209     int N=A.rows();
210     assert( A.columns() == N );
211     assert( P.rows() == N );
212     assert( P.columns() == N );
213     assert( lambda.size() >= N );
214
215     for( int i=0; i<N; i++ ) {
216        for( int j=0; j<N; j++ ) P[i][j] = 0;
217        P[i][i] = 1;
218     }
219
220     // make tridiagnonal
221
222     makeTridiag( A, P );
223
224     // make a more compact data representation:  save only the diagnonal
225     // ans subdiagonal (it has been assumed that A was symmetric).
226
227     Vector<double> subdiag(N-1);
228
229     for( int i=0; i<N;   i++ ) lambda[i]   = A[i][i];
230     for( int i=0; i<N-1; i++ ) subdiag[i] = A[i+1][i];
231
232     int curN = N;
233
234     // limit on loop prevents accidentally going forever
235     for( int step=0; step<10*N; step++ ) {
236
237        int got = oneQRstep( lambda, subdiag, P, curN );
238
239        if( got ) {
240           curN--;
241           if( curN == 0 ) return 0;
242        }
243     }
244
245     // failure to converge ...
246     return 1;
247 }
```

file: QR-test.cpp

```
1   #include <iostream>
2   #include <sstream>
3   #include <iomanip>
4   #include <math.h>
5   #include "Matrix.H"
6   #include "QR.H"
7   using namespace std;
8
9   int main()
10  {
11     cout << scientific << setprecision(4) << showpos;
12     int N=6;
13     Matrix<double> A(N,N), P(N,N);
14     Vector<double> lambda(N);
15
16     stringstream ss( stringstream::in | stringstream::out );
17     ss << "2 -1 0 0 0 0   -1 2 -1 0 0 0 \
18          0 -1 2 -1 0 0   0 0 -1 2 -1 0 \
19          0 0 0 -1 2 -1   0 0 0 0 -1 2";
20     ss >> A;
21
22     cout << "input  A"      << endl << A      << endl;
23
24     QReig( A, P, lambda );
25
26     cout << "tridiagonal A" << endl << A      << endl;
27     cout << "eigenvalues"   << endl << lambda << endl;
28     cout << "eigenvectors"  << endl << P      << endl;
29  }
```

B.6   Singular value decomposition

This SVD example assumes that $m \times n$ matrix **A** has $m \geq n$. The singular value of a matrix with more columns than rows can be made by first transposing the matrix, then transposing the result.

file: SVD.H

```
1   #ifndef __SVD_H__
2   #define __SVD_H__
3   int SVD( Matrix<double>& A,
4            Matrix<double>& U,
5            Matrix<double>& VT,
6            Vector<double>& sigma );
7   #endif
```

file: SVD.cpp

```
1   #include <math.h>
2   #include "Matrix.H"
3   #include "Givens.H"
4
5   #ifndef MAX
```

```
 6   #define MAX(A,B)  (((A)>(B))?(A):(B))
 7   #endif
 8
 9   // A = U D VT
10   //    A is m by n
11   //    U is m by m
12   //    VT is n by n
13
14   //    diag is n
15   //    supdiag is n-1
16
17   void makeBiDiagonal(
18     Matrix<double>&A,
19     Matrix<double>&U,
20     Matrix<double>&VT,
21     Vector<double>&diag ,
22     Vector<double>&supdiag
23   )
24   {
25     int m = A.rows();
26     int n = A.columns();
27     assert( m >= n );
28     Vector<double> w(m);
29     int i,j,l;
30     double len, dot, K;
31
32     // begin with U and VT identity matrices
33     for( i=0; i<m; i++ ) {
34       for( j=0; j<m; j++ ) U[i][j] = 0;
35       U[i][i] = 1;
36     }
37     for( i=0; i<n; i++ ) {
38       for( j=0; j<n; j++ ) VT[i][j] = 0;
39       VT[i][i] = 1;
40     }
41
42     for( j=0; j<n; j++ ) {
43
44       // Householder step to replace column j with k e_1
45
46       len = 0;
47       w[j] = A[j][j];
48       for( i=j+1; i<m; i++ ) {
49         w[i] = A[i][j];
50         len += w[i]*w[i];
51       }
52
53       if( len == 0 ) {
54         // already in correct form — just change representation
55         diag[j] = A[j][j];
56       } else {
57         // if zero, no need to do anything
58
59         K = sqrt( len + w[j]*w[j] );
60         if( w[j] > 0 ) K = -K;
61
62         // K is new diagonal entry.
63         diag[j] = K;
64
65         // Replace column of A with w
66         w[j] -= K;
67
68         len = 1./sqrt(len + w[j]*w[j]);
69         for( i=j; i<m; i++ ) {
70           w[i] *= len;
71         }
72
73         // use w on subsequent columns of A
74         // a := a - 2 w (w.a)  with "a" a column vector
75         for( l=j+1; l<n; l++ ) {
```

```
76            dot = 0.0;
77            for( i=j; i<m; i++ )
78                dot += A[i][l]*w[i];
79            dot *= 2;
80            for( i=j; i<m; i++ )
81                A[i][l] -= dot*w[i];
82         }
83
84         // use w on U, operating from right
85         // u := u - 2 (u.w) w^T with "u" a row vector
86         for( l=0; l<m; l++ ) {
87            dot = 0.0;
88            for( i=j; i<m; i++ )
89                dot += U[l][i]*w[i];
90            dot *= 2;
91            for( i=j; i<m; i++ )
92                U[l][i] -= dot*w[i];
93         }
94      }
95
96      // Householder step to replace row i with k e_1 (shifted)
97
98      if( j+1 < n ) {
99
100        len = 0;
101        w[j+1] = A[j][j+1];
102        for( i=j+2; i<n; i++ ) {
103           w[i] = A[j][i];
104           len += w[i]*w[i];
105        }
106
107        if( len == 0 ) {
108           // already in correct form — just change representation
109           supdiag[j] = A[j][j+1];
110        } else {
111           K = sqrt( len + w[j+1]*w[j+1]);
112           if( w[j+1] > 0 ) K = -K;
113           supdiag[j] = K;
114
115           // Replace row of A with w
116           w[j+1] -= K;
117           len = 1./sqrt(len + w[j+1]*w[j+1]);
118
119           for( i=j+1; i<n; i++ )
120              w[i] *= len;
121
122           // use w on subsequent rows of A
123           // a := a - 2 (a.w) w^T with "a" a row vector
124           for( l=j+1; l<m; l++ ) {
125              dot = 0.0;
126              for( i=j+1; i<n; i++ )
127                 dot += A[l][i]*w[i];
128              dot *= 2;
129              for( i=j+1; i<n; i++ )
130                 A[l][i] -= dot*w[i];
131           }
132
133           // use w on VT, operating from left
134           // v := v - 2 w (w.v) with "v" a column vector
135           for( l=0; l<n; l++ ) {
136              dot = 0.0;
137              for( i=j+1; i<n; i++ )
138                 dot += w[i]*VT[i][l];
139              dot *= 2;
140              for( i=j+1; i<n; i++ )
141                 VT[i][l] -= dot*w[i];
142           }
143        }
144     }
145  }
```

```
146   }
147
148   // if  J  is     d0  s0   0   0
149   //              0  d1  s1   0
150   //              0   0  d2  s2
151   //              0   0   0  d3
152   //
153   // then  J^T  J  is
154   //      d0^2    d0*s0      0            0
155   //      d0*s0   d1^2+s0^2  d1*s1        0
156   //      0       d1*s1      d2^2+s1^2   d2*s2
157   //      0       0          d2*s2       d3^2+s2^2
158   //
159   // operations  are  designed  to  operate  on  J^T  J
160
161   // return  1  if  a  singular  value  has  been  found.
162
163   int  svdstep(
164     Vector<double>&diag,
165     Vector<double>&supdiag,
166     Matrix<double>&U,
167     Matrix<double>&VT,
168     int   nd )
169   {
170     int m = U.columns();
171     int n = VT.rows();
172     assert( n >= nd   );
173     assert( m >= nd );
174     assert( m >= n  );
175     if( nd == 1 ) return 1; // J is a scalar
176
177     double shift, x1, x2, c, s, v, u1, u2, spare;
178     double machine = 1e-16;
179     int    j, k;
180
181     // shift:  JT J has lower 2x2 block
182
183     // diag[nd-2]^2 + supdiag[nd-3]^2    diag[nd-2]*supdiag[nd-2]
184     // diag[nd-2]*supdiag[nd-2]          diag[nd-1]^2 + supdiag[nd-2]^2
185
186     int z = nd-1;
187     do {
188       if( z <= 0 ) break;
189       z--;
190       double a00 = diag[z]*diag[z];
191       if( z > 0 ) a00 += supdiag[z-1]*supdiag[z-1];
192       double a10 = diag[z]*supdiag[z];
193       double a11 = diag[z+1]*diag[z+1]+supdiag[z]*supdiag[z];
194
195       if( fabs(a10) < machine*(fabs(a00)+fabs(a11)) ) {
196         // the matrix is reducible.  set z to be the
197         // "zero" position of last partition.
198         ++z;
199         break;
200       }
201     } while(1);
202
203     if( z > 0 && (fabs(diag[z-1]) < fabs(supdiag[z-1])) ) {
204       // JT J =  x x 0     means J is  x y 0  or   x x 0
205       //         x x 0                 0 x 0       0 0 Y
206       // row z -> 0 0 X                0 0 x       0 0 x
207
208       // element "Y" must be eliminated in order that X can be
209       // stripped of as the square of a singular value.
210       // do this with a succession of Givens operations from left
211
212       spare = supdiag[z-1];
213       supdiag[z-1] = 0;
214       for( int r=z; r<nd; r++ ) {
215         // eliminate spare with diag[r]
```

```
216        x1 = diag[r];
217        x2 = spare;
218        Givens( x1, x2, c, s, v );
219
220        u1 = diag[r];
221        u2 = spare;
222        diag[r] = c*u1 + s*u2;
223
224        if( r<nd-1 ) {
225          u1 = supdiag[r];
226          u2 = 0;
227          supdiag[r] = c*u1 + s*u2;
228          spare      = s*u1 - c*u2;
229        }
230
231        // and fix U
232        for( j=0; j<m; j++ ) {
233          u1 = U[j][r  ];
234          u2 = U[j][z-1];
235          U[j][r  ] = c*u1 + s*u2;
236          U[j][z-1] = s*u1 - c*u2;
237        }
238      }
239    }
240
241    if( z >= nd-1 ) return 1;
242
243    shift = 0;
244
245    double a00 = diag[nd-2]*diag[nd-2];
246    if( nd > 2 ) a00 += supdiag[nd-3]*supdiag[nd-3];
247    double a10 = diag[nd-2]*supdiag[nd-2];
248    double a11 = diag[nd-1]*diag[nd-1]+supdiag[nd-2]*supdiag[nd-2];
249
250    double qa = 1;
251    double qb = -(a00+a11);
252    double qc = a00*a11 - a10*a10;
253    double qs = ((qb<0) ? -1 : 1 );
254
255    double qr = qb*qb - 4*qa*qc;
256
257    // block of JT J is positive semidef, so roots should be real
258    // in exact math, but ...
259    qr = MAX(qr,0.);
260    if( !qb && !qr ) {
261      shift = 0;
262    } else {
263      qr = sqrt(qr);
264      double mu1 = -qs*( fabs(qb)+qr )/( 2 * qa );
265      double mu2 = -qs*( 2 * qc )/( fabs(qb)+qr );
266      shift = mu2;
267      if( fabs(a11-mu1) < fabs(a11-mu2) ) shift = mu1;
268    }
269
270    // first block of  JT J is like
271    // d0^2      d0*s0
272    // d0*s0     d1^2 + s0^2
273    // and if this was QR with implicit shift, then (from right) Givens
274    // reflection would seek to zero (d0*s0) against (d0^2 - k)
275
276    x1 = diag[z]*diag[z] - shift;
277    x2 = diag[z]*supdiag[z];
278    Givens( x1, x2, c, s, v );
279
280    // operate on row z
281    u1 = diag[z];                       // old z,z
282    u2 = supdiag[z];                    // old z,z+1
283    diag[z] = c*u1 + s*u2;              // new z,z
284    supdiag[z] = v*(u1 + diag[z]) - u2; // new z,z+1
285
```

```
286      // operate on row 1
287      u1 = 0;                              // old z+1,z
288      u2 = diag[z+1];                      // old z+1,z+1
289      spare = c*u1 + s*u2;                 // new z+1,z
290      diag[z+1] = v*(u1 + spare) - u2;     // new z+1,z+1
291
292      // "spare" is new element z+1,z — should be zero
293
294      // use this Givens on VT
295      for( j=0; j<n; j++ ) {
296         u1 = VT[z  ][j];
297         u2 = VT[z+1][j];
298         VT[z  ][j] = c*u1 + s*u2;
299         VT[z+1][j] = v*(u1 + VT[z][j]) - u2;
300      }
301
302      // now from left. use element z,z to zero z+1,z (spare)
303      x1 = diag[z];
304      x2 = spare;
305      Givens( x1, x2, c, s, v );
306
307      // column z
308      u1 = diag[z];                        // old z,z
309      u2 = spare;                          // old z+1,z
310      diag[z] = c*u1 + s*u2;               // new z,z
311
312      // column z+1
313      u1 = supdiag[z];                     // old z,z+1
314      u2 = diag[z+1];                      // old z+1,z+1
315      supdiag[z] = c*u1 + s*u2;            // new z,z+1
316      diag[z+1] = v*(u1 + supdiag[z]) - u2; // new z+1,z+1
317
318      // column 2?
319      if( nd > z+2 ) {
320         u1 = 0;                           // old z,z+2
321         u2 = supdiag[z+1];                // old z+1,z+2
322         spare = c*u1 + s*u2;              // new z,z+2
323         supdiag[z+1] = v*(u1 + spare) - u2; // new z+1,z+2
324
325         // spare is new element z,z+2
326      }
327
328      // use this Givens on U
329      for( j=0; j<m; j++ ) {
330         u1 = U[j][z  ];
331         u2 = U[j][z+1];
332         U[j][z  ] = c*u1 + s*u2;
333         U[j][z+1] = v*(u1 + U[j][z]) - u2;
334      }
335
336      for( k=z; k<nd-2; k++ ) {
337
338         // from right. zero spare against supdiag[k]
339         x1 = supdiag[k];
340         x2 = spare;
341         Givens( x1, x2, c, s, v );
342
343         // row k
344         u1 = supdiag[k];                  // old k,k+1
345         u2 = spare;                       // old k,k+2
346         supdiag[k] = c*u1 + s*u2;         // new k,k+1
347
348         // row k+1
349         u1 = diag[k+1];                   // old k+1,k+1
350         u2 = supdiag[k+1];                // old k+1,k+2
351         diag[k+1] = c*u1 + s*u2;          // new k+1,k+1
352         supdiag[k+1] = v*(u1 + diag[k+1]) - u2; // new k+1,k+2
353
354         // row k+2
355         u1 = 0;                           // old k+2,k+1
```

```
356        u2 = diag[k+2];                              // old k+2,k+2
357        spare = c*u1 + s*u2;                         // new k+2,k+1
358        diag[k+2] = v*(u1 + spare) - u2;             // new k+2,k+2
359
360        // use this Givens on VT
361        for( j=0; j<n; j++ ) {
362            u1 = VT[k+1][j];
363            u2 = VT[k+2][j];
364            VT[k+1][j] = c*u1 + s*u2;
365            VT[k+2][j] = v*(u1 + VT[k+1][j]) - u2;
366        }
367
368        // zero spare agains diag[k+1]
369        x1 = diag[k+1];
370        x2 = spare;
371        Givens( x1, x2, c, s, v );
372
373        u1 = diag[k+1];                              // old k+1,k+1
374        u2 = spare;                                  // old k+2,k+1
375        diag[k+1] = c*u1 + s*u2;                     // new k+1,k+1
376
377        u1 = supdiag[k+1];                           // old k+1,k+2
378        u2 = diag[k+2];                              // old k+2,k+2
379        supdiag[k+1] = c*u1 + s*u2;                  // new k+1,k+2
380        diag[k+2] = v*(u1 + supdiag[k+1]) - u2;      // new k+2,k+2
381
382        if( k+2 < nd-1 ) {
383            u1 = 0;                                  // old k+1,k+3
384            u2 = supdiag[k+2];                       // old k+2,k+3
385            spare = c*u1 + s*u2;                     // new k+1,k+3
386            supdiag[k+2] = v*(u1 + spare) - u2;      // new k+2,k+3
387        }
388
389        // use this Givens on U
390        for( j=0; j<m; j++ ) {
391            u1 = U[j][k+1];
392            u2 = U[j][k+2];
393            U[j][k+1] = c*u1 + s*u2;
394            U[j][k+2] = v*(u1 + U[j][k+1]) - u2;
395        }
396    }
397
398    return 0;
399 }
400
401 // input:   "A" is (m x n) with m >= n (more rows than columns).
402 // output, U,VT, and sigma such that A = U*Diag(sigma)*VT
403 // with U and VT unitary and sigma positive.
404 // A is modified
405
406 int SVD( Matrix<double>& A,
407          Matrix<double>& U,
408          Matrix<double>& VT,
409          Vector<double>& sigma )
410 {
411     int i,j;
412     int m = A.rows();
413     int n = A.columns();
414     assert( m >= n );
415     Vector<double>supdiag(n-1);
416     int d = sigma.size();
417
418     // make bi-diagonal.  Store results in "sigma" and "supdiag",
419     // and save tranformation matrices "U" and "VT"
420     makeBiDiagonal(A,U,VT,sigma,supdiag);
421
422     // limit on loop prevents accidentally going forever
423     for( int step=0; step<10*m; step++ ) {
424
425         int got = svdstep( sigma, supdiag, U, VT, d );
```

```
426
427      if ( got ) {
428        d--;
429        if ( d == 0 ) break;
430      }
431    }
432
433    // respect sign convention
434    for ( i=0; i<n; i++ ) {
435      if ( sigma[i] < 0 ) {
436        sigma[i] *= -1;
437        for ( j=0; j<n; j++ ) VT[i][j] *= -1;
438      }
439    }
440
441    return 1;
442  }
```

---

file: SVD-test.cpp

```
1   #include <iostream>
2   #include <sstream>
3   #include <iomanip>
4   #include "Matrix.H"
5   #include "SVD.H"
6   using namespace std;
7
8   int main()
9   {
10    cout << scientific << setprecision(4) << showpos;
11
12    Matrix<double> A(3,3);
13    stringstream ss( stringstream::in | stringstream::out );
14    ss << "1 0 5    -10 2 -10    4 1 0";
15    ss >> A;
16
17    int m = A.rows();
18    int n = A.columns();
19    Vector<double>sigma(n);
20    Matrix<double>U(m,m);
21    Matrix<double>VT(n,n);
22
23    cout << "A before" << endl << A       << endl;
24
25    SVD( A, U, VT, sigma );
26
27    cout << "U"      << endl << U     << endl;
28    cout << "sigma"  << endl << sigma << endl;
29    cout << "VT"     << endl << VT    << endl;
30  }
```

B.7   Conjugate gradient

file: CG.H

```
1   #ifndef _CG_H__
2   #define _CG_H__
3   double cgStep( const Matrix<double>&A,
```

```
4                    Vector<double>& x,
5                    Vector<double>& p,
6                    Vector<double>& r,
7                    Vector<double>& tmp );
8    #endif
```

file: CG.cpp

```
1    #include <iostream>
2    #include <iomanip>
3    #include <math.h>
4    #include "Matrix.H"
5    #include <assert.h>
6
7    // perform 1 step of conjugate gradient method for symmetric
8    // positive definite matrix A.  return the norm of residual.
9
10   // r    residual
11   // p    search direction
12   // x    solution
13   // tmp spare memory
14
15   double cgStep( const Matrix<double>&A,
16                  Vector<double>& x,
17                  Vector<double>& p,
18                  Vector<double>& r,
19                  Vector<double>& tmp )
20   {
21     int n = A.rows();
22     assert( n == A.columns() );
23     assert( n == x.size() );
24     assert( n == p.size() );
25     assert( n == tmp.size() );
26
27     // vector A.p
28     for( int i=0; i<n; i++ ) {
29       tmp[i] = 0;
30       for( int j=0; j<n; j++ ) tmp[i] += A[i][j]*p[j];
31     }
32
33     // alpha, denom of gamma
34     double gamma = 0;
35     double d=0;
36     for( int i=0; i<n; i++ ) {
37       gamma += r[i]*r[i];
38       d     += p[i]*tmp[i];
39     }
40     double alpha = gamma/d;
41
42     // x update and r update
43     d = 0;
44     for( int i=0; i<n; i++ ) {
45       x[i] += alpha*p[i];
46       r[i] -= alpha*tmp[i];
47       d    += r[i]*r[i];
48     }
49
50     // finish gamma
51     gamma = - d/gamma;
52
53     // p
54     d = 0;
55     for( int i=0; i<n; i++ ) {
56       p[i] = r[i] - gamma*p[i];
57       d    += p[i]*p[i];
58     }
59
```

```
60    // L2 norm of p is L2 norm of r
61    return sqrt(d);
62  }
```

---

file: CG-test.cpp

```
1   #include <iostream>
2   #include <iomanip>
3   #include <math.h>
4   #include "Matrix.H"
5   #include "CG.H"
6   using namespace std;
7
8   int main()
9   {
10    Matrix<double> A(2,2);
11    A[0][0] = 75;
12    A[0][1] = A[1][0] = 43;
13    A[1][1] = 26;
14    Vector<double> b(2),x(2),tmp(2),r(2),p(2);
15    b[0] = -15; b[1] = 8;
16
17    x[0] = 0.;
18    x[1] = 0.;
19
20    cout << scientific << setprecision(4) << showpos;
21    int n = x.size();
22
23    for( int i=0; i<n; i++ ) {
24      r[i] = b[i];
25      for( int j=0; j<n; j++ ) r[i] -= A[i][j]*x[j];
26      p[i] = r[i];
27    }
28    cout << "start residual = " << endl << r;
29    cout << "start x = " << endl << x;
30
31    int iter=0;
32    do {
33      ++iter;
34      double resid = cgStep( A, x, p, r, tmp );
35      cout << "L2 of residual = " << resid << " x = " << endl << x;
36    } while(iter<n);
37
38    return 1;
39  }
```

B.8   Jacobi, Gauss–Seidel, and multigrid

This is not a general-purpose subroutine, but a demonstration of some iterative methods for $\mathbf{Ax} = \mathbf{b}$ tailored to Example 4.5.

file: Vcycle-example.cpp

```
1   #include <iostream>
2   #include <math.h>
3   #include "Matrix.H"
4   using std::cout;
5   using std::endl;
```

```
6
7    void prolong( Vector<double>& coarse, Vector<double>& fine )
8    {
9      int nh  = fine.size();
10     int n2h = coarse.size();
11     for( int ih=0; ih<nh; ih++ ) {
12         int i2h = ih/2;
13         if( ih % 2 ) { // odd
14            fine[ih] = coarse[i2h];
15         } else {
16            fine[ih] = 0;
17            if( i2h <  n2h ) fine[ih] += 0.5*coarse[i2h  ];
18            if( i2h >= 1   ) fine[ih] += 0.5*coarse[i2h-1];
19         }
20     }
21   }
22
23   void restrict( Vector<double>& fine, Vector<double>& coarse )
24   {
25     int n2h = coarse.size();
26     for( int i2h=0; i2h<n2h; i2h++ ) {
27        int ih = 2*i2h;
28        coarse[i2h] = 0.25*( fine[ih] + 2.*fine[ih+1] + fine[ih+2] );
29     }
30   }
31
32   void smooth( int nu, const double& omega, const double& scale,
33                Vector<double>& b, Vector<double>& x )
34   {
35     // matrix is scale*( -2, 1, 0, 0, ... ) etc
36     //                 (  1,-2, 1, 0, ... )
37     //                 (  0, 1,-2, 1, ... )
38
39     // GS:
40     // (-2  0  0 ... )           ( 0 -1      ... )
41     // ( 1 -2  0 ... ) * xnew = ( 0  0 -1    ... ) *xold + b/s
42     // ( 0  1 -2 ... )           ( 0  0  0 -1 ... )
43
44     int n = x.size();
45
46     if( omega < 0 ) {
47       // assume Gauss Seidel
48       for( int pass=0; pass<nu; pass++ ) {
49         for( int i=0; i<n; i++ ) {
50           double t = b[i]/scale;
51           if( i < n-1 ) t -= x[i+1];
52           if( i > 0   ) t -= x[i-1];
53           x[i] = -0.5*t;
54         }
55       }
56       return;
57     }
58
59     // Damped Jacobi:
60     // (-2/w   0    0 ...)          (2*(w-1)/w                -1   ...)
61     // (  0  -2/w   0 ...)* xnew =(      -1 2*(w-1)/w         -1   ...)* xold+b/s
62     // (  0    0 -2/w ...)          (         0        -1 2*(w-1)/w  ...)
63
64     // assume damped Jacobi
65     double oldxm1;
66     for( int pass=0; pass<nu; pass++ ) {
67       for( int i=0; i<n; i++ ) {
68         double oldx = x[i];
69         double t = b[i]/scale + 2*(omega-1)*oldx/omega;
70         if( i < n-1 ) t -= x[i+1];
71         if( i > 0   ) t -= oldxm1;
72         x[i] = -0.5*t*omega;
73         oldxm1 = oldx;
74       }
75     }
```

```
76   }
77
78   double residual( const double& scale , Vector<double>& x,
79                      Vector<double>& b, Vector<double> r  )
80   {
81     double r2 = 0;
82     int n = x.size();
83     for( int i=0; i<n; i++ ) {
84       double t = b[i] + 2*scale*x[i];
85       if( i > 0    ) t -= scale*x[i-1];
86       if( i < n-1 ) t -= scale*x[i+1];
87       r[i] = t;
88       r2 += t*t;
89     }
90     return sqrt(r2);
91   }
92
93   double VCycle( int nu, const double& omega, const double& scale ,
94                    Vector<double>&b, Vector<double>&x )
95   {
96     int n = x.size();
97
98     if( n == 1 ) {
99       // bottom solver!
100      x[0] = b[0]/(-2*scale);
101      return 0;
102    }
103
104    // do nu cycles of smoothing
105    smooth( nu, omega, scale, b, x );
106
107    // residual
108    Vector<double> temp1(n);
109    double r2 = residual( scale , x, b, temp1 );
110
111    // restrict residual
112    int nc = ((n+1)/2)-1;
113    Vector<double> temp2(nc);
114    restrict( temp1, temp2 );
115
116    // solve on coarser grid
117    Vector<double> temp3(nc);
118    for( int i=0; i<nc; i++ ) temp3[i] = 0;
119    VCycle( nu, omega, scale/4., temp2, temp3 );
120
121    // prolong
122    prolong( temp3, temp1 );
123
124    // correct
125    for( int i=0; i<n; i++ ) x[i] += temp1[i];
126
127    // and smooth again
128    smooth( nu, omega, scale, b, x );
129
130    r2 = residual( scale , x, b, temp1 );
131    return r2;
132  }
133
134  int main()
135  {
136    double goal = 1.e-4;
137
138    int n = 512-1;
139    Vector<double> b(n);
140    Vector<double> x(n);
141    Vector<double> r(n);
142    for( int i=0; i<n; i++ ) {
143      b[i] = 1;
144      x[i] = 0;
145    }
```

```
146
147      cout << "MULTIGRID WITH DAMPED JACOBI" << endl;
148      int iteration = 0;
149      double r2;
150      do {
151        ++iteration;
152        r2 = VCycle( 3, 2./3., 1., b, x );
153        cout << "iteration " << iteration << "  resid " << r2 << endl;
154      } while( r2 > goal );
155
156      for( int i=0; i<n; i++ ) {
157        b[i] = 1;
158        x[i] = 0;
159      }
160
161      cout << "MULTIGRID WITH GAUSS SEIDEL " << endl;
162      iteration = 0;
163      do {
164        ++iteration;
165        r2 = VCycle( 3, -10., 1., b, x );
166        cout << "iteration " << iteration << "  resid " << r2 << endl;
167      } while( r2 > goal );
168
169      for( int i=0; i<n; i++ ) {
170        b[i] = 1;
171        x[i] = 0;
172      }
173
174      cout << "JUST JACOBI" << endl;
175      iteration = 0;
176      do {
177        ++iteration;
178        smooth( 1, 1., 1., b, x );
179        r2 = residual( 1., x, b, r );
180        if( r2 <= goal ) cout << "iteration " << iteration << "  resid " << r2 << endl;
181      } while( r2 > goal );
182
183      for( int i=0; i<n; i++ ) {
184        b[i] = 1;
185        x[i] = 0;
186      }
187
188      cout << "JUST GAUSS SEIDEL" << endl;
189      iteration = 0;
190      do {
191        ++iteration;
192        smooth( 1, -10., 1., b, x );
193        r2 = residual( 1., x, b, r );
194        if( r2 <= goal ) cout << "iteration " << iteration << "  resid " << r2 << endl;
195      } while( r2 > goal );
196
197      return 1;
198    }
```

B.9   Cooley–Tukey FFT

These examples implement the Cooley–Tukey algorithm in the forward and backward directions. The algorithm stores data in a bit-reversed order, which means that to access it one either needs to re-order the data or provide an address mapping function. Here, the re-ordering algorithm of J. J. Rodríguez [194] is used.

```
file: FFT.H
```

```
1    #ifndef __FFT_H__
2    #define __FFT_H__
3    void fft( Vector<double>& fR, Vector<double>& fI, bool forward );
4    #endif
```

---

file: FFT.cpp

```
1    #include <math.h>
2    #include "Matrix.H"
3    #include <assert.h>
4
5    // re-order data according to bit reversal algorithm of Rodriguez 1988.
6
7    void reverse( Vector<double>& fR,
8                  Vector<double>& fI,
9                  int N, int m )
10   {
11     int i, j, k , NV2, last;
12     double tmpr, tmpi;
13
14     NV2=N>>1;
15     last = (N-1)-(1<<((m+1)>>1));
16     j = 0;
17     for( i=1; i<=last; i++ ) {
18       for( k=NV2; k<=j; k>>=1 ) j-= k;
19       j += k;
20       if( i < j ) {
21         tmpr  = fR[i]; tmpi  = fI[i];
22         fR[i] = fR[j]; fI[i] = fI[j];
23         fR[j] = tmpr;  fI[j] = tmpi;
24       }
25     }
26   }
27
28   // construct the integer address j_0 ,..., j_{p-1}, k_{m-1-p} ,..., k_0
29   // convention:  k_0 is lowest-order bit
30   // alo has bit j_{p-1} clear ,
31   // ahi has bit j_{p-1} set
32
33   inline void address( int k, int j, unsigned int twomml, int p,
34                        int& alo, int& ahi )
35   {
36     unsigned int jmask = 1;
37     unsigned int r=twomml;   // 2^{m-1}
38     alo = ahi = 0;
39
40     // skip the p-1 bit
41     for( int b=0; b<p-1; b++ ) {
42       if(j & jmask) alo |= r;
43       jmask <<= 1;  // shift left 1 = multiply by 2
44       r >>= 1;       // shift right 1 = divide by 2
45     }
46     ahi = r;
47     // now r is 2^(m-1-p)
48     r <<= 1;  // now 2^(m-p)
49     r -= 1;   // now 1 in every place up to and including  m-1-p
50     alo |= ( ((unsigned int)k) & r );
51     ahi |= alo ;
52   }
53
54   // fast fourier transform
55   // if forward=true convert input data to phase amplitutdes
56   // if forward=false convert phase amplitues to data
57
58   void fft( Vector<double>& fR,
59            Vector<double>& fI,
60            bool forward )
```

```
61  {
62    // find "m"
63    int m = 0;
64    int N = fR.size();
65    assert( fR.size() == fI.size() );
66
67    while( N /=2   ) { m++; }
68
69    N = fR.size();
70    int jmax=1;
71    int kmax=N/2;
72    for( int p=0; p<m; p++ ) {
73      for( int j=0; j<jmax; j++ ) {
74        double ang = M_PI*j/((double)jmax);
75        if( forward ) ang = -ang;
76        double c = cos(ang);
77        double s = sin(ang);
78        for( int k=0; k<kmax; k++ ) {
79
80          // addresses
81          // alo has bit j_(p-1) clear
82          // ahi has bit j_(p-1) set
83
84          int alo, ahi;
85          address( k, j, N/2, p+1, alo, ahi );
86
87          double ur = fR[alo];
88          double ui = fI[alo];
89
90          double vr = fR[ahi];
91          double vi = fI[ahi];
92
93          fR[alo] = ur + c*vr - s*vi;
94          fI[alo] = ui + c*vi + s*vr;
95
96          fR[ahi] = ur - c*vr + s*vi;
97          fI[ahi] = ui - c*vi - s*vr;
98        }
99      }
100     jmax *= 2;
101     kmax /= 2;
102   }
103
104   if( forward ) {
105     for( int i=0; i<N; i++ ) {
106       fR[i] /= N;
107       fI[i] /= N;
108     }
109   }
110
111   reverse( fR, fI, N, m );
112 }
```

---

file: FFT-test.cpp

```
1   #include <iostream>
2   #include <iomanip>
3   #include <math.h>
4   #include "Matrix.H"
5   #include <assert.h>
6   #include "FFT.H"
7   using namespace std;
8
9   int main()
10  {
11    cout << scientific << setprecision(4) << showpos;
12
```

```
13     Vector<double> fR(8);
14     Vector<double> fI(8);
15
16     fR[0] = 1; fR[1] = 2; fR[2] = 3; fR[3] = 0;
17     fR[4] = -1; fR[5] =-3; fR[6] =-2; fR[7] =-1;
18
19     fI[0] = 0; fI[1] = 0; fI[2] = 0; fI[3] = 0;
20     fI[4] = 0; fI[5] = 0; fI[6] = 0; fI[7] = 0;
21
22     cout << "start" << endl;
23     cout << "fR" << fR;
24     cout << "fI" << fI;
25
26     fft( fR, fI, true );
27
28     cout << "forward" << endl;
29     cout << "fR" << fR;
30     cout << "fI" << fI;
31
32     // report as coefficients A_ and B_ of cos/sin functions
33     int N = fR.size(); // assumed even
34     assert( N%2 == 0 );
35
36     int M=N/2;
37     cout << "A0 " << 2*fR[0] << 2*fI[0] << "i" << endl;
38
39     for( int j=1; j<M; j++ ) {
40         cout << "A" << noshowpos << j << showpos << " " <<
41         fR[j] + fR[N-j] << fI[j] + fI[N-j] << "i" << endl;
42         cout << "B" << noshowpos << j << showpos << " " <<
43         fI[N-j] - fI[j] << fR[j] - fR[N-j] << "i" << endl;
44     }
45
46     cout << "A" << noshowpos << M << showpos << " "
47         << 2*fR[M] << 2*fI[M] << "i" << endl;
48
49     fft( fR, fI, false );
50
51     cout << "backward" << endl;
52     cout << "fR" << fR;
53     cout << "fI" << fI;
54 }
```

B.10   Variable metric methods

file: GoldenSection.H

```
1   #ifndef __Golden_Section_H__
2   #define __Golden_Section_H__
3   void goldensection( const Vector<double>&x0, const Vector<double>&p,
4                       double& a,   double& b,   double& c,
5                       double& fa,  double& fb,  double& fc,
6                       double eps,  double (*funcp)(const Vector<double>&) );
7   #endif
```

file: GoldenSection.cpp

```
1   #include <math.h>
2   #include "Matrix.H"
3
4   // a<b<c is an ordered set of abscissas, and fa=f(a), etc.
5   // if b is a + (gold-1)*(c-a), or
6   //         c - (gold-1)*(c-a),
7   // then self-similarity will be maintained.
8   // Stop when c-a <= eps.
9
10  void goldensection( const Vector<double>&x0, const Vector<double>&p,
11                      double& a,   double& b,   double& c,
12                      double& fa,  double& fb,  double& fc,
13                      double eps, double (*funcp)(const Vector<double>&) )
14  {
15    double d, fd, gold = (1.+sqrt(5.))/2.;
16    int N = x0.size();
17    Vector<double> x(N);
18
19    // second test stops when c-a cannot be resolved given machine precision:
20    // machine precision ~ ((sqrt(5)-1)/2)^76
21    int count = 0;
22    while( (c-a > eps) && (count++ <= 76) ) {
23
24      // insert "d" in order a,b,d,c
25      if( c-b > b-a ) {
26        d = b + (2.-gold)*(c-b);
27      } else {
28        double tmp = b - (2.-gold)*(b-a);
29        d = b;
30        b = tmp;
31      }
32      for( int i=0; i<N; i++ ) x[i] = x0[i] + d*p[i];
33      fd = (*funcp)( x );
34
35      if( fb < fd ) {
36        c = d; fc = fd;
37      } else {
38        a = b; fa = fb;
39        b = d; fb = fd;
40      }
41    }
42  }
```

---

file: VMM.H

```
1   #ifndef __VMM_H__
2   #define __VMM_H__
3   // 0 for Davidon-Fletcher-Powell
4   // 1 for Broyden-Fletcher-Goldfarb-Shanno
5   // 2 for Oren-Spedicato
6   enum VMMTYPE { DFP=0, BFGS=1, OS=2 };
7
8   // an entry point if Hessian is desired
9
10  int VMM0( Matrix<double>& H,
11            double (*funcp)(const Vector<double>&),
12            void   (*gradp)(const Vector<double>&,Vector<double>&),
13            Vector<double>& x0,
14            VMMTYPE t,
15            int& iteration,
16            const double& thresh,
17            const double& gsthresh );
18
19  // and one if Hessian is not desired
20
21  int VMM( double (*funcp)(const Vector<double>&),
22           void   (*gradp)(const Vector<double>&,Vector<double>&),
```

```
23                Vector<double>& x0,
24                VMMTYPE t,
25                int& iteration,
26                const double& thresh,
27                const double& gsthresh );
28   #endif
```

file: VMM.cpp

```
1    #include "Matrix.H"
2    #include <math.h>
3    #include "VMM.H"
4    #include "GoldenSection.H"
5
6    // leave the Hessian as a global variable to facilitate
7    // interrogating it after an optimization
8
9    void Psi( Matrix<double>& Hnew, const Matrix<double>& Hold,
10             const Vector<double>& y, const Vector<double>& sigma,
11             double beta, double gamma )
12   {
13     int N = Hold.rows();
14     Vector<double> Hy(N);
15     Matrix<double> HysT(N,N);
16
17     // H.y and
18     for( int i=0; i<N; i++ ) {
19       Hy[i] = 0;
20       for( int j=0; j<N; j++ ) Hy[i] += Hold[i][j]*y[j];
21     }
22
23     // H y sigma^T
24     for( int i=0; i<N; i++ ) {
25       for( int j=0; j<N; j++ ) {
26         HysT[i][j] = 0;
27         for( int k=0; k<N; k++ ) HysT[i][j] += Hold[i][k]*y[k];
28         HysT[i][j] *= sigma[j];
29       }
30     }
31
32     // some inner products
33     double yHy = 0, sigy = 0;
34     for( int i=0; i<N; i++ ) {
35       yHy  += y[i]*Hy[i];
36       sigy += sigma[i]*y[i];
37     }
38
39     // some constants
40     double c1 = (1.+gamma*beta*yHy/sigy)/sigy;
41     double c2 = - gamma*(1.-beta)/yHy;
42     double c3 = - gamma*beta/sigy;
43
44     // the update
45     for( int i=0; i<N; i++ ) {
46       for( int j=0; j<N; j++ ) {
47         Hnew[i][j] = gamma*Hold[i][j] + c1*sigma[i]*sigma[j]
48                     + c2*Hy[i]*Hy[j] + c3*(HysT[j][i] + HysT[i][j]);
49       }
50     }
51   }
52
53   int VMM( double (*funcp)(const Vector<double>&),
54            void   (*gradp)(const Vector<double>&,Vector<double>&),
55            Vector<double>& x0,
56            VMMTYPE t,
57            int &iteration,
58            const double& thresh,
```

```
59                const double& gsthresh )
60  {
61     int N = x0.size();
62     Matrix<double> H(N,N);
63     return VMM0( H, funcp, gradp, x0, t, iteration, thresh, gsthresh );
64  }
65
66  int VMM0( Matrix<double>& H,
67            double (*funcp)(const Vector<double>&),
68            void   (*gradp)(const Vector<double>&,Vector<double>&),
69            Vector<double>& x0,
70            VMMTYPE t,
71            int &iteration,
72            const double& thresh,
73            const double& gsthresh )
74  {
75     double gold = (1.+sqrt(5.))/2.;
76     double beta, gamma, ep, tau, s;
77     int N = x0.size();;
78     Vector<double> g0(N), x1(N), g1(N), p(N), y(N), sigma(N), tmp1(N), tmp2(N);
79
80     Matrix<double> Hi(N,N);
81     for( int i=0; i<N; i++ ) {
82       for( int j=0; j<N; j++ ) H[i][j] = Hi[i][j] = 0;
83       H[i][i] = Hi[i][i] = 1;
84     }
85
86     iteration = 0;
87     do {
88
89       double f0 = (*funcp)(x0);
90       (*gradp)(x0,g0);
91       for( int i=0; i<N; i++ ) {
92         p[i] = 0;
93         for( int j=0; j<N; j++ ) p[i] -= H[i][j]*g0[j];
94       }
95
96       // test convergence. df will be theoretical improvement possible.
97       double df=0;
98       for( int i=0; i<N; i++ ) df -= g0[i]*p[i];
99       df /= 2.;
100      if( df < thresh ) return 0;
101
102      // bracket
103      double lambdam, lambda1 = 1;
104      double fm, f1;
105      for( int i=0; i<N; i++ ) x1[i] = x0[i] + lambda1*p[i];
106      f1 = (*funcp)( x1 );
107
108      if( f1 < f0 ) {
109        // look for lambda1,lambdam with greater values
110        do {
111          lambdam = lambda1;
112          fm      = f1;
113          lambda1 *= gold;
114          for( int i=0; i<N; i++ ) x1[i] = x0[i] + lambda1*p[i];
115          f1 = (*funcp)( x1 );
116        } while( fm>f0 || fm>f1 );
117      } else {
118        // look for lambda1,lambdam with lesser values
119        do {
120          lambdam = lambda1/gold;
121          for( int i=0; i<N; i++ ) x1[i] = x0[i] + lambdam*p[i];
122          fm = (*funcp)( x1 );
123          if( fm<f0 && fm<f1 ) break;
124          // lambda ridiculously small is a problem
125          if( lambdam < 1.e-16 ) break;
126          lambda1 = lambdam;
127          f1      = fm;
128        } while(1);
```

```
129         }
130
131         if( lambdam < 1.e−16 ) {
132             // failed to find improvement; inconsistent with df>thresh
133             return −2;
134         }
135
136         double lambda = lambdam;
137         fm = (*funcp)( x1 );
138         double a = 0;
139         goldensection( x0, p, a, lambda, lambda1, f0, fm, f1, gsthresh, funcp );
140
141         // new x, grad, make y & sigma
142         for( int i=0; i<N; i++ ) {
143             sigma[i] = p[i]*lambda;
144             x0[i] += sigma[i];
145         }
146
147         (*gradp)( x0, g1 );
148         for( int i=0; i<N; i++ ) {
149             y[i] = g1[i] − g0[i];
150             g0[i]  = g1[i];
151         }
152
153         switch (t) {
154         case OS:
155
156             // Hy and Hiy
157             for( int i=0; i<N; i++ ) {
158             tmp1[i] = tmp2[i] = 0;
159                 for( int j=0; j<N; j++ ) {
160                     tmp1[i] += H[i][j]*y[j];
161                     tmp2[i] += Hi[i][j]*sigma[j];
162                 }
163             }
164
165             s=0, tau=0, ep=0;
166             for( int i=0; i<N; i++ ) {
167                 s   += y[i]*sigma[i];
168                 tau += y[i]*tmp1[i];
169                 ep  += sigma[i]*tmp2[i];
170             }
171
172             if( (s>0 && ep<=s) || (s<0 && ep>=s) ) {
173                 gamma = ep/s;
174                 beta  = 0;
175             } else if( s >= tau ) {
176                 gamma = s/tau;
177                 beta = 1;
178             } else {
179                 gamma = 1;
180                 beta = s*(ep−s)/(ep*tau−s*s);
181             }
182             break;
183
184         case BFGS:
185             gamma = 1;
186             beta  = 1;
187             break;
188
189         default: // DFP
190             gamma = 1;
191             beta  = 0;
192         }
193
194         Psi( H, H, y, sigma, beta, gamma ); // DFP
195
196         if( t == OS ) {
197             double barb = s*s*(1.−beta);
198             barb = barb/( barb + beta*ep*tau );
```

```
199        Psi( Hi, Hi, sigma, y, barb, 1./gamma ); // DFP
200      }
201
202      ++iteration;
203    } while( iteration <1000);
204    return −1;
205  }
```

---

file: VMM-test.cpp

```
1   #include <iostream>
2   #include <iomanip>
3   #include <math.h>
4   #include "Matrix.H"
5   #include "VMM.H"
6   using namespace std;
7
8   double func( const Vector<double>& x)
9   {
10    // Rosenbrock's banana
11    return 100.*(x[1]−x[0]*x[0])*(x[1]−x[0]*x[0]) + (1.−x[0])*(1.−x[0]);
12  }
13
14  void grad( const Vector<double>& x, Vector<double>& g )
15  {
16    // Rosenbrock's banana
17    g[0] = 2.*( −1. + x[0]*(1. − 200.*x[1] + 200.*x[0]*x[0]));
18    g[1] = 200.*(x[1]−x[0]*x[0]);
19  }
20
21  int main()
22  {
23    Vector<double> x0(2);
24    int iteration;
25    x0[0] = −1.2; x0[1]=1; VMM( func, grad, x0, DFP,  iteration, 1.e−9, 1.e−9 );
26    cout << scientific << setprecision(4)
27       << "DFP:  iterations=" << iteration << " solution=" << endl << x0 << endl;
28
29    x0[0] = −1.2; x0[1]=1; VMM( func, grad, x0, BFGS, iteration, 1.e−9, 1.e−9 );
30    cout << "BFGS: iterations=" << iteration << " solution=" << endl << x0 << endl;
31
32    x0[0] = −1.2; x0[1]=1; VMM( func, grad, x0, OS,   iteration, 1.e−9, 1.e−9 );
33    cout << "OS:   iterations=" << iteration << " solution=" << endl << x0 << endl;
34  }
```

B.11   The simplex method for linear programming

file: Simplex.H

```
1   #ifndef __Simplex_H__
2   #define __Simplex_H__
3   extern double simplex_small;
4   int phase2( const Matrix<double>& A,
5                const Vector<double>& b,
6                Vector<double>& x_J,
7                Vector<int>& J_list,
8                Vector<int>& K_list,
9                const Vector<int>& is_restricted,
```

```
10                    int t_var,
11                    double simplex_small = 1.e-12 );
12   int phase1( Matrix<double>& A,
13               Vector<double>& b,
14               Vector<double>& x_J,
15               Vector<int>& J_list,
16               Vector<int>& K_list,
17               Vector<int>& is_restricted,
18               int t_var,
19               int t_eqn,
20               double simplex_small = 1.e-12 );
21   #endif
```

---

file: Simplex.cpp

```
1    #include <iostream>
2    #include <stdlib.h>
3    #include <stdio.h>
4    #include <math.h>
5    using namespace std;
6
7    #include "Matrix.H"
8    #include "Householder.H"
9
10   // if "MODIFY" is defined, then in Phase 2 matrix modification
11   // will be used instead of reduction.
12   #define MODIFY
13
14   #define MAX(A,B)  (((A)>(B))?(A):(B))
15
16   // anything this big or smaller is close enough to zero.
17
18   // phase 2 of the simplex method.
19   // return 0 on success, not zero if algorithm fails.
20   // x_J is the solution
21
22   int phase2( const Matrix<double>& A,
23               const Vector<double>& b,
24               Vector<double>& x_J,
25               Vector<int>& J_list,
26               Vector<int>& K_list,
27               const Vector<int>& is_restricted,
28               int t_var,
29               double simplex_small )
30   {
31     int i, j, k, r, s, all_neg;
32     double c_k;
33     double r_val, x_J_over_sigma_alpha, sigma;
34     int t_in_J=-1;
35
36     int retval = 0;
37
38     // a bunch of allocated memory
39     int n_var = A.columns();
40     int n_eqn = A.rows();
41     Vector<double> tmp(n_eqn);
42     Vector<double> d(n_eqn);
43     Vector<double> a_bar(n_eqn);
44     Matrix<double> AJ(n_eqn,n_eqn);
45     Matrix<double> AJiAK(n_eqn,n_var-n_eqn);
46     Vector<int> Pl(n_eqn), Pr(n_eqn);
47
48     // index of t in J
49     for( i=0; i<n_eqn; i++ )
50       if( J_list[i] == t_var ) t_in_J = i;
51
52     // first step, compute x_J = Inverse[AJ]*b
```

```
53      for( i=0; i<n_eqn; i++ ) {
54        for( j-0; j<n_eqn; j++ )
55          AJ[i][j] = A[i][J_list[j]];
56        x_J[i] = b[i];
57      }
58      retval = Householder( AJ, Pl, Pr, d, NOTHING );
59      // if there's a problem, return error code.
60      if( retval != 0 ) return retval;
61      solveQR( AJ, Pl, Pr, d, x_J );
62
63      // second, compute Inverse[AJ]*AK
64      for( k=0; k<n_var-n_eqn; k++ ) {
65        for( i=0; i<n_eqn; i++ )
66          tmp[i] = A[i][K_list[k]];
67        solveQR( AJ, Pl, Pr, d, tmp );
68        for( i=0; i<n_eqn; i++ )
69          AJiAK[i][k] = tmp[i];
70      }
71
72      // invalid indices prevent mod/reduction on first pass.
73      r = s = -1;
74
75      do {
76
77        if( r >= 0 && s >= 0 ) {
78
79          // there has been a change to J/K lists.
80
81  #ifdef MODIFY
82
83          // Recompute Inverse(AJ)*AK and Inverse(AJ)*b using MODIFICATION
84
85          // s is position in K where echange happened,
86          // r is position in J where echange happened.
87
88          // "m" is "s" column of AjiAk
89          for( i=0; i<n_eqn; i++ ) tmp[i] = AJiAK[i][s];
90
91          // replace "s" column of AJiAK with e_r
92          for( i=0; i<n_eqn; i++ ) AJiAK[i][s] = 0;
93          AJiAK[r][s] = 1;
94
95          // multiply by M inverse:
96          x_J[r] /= tmp[r];
97          for( i=0; i<n_eqn; i++ )
98            if( i != r ) x_J[i] -= tmp[i]*x_J[r];
99
100         for( j=0; j<n_var-n_eqn; j++ ) {
101           AJiAK[r][j] /= tmp[r];
102           for( i=0; i<n_eqn; i++ )
103             if( i != r ) AJiAK[i][j] -= tmp[i]*AJiAK[r][j];
104         }
105
106  #else
107
108         // Recompute Inverse(AJ)*AK and Inverse(AJ)*b using REDUCTION
109
110         // first step, compute x_J = Inverse[AJ]*b
111         for( i=0; i<n_eqn; i++ ) {
112           for( j=0; j<n_eqn; j++ )
113             AJ[i][j] = A[i][J_list[j]];
114           x_J[i] = b[i];
115         }
116         retval = Householder( AJ, Pl, Pr, d, NOTHING );
117         // if there's a problem, return error code.
118         if( retval != 0 ) return retval;
119         solveQR( AJ, Pl, Pr, d, x_J );
120
121         // second, compute Inverse[AJ]*AK
122         for( k=0; k<n_var-n_eqn; k++ ) {
```

```
123              for( i=0; i<n_eqn; i++ )
124                tmp[i] = A[i][K_list[k]];
125              solveQR( AJ, Pl, Pr, d, tmp );
126              for( i=0; i<n_eqn; i++ )
127                AJiAK[i][k] = tmp[i];
128            }
129
130   #endif
131
132          }
133
134          for( i=0; i<n_eqn; i++ ) {
135            if( is_restricted[J_list[i]] && x_J[i] < -simplex_small ) {
136              cout << "restricted variable #" << i << " " << J_list[i] <<
137                " has illegal value " << x_J[i] << endl;
138              retval = -100;
139              return retval;
140            }
141          }
142
143          // find the coefficients c_k of step 2, p. 260
144          // and do test of step 3.
145          s = -1;
146          for( k=0; k<n_var-n_eqn; k++ ) {
147            c_k = AJiAK[t_in_J][k];
148            if( ( is_restricted[K_list[k]] && c_k < 0 ) ||
149                ( !is_restricted[K_list[k]] && fabs(c_k) > 0) ) {
150              // got a candidate s
151              s = k;
152              sigma = 1;
153              if( c_k > 0 ) sigma = -1;
154              break;
155            }
156          }
157
158          if( s == -1 ) {
159            break; // stop. normal exit.
160          }
161
162          // vector a_bar, of step 4
163          all_neg = 1;
164          for( i=0; i<n_eqn; i++ ) {
165            a_bar[i] = AJiAK[i][s];
166            if( sigma*a_bar[i] > 0 && is_restricted[J_list[i]] )
167              all_neg = 0;
168          }
169
170          if( all_neg ) {
171            cout << "all sigma*a_bar are negative" << endl;
172            retval = -10;
173            break; // abnormal exit.
174          }
175
176          r = -1;
177          r_val = 0;
178          for( i=0; i<n_eqn; i++ ) {
179            if( is_restricted[J_list[i]] ) {
180              if( sigma*a_bar[i] > 0 ) {
181                x_J_over_sigma_alpha = x_J[i]/(sigma*a_bar[i]);
182                if( r < 0 || r_val > x_J_over_sigma_alpha ) {
183                  r_val = x_J_over_sigma_alpha;
184                  r = i;
185                }
186              }
187            }
188          }
189
190          // swap r and s in J/K
191          k = K_list[s];
192          j = J_list[r];
```

```
193          K_list[s] = j;
194          J_list[r] = k;
195
196      } while (1);
197
198      return retval;
199  }
200
201  // phase 1 of the simplex method.
202  // return 0 on success, not zero if algorithm fails.
203  // x_J is the solution
204  // dimensions of all arrays may be changed.
205
206  int phase1( Matrix<double>& A,
207              Vector<double>& b,
208              Vector<double>& x_J,
209              Vector<int>& J_list,
210              Vector<int>& K_list,
211              Vector<int>& is_restricted,
212              int t_var,
213              int t_eqn,
214              double simplex_small )
215  {
216      int i,i2,j2,j;
217      int retval=0;
218
219      // augment:
220      //    rewrite objective equation
221      //    add artifical variables for each eqn other than objective
222
223      int n_eqn = A.rows();
224      int n_var = A.columns();
225
226      // first, all "b" assumed positive
227      for( int i=0; i<n_eqn; i++ ) {
228        if( b[i] < 0 ) {
229          b[i] *= -1;
230          for( int j=0; j<n_var; j++ ) A[i][j] *= -1;
231        }
232      }
233
234      // spare memory to save copy of objective, and for manipulation of lists.
235      Vector<double> objective(n_var);
236      Vector<int>    itmp(n_var+n_eqn);
237
238      // save the objective
239      for( i=0; i<n_var; i++ ) objective[i] = A[t_eqn][i];
240      double b_obj = b[t_eqn];
241
242      // resize A to accommodate new artificial variables
243      A.resize( n_eqn, n_var + n_eqn - 1 );
244
245      // modify A to incorporate artificial variables
246      j2 = n_var;
247      b[t_eqn] = 0;
248      for( i=0; i<n_eqn; i++ ) {
249        if( i == t_eqn ) {
250          // rewrite objective
251          for( j=0; j<n_var; j++ ) A[i][j] = 0;
252          for( j=n_var; j<n_var+n_eqn-1; j++ ) A[i][j] = 1;
253          A[i][t_var] = 1;
254        } else {
255          b[t_eqn] += 2*b[i];
256          for( j=n_var; j<n_var+n_eqn-1; j++ ) A[i][j] = 0;
257          A[i][j2++] = 1;
258        }
259      }
260
261      // modify restricted list
262      is_restricted.resize(n_var+n_eqn-1);
```

```
263        for( i=n_var; i<n_var+n_eqn-1; i++ ) is_restricted[i] = 1;
264
265        // populate J list — size is OK
266        for( i=1; i<n_eqn; i++ )
267          J_list[i] = n_var-1+i;
268        J_list[0] = t_var;
269
270        // K list
271        K_list.resize( n_var - 1 );
272        i2 = 0;
273        for( i=0; i<n_var; i++ )
274          if( i != t_var ) K_list[i2++] = i;
275
276        // call phase 2
277        retval = phase2( A, b, x_J, J_list, K_list, is_restricted, t_var, simplex_small );
278
279        // return error code if there's a problem
280        if( retval != 0 ) return retval;
281
282        // are any augmented variables active?
283        int num_aug_active = 0;
284        double max_aug_value = 0;
285        for( i=0; i<n_eqn; i++ ) {
286          // look for indices that signify augmentation, but omit
287          // augmented objective.
288          if( J_list[i] >= n_var ) {
289            num_aug_active++;
290            max_aug_value = MAX( max_aug_value, fabs(x_J[i]) );
291          }
292        }
293
294        if( num_aug_active ) {
295
296          // we have active augmented variables.
297          if( max_aug_value > simplex_small ) {
298            // value not zero — failure.
299            cout << "phase 1: augmented variables active and nonzero."
300                 << max_aug_value << endl;
301            retval = -20;
302            return retval;
303          }
304
305          // before resizing "A" etc, we need to reorder the artificial variables.
306          Vector<int> newvar(num_aug_active);
307          j = 0;
308          for( i=0; i<n_eqn; i++ )
309            if( J_list[i] >= n_var )
310              newvar[j++] = J_list[i];
311
312          // do the reording
313          for( i=0; i<n_eqn; i++ )
314            for( j=0; j<num_aug_active; j++ )
315              A[i][n_var+j] = A[i][newvar[j]];
316
317          // do the resize
318          A.resize( n_eqn+1, n_var+num_aug_active+1 );
319          b.resize( n_eqn+1 );
320          x_J.resize( n_eqn+1 );
321
322          // restore the objective
323          b[t_eqn] = b_obj;
324          for( i=0; i<n_var; i++ ) A[t_eqn][i] = objective[i];
325          for( i=n_var; i<n_var+num_aug_active+1; i++ ) A[t_eqn][i] = 0;
326
327          // add the new equation that "pins" artificial variables to zero
328          b[n_eqn] = 0;
329          for( i=0; i<n_var; i++ ) A[n_eqn][i] = 0;
330          for( i=n_var; i<n_var+num_aug_active+1; i++ ) A[n_eqn][i] = 1;
331
332          // make coefficient of new "pinning" variable zero except in last eqn
```

```
333        for ( i =0; i<n_eqn; i++ ) A[i][n_var+num_aug_active] = 0;
334
335        // J list — need to renumber the artificial variables to be
336        // compatible with reordering in "A". And, add "pinning" variable.
337        J_list.resize( n_eqn+1 );
338        j = n_var;
339        for ( i =0; i<n_eqn; i++ )
340          if ( J_list[i] >= n_var ) J_list[i] = j++;
341        J_list[n_eqn] = n_var+num_aug_active;
342
343        // K list — reorder, keeping only original variables.
344        j = 0;
345        for ( i =0; i<n_var −1; i++ )
346          if ( K_list[i] < n_var ) K_list[j++] = K_list[i];
347        K_list.resize( (n_var+num_aug_active+1) − (n_eqn+1) );
348
349        // and the restricted list
350        is_restricted.resize( n_var+num_aug_active+1 );
351        is_restricted[n_var+num_aug_active] = 1;
352
353        return retval;
354      }
355
356      // there are no augmented variables in the J list.
357      // they can be removed everywhere.
358
359      // the K list
360      j = 0;
361      for ( i =0; i<n_var −1; i++ )
362        if ( K_list[i] < n_var )
363          K_list[j++] = K_list[i];
364      K_list.resize( n_var − n_eqn );
365
366      // restricted
367      is_restricted.resize( n_var );
368
369      // restore objective
370      b[t_eqn] = b_obj;
371      for ( i =0; i<n_var; i++ ) A[t_eqn][i] = objective[i];
372      for ( i=n_var; i<n_var+num_aug_active+1; i++ ) A[t_eqn][i] = 0;
373
374      // resize A
375      A.resize( n_eqn, n_var );
376
377      return retval;
378    }
```

---

file: Simplex-test.cpp

```
 1   #include <iostream>
 2   #include <sstream>
 3   #include <iomanip>
 4   #include <stdlib.h>
 5   #include <stdio.h>
 6   #include <math.h>
 7   #include "Matrix.H"
 8   #include "Simplex.H"
 9   using namespace std;
10
11   int main( int argc, char ** argv )
12   {
13     cout << scientific << setprecision(4) << showpos;
14     int n_var = 6;
15     int n_eqn = 4;
16
17     // worst case allocation for phase 1
18     int mx_var = n_var + n_eqn;
```

```
19      int mx_eqn = n_eqn + 1;
20
21      // allocate for worst case
22      Matrix<double> A(mx_eqn, mx_var);
23      Vector<double> b(mx_eqn), x(mx_eqn);
24      Vector<int> J_list(mx_eqn), K_list(mx_var-n_eqn), is_restricted(mx_var);
25
26      // resize for standard form problem  (resizing down does not free
27      // memory — so if phase1() needs to resize up, it'll be "free"
28      A.resize(n_eqn, n_var);
29      b.resize(n_eqn);
30      x.resize(n_eqn);
31      J_list.resize(n_eqn);
32      K_list.resize(n_var-n_eqn);
33      is_restricted.resize(n_var);
34
35      // data for a given problem
36      stringstream ss( stringstream::in | stringstream::out );
37      ss << "0.06 0.04 0.08 -1 0 0 1 1 1 0 0 0 \
38           0 0 1 0 1 0 -1 4 0 0 0 1    0 10000 2000 0";
39      ss >> A >> b;
40      ss.clear();
41      ss << "1 1 1 0 1 1";
42      ss >> is_restricted;
43
44      int t_var = 3; // objective variable
45      int t_eqn = 0; // objective equation
46
47      cout << "at start A " << endl << A << endl;
48      cout << "at start b " << endl << b << endl;
49
50      int retval = phase1( A, b, x, J_list, K_list, is_restricted, t_var, t_eqn );
51
52      cout << "phase1 return value " << noshowpos << retval << showpos << endl;
53      cout << "after phase1 A " << endl << A << endl;
54      cout << "after phase1 b " << endl << b << endl;
55      cout << "after phase1 x " << endl << x << endl;
56      cout << "after phase1 J " << endl << noshowpos << J_list << endl;
57      cout << "after phase1 K " << endl << K_list << endl;
58      cout << "after phase1 isrestricted " << endl << is_restricted << showpos << endl;
59
60      if( retval != 0 ) return retval;
61
62      retval = phase2( A, b, x, J_list, K_list, is_restricted, t_var );
63
64      cout << "phase2 return value " << noshowpos << retval << showpos << endl;
65      cout << "after phase2 x " << endl << x << endl;
66      cout << "after phase2 J " << endl << noshowpos << J_list << endl;
67      cout << "after phase2 K " << endl << K_list << showpos << endl;
68
69      return retval;
70  }
```

B.12   Quadratic programming for convex systems

file: ConvexQP.H

```
1   #ifndef __ConvexQP_H__
2   #define __ConvexQP_H__
3
4   int ConvexQP( const Matrix<double>& L,
5                 Vector<double>& x,
```

```
6                   const Matrix<double>& Ce, const Vector<double>& ce,
7                   const Matrix<double>& Ci, const Vector<double>& ci,
8                   double& df,
9                   const double& tolerance );
10
11   #endif
```

file: ConvexQP.cpp

```
1    #include <math.h>
2    #include "Matrix.H"
3    #include "Givens.H"
4
5    #ifndef MIN
6    #define MIN(A,B)  (((A)<(B))?(A):(B))
7    #endif
8
9    // return false of it appears that R is singular.
10   bool qp_modify_append( const Vector<double>& np,
11                          const Vector<double>& d,
12                          Matrix<double>& J,
13                          Matrix<double>& R )
14   {
15     int n = R.rows();
16     int m = R.columns();
17
18     // now fix with a single Householder step
19     Vector<double> w(n);
20
21     double len = 0;
22     w[m] = d[m];
23     for( int i=m+1; i<n; i++ ) {
24       w[i] = d[i];
25       len += d[i]*d[i];
26     }
27     double K = len + d[m]*d[m];
28     if( K <= 0 ) return false;
29     K = sqrt( K );
30     if( d[m] > 0 ) K = -K;
31     w[m] -= K;
32     len = sqrt( len + w[m]*w[m] );
33     for( int i=m; i<n; i++ ) w[i] /= len;
34
35     // fix J by right multiplication by (1-2*w wT)
36     for( int r=0; r<n; r++ ) {
37       double dot = 0;
38       for( int c=m; c<n; c++ ) dot += J[r][c]*w[c];
39       dot *= -2;
40       for( int c=m; c<n; c++ ) J[r][c] += dot*w[c];
41     }
42
43     // and fix R
44     R.resize(n,m+1);
45     for( int i=0; i<m; i++ ) R[i][m] = d[i];
46     R[m][m] = K;
47     for( int i=m+1; i<n; i++ ) R[i][m] = 0;
48
49     return true;
50   }
51
52   void qp_modify_delete( Matrix<double>& J,
53                          Matrix<double>& R,
54                          Vector<int>& A,
55                          Vector<double>& lambda,
56                          int cdel )
57   {
58     for( int i=0; i<A.size(); i++ ) {
```

```
59        if( A[i] == cdel ) {
60          A[i] = -1;
61        } else if( A[i] > cdel ) {\
62          A[i] --;
63        }
64      }
65
66      // fix A and lambda, and begin on R
67      int n = R.rows();
68      int m = R.columns();
69
70      for( int c=cdel+1; c<m; c++ ) {
71        lambda[c-1] = lambda[c];
72        for( int r=0; r<n; r++ ) {
73          R[r][c-1] = R[r][c];
74        }
75      }
76      lambda[m-1] = 0;
77
78      R.resize(n,--m);
79
80      // change from Hessenberg to right triangular with Givens reflections
81      for( int r=cdel; r<m; r++ ) {
82        double cf, sf, vf;
83        Givens( R[r][r], R[r+1][r], cf, sf, vf );
84        for( int c=r; c<m; c++ ) {
85          double y1 = R[r  ][c];
86          double y2 = R[r+1][c];
87          R[r  ][c] = cf*y1 + sf*y2;
88          R[r+1][c] = vf*( R[r][c] + y1 ) - y2;
89        }
90        for( int c=0; c<n; c++ ) {
91          double y1 = J[c][r  ];
92          double y2 = J[c][r+1];
93          J[c][r  ] = cf*y1 + sf*y2;
94          J[c][r+1] = vf*( J[c][r] + y1 ) - y2;
95        }
96      }
97    }
98
99    // the optimum of f(x) = (1/2) x^T G x + b^T x , with G = L L^T positive definite,
100   // subject to constraints Ce^T x + ce =  0, Ce dimension (n x n_equality)
101   // and                    Ci^T x + ci >= 0, Ci dimension (n x n_inequality)
102   // on input, x is the solution to the unconstrained problem.
103   // it is assumed the constraints are independent.
104
105   // return value is 0 if successful, or 1 if no feasible solution exists.
106
107   // df is change to quadratic function upon imposition of constraints.
108
109   // tolerance is the tolerance with which constraints are enforced:
110   // a reasonable value is a few times epsilon * condition number of G,
111   // if available.
112
113   int ConvexQP( const Matrix<double>& L,
114                 Vector<double>& x,
115                 const Matrix<double>& Ce, const Vector<double>& ce,
116                 const Matrix<double>& Ci, const Vector<double>& ci,
117                 double& df,
118                 const double& tolerance )
119   {
120     int n          = x.size();
121     int n_equality  = ce.size();
122     int n_inequality = ci.size();
123     Matrix<double> J(n,n);
124     Matrix<double> R(n,n);
125     Vector<double> np(n);
126     Vector<double> d(n);
127     Vector<double> z(n);
128     Vector<double> r(n);
```

```
129     Vector<double> s(n_inequality);
130     Vector<double> lambda(n);
131
132     // represent "A" list as a mapping of inequality # to column # in matrices N,R
133     Vector<int>   A(n_inequality);
134
135     for( int i=0; i<n; i++ ) lambda[i] = 0;
136     for( int i=0; i<n_inequality; i++ ) A[i] = -1;
137
138     // at start, Q=I and J = L^{-T}
139     // LT J = I
140     // make J
141     for( int c=0; c<n; c++ ) {
142       for( int k=n-1; k>c; k-- ) J[k][c] = 0;
143       J[c][c] = 1./L[c][c];
144       for( int k=c-1; k>=0; k-- ) {
145         double sum = 0;
146         for( int l=k+1; l<n; l++ ) sum += L[l][k]*J[l][c];
147         J[k][c] = -sum/L[k][k];
148       }
149     }
150
151     // initialize R=0
152     for( int r=0; r<n; r++ )
153       for( int c=0; c<n; c++ ) R[r][c] = 0;
154     R.resize(n,0);
155
156     int n_constraints = 0;
157
158     // first pass - add equality constraints
159     for( int p=0; p<n_equality; p++ ) {
160
161       for( int i=0; i<n; i++ ) np[i] = Ce[i][p];
162
163       // d is (J1,J2)^T n_p
164       for( int j=0; j<n; j++ ) {
165         d[j] = 0;
166         for( int i=0; i<n; i++ )
167           d[j] += J[i][j]*np[i];
168       }
169
170       // z = H n_p = J2 * d
171       // step direction in primal (x) space
172       for( int i=0; i<n; i++ ) {
173         z[i] = 0;
174         for( int j=n_constraints; j<n; j++ )
175           z[i] += J[i][j]*d[j];
176       }
177
178       // r = Ndagger n_p = R^-1 d
179       // neg of step direction in dual (lambda) space
180       for( int i=n_constraints-1; i>=0; i-- ) {
181         double sum = 0;
182         for( int j=i+1; j<n_constraints; j++ ) sum += R[i][j]*r[j];
183         r[i] = ( d[i] - sum ) / R[i][i];
184       }
185
186       // t_p is full step length to make constraint np feasible
187       double dot1 = 0, dot2 = 0;
188       for( int i=0; i<n; i++ ) {
189         dot1 += np[i]*x[i];
190         dot2 += np[i]*z[i];
191       }
192       double t_p = - ( dot1 + ce[p] )/dot2;
193
194       // update x assuming full step works
195       for( int i=0; i<n; i++ ) x[i] += t_p*z[i];
196
197       // update lambda, assuming same
198       for( int i=0; i<n_constraints; i++ ) lambda[i] -= t_p*r[i];
```

```
199        lambda[n_constraints] = t_p;
200
201        // change to function value
202        df += 0.5*t_p*t_p*dot2;
203
204        n_constraints++;
205
206        if( !qp_modify_append( np, d, J, R ) ) {
207          // R singular — constraints must be dependent
208          return 1;
209        }
210      }
211
212      // now the inequality constraints
213      do {
214
215        // find a violated inequality constraint
216        double worst_s = 1;
217        int p         = -1;
218        for( int i=0; i<n_inequality; i++ ) {
219          s[i] = ci[i];
220          for( int j=0; j<n; j++ ) s[i] += Ci[j][i]*x[j];
221          if( A[i] < 0 && s[i] < worst_s ) {
222            worst_s = s[i];
223            p       = i;
224          }
225        }
226
227        if( worst_s >= -tolerance ) {
228          // no constraint has been violated. done!.
229          return 0;
230        }
231
232        for( int i=0; i<n; i++ ) np[i] = Ci[i][p];
233        double lambda_p = 0;
234
235        do {
236
237          // d is (J1,J2)^T n_p
238          for( int j=0; j<n; j++ ) {
239            d[j] = 0;
240            for( int i=0; i<n; i++ )
241              d[j] += J[i][j]*np[i];
242          }
243
244          // z = H n_p = J2 * d
245          // step direction in primal (x) space
246          for( int i=0; i<n; i++ ) {
247            z[i] = 0;
248            for( int j=n_constraints; j<n; j++ )
249              z[i] += J[i][j]*d[j];
250          }
251
252          // r = Ndagger n_p = R^-1 d
253          // neg of step direction in dual (lambda) space
254          for( int i=n_constraints-1; i>=0; i— ) {
255            double sum = 0;
256            for( int j=i+1; j<n_constraints; j++ ) sum += R[i][j]*r[j];
257            r[i] = ( d[i] — sum ) / R[i][i];
258          }
259
260          // find a step length
261          bool t_d_bounded = false;
262          bool t_p_bounded = false;
263          double t, t_d, t_p;
264
265          // search for step length in dual (lambda) space.
266          // "l" is first constraint to be infeasible.
267          int l;
268          for( int k=n_equality; k<n_constraints; k++ ) {
```

```
269            if( r[k] > 0.0 ) {
270               if( (!t_d_bounded) || (lambda[k] < r[k]*t_d) ) {
271                  t_d = lambda[k]/r[k];
272                  t_d_bounded = true;
273                  l = k;
274               }
275            }
276         }
277
278         // search for step length in primal (x) space.
279         double dot1 = 0, dot2 = 0;
280         for( int i=0; i<n; i++ ) {
281            dot1 += z[i]*z[i];
282            dot2 += np[i]*z[i];
283         }
284         if( dot1 > 0 && dot2 != 0 ) {
285            t_p = -s[p]/dot2;
286            t_p_bounded = true;
287         }
288
289         if( t_d_bounded ) t = t_d;
290         if( t_p_bounded ) t = t_p;
291         if( t_d_bounded && t_p_bounded ) t = MIN(t_d,t_p);
292
293         // if unbounded, then failure
294         if( !(t_d_bounded || t_p_bounded) ) {
295            return 1;
296            break;
297         }
298
299         if( !t_p_bounded ) {
300            // constraint "l" is infeasible.
301
302            for( int k=0; k<n_constraints; k++ ) lambda[k] -= t*r[k];
303            lambda_p += t;
304
305            qp_modify_delete( J, R, A, lambda, l );
306            n_constraints --;
307
308         } else {
309
310            for( int k=0; k<n; k++ ) x[k] += t*z[k];
311            df += t*dot2*( 0.5*t + lambda_p );
312            for( int k=0; k<n_constraints; k++ ) lambda[k] -= t*r[k];
313            lambda_p += t;
314
315            if( t == t_p ) {
316
317               // a full step was taken.
318               if( !qp_modify_append( np, d, J, R ) ) {
319                  // R singular.
320                  return 1;
321               }
322               A[p] = n_constraints;
323               lambda[n_constraints] = lambda_p;
324               n_constraints++;
325               break;  // break out of do loop
326
327            } else  {
328
329               for( int i=0; i<n_inequality; i++ ) {
330                  s[i] = ci[i];
331                  for( int j=0; j<n; j++ ) s[i] += x[j]*Ci[j][i];
332               }
333               qp_modify_delete( J, R, A, lambda, l );
334               n_constraints --;
335
336            }
337         }
338
```

```
339        } while (1);
340      } while (1);
341      return 0;
342    }
```

file: ConvexQP-test.cpp

```
1    #include <iostream>
2    #include <iomanip>
3    #include "Matrix.H"
4    #include "Cholesky.H"
5    #include "ConvexQP.H"
6    using namespace std;
7
8    int main()
9    {
10
11       Matrix<double> G(2,2), Ci(2,3), Ce;
12       Vector<double> x(2), ci(3), ce;
13
14       // function is 1/2 x^T G x + b^T x
15       // initially store b in x
16       G[0][0] = G[1][1] =  4;
17       G[0][1] = G[1][0] = -2;
18       x[0] = 6;
19       x[1] = 0;
20
21       // constraints
22       Ci[0][0] = 1; Ci[1][0] = 0; ci[0] = 0;
23       Ci[0][1] = 0; Ci[1][1] = 1; ci[1] = 0;
24       Ci[0][2] = 1; Ci[1][2] = 1; ci[2] = -2;
25
26       // convert G to L
27       Cholesky( G );
28
29       // solve for unconstrained optimum
30       solveLLH( G, x );
31       for( int i=0; i<2; i++ ) x[i] = -x[i];
32
33       // impact of constraints
34       double df = 0;
35       int code = ConvexQP( G, x, Ce, ce, Ci, ci, df, 1.e-16 );
36       cout << setprecision(14);
37       cout << "code=" << code << " Convex Quadratic Solution x=" << endl << x << endl;
38       cout << "change to residual " << df << endl;
39    }
```

B.13   Adaptive Simpson's rule integration

file: AdaptiveSimpson.H

```
1    #ifndef __Adaptive_Simpson_H__
2    #define __Adaptive_Simpson_H__
3    double adaptiveSimpson( double a, double b, double tol, double (*func)(double) );
4
5    double adaptiveSimpson2( double& a, double& b,
6                            double& f0, double& f3, double& f6,
7                            double& icoarse, int depth,
8                            double (*func)(double), double iest );
9    #endif
```

---

file: AdaptiveSimpson.cpp

```
1   #include <math.h>
2
3   #ifndef MAX
4   #define MAX(A,B) (((A)>(B))?(A):(B))
5   #endif
6
7   // recursive adaptive Simpson
8   //  inputs:
9   //   a,b:       limits of integration
10  //   f0,f3,f6:  f(a), f( (a+b)/2 ), f(b)
11  //   icoarse:   coarse estimate of integral
12  //   depth:     useful for debugging.
13  //   iest:      estimate of abs of overall integral
14  //  output:  return value of integral over range a,b
15
16  double adaptiveSimpson2( double& a, double& b,
17                          double& f0, double& f3, double& f6,
18                          double& icoarse, int depth,
19                          double (*func)(double), double iest )
20  {
21     double w = (b-a)/3;
22     double x1 = a+w;
23     double x2 = b-w;
24
25     double xl = 0.5*(a+x1);
26     double f1 = (*func)(xl);
27     double f2 = (*func)(x1);
28     double f4 = (*func)(x2);
29     double xr = 0.5*(x2+b);
30     double f5 = (*func)(xr);
31
32     double is1 = w*( f0 + 4*f1 + f2 )/6;
33     double is2 = w*( f2 + 4*f3 + f4 )/6;
34     double is3 = w*( f4 + 4*f5 + f6 )/6;
35     double ifine = is1 + is2 + is3;
36
37     // (1) if the local integral is insignificant relative to the integral estimate.
38     bool stop1 = (iest + (ifine - icoarse) == iest);
39     // (2) if the interval is too small to contain unique representable quad pts.
40     bool stop2 = ( (xl <= a) || (xr >= b) );
41
42     if( stop1 || stop2 ) {
43        return ifine;
44     } else {
45        return
46           adaptiveSimpson2( a,  x1, f0, f1, f2, is1, depth+1, func, iest ) +
47           adaptiveSimpson2( x1, x2, f2, f3, f4, is2, depth+1, func, iest ) +
48           adaptiveSimpson2( x2, b,  f4, f5, f6, is3, depth+1, func, iest );
49     }
50  }
51
52  // a front end to adaptiveSimspon2 to simplify useage
53
54  double adaptiveSimpson( double a, double b, double tolerance,
55                          double (*func)(double) )
56  {
57     double f0 = (*func)(a);
58     double f3 = (*func)(0.5*(a+b));
59     double f6 = (*func)(b);
60     double icoarse = (b-a)*(f0 + 4*f3 + f6)/6;
61
62     double w = (b-a)/3,
63     double x1 = a+w;
64     double x2 = b-w;
```

```
65      double  x1 = 0.5*(a+x1);
66      double  f1 = (*func)(x1);
67      double  f2 = (*func)(x1);
68      double  f4 = (*func)(x2);
69      double  xr = 0.5*(x2+b);
70      double  f5 = (*func)(xr);
71
72      double  is1 = w*( f0 + 4*f1 + f2 )/6;
73      double  is2 = w*( f2 + 4*f3 + f4 )/6;
74      double  is3 = w*( f4 + 4*f5 + f6 )/6;
75      double  ifine = is1 + is2 + is3;
76
77      double  iest = MAX( fabs(icoarse), fabs(ifine) );
78      // rescale per target error
79      double  machine = 1.1e−16;
80      iest *= tolerance/machine;
81
82      // (1) if the local integral is insignificant relative to the integral estimate.
83      bool stop1 = (iest + (ifine − icoarse) == iest);
84      // (2) if the interval is too small to contain unique representable quad pts.
85      bool stop2 = ( (x1 <= a) || (xr >= b) );
86
87      if( stop1 || stop2 ) {
88        return ifine;
89      } else {
90        return
91          adaptiveSimpson2( a,   x1, f0, f1, f2, is1, 2, func, iest ) +
92          adaptiveSimpson2( x1, x2, f2, f3, f4, is2, 2, func, iest ) +
93          adaptiveSimpson2( x2, b,  f4, f5, f6, is3, 2, func, iest );
94      }
95  }
```

file: AdaptiveSimpson-test.cpp

```
1   #include <iostream>
2   #include <iomanip>
3   #include <math.h>
4   #include "AdaptiveSimpson.H"
5   using namespace std;
6
7   // function to integrate
8   double f( double x )
9   {
10    return cos(1./x);
11  }
12
13  int main()
14  {
15    cout << scientific << setprecision(15);
16    double a   = 0.02;
17    double b   = 100;
18    double tol = 1.e−6;
19    double i = adaptiveSimpson( a, b, tol, f );
20    cout << "integral =  " << i << endl;
21  }
```

B.14   Adaptive Runge–Kutta ODE example

file: AdaptiveSS.H

```
1  #ifndef __Adaptive_SS_H__
2  #define __Adaptive_SS_H__
3  void RK4( double x, double h, double y, double& ynew, double (*func)(double,double) );
4
5  void adaptiveSSintegrator( double a, double b, double relerr, double h0,
6                             double (*func)(double,double), int& step, double& y );
7  #endif
```

file: AdaptiveSS.cpp

```
1   #include <math.h>
2
3   #define MIN(A,B)  (((A<B)?(A):(B)))
4   #define MAX(A,B)  (((A>B)?(A):(B)))
5
6   // a single step of the 4th order Runge–Kutta method.
7
8   void RK4( double x, double h, double y, double& ynew, double (*func)(double,double) )
9   {
10    double k1 = (*func)(x,y);
11    double k2 = (*func)(x+h/2, y+h*k1/2);
12    double k3 = (*func)(x+h/2, y+h*k2/2);
13    double k4 = (*func)(x+h, y+h*k3);
14    ynew =  y + h*(k1 + k4 + 2*(k2+k3) )/6;
15  }
16
17  void adaptiveSSintegrator( double a, double b, double relerr, double h0,
18                             double (*func)(double,double), int& step, double& y )
19  {
20    double x, h, ycoarse, yfine, beta;
21    double eps;
22
23    x = a;
24    y = 1./901;
25    h = h0;
26    step = 0;
27    do {
28      step++;
29
30      do {
31        eps = relerr*fabs(y);
32
33        // coarse calculation
34        RK4( x, h, y, ycoarse, func );
35
36        // fine calculation
37        RK4( x,     h/2, y,     yfine, func );
38        RK4( x+h/2, h/2, yfine, yfine, func );
39
40        // find new h
41        beta = pow( eps*15/MAX(fabs(yfine-ycoarse),1.e-16), 0.2 );
42        beta = MIN(beta,2); // prevent too big of a step
43
44        if( beta > 0.5 ) {
45          y = yfine;
46          x += h;
47          h = h*beta;
48          break;
49        }
50        h = h*beta;
51      } while(1);
52      if( b-x > h ) h = b-x;
53    } while( x < b );
54  }
```

file: AdaptiveSS-test.cpp

```
1   #include <iostream>
2   #include <iomanip>
3   #include <math.h>
4   #include "AdaptiveSS.H"
5   using namespace std;
6
7   double f( double x, double y )
8   {
9     return - 200*x*y*y;
10  }
11
12  int main()
13  {
14    double a  = -3;
15    double b  =  0;
16    double h0 = 0.1;
17    double y  = 1./901;
18    double relerr = 2.e-16;
19    int    step;
20    adaptiveSSintegrator( a, b, relerr, h0, f, step, y );
21    cout << scientific << setprecision(15);
22    cout << "end   y " << y << " error " << y-1 << endl;
23    cout << "number of steps " << step << endl;
24  }
```

B.15   Adaptive multistep ODE example

file: AdaptiveMS.H

```
1   #ifndef __Adaptive_MS_H__
2   #define __Adaptive_MS_H__
3
4   void makeGs( Vector<double>& abscissas,
5                Vector<double>& gpred,
6                Vector<double>& gcorr );
7
8   void NewtonLastDiag( double& xnew, double& fnew,
9                        Vector<double>& abscissas,
10                       Vector<double>& tableau );
11
12  void rollup( Vector<double>& abscissas, Vector<double>& tableau );
13
14  void predictor( int p,
15                  const Vector<double>& abscissas,
16                  const Vector<double>& gpred,
17                  const Vector<double>& tableau,
18                  double& y );
19
20  void corrector( int p,
21                  const Vector<double>& abscissas,
22                  const Vector<double>& gcorr,
23                  const Vector<double>& tableau,
24                  double& y );
25
26  void adaptiveMSintegrator( double a, double b, double relerr,
27                             int maxallocorder, double h0,
28                             double (*func)(double,double),
29                             int& maxorder, int& step, double& y );
30  #endif
```

---

file: AdaptiveMS.cpp

```
1   #include <math.h>
2   #include "Matrix.H"
3   #define MIN(A,B) (((A<B)?(A):(B)))
4   #define MAX(A,B) (((A>B)?(A):(B)))
5
6   // the integrals of factors (x-x_k)*(x-x_(k-1)) etc.
7   // "gpred" are integrals used for predictor,
8   // and "gcorr" are integrals for corrector.
9   // recults include those from recursion, and the trivial cases.
10
11  void makeGs( Vector<double>& abscissas ,
12               Vector<double>& gpred ,
13               Vector<double>& gcorr )
14  {
15    // assume abscissas is equipped for variable order, so
16    int s = abscissas.size();
17    gpred.resize(s+1);
18    gcorr.resize(s+1);
19    Vector<double> tmp(s);
20
21    // make g1j. assume most recent x is last
22    double dx = abscissas[s-2] - abscissas[s-1];
23    double t = dx*dx;
24    for( int j=1; j<=s; j++ ) {
25      tmp[j-1] = t/(j*(j+1.));
26      t *= dx;
27    }
28
29    // the trivial cases, not from recursion
30    gpred[0] = -dx;
31    gcorr[0] = -dx;
32    gcorr[1] = -0.5*dx*dx;
33
34    // do recursion
35    for( int i=1; i<s; i++ ) {
36      if( i != 1 ) {
37        for( int j=1; j<=s+1-i; j++ ) {
38          tmp[j-1] *= (abscissas[s-1]-abscissas[s-1-i]);
39          tmp[j-1] += tmp[j];
40        }
41      }
42      gpred[i] = tmp[0];
43      if( i<s-1 ) gcorr[i+1] = tmp[1];
44    }
45  }
46
47  // construct only the bottom diagonal of the Newton tableau —
48  // assuming the rest have already been calculated.
49
50  void NewtonLastDiag( double& xnew, double& fnew,
51                       Vector<double>& abscissas ,
52                       Vector<double>& tableau )
53  {
54    int n = abscissas.size();
55    int i = n-1;
56    abscissas[i] = xnew;
57    int j = (i*(i+1))/2;
58    tableau[j] = fnew;
59    for( int k=j+1; k<=j+i; k++ ) {
60      tableau[k] = (tableau[k-1]-tableau[k-i-1])/(abscissas[i]-abscissas[i+j-k]);
61    }
62  }
63
64  // roll data "up" in the Newton tableau so the last diagonal is empty
```

```
65    // for the next iteration.
66
67    void rollup( Vector<double>& abscissas, Vector<double>& tableau )
68    {
69      int rows = abscissas.size();
70      int k = 0;
71      for( int r=0; r<rows-1; r++ ) {
72        abscissas[r] = abscissas[r+1];
73        int k2 = ((r+1)*(r+2))/2;
74        for( int c=0; c<=r; c++ ) {
75          tableau[k++] = tableau[k2++];
76        }
77      }
78    }
79
80    // compute the predictor. on input, y is previous value.
81    // on output, next value by predictor routine of order "p"
82
83    void predictor( int p,
84                    const Vector<double>& abscissas,
85                    const Vector<double>& gpred,
86                    const Vector<double>& tableau,
87                    double& y )
88    {
89      int s = abscissas.size();
90      int k = ((s-2)*(s-1))/2;
91      for( int j=0; j<p; j++ )
92        y += tableau[k+j]*gpred[j];
93    }
94
95    // compute the corrector. on input, y is previous value.
96    // on output, next value by corrector routine of order "p"
97
98    void corrector( int p,
99                    const Vector<double>& abscissas,
100                   const Vector<double>& gcorr,
101                   const Vector<double>& tableau,
102                   double& y )
103   {
104     int s = abscissas.size();
105     int k = (s*(s-1))/2;
106     for( int j=0; j<p; j++ )
107       y += tableau[k+j]*gcorr[j];
108   }
109
110   // input:
111   //    a,b            limits of x
112   //    y              y(a)
113   //    maxallocorder  max order method that can be used
114   //    h0             input guess for steps size
115   //    func           dy/dx = func(x,y)
116   // output:
117   //    y              y(b)
118   //    maxorder       max order actually used
119   //    step           number of steps taken
120
121   void adaptiveMSintegrator(  double a, double b, double relerr,
122                               int maxallocorder, double h0,
123                               double (*func)(double,double),
124                               int& maxorder, int& step, double& y )
125   {
126
127     double hold, ypred, ycorr, xp, beta, betam, betap;
128     double steperr, err;
129     Vector<double> abscissas(2), tableau(3), gpred, gcorr;
130     double h = h0;                  // an initial value
131     double x = a;                   // starting value
132     int    trial=0;
133     int    allocorder = 1;          // tableau has memory allocated for this order
134     int    order = 1;               // initial order
```

```
135
136     maxorder = 1;
137     step     = 0;
138
139     // put starting info into tableau.
140     // NOTE: we will arrange it so the most recent value of the absicssa is
141     // always zero — to avoid subtracting large numbers.
142     abscissas[0] = 0;
143     tableau[0]   = (*func)(x,y);
144
145     do {
146       trial = 1;
147       if( order > maxorder ) maxorder = order;
148       do {
149         err = relerr * y;
150         h = MIN(b − x, h);
151         hold = h;
152         xp = x+h;
153         abscissas[abscissas.size()−1] = h;
154         makeGs( abscissas, gpred, gcorr );
155
156         // predictor
157         ypred = y;
158         predictor( order, abscissas, gpred, tableau, ypred );
159
160         // fill out tableau
161         double fp = (*func)(xp,ypred);
162         NewtonLastDiag( h, fp, abscissas, tableau );
163
164         // corrector
165         ycorr = y;
166         corrector( order, abscissas, gcorr, tableau, ycorr );
167
168         // compute new step size, see if order change indicated
169         int k = abscissas.size();
170         k = (k*(k−1))/2;
171         steperr = fabs(gcorr[order−1]*tableau[k+order]);
172         beta = pow(err/steperr, 1./(order+1.));
173         if( beta > 0.5 ) {
174           betam = −1, betap = −1;
175           if( order > 1 ) {
176             steperr = fabs(gcorr[order−2]*tableau[k+order−1]);
177             betam = pow(err/steperr, 1./order);
178           }
179           if( order < allocorder ) {
180             steperr = fabs(gcorr[order]*tableau[k+order+1]);
181             betap = pow(err/steperr, 1./(order+2));
182           }
183           if( betam > MAX(beta,betap) ) {
184             // drop order
185             h = h*MAX(MIN(16.,betam),1./16.);
186             order −−;
187           } else if( betap > MAX(beta,betam) ) {
188             // raise order
189             h = h*MAX(MIN(16.,betap),1./16.);
190             order++;
191           } else {
192             h = h*MAX(MIN(16.,beta),1./16.);
193           }
194           break;
195         }
196         // the 16 and 1/16 are limits to control crazy steps — safety factors.
197         h = h*MAX(MIN(16.,beta),1./16.);
198       } while(++trial <20);
199       ++step;
200       double fp = (*func)(xp,ycorr);
201       NewtonLastDiag( hold, fp, abscissas, tableau );
202
203       // reset the abscissas in the tableau in order to mitigate
204       // cancellation errros.
```

```
205        for( int i=0; i<abscissas.size(); i++ ) abscissas[i] -= hold;
206
207        x = xp;
208        y = ycorr;
209
210        if( (order+1 > allocorder) && (allocorder != maxallocorder) ) {
211          // increase tableau size
212          int s = abscissas.size();
213          abscissas.resize(s+1);
214          tableau.resize(((s+1)*(s+2))/2);
215          allocorder++;
216        } else {
217          // keep size contant, roll values up
218          rollup( abscissas, tableau );
219        }
220      } while(x<b);
221  }
```

file: AdaptiveMS-test.cpp

```
1   #include <iostream>
2   #include <iomanip>
3   #include "Matrix.H"
4   #include "AdaptiveMS.H"
5   using namespace std;
6
7   #define MIN(A,B) (((A<B)?(A):(B)))
8   #define MAX(A,B) (((A>B)?(A):(B)))
9
10  double f( double x, double y )
11  {
12    return - 200*x*y*y;
13  }
14
15  int main()
16  {
17    double a = -3;
18    double b =  0;
19    double y =  1./901;
20    double h0 = 0.0001;
21    double relerr = 1.e-15;
22    int    maxallocorder = 20;
23    int    maxorder, step;
24    adaptiveMSintegrator( a, b, relerr, maxallocorder, h0, f, maxorder, step, y );
25    cout << scientific << setprecision(15);
26    cout << "end  y " << y << " error " << y-1 << endl;
27    cout << "maximum order " << maxorder << endl;
28    cout << "number of steps " << step << endl;
29  }
```

B.16   Stochastic integration and testing

file: SDEintegrate.cpp

```
1   #include <iostream>
2   #include <iomanip>
3   #include <math.h>
4   #include "dSFMT.h"
```

```
 5   #include "Matrix.H"
 6   using namespace std;
 7   #ifdef USEMPI
 8   #include <mpi.h>
 9   int rank, nproc;
10   #endif
11   bool doio;
12
13   #ifndef MIN
14   #define MIN(A,B) (((A)<(B))?(A):(B))
15   #endif
16
17   // initialize the random number generator. Here, dSFMT
18   // see http://www.math.sci.hiroshima-u.ac.jp/~m-mat/MT/emt.html
19
20   dsfmt_t internal_state;
21
22   void setup(uint32_t seed)
23   {
24     dsfmt_init_gen_rand( &internal_state, seed );
25   }
26
27   // Box-Muller transform to make gaussian normal deviates
28   // from uniform deviates.
29
30   double normaldeviate()
31   {
32     static int got_one = 0;
33     static double extra_value;
34     double v1, v2, tmp, rsq;
35
36     if( got_one ) {
37       got_one = 0;
38       return extra_value;
39     } else {
40       do {
41         // get 2 uniform deviates on [-1,1]
42         // dsfmt_genrand_close_open() returns a number on [0,1)
43         v1 = 2.*dsfmt_genrand_close_open(&internal_state)-1.;
44         v2 = 2.*dsfmt_genrand_close_open(&internal_state)-1.;
45         rsq = v1*v1 + v2*v2;
46       } while( rsq==0 || rsq >= 1 );
47       tmp = sqrt( - 2.*log(rsq)/rsq );
48       extra_value = tmp*v2;
49       got_one = 1;
50       return tmp*v1;
51     }
52   }
53
54   void makeWZ( int nstep, double& h,
55                Vector<double>& DW, Vector<double>& DZ )
56   {
57     // make npath*nstep Wiener steps DW, and consistent DZ
58     double sqrth = sqrt(h);
59     double h32 = h*sqrth;
60     double s3     = sqrt(3.);
61     double U1, U2;
62
63     for( int n=0; n<nstep; n++ ) {
64       U1 = normaldeviate();
65       U2 = normaldeviate();
66       DW[n] = sqrth*U1;
67       DZ[n] = 0.5*h32*(U1 + U2/s3);
68     }
69   }
70
71   void coarsenby2( int nstepfine, double& hfine,
72                    Vector<double>& DW, Vector<double>& DZ )
73   {
74     int nstepcoarse = nstepfine/2;
```

```
75      int  nfine = 0;
76      for( int n=0; n<nstepcoarse; n++ ) {
77        // the fine steps in corresponding coarse step
78        double dWf1 = DW[ nfine ];
79        double dZf1 = DZ[ nfine ];
80        nfine ++;
81        double dWf2 = DW[ nfine ];
82        double dZf2 = DZ[ nfine ];
83        nfine ++;
84
85        // the coarse step
86        DW[n] = dWf1 + dWf2;
87        DZ[n] = dZf1 + dZf2 + hfine*dWf1;
88      }
89    }
90
91    void integrate( int nstep, double& h,
92                        Vector<double>& DW,  Vector<double>& DZ,
93                        double& result )
94    {
95      // Ito−Taylor expansion of dx = f*dt + g*dW with
96      //  f = (x − x^3 + A*cos[2*Pi*t])/eps
97      //  g = sig/sqrt(eps)
98      //  after Berglund and Gentz 2002
99
100     double eps = 0.0025;
101     double sqe = sqrt(eps);
102     double a0  = 0.005;
103     double sig = 0.065;
104     double x0  = 1.15303;
105     double A   = 2./(3.*sqrt(3.)) − a0;
106
107     // IC
108     double x = x0;
109     double t= 0;
110     // integrate
111     for( int n=0; n<nstep; n++ ) {
112       double ac = A*cos(2.*M_PI*t);
113       double dac = −2*A*M_PI*sin(2.*M_PI*t);
114       x += (sig/sqe)*DW[n] + (x*(1.−x*x) + ac)*h/eps
115         + 0.5*h*h*( x*((1.−3*sig*sig) + x*x*(−4.+3.*x*x))
116           + ac*(1.−3.*x*x) + eps*dac )/(eps*eps)
117           + (1.−3.*x*x)*sig*DZ[n]/(eps*sqe);
118       t += h;
119     }
120     result = x;
121   }
122
123   void analyze( double hfine, int nstep, int nrefine, int npath, int nmom )
124   {
125     Vector<double> DW(nstep);
126     Vector<double> DZ(nstep);
127     Vector<double> refres(nrefine);
128     Vector<double> strong_err(nrefine −1);
129     Vector<double> weak_err(nrefine −1);
130     Matrix<double> weak_expect(nrefine ,nmom);
131
132     for( int p=0; p<npath; p++ ) {
133       makeWZ( nstep, hfine, DW, DZ );
134       double resh = hfine;
135       int resn = nstep;
136       for( int r=0; r<nrefine; r++ ) {
137         double result;
138         integrate( resn, resh, DW, DZ, result );
139         refres[r] = result;
140         if( r != nrefine−1 ) {
141           coarsenby2( resn, resh, DW, DZ );
142           resn /= 2;
143           resh *= 2;
144         }
```

```
145        }
146        // accumulate strong error
147        for( int r=0; r<nrefine -1; r++ ) {
148          double e = refres [r] - refres [r+1];
149          strong_err [r] += e*e;
150        }
151        // accumulate weak expectation
152        for( int r=0; r<nrefine; r++ ) {
153          for( int m=1; m<=nmom; m++ ) {
154            weak_expect [r][m-1] += pow( refres [r], (double) m );
155          }
156        }
157      }
158
159      // accumulate across processors
160      long long allnpath = npath;
161   #ifdef USEMPI
162      double *mydat = new double[nrefine];
163      double *alldat = new double[nrefine];
164
165      // sum strong errors
166      for( int i=0; i<nrefine -1; i++ ) mydat[i] = strong_err[i];
167      MPI_Allreduce( mydat, alldat, nrefine -1, MPI_DOUBLE, MPI_SUM, MPI_COMM_WORLD );
168      for( int i=0; i<nrefine -1; i++ ) strong_err[i] = alldat[i];
169
170      // weak expectations
171      for( int m=1; m<=nmom; m++ ) {
172        for( int i=0; i<nrefine; i++ ) mydat[i] = weak_expect[i][m-1];
173        MPI_Allreduce( mydat, alldat, nrefine, MPI_DOUBLE, MPI_SUM, MPI_COMM_WORLD );
174        for( int i=0; i<nrefine; i++ ) weak_expect[i][m-1] = alldat[i];
175      }
176
177      // count
178      long long mynpath = npath;
179      MPI_Allreduce( &mynpath, &allnpath, 1, MPI_LONG_LONG, MPI_SUM, MPI_COMM_WORLD );
180
181      delete [] alldat;
182      delete [] mydat;
183   #endif
184
185      // and generate convergence results
186      for( int r=0; r<nrefine -1; r++ ) {
187        strong_err[r] /= allnpath;
188        strong_err[r] = sqrt(strong_err[r]);
189        if( doio )
190          cout << "strong error res=" << r << " " << strong_err[r] << endl;
191      }
192      if( doio ) {
193        for( int r=0; r<nrefine -2; r++ ) {
194          cout << "strong order " << log(strong_err[r+1]/strong_err[r])/log(2.) << endl;
195        }
196      }
197
198      for( int m=1; m<=nmom; m++ ) {
199        for( int r=0; r<nrefine; r++ )
200          weak_expect[r][m-1] /= allnpath;
201        for( int r=0; r<nrefine -1; r++ ) {
202          weak_err[r] = fabs(weak_expect[r][m-1] - weak_expect[r+1][m-1]);
203          if( doio )
204            cout << "m=" << m << " weak error res=" << r << " " << weak_err[r] << endl;
205        }
206        if( doio ) {
207          for( int r=0; r<nrefine -2; r++ ) {
208            cout << "m=" << m << " weak order "
209                 << log(weak_err[r+1]/weak_err[r])/log(2.) << endl;
210          }
211        }
212      }
213    }
214
```

```
215   int main(int argc, char *argv[])
216   {
217   #ifdef USEMPI
218     MPI_Init(&argc,&argv);
219     MPI_Comm_rank(MPI_COMM_WORLD,&rank);
220     MPI_Comm_size(MPI_COMM_WORLD,&nproc);
221     doio = (rank==0);
222   #else
223     rank = 0;
224     dioo = true;
225   #endif
226     cout << scientific << setprecision(6) << endl;
227     setup(rank);
228     int npath = 1000000;
229     int nrefine = 8;
230     int nstep = 16384;
231     double hfine = 1.e-3/nstep;
232     int nmom = 5; // necessary to prove 2nd order
233     analyze( hfine, nstep, nrefine, npath, nmom );
234   #ifdef USEMPI
235     MPI_Finalize();
236   #endif
237     return 0;
238   }
```

B.17   Big example: Hartree–Fock–Roothaan

The following code computes fits and interpolation coefficients for the $n = 4$ Rys polynomials. It creates the file "rysfit.H" which is used in the Hartree–Fock–Roothaan code.

file: Rysfit.cpp

```
1    #include <iostream>
2    #include <fstream>
3    #include <iomanip>
4    #include <math.h>
5    #include "Matrix.H"
6    #include "QR.H"
7    #include "AdaptiveSimpson.H"
8    #include "Householder.H"
9    #include "HQsort.H"
10   using namespace std;
11
12   fstream io;
13
14   // Stieltjes values are global for benefit of integrator
15   Vector<double> s_delta;
16   Vector<double> s_gammasq;
17
18   // more global info for integrator
19   int     s_pwr;
20   bool    s_extra_x;
21   double  g_alpha;
22
23   // evaluate polynomial degree "deg" with Stieltjes recursion
24   double polyfn( double x, int deg )
25   {
26     double pm2=0, pm1=0, p=1;
27     for( int i=1; i<=deg; i++ ) {
28       pm2 = pm1;
29       pm1 = p;
30       p *= x-s_delta[i-1];
```

```
31        if( i!=1 ) p -= s_gammasq[i-2]*pm2;
32      }
33      return p;
34   }
35
36   // the function to be evaluated for Stieltjes form
37   double orth_rys_f( double x )
38   {
39      double pi = polyfn(x*x,s_pwr);
40      double retval = pi*pi*exp(-g_alpha*x*x);
41      if( s_extra_x ) retval *= x*x;
42      return retval;
43   }
44
45   void coefficients_from_Stieltjes( int maxorder, double& p0p0 )
46   {
47      s_delta.resize(maxorder);        // in case
48      s_gammasq.resize(maxorder-1);
49
50      double oldIpipi;
51      for( int i=0; i<maxorder; i++ ) {
52        s_pwr    = i;       // global variable passes degree to orth_rys_f
53        s_extra_x = false;
54        double Ipipi = adaptiveSimpson( 0., 1., 1.e-15, orth_rys_f );
55        s_extra_x = true;
56        double Ixpipi = adaptiveSimpson( 0., 1., 1.e-15, orth_rys_f );
57
58        if( i == 0 ) p0p0 = Ipipi;
59        s_delta[i] = Ixpipi/Ipipi;
60        if( i != 0 ) s_gammasq[i-1] = Ipipi/oldIpipi;
61        oldIpipi = Ipipi;
62      }
63   }
64
65   // global power for benefit of moment integrator
66   int f_pwr;
67
68   // integrand for moment calculation
69   void calcRysData( const int& order, const double& a_alpha,
70                     Vector<double>& x, Vector<double>& W )
71   {
72      g_alpha = a_alpha;
73      double p0p0;
74      Vector<double> delta(order), gamma(order-1);
75
76      coefficients_from_Stieltjes( order, p0p0 );
77      // copy from global array
78      for( int i=0; i<order-1; i++ ) {
79        delta[i] = s_delta[i];
80        gamma[i] = sqrt(s_gammasq[i]);
81      }
82      delta[order-1] = s_delta[order-1];
83
84      // J
85      Matrix<double> J(order,order);
86      for( int i=0; i<order; i++ ) {
87        for( int j=0; j<order; j++ ) J[i][j] = 0;
88        J[i][i] = delta[i];
89      }
90      for( int i=0; i<order-1; i++ )
91        J[i][i+1] = J[i+1][i] = gamma[i];
92
93      // eigenvalues and vectors
94      Matrix<double> V(order,order);
95      QReig( J, V, x );
96
97      // weights
98      for( int i=0; i<order; i++ ) {
99        double len = 0;
100       for( int j=0; j<order; j++ ) len += V[j][i]*V[j][i];
```

```
101        W[i] = (p0p0/len)*V[0][i]*V[0][i];
102      }
103
104      // sort
105      Vector<int> perm(order);
106      Vector<double> wtmp(order);
107      hqsort0( x, perm );
108      for( int i=0; i<order; i++ ) wtmp[i] = W[perm[i]];
109      for( int i=0; i<order; i++ ) W[i] = wtmp[i];
110    }
111
112    void Chebyfit( const double& a, const double& b, int npts, const string& prefix )
113    {
114      io << showpos << scientific << setprecision(15);
115      io << "// a=" << a << " b=" << b << endl;
116      int order = 4;
117      Vector<double> t(order), W(order), bj(npts);
118      Matrix<double> Ws(order,npts), ts(order,npts);
119
120      for( int i=0; i<npts; i++ ) {
121        double xk = cos(M_PI*(2.*i+1)/(2.*npts));
122        double alpha = (b+a)/2 + xk*(b-a)/2;
123        calcRysData( order, alpha, t, W );
124        for( int j=0; j<order; j++ ) {
125          Ws[j][i] = W[j];
126          ts[j][i] = t[j];
127        }
128      }
129      for( int wt=0; wt<2; wt++ ) {
130        char c = wt ? 'W' : 't';
131        for( int o=0; o<order; o++ ) {
132          for( int j=0; j<npts; j++ ) bj[j] = 0;
133          for( int k=0; k<npts; k++ ) {
134            double xk = cos(M_PI*(2.*k+1)/(2.*npts));
135            double Tjold=0, Tj=1;
136            double val = (wt ? Ws[o][k] : ts[o][k] );
137            for( int j=0; j<npts; j++ ) {
138              bj[j] += Tj*val;
139              if( j==0 ) {
140                Tj = xk;
141                Tjold = 1;
142              } else {
143                double foo = Tj;
144                Tj = 2*xk*Tj - Tjold;
145                Tjold = foo;
146              }
147            }
148          }
149          for( int j=0; j<npts; j++ )
150            bj[j] *= ((j==0)?1.:2.)/npts;
151          io << "static double " << prefix << c << noshowpos << o
152             << "[" << npts << showpos << "]={" << endl;
153          for( int i=0; i<npts; i++ ) {
154            io << bj[i];
155            if( i != npts-1 ) io << ", ";
156            if( i%3 == 2 && i != npts-1 ) io << endl;
157          }
158          io << "};" << endl;
159        }
160      }
161    }
162
163    void LaurentFit( const double& xmin, const double& xmax,
164                     int npts, int kmin, int kmax )
165    {
166      static double rHermite[4] = {
167        2.9306374202572440192235027052 4, 1.98165675669584292585463063976,
168        1.15719371244678019472076577906, 0.38118699020732211685471888558 };
169
170      static double wHermite[4] = {
```

```
171      0.000199604072211367619206090452 5, 0.01707798300741347545620305643 64,
172      0.207802325814891879543258620285 7, 0.661147012558241291030415974495 8 };
173
174    int order = 4;
175    int nk = kmax+1-kmin;
176    Vector<double> t(order), W(order), xpts(npts), b(npts), diag(nk);
177    Matrix<double> Ws(order,npts), ts(order,npts), A(npts,nk);
178    Vector<int> Pl(npts), Pr(nk);
179
180    io << showpos << scientific << setprecision(15);
181    io << "// a=" << xmin << " b=" << xmax << endl;
182
183    int n1 = npts/3;
184    int n2 = npts*2/3;
185    double x1 = xmin + 0.1*(xmax-xmin);
186    double x2 = x1 + 0.25*(xmax-x1);
187    for( int i=0; i<n1; i++ ) {
188      xpts[i] = xmin + i*(x1-xmin)/n1;
189    }
190    for( int i=n1; i<n2; i++ ) {
191      xpts[i] = x1 + (i-n1)*(x2-x1)/(n2-n1);
192    }
193    for( int i=n2; i<npts; i++ ) {
194      xpts[i] = x2 + (i-n2)*(xmax-x2)/(npts - n2 -1);
195    }
196
197    for( int i=0; i<npts; i++ ) {
198      calcRysData( order, xpts[i], t, W );
199      for( int j=0; j<order; j++ ) {
200        Ws[j][i] = W[j];
201        ts[j][i] = t[j];
202      }
203    }
204
205    for( int wt=0; wt<2; wt++ ) {
206      char c = wt ? 'W' : 't';
207      for( int o=0; o<order; o++ ) {
208
209        for( int i=0; i<npts; i++ ) {
210          for( int j=0; j<nk; j++ ) A[i][j] = 0;
211          b[i] = 0;
212        }
213
214        for( int i=0; i<npts; i++ ) {
215          double x = xpts[i];
216          double xs = x/xmax; // use scaled variable
217          double dat = wt ? Ws[o][i] : ts[o][i];
218          if( wt ) {
219            dat -= wHermite[o]/sqrt(x);
220          } else {
221            double rsq = rHermite[o]; rsq *= rsq;
222            dat -= rsq/x;
223          }
224          b[i] = dat;
225          double xk = pow(xs,(double)kmin);
226          for( int j=0; j<nk; j++ ) {
227            // weight by exp(-x)
228            A[i][j] = xk*exp(-x);
229            // A[i][j] = xk;
230            xk *= xs;
231          }
232        }
233
234        Householder( A, Pl, Pr, diag, POWELLREID );
235        solveQR( A, Pl, Pr, diag, b );
236
237        io << "static double " << c << noshowpos << o << "coef[" << nk
238          << showpos << "]={" << endl;
239        for( int i=0; i<nk; i++ ) {
240          io << b[i];
```

```
241              if ( i != nk-1 ) io << ", ";
242              if ( i%3 == 2 && i != nk-1 ) io << endl;
243          }
244          io << "};" << endl;
245      }
246    }
247  }
248
249  int main()
250  {
251    io.open("rysfit.H", fstream::out );
252    // fit to Laurent series 20,50
253    LaurentFit( 20, 50, 150, -3, 7 );
254    // fit to Chebyshev in range 0,1
255    Chebyfit( 0.0, 1.0, 32, "s1" );
256    // fit to Chebyshev in range 1,5
257    Chebyfit( 1.0, 5.0, 32, "s2" );
258    // fit to Chebyshev in range 5,10
259    Chebyfit( 5.0, 10., 32, "s3" );
260    // fit to Chebyshev in range 10,15
261    Chebyfit( 10., 15., 32, "s4" );
262    // fit to Chebyshev in range 15,20
263    Chebyfit( 15., 20., 32, "s5" );
264    io.close();
265  }
```

This is the big example: optimize the structure of H_2O using the Hartree–Fock–Roothaan self-consistent LCAO approach.

file: HFRopt.cpp

```
1   #include <iostream>
2   #include <iomanip>
3   #include <sstream>
4   #include <math.h>
5   #include "Matrix.H"
6   #include "SVD.H"
7   #include "QR.H"
8   #include "HQsort.H"
9   #include "VMM.H"
10  #include "rysfit.H"
11  using namespace std;
12
13  // McMurchie-Davidson d_0^{nmu,nlambda} factors for overlap and KE.
14
15  double MD( const double& alpha,
16            const int&    a_nmu,    const int&    a_nlambda,
17            const double& a_dx_amu, const double& a_dx_alambda )
18  {
19    // order so what we call mu has the bigger n
20    int nmu, nlambda;
21    double dx_amu, dx_alambda;
22    if ( a_nmu >= a_nlambda ) {
23      nmu    = a_nmu;     dx_amu     = a_dx_amu;
24      nlambda = a_nlambda; dx_alambda = a_dx_alambda;
25    } else {
26      nmu    = a_nlambda; dx_amu     = a_dx_alambda;
27      nlambda = a_nmu;    dx_alambda = a_dx_amu;
28    }
29
30    Vector<double> d(nmu+nlambda+1);
31    d[0] = 1;
32    for ( int n=1; n<=nmu+nlambda; n++ ) {
33      d[n] = dx_amu*d[n-1];
34      if ( n>=2 ) d[n] += (n-1)*d[n-2]/(2*alpha);
35    }
36    for ( int m=1; m<=nlambda; m++ ) {
```

```
37        for( int n=0; n<=nmu+nlambda-m; n++ ) {
38          d[n] = (dx_alambda-dx_amu)*d[n] + d[n+1];
39        }
40      }
41      return d[nmu];
42    }
43
44    double Rys_A( const int& nmu,    const int& nlam,
45                  const double& tsq, const double& alpha,
46                  const double& xmu, const double& xlam,
47                  const double& xa,  const double& xi )
48    {
49      Vector<double>A (nmu+nlam+1);
50      A[0] = 1;
51      for( int n=1; n<=nmu+nlam; n++ ) {
52        A[n] = ( tsq*(xi-xa)  + (xa-xmu) )*A[n-1];
53        if( n >= 2 ) A[n] += (n-1)*(1-tsq)*A[n-2]/(2.*alpha);
54      }
55      for( int m=1; m<= nlam; m++ ) {
56        for( int n=0; n<=nmu+nlam-m; n++ ) {
57          A[n] = A[n]*(xmu-xlam) + A[n+1];
58        }
59      }
60      return A[nmu];
61    }
62
63    // Rys Dupuis King "G" factors, relative to G_00 = 1
64
65    double RDK_G( const int& na, const int& nb,
66                  const double& alpha, const double& beta, const double& tsq,
67                  const double& xa, const double& xb,
68                  const double& xi, const double& xk )
69    {
70      double b00 = tsq/(2.*(alpha+beta));
71      double b10 = 1./(2.*alpha)  - beta*b00/alpha;
72      double c00 = (xa-xi) + beta*tsq*(xb-xa)/(alpha+beta);
73      double cprime = (xb-xk) + alpha*tsq*(xa-xb)/(alpha+beta);
74      double bprime = 1./(2.*beta)  - alpha*b00/beta;
75      Matrix<double> G(na+1,nb+1);
76      for( int ia=0; ia<=na; ia++ )
77        for( int ib=0; ib<=nb; ib++ ) {
78          if( ib==0 && ia==0 ) {
79            G[ia][ib] = 1;
80          } else if( ib==0 ) {
81            G[ia][ib] = c00*G[ia-1][ib];
82            if( ia > 1 ) G[ia][ib] += (ia-1)*b10*G[ia-2][ib];
83    //      if( ib > 0 ) G[ia][ib] += ib*b00*G[ia-1][ib-1];
84          } else {
85            G[ia][ib] = cprime*G[ia][ib-1];
86            if( ib > 1 ) G[ia][ib] += (ib-1)*bprime*G[ia][ib-2];
87            if( ia > 0 ) G[ia][ib] += ia*b00*G[ia-1][ib-1];
88          }
89        }
90      }
91      return G[na][nb];
92    }
93
94    // mu,lambda on electron 1; nu,sigma on electron 2
95
96    double RDK_Gxfer( const int& nmu,       const int& nnu,
97                      const int& nlambda,   const int& nsigma,
98                      const double& A,      const double& B,        const double& tsq,
99                      const double& xa,     const double& xb,
100                     const double& xmu,    const double& xnu ,
101                     const double& xlambda, const double& xsigma )
102   {
103     Matrix<double> Ii(nlambda+1,nlambda+1);
104     Matrix<double> Ik(nsigma+1,nsigma+1);
105     for( int n2=nnu; n2<=nnu+nsigma; n2++ ) {
106       // for each "n" on electron 2 ...
```

```
107          // find the G_{n1,n2} factor
108          for( int n1=nmu; n1<=nmu+nlambda; n1++ ) {
109            // first index: related to n1
110            // second index: nlambda, initially zero
111            Ii[n1-nmu][0] = RDK_G( n1, n2, A, B, tsq, xa, xb, xmu, xnu );
112          }
113          // transfer momentum from mu to lambda
114          for( int j=1; j<=nlambda; j++ ) {
115            for( int n=nmu; n<=nmu+nlambda-j; n++ ) {
116              Ii[n-nmu][j] = Ii[n-nmu+1][j-1] + (xmu-xlambda)*Ii[n-nmu][j-1];
117            }
118          }
119          // save result in matrix Ik for nu/sigma transfer.
120          Ik[n2-nnu][0] = Ii[0][nlambda];
121        }
122        // transfer from nu to sigma
123        for( int n2=1; n2<=nsigma; n2++ ) {
124          for( int n1=nnu; n1<=nnu+nsigma-n2; n1++ ) {
125            Ik[n1-nnu][n2] = Ik[n1-nnu+1][n2-1] + (xnu-xsigma)*Ik[n1-nnu][n2-1];
126          }
127        }
128        return Ik[0][nsigma];
129      }
130
131      // use interpolations/fit to calculate n=4 Rys quadrature data
132
133      void Rys( const double& gamma, Vector<double>& rys_t, Vector<double>& rys_W )
134      {
135        static double rHermite[4] = {
136          2.9306374202572440192235027052 4, 1.981656756695842925854630639 76,
137          1.15719371244678019472076577906, 0.38118699020732211685471888558 };
138
139        static double wHermite[4] = {
140          0.0001996040722113676192060904525, 0.0170779830074134754562030564364,
141          0.2078023258148918795432586202857, 0.66114701255824129103041597449 58 };
142
143        int     order = 4;
144        double *tp[4], *wp[4], m, hw;
145        if( gamma <= 1 ) {
146          m=0.5; hw=0.5;
147          tp[0] = s1t0; tp[1] = s1t1; tp[2] = s1t2; tp[3] = s1t3;
148          wp[0] = s1W0; wp[1] = s1W1; wp[2] = s1W2; wp[3] = s1W3;
149        } else if( gamma <= 5 ) {
150          m=3; hw=2;
151          tp[0] = s2t0; tp[1] = s2t1; tp[2] = s2t2; tp[3] = s2t3;
152          wp[0] = s2W0; wp[1] = s2W1; wp[2] = s2W2; wp[3] = s2W3;
153        } else if( gamma <= 10 ) {
154          m = 7.5; hw=2.5;
155          tp[0] = s3t0; tp[1] = s3t1; tp[2] = s3t2; tp[3] = s3t3;
156          wp[0] = s3W0; wp[1] = s3W1; wp[2] = s3W2; wp[3] = s3W3;
157        } else if( gamma <= 15 ) {
158          m = 12.5; hw=2.5;
159          tp[0] = s4t0; tp[1] = s4t1; tp[2] = s4t2; tp[3] = s4t3;
160          wp[0] = s4W0; wp[1] = s4W1; wp[2] = s4W2; wp[3] = s4W3;
161        } else if( gamma <= 20 ) {
162          m = 17.5; hw=2.5;
163          tp[0] = s5t0; tp[1] = s5t1; tp[2] = s5t2; tp[3] = s5t3;
164          wp[0] = s5W0; wp[1] = s5W1; wp[2] = s5W2; wp[3] = s5W3;
165        }
166
167        if( gamma <= 20 ) {
168          int nc=20; // number of terms to include
169          // Clenshaw algorithm
170          Vector<double> y(nc+2);
171          y[nc+1] = y[nc] = 0;
172          // put in range -1,1
173          double gammam = (gamma-m)/hw;
174          for( int o=0; o<order; o++ ) {
175            for( int k=nc-1; k>=0; k-- ) {
176              int f = (k==0)?1:2;
```

```
177          y[k] = f*gammam*y[k+1] − y[k+2] + tp[o][k];
178        }
179        rys_t[o] = y[0];
180      }
181      for( int o=0; o<order; o++ ) {
182        for( int k=nc−1; k>=0; k− ) {
183          int f = (k==0)?1:2;
184          y[k] = f*gammam*y[k+1] − y[k+2] + wp[o][k];
185        }
186        rys_W[o] = y[0];
187      }
188    } else { // a Laurent expansion fit
189      tp[0] = t0coef; tp[1] = t1coef; tp[2] = t2coef; tp[3] = t3coef;
190      wp[0] = W0coef; wp[1] = W1coef; wp[2] = W2coef; wp[3] = W3coef;
191      double gammamax = 50.;
192      int kmin = −3;
193      int nk = 11;
194      for( int o=0; o<order; o++ ) {
195        double q=0;
196        for( int j=nk−1; j>=0; j− ) {
197          q *= gamma/gammamax;
198          q += tp[o][j];
199        }
200        q *= pow(gamma/gammamax,(double)kmin);
201        double rsq = rHermite[o]; rsq *= rsq;
202        rys_t[o] = rsq/gamma + exp(−gamma)*q;
203      }
204      for( int o=0; o<order; o++ ) {
205        double q=0;
206        for( int j=nk−1; j>=0; j− ) {
207          q *= gamma/gammamax;
208          q += wp[o][j];
209        }
210        q *= pow(gamma/gammamax,(double)kmin);
211        rys_W[o] = wHermite[o]/sqrt(gamma) + exp(−gamma)*q;
212      }
213    }
214  }
215
216  void integrals( Vector<double>& fourcenter , Matrix<double>& H1,
217                  Matrix<double>& S, Matrix<double>& nucleix ,
218                  Vector<int>& nucleiZ , Matrix<double>& basisx ,
219                  Matrix<int>& basisn , Vector<double>& basisa )
220  {
221    int Nnuclei = nucleiZ.size();
222    int Nbasis = basisa.size();
223    int N2 = (Nbasis*(Nbasis+1))/2;
224    int N4 = (N2*(N2+1))/2;
225    Vector<double> xa(3), xb(3), rys_t(4), rys_W(4);
226
227    // the H1 terms and overlap
228    for( int mu=0; mu<Nbasis; mu++ ) {
229      double amu = basisa[mu];
230      const Vector<double>& xmu = basisx[mu];
231      for( int lambda=0; lambda<=mu; lambda++ ) {
232        double alambda = basisa[lambda];
233        const Vector<double>& xlambda = basisx[lambda];
234        double alpha = amu+alambda;
235        double rho = amu*alambda/alpha;
236        double dxsq=0;
237        for( int d=0; d<3; d++ ) {
238          double tmp = xmu[d]−xlambda[d];
239          dxsq += tmp*tmp;
240          xa[d] = (amu*xmu[d] + alambda*xlambda[d])/alpha;
241        }
242
243        // overlap: this is the S−orbital result
244        double S_olap = M_PI/alpha;
245        S_olap *= sqrt(S_olap)*exp(−rho*dxsq);
246
```

```
247        // McMurchie Davidson —— modify "S" overlap to generate other L states
248        double md_olap = 1;
249        for( int dir=0; dir<3; dir++ ) {
250          int nmu = basisn[mu][dir];
251          int nlambda = basisn[lambda][dir];
252          double d_a_mu = xa[dir] - xmu[dir];
253          double d_m_lambda = xa[dir] - xlambda[dir];
254          double d = MD( alpha,nmu,nlambda,d_a_mu,d_m_lambda );
255          md_olap *= d;
256        }
257        md_olap *= S_olap;
258        S[mu][lambda] = S[lambda][mu] = md_olap;
259
260        // kinetic energy.  If just S, KE is olap*rho*(3 - 2*rho*dxsq);
261        Vector<double> td(3);
262        td[0]=td[1]=td[2]=1;
263        for( int dir=0; dir<3; dir++ ) {
264          int nmu = basisn[mu][dir];
265          int nlambda = basisn[lambda][dir];
266          double d_a_mu = xa[dir] - xmu[dir];
267          double d_m_lambda = xa[dir] - xlambda[dir];
268          double tdd = 0;
269          tdd += 2.*amu*alambda*MD( alpha, nmu+1, nlambda+1, d_a_mu, d_m_lambda );
270          if( nmu > 0 ) {
271            tdd -= nmu*alambda*MD( alpha, nmu-1, nlambda+1, d_a_mu, d_m_lambda );
272          }
273          if( nlambda > 0 ) {
274            tdd -= nlambda*amu*MD( alpha, nmu+1, nlambda-1, d_a_mu, d_m_lambda );
275          }
276          if( nmu > 0 && nlambda > 0 ) {
277            tdd += 0.5*nmu*nlambda*MD( alpha, nmu-1, nlambda-1, d_a_mu, d_m_lambda );
278          }
279          td[dir] *= tdd;
280          double d = MD( alpha, nmu, nlambda, d_a_mu, d_m_lambda );
281          td[(dir+1)%3] *= d;
282          td[(dir+2)%3] *= d;
283        }
284        double md_ke = 0;
285        for( int dir=0; dir<3; dir++ ) md_ke += td[dir];
286        md_ke *= S_olap;
287        Hl[mu][lambda] = Hl[lambda][mu] =  md_ke;
288
289        // nuclear attraction
290        double total_rys_nuc =0;
291        for( int k=0; k<Nnuclei; k++ ) {
292          double rys_nuc = 0;
293          const Vector<double>& xn = nucleix[k];
294          int Z = nucleiZ[k];
295          double xaisq=0;
296          double xmlsq=0;
297          for( int dir=0; dir<3; dir++ ) {
298            double tmp = xmu[dir]-xlambda[dir];
299            xmlsq += tmp*tmp;
300            tmp = xa[dir] - xn[dir];
301            xaisq += tmp*tmp;
302          }
303          Rys( alpha*xaisq, rys_t, rys_W );
304          for( int root=0; root<4; root++ ) {
305            double tsq = rys_t[root];
306            double P = 1;
307            for( int dir=0; dir<3; dir++ ) {
308              int nmu = basisn[mu][dir];
309              int nlam = basisn[lambda][dir];
310              double A = Rys_A( nmu, nlam, tsq, alpha,
311                  xmu[dir], xlambda[dir], xa[dir], xn[dir] );
312              P *= A;
313            }
314            rys_nuc += P*rys_W[root];
315          }
316          rys_nuc *= 2*M_PI*exp(-amu*alambda*xmlsq/alpha)/alpha;
```

```
317              total_rys_nuc  -= Z*rys_nuc;
318          }
319
320          Hl[mu][lambda] += total_rys_nuc;
321          if( mu!=lambda ) Hl[lambda][mu] += total_rys_nuc;
322      }
323  }
324
325  // the four-center Coulomb/exchange terms
326  // packed in funny way to exploit symmetry
327  int mu=0, nu=0, lambda=0, sigma=0, eone=0, etwo=0;
328  for( int indx=0; indx<N4; indx++ ) {
329      const Vector<double>& xmu    = basisx[mu];
330      const Vector<double>& xnu    = basisx[nu];
331      const Vector<double>& xlambda = basisx[lambda];
332      const Vector<double>& xsigma = basisx[sigma];
333      double amu     = basisa[mu];
334      double anu     = basisa[nu];
335      double alambda = basisa[lambda];
336      double asigma  = basisa[sigma];
337      double alpha = amu+alambda;
338      double beta  = anu+asigma;
339      double tmp, dxaasq=0, dxbbsq=0, dxabsq=0;
340      for( int d=0; d<3; d++ ) {
341          xa[d] = (amu*xmu[d] + alambda*xlambda[d])/alpha;
342          xb[d] = (anu*xnu[d] + asigma*xsigma[d])/beta;
343          tmp = xmu[d] - xlambda[d];
344          dxaasq += tmp*tmp;
345          tmp = xnu[d] - xsigma[d];
346          dxbbsq += tmp*tmp;
347          tmp = xa[d] - xb[d];
348          dxabsq += tmp*tmp;
349      }
350      double Ca = exp( - amu*alambda*dxaasq/alpha );
351      double Cb = exp( - anu*asigma*dxbbsq/beta );
352      double rho = alpha*beta/(alpha+beta);
353
354      double rys_rep = 0;
355      Rys( rho*dxabsq, rys_t, rys_W );
356      for( int root=0; root<4; root++ ) {
357          double tsq = rys_t[root];
358          double P = 1;
359          for( int dir=0; dir<3; dir++ ) {
360              int nmu    = basisn[mu][dir];
361              int nlambda = basisn[lambda][dir];
362              int nnu    = basisn[nu][dir];
363              int nsigma = basisn[sigma][dir];
364              P *= RDK_Gxfer( nmu, nnu, nlambda, nsigma, alpha, beta, tsq,
365                      xa[dir], xb[dir], xmu[dir], xnu[dir], xlambda[dir], xsigma[dir] );
366          }
367          rys_rep += P*rys_W[root];
368      }
369      rys_rep *= 2*sqrt(M_PI)*M_PI*M_PI*Ca*Cb;
370      rys_rep /= alpha*beta*sqrt(alpha+beta);
371      fourcenter[indx] = rys_rep;
372
373      if( eone == etwo ) {
374          etwo = nu = sigma = 0;
375          if( lambda == mu ) { mu++; lambda=0; } else { lambda++; }
376          eone++;
377      } else {
378          if( sigma == nu ) { nu++; sigma=0; } else { sigma++; }
379          etwo++;
380      }
381  }
382  }
383
384  // With Lowdin decomposition of S, U D^2 U^T
385  // orthogonality is (D U^T C)^T (D U^T C) = 1.
386  // Use Gram-Schmidt to make C compatible with this.
```

```
387
388    void orthogonalize( Matrix<double>& occOrbCoef,
389                        Matrix<double>& U,
390                        Vector<double>& sigma )
391    {
392      int Nbasis = occOrbCoef.columns();
393      int Npairs = occOrbCoef.rows();
394      Matrix<double> tmp(Npairs,Nbasis);
395
396      for( int i=0; i<Npairs; i++ ) {
397        for( int j=0; j<Nbasis; j++ ) {
398          tmp[i][j] = 0;
399          for( int k=0; k<Nbasis; k++ )
400            tmp[i][j] += sigma[j]*U[k][j]*occOrbCoef[i][k];
401        }
402      }
403
404      double d;
405      for( int i=0; i<Npairs; i++ ) {
406        for( int j=0; j<i; j++ ) {
407          d = 0;
408          for( int k=0; k<Nbasis; k++ )
409            d += tmp[i][k]*tmp[j][k];
410          for( int k=0; k<Nbasis; k++ )
411            tmp[i][k] -= d*tmp[j][k];
412        }
413        d = 0;
414        for( int k=0; k<Nbasis; k++ )
415          d += tmp[i][k]*tmp[i][k];
416        d = 1./sqrt(d);
417        for( int k=0; k<Nbasis; k++ )
418          tmp[i][k] *= d;
419      }
420
421      for( int i=0; i<Npairs; i++ ) {
422        for( int j=0; j<Nbasis; j++ ) {
423          occOrbCoef[i][j] = 0;
424          for( int k=0; k<Nbasis; k++ )
425            occOrbCoef[i][j] += U[j][k]*tmp[i][k]/sigma[k];
426        }
427      }
428    }
429
430    void setupH2O( const double& angle, const double& length1, const double& length2,
431                   Matrix<double>& nucleix, Vector<int>& nucleiZ,
432                   Matrix<double>& basisx, Matrix<int>& basisn, Vector<double>& basisa )
433    {
434      int Nbasis = 24;
435      int Nnuclei = 3;
436      nucleix.resize(Nnuclei,3);
437      nucleiZ.resize(Nnuclei);
438      basisx.resize(Nbasis,3);
439      basisn.resize(Nbasis,3);
440      basisa.resize(Nbasis);
441
442      // oxygen
443      nucleix[0][0] = nucleix[0][1] = nucleix[0][2] = 0;
444      nucleiZ[0] = 8;
445      // hydrogen
446      nucleix[1][0]= length1*cos(angle/2);
447      nucleix[1][1]= length1*sin(angle/2);
448      nucleix[1][2]= 0;
449      nucleiZ[1] = 1;
450      // hydrogen
451      nucleix[2][0]= length2*cos(angle/2);
452      nucleix[2][1]= -length2*sin(angle/2);
453      nucleix[2][2]= 0;
454      nucleiZ[2] = 1;
455
456      // STO-3G components for hydrogen 1
```

```
457     for( int b=0; b<3; b++ ) {
458       for( int i=0; i<3; i++ ) {
459         basisx[b][i] = nucleix[1][i];
460         basisn[b][i] = 0;
461       }
462     }
463     basisa[0] = 0.16885540; basisa[1] = 0.62391373; basisa[2] = 3.42525091;
464
465     // STO-3G components for hydrogen 2
466     for( int b=3; b<6; b++ ) {
467       for( int i=0; i<3; i++ ) {
468         basisx[b][i] = nucleix[2][i];
469         basisn[b][i] = 0;
470       }
471     }
472     basisa[3] = 0.16885540; basisa[4] = 0.62391373; basisa[5] = 3.42525091;
473
474     // STO-3G components for oxygen
475     for( int b=6; b<21; b++ ) {
476       for( int i=0; i<3; i++ ) {
477         basisx[b][i] = nucleix[0][i];
478       }
479     }
480     // S
481     for( int b=6; b<15; b++ ) {
482       for( int i=0; i<3; i++ ) {
483         basisn[b][i] = 0;
484       }
485     }
486     // Px
487     for( int b=15; b<18; b++ ) {
488       basisn[b][0] = 1; basisn[b][1] = basisn[b][2] = 0;
489     }
490     // Py
491     for( int b=18; b<21; b++ ) {
492       basisn[b][1] = 1; basisn[b][0] = basisn[b][2] = 0;
493     }
494     // Pz
495     for( int b=21; b<24; b++ ) {
496       basisn[b][2] = 1; basisn[b][0] = basisn[b][1] = 0;
497     }
498     // 1S
499     basisa[6 ] = 0.16885540; basisa[7 ] = 0.62391373; basisa[8 ] = 3.42525091;
500     // 2S
501     basisa[9 ] = 0.3803890;  basisa[10] = 1.1695961;  basisa[11] = 5.0331513;
502     basisa[12] = 6.4436083;  basisa[13] = 23.808861;  basisa[14] = 130.70932;
503     // 2P
504     basisa[15] = 0.3803890;  basisa[16] = 1.1695961;  basisa[17] = 5.0331513;
505     // 2P
506     basisa[18] = 0.3803890;  basisa[19] = 1.1695961;  basisa[20] = 5.0331513;
507     // 2P
508     basisa[21] = 0.3803890;  basisa[22] = 1.1695961;  basisa[23] = 5.0331513;
509   }
510
511   double solveHFR( const double& damping,    Matrix<double>& occOrbCoef,
512                    Matrix<double>& nucleix,  Vector<int>& nucleiZ,
513                    Matrix<double>& basisx,   Matrix<int>& basisn,
514                    Vector<double>& basisa )
515   {
516     double small = 1.e-7; // convergence criterion
517     int Nbasis  = basisx.rows();
518     int Nnuclei = nucleix.rows();
519     int Npairs  = occOrbCoef.rows();
520     int N2 = (Nbasis*(Nbasis+1))/2;
521     int N4 = (N2*(N2+1))/2;
522
523     Vector<int>     perm(Nbasis);
524     Vector<double> fourcenter(N4), sigma(Nbasis), lambda(Nbasis);
525     Matrix<double> H1(Nbasis,Nbasis), S(Nbasis,Nbasis), J(Nbasis,Nbasis),
526                    K(Nbasis,Nbasis), U(Nbasis,Nbasis), VT(Nbasis,Nbasis),
```

```
527                         P(Nbasis,Nbasis), F(Nbasis,Nbasis), X(Nbasis,Nbasis),
528                         newOccOrbCoef(Npairs,Nbasis);
529
530      double Etot=0, deltaE, deltaC;
531
532      integrals( fourcenter, H1, S, nucleix, nucleiZ, basisx, basisn, basisa );
533      SVD( S, U, VT, sigma );
534      for( int i=0; i<Nbasis; i++ ) sigma[i] = sqrt(sigma[i]);
535
536      int iteration = 0;
537      double Etotal;
538      do {
539        ++iteration;
540
541        // needed if damping was employed, or for bad IC
542        if( iteration==1 || damping != 0 ) orthogonalize( occOrbCoef, U, sigma );
543
544        // P matrix
545        for( int i=0; i<Nbasis; i++ ) {
546          for( int j=0; j<=i; j++ ) {
547            double  t = 0;
548            for( int k=0; k<Npairs; k++ ) {
549              t += occOrbCoef[k][i]*occOrbCoef[k][j];
550            }
551            P[i][j] = P[j][i] = t;
552          }
553        }
554
555        for( int i=0; i<Nbasis; i++ )
556          for( int j=0; j<Nbasis; j++ ) J[i][j] = K[i][j] = 0;
557
558        int l1=0, l2=0, r1=0, r2=0, eone=0, etwo=0;
559        for( int indx=0; indx<N4; indx++ ) {
560          double fc = fourcenter[indx];
561          J[l1][r1] +=    fc*P[l2][r2]; K[l1][r2] += fc*P[l2][r1];
562          if( l1!=r1 ) {
563            J[r1][l1] +=    fc*P[l2][r2]; K[r1][r2] += fc*P[l2][l1];
564          }
565          if( l2!=r2 ) {
566            J[l1][r1] +=    fc*P[r2][l2]; K[l1][l2] += fc*P[r2][r1];
567          }
568          if( l1!=r1 && l2!=r2 ) {
569            J[r1][l1] +=    fc*P[r2][l2]; K[r1][l2] += fc*P[r2][l1];
570          }
571          if( eone != etwo ) {
572            J[l2][r2] +=    fc*P[l1][r1]; K[l2][r1] += fc*P[l1][r2];
573            if( l2!=r2 ) {
574              J[r2][l2] +=    fc*P[l1][r1]; K[r2][r1] += fc*P[l1][l2];
575            }
576            if( l1!=r1 ) {
577              J[l2][r2] +=    fc*P[r1][l1]; K[l2][l1] += fc*P[r1][r2];
578            }
579            if( l2!=r2 && l1!=r1 ) {
580              J[r2][l2] +=    fc*P[r1][l1]; K[r2][l1] += fc*P[r1][l2];
581            }
582          }
583
584          if( eone == etwo ) {
585            etwo = l2 = r2 = 0;
586            if( r1 == l1 ) { l1++; r1=0; } else { r1++; }
587            eone++;
588          } else {
589            if( r2 == l2 ) { l2++; r2=0; } else { r2++; }
590            etwo++;
591          }
592        }
593
594        for( int i=0; i<Nbasis; i++ )
595          for( int j=0; j<Nbasis; j++ ) F[i][j] = H1[i][j] + 2*J[i][j] − K[i][j];
596
```

```
597        // evaluate nuclear-nuclear repulsion energy
598        double Enn = 0;
599        for( int i=0; i<Nnuclei; i++ ) {
600          const Vector<double>& xi = nucleix[i];
601          int zi = nucleiZ[i];
602          for( int j=0; j<i; j++ ) {
603            const Vector<double>& xj = nucleix[j];
604            int zj = nucleiZ[j];
605            double r = 0;
606            for( int dir=0; dir<3; dir++ )
607              r += (xi[dir]-xj[dir])*(xi[dir]-xj[dir]);
608            Enn += zi*zj/sqrt(r);
609          }
610        }
611
612        // 1/sqrt(sigma) UT F U 1/sqrt(sigma) sqrt(sigma) UT C = E sqrt(sigma) UT C
613
614        // replace VT with F U/sqrt(sigma)
615        for( int i=0; i<Nbasis; i++ ) {
616          for( int j=0; j<Nbasis; j++ ) {
617            VT[i][j] = 0;
618            for( int k=0; k<Nbasis; k++ ) VT[i][j] += F[i][k]*U[k][j]/sigma[j];
619          }
620        }
621
622        // replace F with 1/sqrt(sigma) UT VT
623        for( int i=0; i<Nbasis; i++ ) {
624          for( int j=0; j<Nbasis; j++ ) {
625            F[i][j] = 0;
626            for( int k=0; k<Nbasis; k++ ) F[i][j] += U[k][i]*VT[k][j]/sigma[i];
627          }
628        }
629
630        // eigenvalue problem
631        QReig( F, X, lambda );
632
633        // X = sqrt(sigma) UT C
634        // C = U 1/sqrt(sigma) X
635
636        // but collect only lowest energy pairs
637        // routine uses descending order ...
638        hqsort0( lambda, perm );
639
640        // sort C in place: highest to lowest
641        for( int i=Nbasis-1; i>=0; i-- ) {
642          int k = perm[i];
643          while( k > i ) k = perm[k];
644          if( k != i ) {
645            for( int j=0; j<Nbasis; j++ ) {
646              double d=X[j][k]; X[j][k]=X[j][i]; X[j][i]=d;
647            }
648          }
649        }
650
651        double Eorb = 0;
652        for( int i=0; i<Npairs; i++ ) Eorb += 2*lambda[Nbasis-1-i];
653        double Ej = 0, Ek = 0;
654        for( int p=0; p<Nbasis; p++ ) {
655          for( int q=0; q<Nbasis; q++ ) {
656            Ej += P[p][q]*J[p][q];
657            Ek += P[p][q]*K[p][q];
658          }
659        }
660        Etotal = Eorb -(2*Ej-Ek) + Enn;
661        cout << "iteration: " << iteration << " E total " << Etotal << endl;
662
663        // for convergence testing
664        deltaE = fabs(Etotal-Etot);
665        Etot = Etotal;
666
```

```
667        for( int i=0; i<Npairs; i++ ) {
668          for( int j=0; j<Nbasis; j++ ) {
669            newOccOrbCoef[i][j] = 0;
670            for( int k=0; k<Nbasis; k++ )
671              newOccOrbCoef[i][j] += U[j][k]*X[k][Nbasis-1-i]/sigma[k];
672          }
673        }
674
675        // don't use damping on first pass since IC may be bad
676        double idamp = (iteration==1) ? 0 : damping;
677        deltaC = 0;
678        for( int i=0; i<Npairs; i++ ) {
679          for( int j=0; j<Nbasis; j++ ) {
680            double tmp = (1-idamp)*newOccOrbCoef[i][j] + idamp*occOrbCoef[i][j];
681            double diff = occOrbCoef[i][j] - tmp;
682            deltaC += diff*diff;
683            occOrbCoef[i][j] = tmp;
684          }
685        }
686        deltaC = sqrt(deltaC);
687
688        cout << "deltaE " << deltaE << " deltaC " << deltaC << endl;
689      } while( (iteration < 2) || deltaE > small || deltaC > small );
690      return Etotal;
691    }
692
693    Matrix<double> nucleix, basisx;
694    Matrix<int>    basisn;
695    Vector<double> basisa;
696    Vector<int>    nucleiZ;
697    Matrix<double> occOrbCoef;
698
699    double func( const Vector<double>& params )
700    {
701      extern Matrix<double> nucleix, basisx, occOrbCoef;
702      extern Matrix<int>    basisn;
703      extern Vector<double> basisa;
704      extern Vector<int>    nucleiZ;
705      double len1 = params[0];
706      double len2 = params[1];
707      double ang  = params[2];
708      setupH2O( ang, len1, len2, nucleix, nucleiZ, basisx, basisn, basisa );
709      return solveHFR( 0.0, occOrbCoef, nucleix, nucleiZ, basisx, basisn, basisa );
710    }
711
712    void grad( const Vector<double>& params, Vector<double>& g )
713    {
714      double epsilon = 1.e-3;
715
716      Vector<double> p2(3);
717      for( int i=0; i<3; i++ ) p2[i] = params[i];
718
719      p2[0] = params[0] + epsilon;
720      double Ep = func(p2);
721      p2[0] = params[0] - epsilon;
722      double Em = func(p2);
723      g[0] = (Ep-Em)/(2*epsilon);
724      p2[0] = params[0];
725
726      p2[1] = params[1] + epsilon;
727      Ep = func(p2);
728      p2[1] = params[1] - epsilon;
729      Em = func(p2);
730      g[1] = (Ep-Em)/(2*epsilon);
731      p2[1] = params[1];
732
733      p2[2] = params[2] + epsilon;
734      Ep = func(p2);
735      p2[2] = params[2] - epsilon;
736      Em = func(p2);
```

```
737      g[2] = (Ep-Em)/(2*epsilon);
738      p2[2] = params[2];
739    }
740
741    int main()
742    {
743      int Npairs=5;
744      int Nbasis=24;
745      extern Matrix<double> occOrbCoef;
746
747      // begin things with a wild guess.  Each HFR evaluation will take
748      // the previous "C" matrix output as input.
749      occOrbCoef.resize(Npairs,Nbasis);
750      for( int p=0; p<Npairs; p++ )
751        for( int i=0; i<Nbasis; i++ ) occOrbCoef[p][i] = 0;
752      // 1S of O
753      occOrbCoef[0][7] = 1;
754      // 2S of O
755      occOrbCoef[1][10] = 1;
756      // Px-H
757      occOrbCoef[2][16] = 0.1;
758      occOrbCoef[2][18] = 0.5;
759      occOrbCoef[2][1]  = 0.2;
760      // Px-H
761      occOrbCoef[3][16] = 0.1;
762      occOrbCoef[3][18] = -0.5;
763      occOrbCoef[3][4]  = 0.2;
764      // Pz
765      occOrbCoef[4][21] = 1;
766
767      cout << scientific << setprecision(14);
768      Vector<double> params(3);
769      // parameters 0,1 are bond lengths;
770      //            2 is the angle.
771      params[0] = 1.8;
772      params[1] = 1.8;
773      params[2] = 105.*M_PI/180.;
774      int iteration;
775      VMM( func, grad, params, DFP, iteration, 1.e-5, 1.e-5 );
776      cout << "DPF:  " << iteration << " iterations" << endl;
777      cout << "optimum energy " << func(params) << endl;
778      cout << "optimum len1  " << params[0] << endl;
779      cout << "optimum len2  " << params[1] << endl;
780      cout << "optimum angle " << params[2]*180/M_PI << endl;
781    }
```

Solutions

Chapter 1

1.1 The following C++ program calculates machine precision. Note that it assumes "round ties to even." A more robust program that makes no assumptions is William Kahan's `paranoia.c`, currently available at `\tthttp://www.netlib.org/paranoi`

```
1   #include <iostream>
2   using namespace std;
3
4   // Note:  this is probably machine dependent.
5   // It "works" with IEEE-754 (1985 & 2008) with the "Round to ties to even"
6   // rounding mode.
7
8   template <class T> void precision()
9   {
10    int t = 1;
11    T   test, half = 0.5,  epsilon = 0.5;
12
13    do {
14      test = half + epsilon;
15      if( test == half ) {
16
17        // this can only happen if round(half+epsilon) = half,
18        // which means epsilon is 2^{-(t+2)}:
19        //
20        //    if epsilon = 2^{-t}, half+epsilon would be machine
21        //    representable.
22        //
23        //    if epsilon = 2^{-(t+1)}, then half+epsilon is not
24        //    machine representable, and will not round up to
25        //    something other than half in round-to-nearest mode.
26
27        //    if epsilon 2^{-(t+1)}  and half+epsilon rounds up,
28        //    then change the following two lines to
29        //    t-=2; and epsilon *=4;
30
31        t --;
32        epsilon *= 2;
33
34        // done
35
36        cout << "machine epsilon is " << epsilon
37             << " (2^" << -t << ")"
38             << " with " << sizeof(T) << " byte precision" << endl;
39        return;
40      }
41
42      // machine precision not yet found:  make epsilon smaller ...
43      epsilon /= 2;
44      t ++;
45
```

```
46      } while (1);
47  }
48
49  int main(int argc , char **argv)
50  {
51      precision <double >();
52      precision <float >();
53  }
```

with output:

```
machine epsilon is 1.11022e-16 (2^-53) with 8 byte precision
machine epsilon is 5.96046e-08 (2^-24) with 4 byte precision
```

1.2

	f	$\frac{x}{f}\frac{\partial f}{\partial x}$	$\frac{y}{f}\frac{\partial f}{\partial y}$
(i)	$\cos(x)$	$-x\tan(x)$	
(ii)	$\sin(x)$	$x\cot(x)$	
(iii)	$\tan(x)$	$x\sec(x)\csc(x)$	
(iv)	$\cot(x)$	$-x\sec(x)\csc(x)$	
(v)	$\arccos(x)$	$-\frac{\cos(f)}{f\sin(f)}$	
(vi)	$\arcsin(x)$	$\frac{\tan(f)}{f}$	
(vii)	$\arctan(x)$	$\frac{\sin(f)\cos(f)}{f}$	
(viii)	$\arctan(y/x)$	$-\frac{\sin(f)\cos(f)}{f}$	$+\frac{\sin(f)\cos(f)}{f}$
(ix)	e^x	x	
(x)	x^α	α	
(xi)	x^y	y	$y\ln(x)$
(xii)	$\ln(x)$	$\frac{1}{\ln(x)}$	

1.3 For the new approximation we have the trivial algorithm $y_a = x$, the only numerical error is propagated input error:

$$\frac{\Delta y}{y_a} = \epsilon_x.$$

Combining the approximation error, and adjusting the denominator to reflect the exact solution,

$$\frac{\Delta y}{y_e} = \epsilon_x \frac{x}{e^x - 1} + \frac{x^2 e^\xi}{2}\frac{1}{e^x - 1}.$$

The threshold where the maximum errors are equal comes from evaluation of

$$\left|\frac{x}{e^x - 1}\epsilon + \frac{x^2 e^\xi}{2(e^x - 1)}\right| = \left(1 + \left|\frac{xe^x}{e^x - 1}\right| + \left|\frac{e^x}{e^x - 1}\right|\right)\epsilon$$

using the analysis of the exact method from (1.17).

If $x > 0$, we have $e^x - 1 > 0$, and $\xi \in [0, x]$ gives a maximum remainder error when $\xi = x$, so

$$\frac{x}{e^x - 1}\epsilon + \frac{x^2 e^x}{2(e^x - 1)} = \left(1 + \frac{xe^x}{e^x - 1} + \frac{e^x}{e^x - 1}\right)\epsilon$$

$$x^2 = 2\epsilon e^{-x}(e^x(2 + x) - 1 - x)$$

$$x \approx \sqrt{2\epsilon} \approx 1.5 \times 10^{-8}.$$

When $x < 0$, $e^x - 1 < 0$, and $\xi \in [x, 0]$ gives a maximum remainder error with $\xi = 0$:

$$\frac{x}{e^x - 1}\epsilon - \frac{x^2}{2(e^x - 1)} = \left(1 + \frac{xe^x}{e^x - 1} - \frac{e^x}{e^x - 1}\right)\epsilon$$

$$x^2 = 2\epsilon(1 + x - xe^x)$$

$$x \approx -\sqrt{2\epsilon} \approx -1.5 \times 10^{-8}.$$

So, to a first approximation x is more numerically trustworthy than $e^x - 1$ for $x \in [-x_0, x_0]$, $x_0 \approx 1.5 \times 10^{-8}$.

The observed errors are qualitatively consistent with the analysis (Figure S.1).

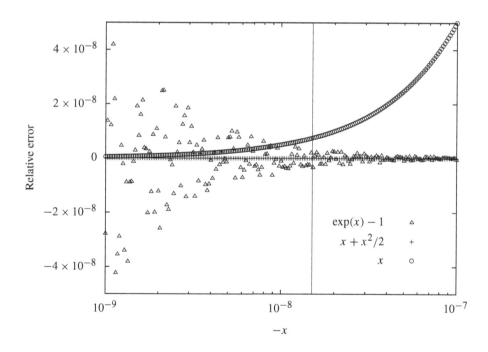

Figure S.1 $(e^x - 1)$ vs x and $x^2/2 + x$; Problem 1.3.

1.4 For the given formula, an algorithm is $\eta_1 = \exp(x)$, $\eta_2 = \ln \eta_1$, $\eta_3 = \eta_2/x$, $\eta_4 = \eta_1 - 1$, $\eta_5 = \eta_4/\eta_3$. The Bauer graph of this algorithm is given in Figure S.2.

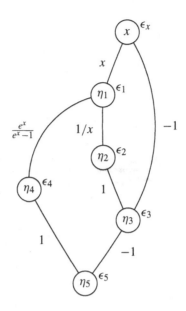

Figure S.2 Bauer graph for $(x/\ln\exp x)(e^x - 1)$; Problem 1.4.

From this graph the error is

$$\frac{\Delta\eta_5}{\eta_5} = \frac{xe^x}{e^x - 1}\epsilon_x + \left(\frac{e^x}{e^x - 1} - \frac{1}{x}\right)\epsilon_1 - \epsilon_2 - \epsilon_3 + \epsilon_4 + \epsilon_5$$

$$\left|\frac{\Delta\eta_5}{\eta_5}\right| \le \left(4 + \left|\frac{e^x}{e^x - 1} - \frac{1}{x}\right| + \left|\frac{xe^x}{e^x - 1}\right|\right)\epsilon.$$

Stripping off the inherent error, we have

$$\left(3 + \left|\frac{e^x}{e^x - 1} - \frac{1}{x}\right|\right)\epsilon$$

versus

$$\left|\frac{e^x}{e^x - 1}\right|\epsilon,$$

for $e^x - 1$, cf. (1.17).

If $x > 0$, then $e^x/(e^x - 1) > 1/x$, so

$$\left(3 + \left|\frac{e^x}{e^x - 1} - \frac{1}{x}\right|\right)\epsilon < \left|\frac{e^x}{e^x - 1}\right|\epsilon$$

$$3 - \frac{1}{x} < 0$$

$$x < \frac{1}{3},$$

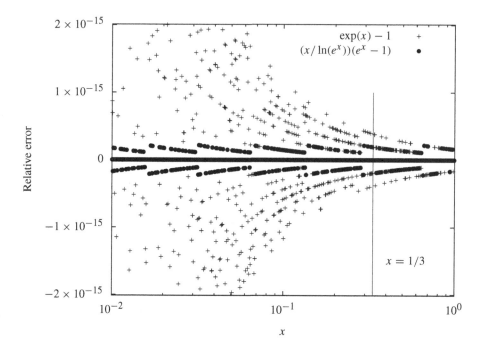

Figure S.3 $(x/\ln(e^x))(e^x - 1)$; Problem 1.4.

so the given formula is more numerically stable for all $0 < x < \frac{1}{3}$. If $x < 0$, then

$$3 + \left| \frac{e^x}{e^x - 1} - \frac{1}{x} \right| < \left| \frac{e^x}{e^x - 1} \right|$$

$$3 + \frac{e^x}{e^x - 1} - \frac{1}{x} < - \frac{e^x}{e^x - 1}$$

$$5xe^x - 3x < e^x - 1$$

$$x \gtrsim -0.252\,657.$$

See Figure S.3.

When $|x| \lesssim \epsilon$, one will find round(ln round(exp round(x))) = 0, which will lead to a division by zero error. In this case, the approximation $y \approx x$ should be used.

1.5 The algorithm

$$\phi_1 = 1 + x$$

$$\phi_2 = \ln \phi_1$$

$$\phi_3 = \phi_1 - 1$$

$$\phi_4 = x/\phi_3$$

$$\phi_5 = \phi_4 \phi_2$$

has errors

$$\frac{\Delta\phi_1}{\phi_1} = \frac{x}{1+x}\frac{\Delta x}{x} + \epsilon_1$$

$$\frac{\Delta\phi_2}{\phi_2} = \frac{1}{\ln(1+x)}\frac{\Delta\phi_1}{\phi_1} + \epsilon_2$$

$$\frac{\Delta\phi_3}{\phi_3} = \frac{1+x}{x}\frac{\Delta\phi_1}{\phi_1} + \epsilon_3$$

$$\frac{\Delta\phi_4}{\phi_4} = \frac{\Delta x}{x} - \frac{\Delta\phi_3}{\phi_3} + \epsilon_4$$

$$\frac{\Delta\phi_5}{\phi_5} = \frac{\Delta\phi_2}{\phi_2} + \frac{\Delta\phi_4}{\phi_4} + \epsilon_5$$

graphed in Figure S.4.

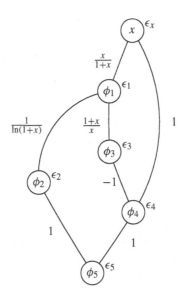

Figure S.4 Bauer graph for $(x/((1+x)-1))\ln(1+x)$; Problem 1.5.

The relative error of $y = \ln(1+x)$ is given by $\Delta\phi_2/\phi_2$:

$$\left|\frac{\Delta\phi_2}{\phi_2}\right| \leq \underbrace{\left(1 + \left|\frac{x}{(1+x)\ln(1+x)}\right|\right)\epsilon}_{\text{inherent}} + \underbrace{\left|\frac{1}{\ln(1+x)}\right|\epsilon}_{\text{algorithm}},$$

where the algorithm error is due to ϵ_1 amplified by $1/\ln(1+x)$.

The relative error of the compensated algorithm is

$$\left|\frac{\Delta\phi_5}{\phi_5}\right| \leq \underbrace{\left(1 + \left|\frac{x}{(1+x)\ln(1+x)}\right|\right)\epsilon}_{\text{inherent}} + \underbrace{\left(3 + \left|\frac{1}{\ln(1+x)} - \frac{1+x}{x}\right|\right)\epsilon}_{\text{algorithm}}.$$

The compensated algorithm is more numerically trustworthy for $|x| \ll 1$ because

$$\ln(1+x) = x - \frac{x^2}{2} + \frac{x^3}{3} - \frac{x^4}{4} + \cdots \quad \text{for } -1 < x \le 1,$$

which means

$$\lim_{x \to 0} \frac{1}{\ln(1+x)} \sim \lim_{x \to 0} \frac{1}{x} \to \infty$$

$$\lim_{x \to 0} \left(\frac{1}{\ln(1+x)} - \frac{1+x}{x} \right) \to -\frac{1}{2}.$$

The error bounds for the two methods are equal when

$$3 + \left| \frac{1}{\ln(1+x)} - \frac{1+x}{x} \right| = \left| \frac{1}{\ln(1+x)} \right|.$$

If $x \ge 0$ then $\ln(1+x) \ge 0$ and for $|x| \ll 1$,

$$3 - \frac{1}{\ln(1+x)} + \frac{1+x}{x} = \frac{1}{\ln(1+x)}$$

$$1 + 4x = \frac{2x}{\ln(1+x)} \sim 2 + x - \frac{x^2}{6} + \frac{x^3}{12} - \cdots$$

$$x \sim 1/3.$$

A more accurate result using Newton–Raphson is $x \sim 0.328\,166$. If $x \le 0$, then $\ln(1+x) \le 0$ and for $|x| \ll 1$,

$$3 - \frac{1}{\ln(1+x)} + \frac{1+x}{x} = -\frac{1}{\ln(1+x)}$$

$$x = -1/4.$$

So the compensated formula $x \ln(1+x)/((1+x) - 1)$ is more numerically trustworthy in the range $[-1/4, 0.328\,166]$. But, when round(round(1 + round(x)) − 1) = 0 this formula will incur division by zero, and this will occur when $|x| \le \epsilon$. In this case one should use the approximation $y \approx x$. See Figure S.5.

1.6 The inherent error is independent of algorithm for mathematically identical equations. All numerical error associated with y_1 is inherent. Therefore, the error bound associated with y_2 cannot ever be smaller.

To show this the long way, we construct the algorithm: $\eta_1 = a+b$, $\eta_2 = a^2$, $\eta_3 = b^2$, $\eta_4 = \eta_2 - \eta_3$, $y = \eta_4/\eta_1$. The Bauer graph for this algorithm is shown in Figure S.6. The error is

$$\frac{\Delta y}{y} = \left(\frac{2a^2}{a^2 - b^2} - \frac{a}{a+b} \right) \epsilon_a + \left(-\frac{2b^2}{a^2 - b^2} - \frac{b}{a+b} \right) \epsilon_b$$

$$- \epsilon_1 + \frac{a^2}{a^2 - b^2} \epsilon_2 - \frac{b^2}{a^2 - b^2} \epsilon_3 + \epsilon_4 + \epsilon_5$$

$$= \underbrace{\frac{a}{a-b} \epsilon_a - \frac{b}{a-b} \epsilon_b + \epsilon_5}_{\text{inherent}} - \epsilon_1 + \frac{a^2}{a^2 - b^2} \epsilon_2 - \frac{b^2}{a^2 - b^2} \epsilon_3 + \epsilon_4.$$

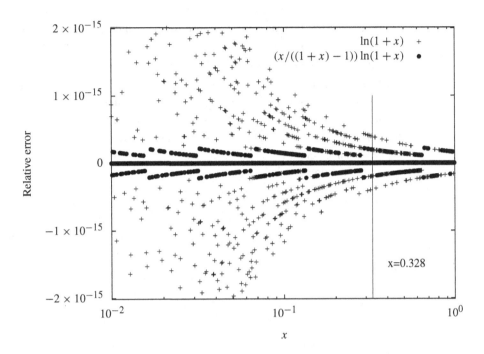

Figure S.5 $(x/((1 + x) - 1)) \ln(1 + x)$; Problem 1.5.

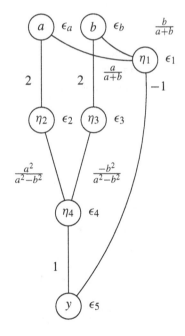

Figure S.6 Bauer graph for Problem 1.6.

So, there is no way to avoid cancellation. In fact, $(a + b)(a - b)$ is more numerically trustworthy than $a^2 - b^2$ (because the bad amplification factor associated with subtraction affects only input error, not algorithm error). So, the best way to evaluate y_2 is $(a + b)(a - b)/(a + b)$ which is clearly pointless.

1.7 This formula represents Archimedes' (second century BC) method for estimating π from the circumference of a polygon. The initial value $2\sqrt{2}$ is the circumference in the top half-plane of an inscribed cube. The next value is the circumference in the top half-plane of an inscribed octagon, etc. (Figure S.7). The formula is derived using the Pythagorean theorem to calculate the length of the chords at each iteration from the lengths of the previous iteration.

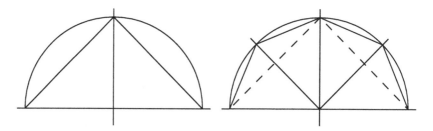

Figure S.7 Archimedes' determination of π; Problem 1.7.

A more numerically stable representation is

$$p^{(n+1)} = \frac{2p^{(n)}}{\sqrt{2 + 2\sqrt{1 - 2^{-2(n+1)}(p^{(n)})^2}}}$$

which comes from the same rearrangement used to improve solution of the quadratic formula. The original formula had two problems. First, there was an extra subtraction that introduced unnecessary numerical error. Second, the factor 2^{n+1} cannot be calculated for every n because of exponential overflow. In the improved version one needs only worry about $2^{-2(n+1)}$, which will also fail (by exponential underflow). Fortunately, underflow rounds to 0 which is representable. One subtraction remains in the improved solution.

1.8 We will analyze the errors of the following equations:

$$y_0 = \cos x - 1$$
$$y_1 = -2 \sin^2 \frac{x}{2}$$
$$y_2 = -\frac{\sin^2(x)}{1 + \cos(x)}$$
$$y_3 = \left(\frac{x}{\arccos(\cos(x))}\right)^2 (\cos(x) - 1)$$
$$y_4 = -\frac{x^2}{2},$$

where y_0 is the original, y_1, y_2, and y_3 are mathematically identical alternatives, and y_4 is an approximation.

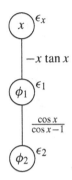

Figure S.8 Bauer graph for $y_0 = \cos x - 1$; Problem 1.8.

For y_0 an algorithm is

$$\phi_1 = \cos x$$
$$\phi_2 = \phi_1 - 1,$$

and the graph is shown in Figure S.8. The errors are:

$$\frac{\Delta\phi_1}{\phi_1} = -x \tan x \frac{\Delta x}{x} + \epsilon_1$$
$$\frac{\Delta\phi_2}{\phi_2} = \frac{\cos x}{\cos x - 1} \frac{\Delta\phi_1}{\phi_1} + \epsilon_2,$$

so with $\Delta x / x = \epsilon_x$

$$\frac{\Delta\phi_2}{\phi_2} = \underbrace{-\frac{x \sin x}{\cos_x - 1}\epsilon_x + \epsilon_2}_{\text{inherent}} + \frac{\cos x}{\cos x - 1}\epsilon_1.$$

The problem here is that the coefficient of ϵ_1 increases without bound as $x \to 0$. The other terms are finite. Note

$$\lim_{x \to 0} \frac{x \sin x}{\cos_x - 1} = -2,$$

so the inherent error bound is 3ϵ.

For y_1 an algorithm is

$$\eta_1 = x/2$$
$$\eta_2 = \sin \eta_1$$
$$\eta_3 - \eta_2^2$$
$$\eta_4 = -2\eta_3,$$

with errors

$$\frac{\Delta \eta_1}{\eta_1} = \frac{\Delta x}{x}$$

$$\frac{\Delta \eta_2}{\eta_2} = \frac{x}{2} \cot \frac{x}{2} \frac{\Delta \eta_1}{\eta_1} + \epsilon_2$$

$$\frac{\Delta \eta_3}{\eta_3} = 2 \frac{\Delta \eta_2}{\eta_2} + \epsilon_3$$

$$\frac{\Delta \eta_4}{\eta_4} = \frac{\Delta \eta_3}{\eta_3}.$$

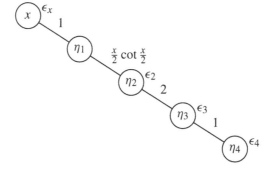

Figure S.9 Bauer graph for $y_1 = -2 \sin^2(x/2)$; Problem 1.8.

See Figure S.9. The last step has no error ϵ_4 since only the sign and exponent bits are affected. All together,

$$\frac{\Delta \eta_4}{\eta_4} = \underbrace{\epsilon_3 + \frac{x}{2} \cot \frac{2}{2} \epsilon_x}_{\text{inherent}} + 2 \epsilon_2.$$

The coefficient of ϵ_x here is the same as for algorithm y_0 – the inherent error bounds are independent of algorithm for mathematically identical formulas. The algorithm-dependent error here is bounded by 2ϵ which is numerically harmless as $x \to 0$.

For y_2

$$\psi_1 = \sin x$$

$$\psi_2 = \psi_1^2$$

$$\psi_3 = \cos x$$

$$\psi_4 = \psi_3 + 1$$

$$\psi_5 = \frac{\psi_2}{\psi_4}$$

$$\psi_6 = -\psi_5,$$

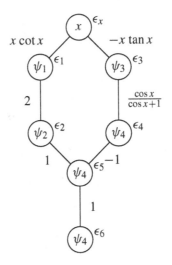

Figure S.10 Bauer graph for $y_2 = -\sin^2 x/(1 + \cos x)$; Problem 1.8.

with the graph shown in Figure S.10. The errors are

$$\frac{\Delta \psi_1}{\psi_1} = x \cot x \frac{\Delta x}{x} + \epsilon_1$$

$$\frac{\Delta \psi_2}{\psi_2} = 2 \frac{\Delta \psi_1}{\psi_1} + \epsilon_2$$

$$\frac{\Delta \psi_3}{\psi_3} = -x \tan x \frac{\Delta x}{x} + \epsilon_3$$

$$\frac{\Delta \psi_4}{\psi_4} = \frac{\cos x}{\cos x + 1} \frac{\Delta \psi_3}{\psi_3} + \epsilon_4$$

$$\frac{\Delta \psi_5}{\psi_5} = \frac{\Delta \psi_2}{\psi_2} - \frac{\Delta \psi_4}{\psi_4} + \epsilon_5$$

$$\frac{\Delta \psi_6}{\psi_6} = \frac{\Delta \psi_5}{\psi_5}.$$

There is no ϵ_6 since only the sign bit is affected. All together,

$$\frac{\Delta \psi_6}{\psi_6} = \epsilon_5 + \underbrace{\left(2x \tan x + \frac{x \sin x}{\cos x + 1} \right) \epsilon_x}_{\text{inherent}} + \epsilon_2 + 2\epsilon_1 - \epsilon_4 - \frac{\cos x}{\cos x + 1} \epsilon_3.$$

The algorithm error here is $\frac{9}{2}\epsilon$ as $x \to 0$, which is numerically harmless, but less numerically trustworthy than y_1.

For y_3 (see Figure S.11),

$$\mu_1 = \cos x$$

$$\mu_2 = \mu_1 - 1$$

$$\mu_3 = \arccos \mu_1$$

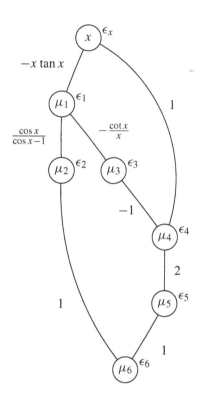

Figure S.11 Bauer graph for $y_3 = (x / \arccos \cos x)^2 (\cos x - 1)$; Problem 1.8.

$$\mu_4 = x / \mu_3$$
$$\mu_5 = \mu_4^2$$
$$\mu_6 = \mu_5 \mu_2$$

$$\frac{\Delta \mu_1}{\mu_1} = -x \tan x \frac{\Delta x}{x} + \epsilon_1$$

$$\frac{\Delta \mu_2}{\mu_2} = \frac{\cos x}{\cos x - 1} \frac{\Delta \mu_1}{\mu_1} + \epsilon_2$$

$$\frac{\Delta \mu_3}{\mu_3} = -\frac{\cot x}{x} \frac{\Delta \mu_1}{\mu_1} + \epsilon_3$$

$$\frac{\Delta \mu_4}{\mu_4} = \frac{\Delta x}{x} - \frac{\Delta \mu_3}{\mu_3} + \epsilon_4$$

$$\frac{\Delta \mu_5}{\mu_5} = 2 \frac{\Delta \mu_4}{\mu_4} + \epsilon_5$$

$$\frac{\Delta \mu_6}{\mu_6} = \frac{\Delta \mu_2}{\mu_2} + \frac{\Delta \mu_5}{\mu_5} + \epsilon_6$$

$$\frac{\Delta \mu_6}{\mu_6} = \epsilon_6 - \underbrace{\frac{x \sin x}{\cos x - 1} \epsilon_x}_{\text{inherent}} + \epsilon_5 + \epsilon_2 + 2\epsilon_4 - 2\epsilon_3 + \left(\frac{\cos x}{\cos x - 1} + \frac{2 \cot x}{x} \right) \epsilon_1.$$

Because

$$\lim_{x \to 0} \left(\frac{\cos x}{\cos x - 1} + \frac{2 \cot x}{x} \right) = \frac{1}{6}$$

the algorithm error is $6\frac{1}{6}\epsilon$, which is again numerically harmless but less numerically trustworthy than y_1 and y_2. Note that for $x \lesssim 10^{-8}$,

$$\text{round}(\arccos \text{round}(\cos \text{round}(x))) = 0.$$

When this occurs, $y \approx -x^2/2$, i.e., y_4, should be used.

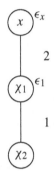

Figure S.12 Bauer graph for $y_4 = -x^2/2$; Problem 1.8.

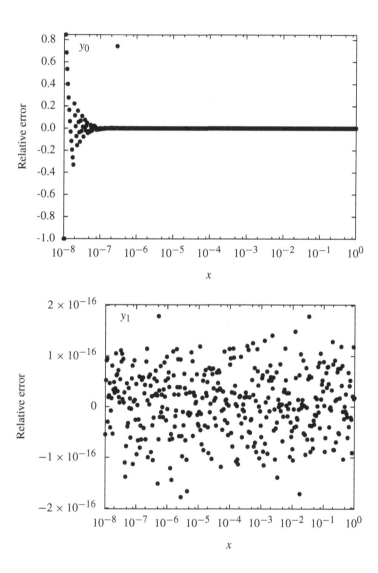

Figure S.13 Relative errors for $\cos x - 1$; Problem 1.8.

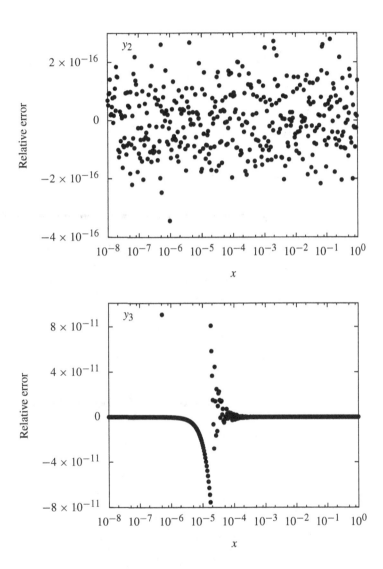

Figure S.13 continued.

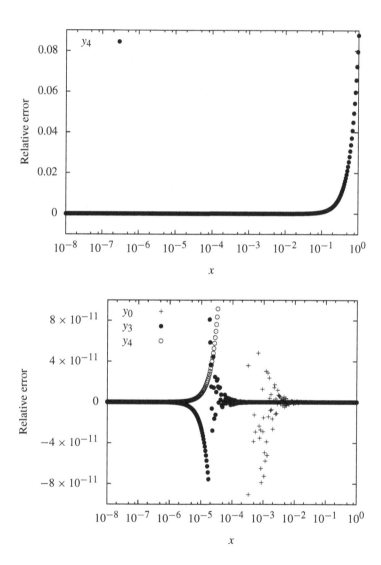

Figure S.13 continued.

Finally, y_4 (Figure S.12):

$$\chi_1 = x^2$$
$$\chi_2 = -\chi_1/2,$$

with numerical errors

$$\frac{\Delta\chi_1}{\chi_1} = 2\frac{\Delta x}{x} + \epsilon_1$$
$$\frac{\Delta\chi_2}{\chi_2} = \frac{\Delta\chi_1}{\chi_1}.$$

There is no ϵ_2 because only the sign and mantissa are affected in the last step. There is also an approximation error from the Taylor series,

$$y_0 = y_4 + \frac{\xi^4}{24}$$

for some $\xi \in [0, x]$. An error expression that embraces both numerical and approximation error is

$$\frac{\tilde{y}_4 - y_4}{y_4} = \frac{\Delta\chi_2}{\chi_2} = \epsilon_1 + 2\epsilon_x$$

$$\frac{\tilde{y}_4 - y_0}{y_0} = (\epsilon_1 + 2\epsilon_x)\frac{x^2}{2(1 - \cos x)} - \frac{\xi^4}{24(\cos x - 1)}.$$

When considering the limit $x \to 0$, one must keep in mind that ξ depends on x. The maximum error bound comes with the assumption $\xi = x$:

$$\left|\frac{\tilde{y}_4 - y_0}{y_0}\right| \le (|\epsilon_1| + 2|\epsilon_x|)\frac{x^2}{2(1 - \cos x)} + \frac{x^4}{24(1 - \cos x)},$$

so in the limit $x \to 0$ the error bound is 3ϵ including the inherent error (which itself is 3ϵ). This is the most numerically trustworthy choice, but only on a restricted range of x near zero.

The relative error of these methods is shown in Figure S.13. Note that y_0 has 100% relative errors as $x \to 10^{-8}$, and the other methods are bounded in that limit. The sixth panel of the figure shows y_0, y_3, and y_4 on a uniform scale where the the relative strengths of the methods are better appreciated for modest values of x.

1.9 To show this use (i)

$$\lim_{m\to\infty}\left(1 + \frac{x}{m}\right)^m = e^x,$$

and $(1 + x/m)^m$ increases monotonically with m for $|x| < 1$; and (ii)

$$\exp(1/10)^x - 1 \le (\exp(1/10) - 1)x \qquad\qquad \text{if } 0 \le x \le 1$$
$$\exp(1/10)^x - 1 \ge (\exp(1/10) - 1)x \qquad\qquad \text{if } -1 \le x \le 0,$$

which can be shown graphically (Figure S.14). We will be concerned with case $x = \pm 10\epsilon n$. Then,

$$\overbrace{(1 + \epsilon)^n - 1 = (1 + \epsilon)^{\frac{n\epsilon}{\epsilon}} - 1 < e^{\epsilon n} - 1}^{\text{with (i)}} = \underbrace{e^{\frac{10\epsilon n}{10}} - 1 \le 10(e^{\frac{1}{10}} - 1)\epsilon n}_{\text{with (ii)}} < 1.06\epsilon n$$

and

$$(1 - \epsilon)^n - 1 = (1 - \epsilon)^{\frac{n\epsilon}{\epsilon}} - 1 > e^{-\epsilon n} - 1 = e^{\frac{-10\epsilon n}{10}} - 1 \ge -10(e^{\frac{1}{10}} - 1)\epsilon n > -1.06\epsilon n.$$

Here we are treating ϵ like $1/m$ in the first terms, and using the monotonicity property to assign the first inequality.

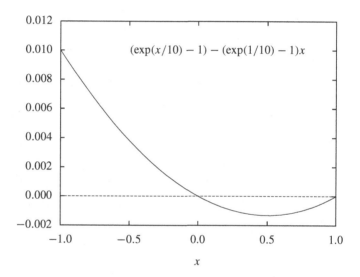

Figure S.14 An inequality for Wilkinson's bound for $(1 \pm \epsilon)^n$; Problem 1.9.

A non-graphical proof of (ii) follows: let $g > 1$ (the specific case above is $g = \exp(1/10)$, but we can be more general), and consider

$$y(x) = g^x - 1 - (g - 1)x.$$

Clearly $y(0) = 0$, and $y(1) = 0$. The derivative

$$y'(x) = g^x \ln g - g + 1$$

has one zero,

$$\xi = \frac{\ln \frac{g-1}{\ln g}}{\ln g},$$

and $0 \le \xi \le 1$ is required by Rolle's theorem. As $y'(0) < 0$, so $y(x) < 0$ in the interval $(0, 1)$, and $y(x) > 0$ for $x < 0$ and $x > 1$. This proves (ii) for the general case, e.g., for the circumstance $n\epsilon < 1/100$, which permits some general comments.

The 6% error implied by omission of terms of order ϵ^2 occurs for $n \approx 1/(10\epsilon) \approx 9 \times 10^{14}$ in double precision (which is really enormous; comparable to the Earth–Sun distance measured in human hair diameters), or $n \approx 1.7 \times 10^6$ in single precision (which is more conceivable, but still huge: the number of hours one would have to drive at 55 mph to cover the Earth–Sun distance). If $\epsilon n < 1/500$ ($n \approx 1.8 \times 10^{13}$ in double precision) the omission of terms of order ϵ^2 results in an error of only 0.1%. For any $n > 1$ we commit an error in neglecting terms of order ϵ^2, but when double precision is used these are negligible in practical terms for all but exceptional cases.

1.10 (i) Using the Taylor series,

$$f(x \pm h) = f(x) \pm hf'(x) + \frac{h^2}{2}f''(x) \pm \frac{h^3}{6}f'''(\xi_\pm) \quad \xi_\pm \in [x, x+h]$$

$$\frac{f(x+h) - f(x-h)}{2h} = f'(x) + \frac{h^2}{12}f'''(\xi_+) + \frac{h^2}{12}f'''(\xi_-)$$

$$\frac{f(x+h) - f(x-h)}{2h} = f'(x) + \frac{h^2}{6}f'''(\bar{\xi}) \quad \bar{\xi} \in [x - h, x + h],$$

so this is an $\mathcal{O}(h^2)$ approximation.
(ii) Knowing nothing about f, an algorithm is

$$\phi_1 = x + h$$
$$\phi_2 = x - h$$
$$\phi_3 = f(\phi_1)$$
$$\phi_4 = f(\phi_2)$$
$$\phi_5 = \phi_3 - \phi_4$$
$$\phi_6 = \phi_5/2h,$$

with errors

$$\frac{\Delta\phi_1}{\phi_1} = \frac{x}{x+h}\frac{\Delta x}{x} + \epsilon_1$$

$$\frac{\Delta\phi_2}{\phi_2} = \frac{x}{x-h}\frac{\Delta x}{x} + \epsilon_2$$

$$\frac{\Delta\phi_3}{\phi_3} = \frac{(x+h)f'(x+h)}{f(x+h)}\frac{\Delta\phi_1}{\phi_1} + \epsilon_3$$

$$\frac{\Delta\phi_4}{\phi_4} = \frac{(x-h)f'(x-h)}{f(x-h)}\frac{\Delta\phi_2}{\phi_2} + \epsilon_4$$

$$\frac{\Delta\phi_5}{\phi_5} = \frac{f(x+h)}{f(x+h) - f(x-h)}\frac{\Delta\phi_3}{\phi_3} - \frac{f(x-h)}{f(x+h) - f(x-h)}\frac{\Delta\phi_4}{\phi_4} + \epsilon_5$$

$$\frac{\Delta\phi_6}{\phi_6} = \frac{\Delta\phi_5}{\phi_5},$$

where there is no error due to division by $2h$ assuming h to be a power of 2. The relative errors ϵ_3 and ϵ_4 are not necessarily bounded by ϵ since f is not necessarily an elementary operation. Figure S.15 presents the associated Bauer graph.

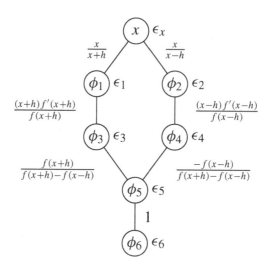

Figure S.15 Numerical error for centered differences; Problem 1.10.

The numerical error is

$$
\frac{\Delta\phi_6}{\phi_6} = \epsilon_6 + \epsilon_5 + \frac{f(x+h)}{f(x+h) - f(x-h)}\epsilon_3 + \frac{(x+h)f'(x+h)}{f(x+h) - f(x-h)}\epsilon_1
$$
$$
- \frac{f(x-h)}{f(x-h) - f(x-h)}\epsilon_4 - \frac{(x-h)f'(x-h)}{f(x+h) - f(x-h)}\epsilon_2
$$
$$
+ \frac{x(f'(x+h) - f'(x-h))}{f(x+h) - f(x-h)}\epsilon_x.
$$

(iii) We might estimate the optimum h as the value for which the upper bound on the approximation error equals the upper bound on the numerical error. First, express both errors in an absolute fashion (not relative),

$$
\left|\frac{h^2}{6}f'''(\bar\xi)\right| = \left|\epsilon_6\phi_6 + \epsilon_5\phi_6 + \frac{f(x+h)}{2h}\epsilon_3 + \frac{(x+h)f'(x+h)}{2h}\epsilon_1 \right.
$$
$$
\left. - \frac{f(x-h)}{2h}\epsilon_4 - \frac{(x-h)f'(x-h)}{2h}\epsilon_2 + \frac{x(f'(x+h) - f'(x-h))}{2h}\epsilon_x\right|.
$$

Next, anticipate the circumstance that $h \to 0$, so

$$
\left|\frac{h^2}{6}f'''(x)\right| = \left|2\epsilon|f'(x)| + \frac{|f(x)|}{h}\epsilon' + \frac{|xf'(x)|}{h}\epsilon + |xf''(x)|\epsilon\right|,
$$

where ϵ' is the upper bound for ϵ_3, ϵ_4. The right-hand side terms proportional to h^{-1} will dominate, so

$$
\left|\frac{h^2}{6}f'''(x)\right| = \left(\frac{|f(x)|}{h}\epsilon' + \frac{|xf'(x)|}{h}\epsilon\right)
$$

$$
h = \sqrt[3]{\frac{6(|f(x)|\epsilon' + |xf'(x)|\epsilon)}{|f'''(x)|}}.
$$

A commonly used rule of thumb is

$$h = \sqrt[3]{\frac{6|f(x)|\epsilon}{|f'''(x)|}},$$

which assumes that $f(x+h)$ and $f(x-h)$ are evaluated as well as possible, with only one rounding error each. All other errors are neglected.

(iv) With $f = \sin(x)$ and $x = \pi/4$, and assuming double precision, the h value where approximation error balances numerical error is calculated to be $h \approx 1.1 \times 10^{-5}$. This compares favorably with computed values (Figure S.16).

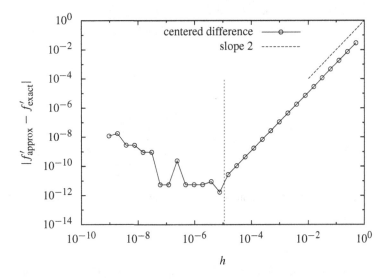

Figure S.16 Centered difference error for $\sin'$; Problem 1.10.

Chapter 2

2.1 (i)

$$\text{lub}_2(\mathbf{A}) = \max_{\mathbf{x}} \frac{\|\mathbf{A}\mathbf{x}\|_2}{\|\mathbf{x}\|_2}$$
$$\leq \max_{\mathbf{x}} \frac{\sqrt{n}\|\mathbf{A}\mathbf{x}\|_\infty}{\|\mathbf{x}\|_\infty}$$
$$= \sqrt{n}\ \text{lub}_\infty(\mathbf{A})$$

using (2.16) in the denominator and (2.17) in the numerator. It writing $\mathbf{A}^{-1}$ for $\mathbf{A}$, one gets also $\text{lub}_2(\mathbf{A}^{-1}) \leq \sqrt{n}\ \text{lub}_\infty(\mathbf{A}^{-1})$, and therefore

$$\text{cond}_2(\mathbf{A}) \leq n\ \text{cond}_\infty(\mathbf{A}).$$

(ii)

$$\text{lub}_\infty(\mathbf{A}) = \max_{\mathbf{x}} \frac{\|\mathbf{Ax}\|_\infty}{\|\mathbf{x}\|_\infty}$$

$$= \max_{\mathbf{x}} \frac{\max_i \left|\sum_k A_{ik} x_k\right|}{\max_i |x_i|}$$

$$= \max_i \sum_k |A_{ik}|.$$

Here, the denominator will be $\max_i |x_i|$, the largest element of $\mathbf{x}$. Let x_k be the element of $\mathbf{x}$ largest in absolute value. Then, $\mathbf{x} = (\text{sign}(A_{i1})x_k, \ \text{sign}(A_{i2})x_k, \ ...)^T$ is a vector with $\|\mathbf{x}\| = |x_k|$, and with $\|\mathbf{Ax}\| \le |x_k| \sum_k |A_{ik}|$. The largest least upper bound norm is obtained from an $\mathbf{x}$ of this sort, with the particular index i that gives the greatest value.

2.2 If

$$\mathbf{A}^{(d)} = \mathbf{GPA}^{(d-1)}$$

then

$$\text{cond}(\mathbf{A}^{(d)}) = \overbrace{\text{lub}(\mathbf{G})\text{lub}(\mathbf{G}^{-1})}^{\text{cond}(\mathbf{G})} \underbrace{\text{lub}(\mathbf{P})}_{1} \overbrace{\text{lub}(\mathbf{P}^{-1})}^{1} \text{cond}(\mathbf{A}^{d-1}),$$

where we recognize that permutation matrices have unit norm. To bound $\text{lub}(\mathbf{G})$, we have

$$\text{lub}_\infty(\mathbf{G}) = \max_{i=d+1,n} (1 + |f_i|) \le 2.$$

Here we use the form of the Frobenius matrix (2.3), with the condition $|f_i| \le 1$ that comes from partial pivoting. Similarly, for $\text{lub}(\mathbf{G}^{-1})$ we have

$$\text{lub}_\infty(\mathbf{G}^{-1}) = \max_{i=d+1,n} (1 + |f_i|) \le 2,$$

using the form (2.8). Together,

$$\text{cond}_\infty(\mathbf{G}) \le 4$$
$$\text{cond}_2(\mathbf{G}) \le 4n,$$

so

$$\text{cond}_\infty(\mathbf{A}^{(d)}) \le 4 \, \text{cond}_\infty(\mathbf{A}^{(d-1)})$$
$$\text{cond}_2(\mathbf{A}^{(d)}) \le 4n \, \text{cond}_2(\mathbf{A}^{(d-1)}).$$

2.3 The results are displayed in the table below. For this very poorly conditioned family of matrices one sees a tendency for pivoting strategies to have their advertised benefits, and for Householder to be slightly advantageous relative to Gaussian elimination, at least when n is large.

	$n = 4$	$n = 8$	$n = 12$	$n = 16$
GE				
trivial	4.51×10^{-15}	1.03×10^{-10}	9.16×10^{-7}	6.59×10^{-3}
partial	4.48×10^{-15}	4.33×10^{-11}	1.13×10^{-7}	8.92×10^{-4}
scaled	4.32×10^{-15}	2.01×10^{-11}	5.49×10^{-9}	4.73×10^{-4}
rook	4.40×10^{-15}	3.83×10^{-11}	6.48×10^{-8}	5.87×10^{-4}
complete	4.40×10^{-15}	3.96×10^{-11}	6.48×10^{-8}	4.36×10^{-4}
QR				
trivial	4.07×10^{-15}	1.08×10^{-11}	2.23×10^{-8}	5.73×10^{-4}
Golub	1.06×10^{-14}	1.45×10^{-11}	2.09×10^{-8}	3.63×10^{-4}
Powell–Reid	1.37×10^{-14}	3.82×10^{-12}	3.16×10^{-9}	4.83×10^{-4}
Björck	1.06×10^{-14}	1.45×10^{-11}	2.09×10^{-8}	3.63×10^{-4}

```
1   #include <iostream>
2   #include <iomanip>
3   #include <math.h>
4   #include "Matrix.H"
5   #include "GE.H"
6   #include "Householder.H"
7   using namespace std;
8
9   void makeVandermonde( int n, Matrix<double>& A, Vector<double>& b )
10  {
11    for( int i=0; i<n; i++ ) {
12      b[i] = 0;
13      double xi = (n-1.-i)/(n-1.);
14      double xij = 1;
15      for( int j=0; j<n; j++ ) {
16        b[i] += xij;
17        A[i][j] = xij;
18        xij *= xi;
19      }
20    }
21  }
22
23  int main()
24  {
25    cout << scientific << setprecision(2);
26    Matrix<double> A(16,16);
27    Vector<double> b(16), diag(16);
28    Vector<int>    pl(16), pr(16);
29    int code;
30
31    string getypes[5]={"trivial","partial","implicit scaled","rook","complete"};
32    string qrtypes[4]={"trivial","Golub","Powell-Reid","Bjorck"};
33
34    for( int n=4; n<=16; n+=4 ) {
35      cout << "n=" << n << endl;
36      A.resize(n,n);
37      b.resize(n);
38      diag.resize(n);
39      pl.resize(n);
40      pl.resize(n);
41
42      for( int type=0; type<5; type++ ) {
43        makeVandermonde(n, A, b);
44        code = ge( A, pl, pr, (GEPivotType)type );
45        cout << "  GE with " << getypes[type] << " pivoting ("
46             << (code? "fail" : "ok") << ")" << endl;
47        if( !code ) {
48          solveLR(A, pl, pr, b );
```

```
49              double  norm  =  0;
50              for ( int  i =0;  i<n;  i++ )  norm  +=  (b[i]-1.)*(b[i]-1.);
51              norm  =  sqrt(norm);
52              cout << "     norm = " << norm << endl;
53            }
54          }
55
56          for( int  type =0;  type <4;  type++ )  {
57            makeVandermonde(n, A, b);
58            code  =  Householder( A,  pl,  pr,  diag,  (QRPivotType)type );
59            cout << "  Householder  with " << qrtypes[type] << "  pivoting  ("
60                 << (code? "fail" : "ok") << ")" << endl;
61            if ( !code ) {
62              solveQR( A,  pl,  pr,  diag,  b );
63              double  norm  =  0;
64              for ( int  i =0;  i<n;  i++ )  norm  +=  (b[i]-1.)*(b[i]-1.);
65              norm  =  sqrt(norm);
66              cout << "     norm = " << norm << endl;
67            }
68          }
69        }
70    }
```

2.4 In the first step of Gaussian elimination, the given condition on **A** guarantees that pivot a_{11} is not zero, so pivoting is not required. The result of the first elimination step would be

$$\begin{pmatrix} a_{11} & a_{12} & \cdots & a_{1j} & \cdots \\ 0 & a_{22} - \frac{a_{21}}{a_{11}}a_{12} & \cdots & a_{2j} - \frac{a_{21}}{a_{11}}a_{1j} & \cdots \\ 0 & a_{32} - \frac{a_{31}}{a_{11}}a_{12} & \cdots & a_{3j} - \frac{a_{31}}{a_{11}}a_{1j} & \cdots \\ \vdots & & & & \end{pmatrix}.$$

From $|a_{11}| > \sum_{j \neq 1} |a_{1j}|$ one has

$$1 > \sum_{j \neq 1} \left| \frac{a_{1j}}{a_{11}} \right|,$$

and so for $k \geq 2$

$$|a_{kk}| > |a_{k1}| + \sum_{j \neq 1,k} |a_{kj}|$$

$$> |a_{k1}| \left| \frac{a_{1k}}{a_{11}} \right| + |a_{k1}| \sum_{j \neq 1,k} \left| \frac{a_{1j}}{a_{11}} \right| + \sum_{j \neq 1,k} |a_{kj}|$$

$$= \left| \frac{a_{k1}}{a_{11}} \right| |a_{1k}| + \sum_{j \neq 1,k} \left| \frac{a_{k1}}{a_{11}} \right| |a_{1j}| + \sum_{j \neq 1,k} |a_{kj}|$$

$$\left| a_{kk} - \frac{a_{k1}}{a_{11}}a_{1k} \right| \geq |a_{kk}| - \left| \frac{a_{k1}}{a_{11}}a_{1k} \right| \geq \sum_{j \neq 1,k} \left(|a_{kj}| + \left| \frac{a_{k1}}{a_{11}}a_{1j} \right| \right) \geq \sum_{j \neq 1,k} \left| a_{kj} - \frac{a_{k1}}{a_{11}}a_{1j} \right|,$$

and therefore in the lower-right $(n-1) \times (n-1)$ block of the above matrix $\mathbf{A}'$, the same condition holds: $|a'_{ii}| \geq \sum_{j \neq i} |a_{ij}|$. By induction, pivoting will never be required to avoid a division by zero.

2.5 Define $\mathbf{x}$ as the solution to $\mathbf{A}\mathbf{x} = \mathbf{b}$, then $\mathbf{e} = \mathbf{y} - \mathbf{x} = \Delta\mathbf{x}$ and

$$\frac{\rho_x}{\rho_r} = \frac{\|\Delta\mathbf{x}\|}{\|\mathbf{x}\|} \frac{\|\mathbf{b}\|}{\|\mathbf{A}\Delta\mathbf{x}\|}.$$

Write

$$\mathbf{b} = \mathbf{A}\mathbf{x}$$
$$\|\mathbf{b}\| \leq \|\mathbf{A}\|\|\mathbf{x}\|,$$

and

$$\Delta\mathbf{x} = \mathbf{A}^{-1}\mathbf{A}\Delta\mathbf{x}$$
$$\|\Delta\mathbf{x}\| \leq \|\mathbf{A}^{-1}\|\|\mathbf{A}\Delta\mathbf{x}\|.$$

Substituting these inequalities into the numerator of ρ_x/ρ_r gives

$$\frac{\rho_x}{\rho_r} \leq \frac{\|\mathbf{A}^{-1}\|\|\mathbf{A}\Delta\mathbf{x}\|}{\|\mathbf{x}\|} \frac{\|\mathbf{A}\|\|\mathbf{x}\|}{\|\mathbf{A}\Delta\mathbf{x}\|} = \text{cond}(\mathbf{A}).$$

Also

$$\mathbf{x} = \mathbf{A}^{-1}\mathbf{b}$$
$$\|\mathbf{x}\| \leq \|\mathbf{A}^{-1}\|\|\mathbf{b}\|,$$

and

$$\|\mathbf{A}\Delta\mathbf{x}\| \leq \|\mathbf{A}\|\|\Delta\mathbf{x}\|.$$

Substitute these in the denominator of ρ_x/ρ_r to give

$$\frac{\rho_x}{\rho_r} \geq \frac{\|\Delta\mathbf{x}\|}{\|\mathbf{A}^{-1}\|\|\mathbf{b}\|} \frac{\|\mathbf{b}\|}{\|\mathbf{A}\|\|\Delta\mathbf{x}\|} = \frac{1}{\text{cond}(\mathbf{A})}.$$

There exist particular values of $\mathbf{b}$ and $\mathbf{y}$ such that each of the bounds on ρ_x/ρ_r can be satisfied as equalities.

Chapter 3

3.1 First calculate the $\mathbf{Q}$ matrix: find k from (2.37); find $\mathbf{w}$ from (2.36); then compute $\mathbf{Q}$ from (2.33):

$$\mathbf{Qx} = (\mathbf{I} - 2\mathbf{ww}^{\mathrm{T}}) \begin{pmatrix} x_1 \\ x_2 \end{pmatrix} = \begin{pmatrix} k \\ 0 \end{pmatrix}$$

$$k = -\operatorname{sign}(x_1)\sqrt{x_1^2 + x_2^2}$$

$$\mathbf{w} = \begin{pmatrix} x_1 + \operatorname{sign}(x_1)\sqrt{x_1^2 + x_2^2} \\ x_2 \end{pmatrix} \frac{1}{\sqrt{2[x_1^2 + x_2^2 + |x_1|\sqrt{x_1^2 + x_2^2}]}}$$

$$\mathbf{ww}^{\mathrm{T}} = \frac{1}{2[x_1^2 + x_2^2 + |x_1|\sqrt{x_1^2 + x_2^2}]}$$
$$\times \begin{pmatrix} 2x_1^2 + x_2^2 + 2|x_1|\sqrt{x_1^2 + x_2^2} & x_1 x_2 + \operatorname{sign}(x_1)x_2\sqrt{x_1^2 + x_2^2} \\ x_1 x_2 + \operatorname{sign}(x_1)x_2\sqrt{x_1^2 + x_2^2} & x_2^2 \end{pmatrix}$$

$$\mathbf{Q} = \frac{1}{[x_1^2 + x_2^2 + |x_1|\sqrt{x_1^2 + x_2^2}]}$$
$$\times \begin{pmatrix} -x_1^2 - |x_1|\sqrt{x_1^2 + x_2^2} & -x_1 x_2 - \operatorname{sign}(x_1)x_2\sqrt{x_1^2 + x_2^2} \\ -x_1 x_2 - \operatorname{sign}(x_1)x_2\sqrt{x_1^2 + x_2^2} & x_1^2 + |x_1|\sqrt{x_1^2 + x_2^2} \end{pmatrix}.$$

Multiply numerator and denominator by $x_1^2 + x_2^2 - |x_1|\sqrt{x_1^2 + x_2^2}$ and simplify

$$\mathbf{Q} = \begin{pmatrix} \dfrac{-|x_1|}{\sqrt{x_1^2 + x_2^2}} & -\dfrac{x_2 \operatorname{sign}(x_1)}{\sqrt{x_1^2 + x_2^2}} \\ -\dfrac{x_2 \operatorname{sign}(x_1)}{\sqrt{x_1^2 + x_2^2}} & \dfrac{|x_1|}{\sqrt{x_1^2 + x_2^2}} \end{pmatrix}.$$

Next, compute $\mathbf{G}$ from (3.30a–3.30c) and (3.29):

$$k = \operatorname{sign}(x_1)\sqrt{x_1^2 + x_2^2}$$

$$c = \frac{|x_1|}{\sqrt{x_1^2 + x_2^2}}$$

$$s = \frac{x_2 \operatorname{sign}(x_1)}{\sqrt{x_1^2 + x_2^2}}$$

$$\mathbf{G} = \begin{pmatrix} \dfrac{|x_1|}{\sqrt{x_1^2 + x_2^2}} & \dfrac{x_2 \operatorname{sign}(x_1)}{\sqrt{x_1^2 + x_2^2}} \\ \dfrac{x_2 \operatorname{sign}(x_1)}{\sqrt{x_1^2 + x_2^2}} & -\dfrac{|x_1|}{\sqrt{x_1^2 + x_2^2}} \end{pmatrix}.$$

Therefore, $\mathbf{G} = -\mathbf{Q}$. $\mathbf{Q}$ is a reflection matrix, just with different sign conventions.

3.2

$$\mathbf{A}^{(i+1)} = \mathbf{R}^{(i)}\mathbf{Q}^{(i)} + \mu^{(i)}\mathbf{I}$$
$$= \mathbf{R}^{(i)}\mathbf{Q}^{(i)} + \mu^{(i)}\mathbf{Q}^{(i)H}\mathbf{Q}^{(i)}$$
$$= \mathbf{Q}^{(i)H}\mathbf{Q}^{(i)}\mathbf{R}^{(i)}\mathbf{Q}^{(i)} + \mu^{(i)}\mathbf{Q}^{(i)H}\mathbf{Q}^{(i)}$$
$$= \mathbf{Q}^{(i)H}\left(\mathbf{A}^{(i)} - \mu^{(i)}\mathbf{I}\right)\mathbf{Q}^{(i)} + \mathbf{Q}^{(i)H}\left(\mu^{(i)}\mathbf{I}\right)\mathbf{Q}^{(i)}$$
$$= \mathbf{Q}^{(i)H}\mathbf{A}^{(i)}\mathbf{Q}^{(i)},$$

which is a similarity transformation.

3.3 The goal is to find the smallest eigenvalue, which should immediately suggest inverse vector iteration. We can expect the accuracy of the result to improve with increasing N, which means we should think of appropriate methods for large tridiagonal systems. The given matrix can be made triangular with $\mathcal{O}(N)$ Givens reflections. The fastest way to accurately solve this problem will be to seek an orthogonal QR decomposition, where the unitary matrix Q is not Householder but Givens. Inverse vector iteration with this approach is used in the following example.

```
1    #include <iostream>
2    #include <iomanip>
3    #include "Matrix.H"
4    #include "Givens.H"
5    using namespace std;
6
7    double solve( int N )
8    {
9      double u1, u2, c, s, v, delta, lambda=1.e6;
10     int iterations;
11
12     // matrix by row, original and triangular
13     Vector<double>diag(N);
14     Vector<double>subdiag(N);
15     Vector<double>supdiag(N);
16     Vector<double>supsupdiag(N);
17
18     // eigenvector
19     Vector<double>z(N);
20
21     // transformation
22     Vector<double>givS(N);
23     Vector<double>givC(N);
24     Vector<double>givV(N);
25
26     // create representation of matrix
27     double L = 1;
28     double h = L/(N+1.);
29     double hsq = h*h;
30
31     for( int row=0; row<N; row++ ) {
32       double x = (row+1)*h;
33       double c = 3 + x*x/(L*L);
34       double csq = c*c;
35       double f = csq/hsq;
36       diag[row] = 2.*f;
37       if( row > 0 ) subdiag[row] = -f;
38       if( row < N-1 ) supdiag[row] = -f;
39       if( row < N-2 ) supsupdiag[row] = 0;
40     }
41
42     // find c,s Givens pairs to make triangular.
43     for( int row=1; row<N; row++ ) {
```

```
44          Givens ( diag [row −1], subdiag [row], c, s, v );
45          givC [row]  = c ;
46          givS [row]  = s ;
47          givV [row]  = v ;
48          u1 = diag [row −1];
49          u2 = subdiag [row ];
50          diag [row −1] = c*u1 + s*u2 ;
51          subdiag [row]  = 0;
52          u1 = supdiag [row −1];
53          u2 = diag [row ];
54          supdiag [row −1] = c*u1 + s*u2 ;
55          diag [row ] = v*(u1 + supdiag [row −1]) − u2 ;
56          u1 = supsupdiag [row −1];
57          u2 = supdiag [row ];
58          supsupdiag [row −1] = c*u1 + s*u2 ;
59          supdiag [row ] = v*(u1 + supsupdiag [row −1]) − u2 ;
60        }
61
62        // "random" eigenvector
63        for ( int row=0; row<N; row++ ) z[row] = 1/ sqrt (( double )N);
64
65        // iterate until relative change in lambda is a few times machine precision
66        iterations = 0;
67        do {
68          // z −> G*z
69          for ( int row=1; row<N; row++ ) {
70            c = givC [row ];
71            s = givS [row ];
72            v = givV [row ];
73            u1 = z[row −1];
74            u2 = z[row ];
75            z[row −1] = c*u1 + s*u2 ;
76            z[row ]    = v*(u1 + z[row −1]) − u2 ;
77          }
78
79          // back substitution. assemble length for renormalization
80          double len = 0;
81          for ( int row=N−1; row >=0; row— ) {
82            if ( row<N−1 ) z[row] −= supdiag [row]*z[row +1];
83            if ( row<N−2 ) z[row] −= supsupdiag [row]*z[row +2];
84            z[row ] /= diag [row ];
85            len += z[row]*z[row ];
86          }
87          // z grows by 1/lambda so len proportional to 1/lambda^2
88          len = 1./ sqrt (len );
89          // now proportional to lambda
90          delta = (len − lambda )/ lambda ;
91          lambda = len ;
92          for ( int row=0; row<N; row++ ) z[row] *= len ;
93
94          ++iterations ;
95        } while ( delta < 1e−15 && iterations < 10);
96
97        if ( iterations >= 10 ) {
98          cout << "inverse vector iteration failed to converge" << endl;
99        }
100
101       return lambda ;
102     }
103
104     int main ()
105     {
106       cout << scientific << setprecision (15);
107
108       int N = 10;
109       for ( int i=0; i <6; i++ ) {
110         double lambda = solve (N);
111         cout << "N=" << N << " omega = " << sqrt (lambda) << endl;
112         N *= 10;
113       }
```

114   return 1;
115   }

The code converges rapidly and shows convergence to $\omega \approx 10.2468$ for $N \geq 1000$.

3.4

$$\mathbf{Ax} = \mathbf{b} - \mathbf{r}$$

$$\mathbf{A\bar{x}} = \mathbf{b}$$

$$\mathbf{Ae} = \mathbf{A}(\mathbf{x} - \mathbf{\bar{x}}) = -\mathbf{r}.$$

Now substitute the SVD expansion of $\mathbf{A}$, and expand $\mathbf{e}$ in columns of $\mathbf{V}$, i.e.,

$$\mathbf{e} = c_1(\mathbf{Ve}_1) + c_2(\mathbf{Ve}_2) + \cdots$$

$$\mathbf{Vc} = \mathbf{e}.$$

Then,

$$\mathbf{U\Sigma V}^T\mathbf{Vc} = -\mathbf{r}$$

$$\mathbf{\Sigma c} = -\mathbf{U}^T\mathbf{r}$$

$$\|\mathbf{\Sigma c}\|_2 = \|\mathbf{r}\|_2.$$

$$\|\mathbf{r}\|_2 = \sqrt{c_1^2\sigma_1^2 + c_2^2\sigma_2^2 + \cdots + c_n^2\sigma_n^2}.$$

If $\mathbf{e}$ is proportional to a column of $\mathbf{V}$ with singular value less than 1, then $\|\mathbf{r}\|_2 < \|\mathbf{e}\|_2$. On the other hand, if $\mathbf{e}$ is proportional to a column of $\mathbf{V}$ with singular value greater than 1, then $\|\mathbf{r}\|_2 > \|\mathbf{e}\|_2$.

3.5

$$x_5 = 1$$

$$x_4 = (a_5 + \lambda)x_5/1 = a_5 + \lambda$$

$$x_3 = (a_4x_5 + \lambda x_4)/1 = a_4 + \lambda a_5 + \lambda^2$$

$$x_2 = (a_3x_5 + \lambda x_3)/1 = a_3 + \lambda a_4 + \lambda^2 a_5 + \lambda^3$$

$$x_1 = (a_2x_5 + \lambda x_2)/1 = a_2 + \lambda a_3 + \lambda^2 a_4 + \lambda^3 a_5 + \lambda^4$$

$$x_0 = (a_1x_5 + \lambda x_1)/1 = a_1 + \lambda a_2 + \lambda^2 a_3 + \lambda^3 a_4 + \lambda^4 a_5 + \lambda^5$$

$$b = -\lambda x_0 - a_0x_5 = -a_0 - \lambda a_1 - \lambda^2 a_2 - \lambda^3 a_3 - \lambda^4 a_4 - \lambda^5 a_5 - \lambda^6.$$

Therefore, $b(\lambda)$ is proportional to the characteristic equation (A.22), as advertised.

3.6 The estimated singular values range from 6.66 to 5.77×10^{-16}, so the condition number is calculated to be 1.15×10^{16}. Using method of Lawson and Hanson, let $\mathbf{x}_k$ be the solution $\mathbf{x}$ computed with the greatest k singular values – the others being set to zero. A plot of $\|\mathbf{r}_k\|$ vs. $\|\mathbf{x}_k\|$ suggest that the optimum solution occurs when $k \approx 12$: Figure S.17.

Since we have the exact solution $\mathbf{x} = \mathbf{1}$ for this problem, we can check the estimated result. A plot of solution error versus $\|\mathbf{x}_k\|$ verifies the prediction (see Figure S.18). The solution error is 1.7×10^{-12} when $k \approx 12$.

Compare this with Solution 2.3. Direct methods of solution carry much larger errors for better conditioned Vandermonde matrices.

```
1   #include <iostream>
2   #include <iomanip>
3   #include <math.h>
4   #include "Matrix.H"
5   #include "SVD.H"
6   using namespace std;
7
8   template <class T> void qsort( Vector<T>& list, int left, int right )
9   {
10    if( left < right ) {
11      int    pivot = (left+right)/2;
12      T      pval  = list[pivot];
13      int lo=left;
14      int hi=right;
15      while( lo <= hi ) {
16        while( list[lo] > pval ) lo++;
17        while( list[hi] < pval ) hi--;
18        if( lo <= hi ) {
19          double d=list[lo]; list[lo]=list[hi]; list[hi]=d;
20          lo++;
21          hi--;
22        }
23      };
24      if( left < hi   ) qsort<T>( list, left, hi );
25      if( lo < right ) qsort<T>( list, lo, right );
26    }
27  }
28
29  void makeVandermonde( int n, Matrix<double>& A, Vector<double>& b )
30  {
31    for( int i=0; i<n; i++ ) {
32      b[i] = 0;
33      double xi = (n-1.-i)/(n-1.);
34      double xij = 1;
35      for( int j=0; j<n; j++ ) {
36        b[i] += xij;
37        A[i][j] = xij;
38        xij *= xi;
39      }
40    }
41  }
42
43  int main()
44  {
45    cout << scientific << setprecision(10);
46    int n = 20;
47    Matrix<double> A(n,n), Acpy(n,n), U(n,n), VT(n,n);
48    Vector<double> b(n),    bcpy(n), sigma(n), orderedlist(n), UTb(n);
49
50    // make test matrix and rhs, make a copy of rhs
51    makeVandermonde(n, A, b);
52    makeVandermonde(n, Acpy, bcpy);
53
54    // perform SVD
55    SVD( A, U, VT, sigma );
56    for( int i=0; i<n; i++ ) orderedlist[i]=sigma[i];
57    qsort<double>( orderedlist, 0, n-1 );
58    cout << "singular values" << endl << orderedlist << endl;
59
60    // min and max singular values; condition number
61    cout << "condition number estimate in L2 " << orderedlist[0]/orderedlist[n-1] << endl;
62
63    for( int k=0; k<=n; k++ ) {
64      // include k singular values
65
66      // solve U sigma VT x = b;    x = V 1/sigma UT b
67      for( int i=0; i<n; i++ ) {
```

```
68        UTb[i] = 0;
69        for( int j=0; j<n; j++ ) UTb[i] += U[j][i]*bcpy[j];
70      }
71      // divide by sigma only where sigma is greater than cutoff
72      for( int i=0; i<n; i++ ) {
73        if( k && sigma[i] >= orderedlist[k-1] ) {
74          UTb[i] /= sigma[i];
75        } else {
76          UTb[i] = 0;
77        }
78      }
79      for( int i=0; i<n; i++ ) {
80        b[i] = 0;
81        for( int j=0; j<n; j++ ) b[i] += VT[j][i]*UTb[j];
82      }
83      // solution error norm, residual norm, solution norm
84      double snorm=0, rnorm = 0, xnorm=0;
85      for( int i=0; i<n; i++ ) {
86        double r=bcpy[i];
87        for( int j=0; j<n; j++ ) r -= Acpy[i][j]*b[j];
88        rnorm += r*r;
89        snorm += (b[i]-1.)*(b[i]-1.);
90        xnorm += b[i]*b[i];
91      }
92      rnorm = sqrt(rnorm);
93      snorm = sqrt(snorm);
94      xnorm = sqrt(xnorm);
95      cout << xnorm << " " << rnorm << " " << snorm << " \"" << k << "\"" <<endl;
96    }
97  }
```

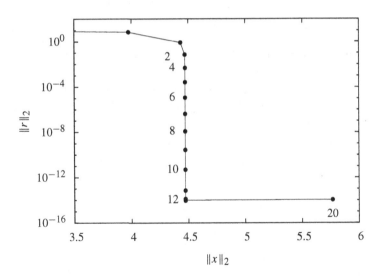

Figure S.17 Residual norm versus solution length, Problem 3.6.

3.7 Let $B = USV^T$, with U and V from the SVD of A. Since B has no *a priori* relation to A, S cannot be expected to be diagonal. It is simply given by $S = U^T B V$. The desired B comes from minimizing

$$x^2 = (\|A - B\|_2)^2 = \operatorname{tr} A^T A - 2\operatorname{tr} A^T B + \operatorname{tr} B^T B$$

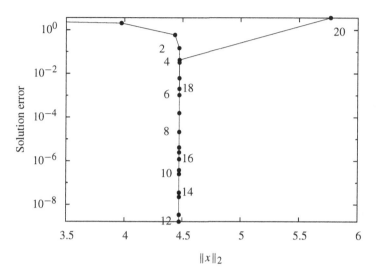

Figure S.18 Solution error versus solution length, Problem 3.6.

$$= \sum_i \sigma_i^2 - 2 \operatorname{tr}(\Sigma S) + \operatorname{tr}\left(S^T S\right).$$

This uses the fact that the trace is the sum of eigenvalues, and $B^T B = U^T S^T S U$ is a similarity transformation.

First consider off-diagonal elements of S. The derivative of x^2 with respect to some off diagonal element S_{ij}, $i \neq j$, will be $2S_{ij}$. Therefore, for x^2 to be a minimum, S must be diagonal.

Minimization of x^2 with respect to the diagonal elements of S requires that $S_{ii} = \sigma_i$. But, this would give $B = A$, and the rank of B will not be $s < k$ as desired. The minimum of

$$\sum_{i=1}^{s}(\sigma_i - S_{ii})^2 + \sum_{i=s+1}^{k}(\sigma_i - S_{ii})^2$$

subject to S having rank s, is

$$S_{ii} = \begin{cases} \sigma_i & i \leq s \\ 0 & \text{otherwise.} \end{cases}$$

Therefore, when using truncated SVD to solve a poorly conditioned linear system, one is using a lower-rank least squares approximation to A.

Chapter 4

4.1 Consider the 5×5 example

$$
L = \begin{pmatrix} 0 & & & & \\ * & 0 & & & \\ * & * & 0 & & \\ * & * & * & 0 & \\ * & * & * & * & 0 \end{pmatrix} \qquad
L^2 = \begin{pmatrix} 0 & & & & \\ 0 & 0 & & & \\ * & 0 & 0 & & \\ * & * & 0 & 0 & \\ * & * & * & 0 & 0 \end{pmatrix}
$$

$$
L^3 = \begin{pmatrix} 0 & & & & \\ 0 & 0 & & & \\ 0 & 0 & 0 & & \\ * & 0 & 0 & 0 & \\ * & * & 0 & 0 & 0 \end{pmatrix} \qquad
L^4 = \begin{pmatrix} 0 & & & & \\ 0 & 0 & & & \\ 0 & 0 & 0 & & \\ 0 & 0 & 0 & 0 & \\ * & 0 & 0 & 0 & 0 \end{pmatrix}
$$

$$ L^5 = 0. $$

Therefore, for $n \times n$ L, $L^n = 0$. This has nothing to do with the sign of the elements, so also $|L|^n = 0$.

$$ (I - L)(I + L + \cdots + L^{n-1}) = (I + L + \cdots + L^{n-1}) - (L + \cdots + L^n) = I $$
$$ (I + L + \cdots + L^{n-1}) = (I - L)^{-1} $$
$$ (I + |L| + \cdots + |L^{n-1}|) = (I - |L|)^{-1}. $$

Again, this has nothing to do with the signs of the elements of L so it is true also for $|L|$.

Termwise, the expansion of $(I - |L|)^{-1}$ is $\geq$ the expansion of $(I - L)^{-1}$, and clearly the former is $\geq I$.

4.2 The matrix B is trivial for Jacobi ($B = I$), and a very simple Frobenius matrix in the Gauss–Seidel case. As such, its inverse is trivial (2.8). The J and H matrices are therefore easy to construct. The sparseness of these matrices makes their characteristic equations easy to evaluate by expansion in cofactors:

$$
J = \begin{pmatrix} 0 & & & -a \\ -1 & \ddots & & \\ & \ddots & \ddots & \\ -b & & -1 & 0 \end{pmatrix}
$$

$$ \det(J - \lambda I) = (-1)^n \lambda^n - (-1)^n ab\lambda^{n-2} - a $$

$$\mathbf{H} = \begin{pmatrix} 0 & & & & & -a \\ & & & & & a \\ & & & & & -a \\ & & & & & a \\ & & & & & \vdots \\ & & & & & a((-1)^n + b) \end{pmatrix}$$

$$\det(\mathbf{H} - \lambda\mathbf{I}) = \lambda^{n-1}\left((-1)^n\lambda - (-1)^n ab - a\right).$$

If $a = (-1)^{n+1}/2$ and $b = (-1)^n$, then

$$\psi_J = \det(\mathbf{J} - \lambda\mathbf{I}) = (-1)^n\left[\lambda^n + \frac{1}{2}\lambda^{n-2} + \frac{1}{2}\right]$$

$$\det(\mathbf{H} - \lambda\mathbf{I}) = (-1)^n\lambda^{n-1}\left[\lambda + 1\right].$$

Clearly $\rho(\mathbf{H}) = -1$, so Gauss–Seidel does not converge. For Jacobi, let $f(\lambda) = \lambda^n + 1/2$ and $g(\lambda) = \lambda^{n-2}/2$. $|f| > |g|$ on the unit circle $|\lambda| = 1$ except when n is odd and $\lambda = -1$, in which case $|f| = |g| = 1/2$. However, -1 is not a root of the characteristic equation, so we may use a deformed contour: a unit circle in the complex plane with a small dimple at $\lambda = -1$, Figure S.19. f has n zeros in the contour, Rouché's theorem applies on the contour, therefore $\psi_J = f + g$ has all zeroes in the contour. Since $\lambda \neq -1$, we can take the radius of the deformation to zero and be assured that no root $|\lambda| \geq 1$ exists.

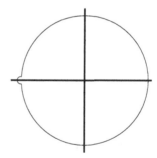

Figure S.19 Deformed contour in complex plane; Problem 4.2.

If $a \geq 1$ and $b = (-1)^{n+1}$

$$\det(\mathbf{J} - \lambda\mathbf{I}) = (-1)^n\lambda^n + a\lambda^{n-2} - a$$

$$\det(\mathbf{H} - \lambda\mathbf{I}) = (-1)^n\lambda^n.$$

Clearly $\rho(\mathbf{H}) = 0$, so Gauss–Seidel converges. For Jacobi, expressing the characteristic equation in the form

$$(-1)^n(\lambda - \lambda_0)(\lambda - \lambda_1)\cdots(\lambda - \lambda_{n-1})$$

shows that

$$(-1)^n\prod_i\lambda_i = a > 1,$$

which cannot be true unless some eigenvalue has magnitude greater than one.

4.3 The discrete Laplace operator is

$$(\Delta u)_{ij} = \frac{u_{i-1,j} + u_{i+1,j} + u_{i,j-1} + u_{i,j+1} - 4u_{ij}}{h^2}.$$

We will use this discretization at each node of the grid. Where a value is referenced that lies on the domain boundary, we substitute the boundary condition and move the term to the right-hand side. When $N = 4$ the result is

$$\frac{1}{h^2}
\left(\begin{array}{ccc|ccc|ccc}
-4 & 1 & & 1 & & & & & \\
1 & -4 & 1 & & 1 & & & & \\
 & 1 & -4 & & & 1 & & & \\
\hline
1 & & & -4 & 1 & & 1 & & \\
 & 1 & & 1 & -4 & 1 & & 1 & \\
 & & 1 & & 1 & -4 & & & 1 \\
\hline
 & & & 1 & & & -4 & 1 & \\
 & & & & 1 & & 1 & -4 & 1 \\
 & & & & & 1 & & 1 & -4
\end{array}\right)
\left(\begin{array}{c}
u_{11} \\ u_{12} \\ u_{13} \\ u_{21} \\ u_{22} \\ u_{23} \\ u_{31} \\ u_{32} \\ u_{33}
\end{array}\right)
= -\frac{1}{h^2}
\left(\begin{array}{c}
-\frac{1}{h^2} + \frac{1}{h^2} \\ +\frac{4}{h^2} \\ 1 - \frac{1}{h^2} + \frac{9}{h^2} \\ \hline -\frac{4}{h^2} \\ 1 - \frac{4}{h^2} \\ -\frac{9}{h^2} + \frac{1}{h^2} - 1 \\ \hline 1 - \frac{9}{h^2} + \frac{4}{h^2} - 1 \\ +\frac{9}{h^2} - 1
\end{array}\right).$$

This matrix does not obey the strong row sum criterion because there are rows where the magnitude of the diagonal $|4/h^2|$ is equal to the sum of off-diagonal entries.

To apply the weak row sum criterion, we first need to assess whether or not the matrix is irreducible. Its graph is shown below.

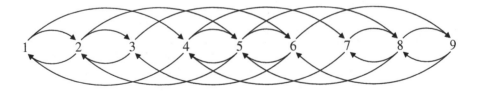

This graph is fully connected: one may pass from any node to any other node by following the curves in the direction of the arrows. Therefore, this matrix is irreducible and the weak row sum criterion may be applied. The matrix passes the weak row sum criterion because there are rows where the magnitude of the diagonal is greater than the sum of off-diagonals, and there are no rows where the diagonal is less than the sum of off-diagonals. This assures us that Jacobi iteration will converge.

Finally, since the diagonal entries are all negative, and the off-diagonal entries are all positive (or zero), **J** will be positive. Therefore, by the analysis of Stein and Rosenberg Gauss–Seidel will converge more rapidly than will Jacobi.

4.4, 4.5, and 4.6 For convenience, these problems are solved together in the single code displayed below. The desired stopping criteria are met after 147 iterations of conjugate gradient alone, 9 individual multigrid V cycles (using 3 iterations of damped Jacobi as a smoother), and 5 iterations of the multigrid-preconditioned conjugate gradient approach. The solution is displayed in Figure S.20.

method	iterations	norm
CG	147	$\|\mathbf{p}\|_2 = 0.000\,930\,276$
MG	9	$\|\mathbf{r}\|_2 = 0.000\,208\,228$
MGCG	5	$\|\mathbf{p}\|_2 = 0.000\,048\,647$

```cpp
1   #include <Matrix.H>
2   #include <iostream>
3   #include <iomanip>
4   #include <math.h>
5   #include <stdlib.h>
6   using namespace std;
7
8   void prolong( const Vector<double>& coarse, Vector<double>& fine, int Nh )
9   {
10    int nvh = fine.size();
11    int N2h = Nh/2;
12    for( int kh=0; kh<nvh; kh++ ) {
13      int jh = kh/(Nh-1);
14      int ih = kh%(Nh-1);
15      // node i2h,j2h may be fictitious
16      // h:    0,2   1,2   2,2    c:   .    .    .
17      //       0,1   1,1   2,1         .   0,0   .
18      //       0,0   1,0   2,0         .    .    .
19      int i2h = ih/2;
20      int j2h = jh/2;
21      int k2h = (N2h-1)*j2h + i2h;
22      fine[kh] = 0;
23      if( ih%2 && jh%2 ) {
24        // odd in both indices — fine lies on top of coarse
25        fine[kh] = coarse[k2h];
26      } else if( ih%2 ) {
27        // aligned in x direction, not y
28        if( j2h > 0     ) fine[kh] += 0.5*coarse[k2h-N2h+1];   // below
29        if( j2h < N2h-1 ) fine[kh] += 0.5*coarse[k2h      ];   // above
30      } else if( jh%2 ) {
31        // aligned in y direction, not x
32        if( i2h > 0     ) fine[kh] += 0.5*coarse[k2h-1];   // left
33        if( i2h < N2h-1 ) fine[kh] += 0.5*coarse[k2h ];   // right
34      } else {
35        // not aligned
36        if( i2h > 0 ) {
37          if( j2h > 0     ) fine[kh] += 0.25*coarse[k2h-N2h];   // lower left
38          if( j2h < N2h-1 ) fine[kh] += 0.25*coarse[k2h-1  ];   // upper left
39        }
40        if( i2h < N2h-1 ) {
41          if( j2h > 0     ) fine[kh] += 0.25*coarse[k2h-N2h+1]; // lower right
42          if( j2h < N2h-1 ) fine[kh] += 0.25*coarse[k2h      ]; // upper right
43        }
```

```
44       }
45     }
46   }
47
48   void restrict( const Vector<double> fine, Vector<double>& coarse, int N2h )
49   {
50     int nv2h = coarse.size();
51     int Nh = N2h*2;
52
53     for( int k2h=0; k2h<nv2h; k2h++ ) {
54       int j2h = k2h/(N2h-1);
55       int i2h = k2h%(N2h-1);
56       // this is the fine cell under the coarse cell
57       int ih = i2h*2+1;
58       int jh = j2h*2+1;
59       int kh = (Nh-1)*jh + ih;
60       // fine cell beneath coarse
61       coarse[k2h] = fine[kh]/4.;
62       // upper/lower left/right
63       coarse[k2h] += ( fine[kh-1-(Nh-1)] + fine[kh-1+(Nh-1)] +
64                        fine[kh+1-(Nh-1)] + fine[kh+1+(Nh-1)] )/16.;
65       // lower, upper, left, right
66       coarse[k2h] += ( fine[kh-(Nh-1)]   + fine[kh+(Nh-1)] +
67                        fine[kh-1     ]   + fine[kh+1     ] )/8.;
68     }
69   }
70
71   void smooth( int nu, const double& scale, int N,
72                const Vector<double>& b, Vector<double>& x )
73   {
74     int nv = x.size();
75     // for Jacobi need a spare copy
76     Vector<double> s(nv);
77     for( int c=0; c<nu; c++ ) {
78       for( int k=0; k<nv; k++ ) s[k] = x[k];
79       for( int k=0; k<nv; k++ ) {
80         x[k] = b[k]/scale;
81         // -2 is (-4/(2/3))+4
82         x[k] -= 2*s[k];
83         int j = k/(N-1);
84         int i = k%(N-1);
85         if( i!=0   ) x[k] -= s[k-1];
86         if( i!=N-2 ) x[k] -= s[k+1];
87         if( j!=0   ) x[k] -= s[k-(N-1)];
88         if( j!=N-2 ) x[k] -= s[k+(N-1)];
89         // -6 is -4/(2/3))
90         x[k] /= -6;
91       }
92     }
93   }
94
95   double residual( const double& scale, int N, const Vector<double>&x,
96                    const Vector<double>&b, Vector<double>&r )
97   {
98     int nv = x.size();
99     double r2 = 0;
100    for( int k=0; k<nv; k++ ) {
101      r[k] = b[k] + 4*scale*x[k];
102      int j = k/(N-1);
103      int i = k%(N-1);
104      if( i!=0   ) r[k] -= scale*x[k-1];
105      if( i!=N-2 ) r[k] -= scale*x[k+1];
106      if( j!=0   ) r[k] -= scale*x[k-(N-1)];
107      if( j!=N-2 ) r[k] -= scale*x[k+(N-1)];
108      r2 += r[k]*r[k];
109    }
110    return sqrt(r2);
111  }
112
113  double Vcycle( int nu, int Nh, const double& scale, Vector<double>&b,
```

```
114                    Vector<double>&x  )
115  {
116     int nvh = (Nh-1)*(Nh-1);
117     if( Nh==2 ) {
118        // one point ... trivial bottom solver
119        x[0] = b[0]/(-4*scale);
120        return 0;
121     }
122
123     // do nu cycles of smoothing
124     smooth( nu, scale, Nh, b, x );
125
126     // residual
127     Vector<double> tmp1( nvh );
128     double r2 = residual( scale, Nh, x, b, tmp1 );
129
130     // restrict
131     int N2h = Nh/2;
132     int nv2h = (N2h-1)*(N2h-1);
133     Vector<double> tmp2( nv2h );
134     restrict( tmp1, tmp2, N2h );
135
136     // solve on coarser grid
137     Vector<double> tmp3( nv2h );
138     for( int i=0; i<nv2h; i++ ) tmp3[i] = 0;
139     Vcycle( nu, N2h, scale/4., tmp2, tmp3 );
140
141     // prolong
142     prolong( tmp3, tmp1, Nh );
143
144     // correct
145     for( int i=0; i<nvh; i++ ) x[i] += tmp1[i];
146
147     // smooth again
148     smooth( nu, scale, Nh, b, x );
149
150     r2 = residual( scale, Nh, x, b, tmp1 );
151     return r2;
152  }
153
154  // compute p = Ax without storing A
155  void Ax( Vector<double>& p, const Vector<double>& x, int N, double& hsq )
156  {
157     int nv = x.size();
158     for( int k=0; k<nv; k++ ) {
159        int j = k/(N-1);
160        int i = k%(N-1);
161        p[k] = -4.*x[k]/hsq;
162        if( i!=0   ) p[k] += x[k-1]/hsq;
163        if( i!=N-2 ) p[k] += x[k+1]/hsq;
164        if( j!=0   ) p[k] += x[k-(N-1)]/hsq;
165        if( j!=N-2 ) p[k] += x[k+(N-1)]/hsq;
166     }
167  }
168
169  int main()
170  {
171     int N = 64;
172     int nv = (N-1)*(N-1);
173     double h = 1./N;
174     double hsq = h*h;
175
176     Vector<double> x(nv), q(nv), b(nv), p(nv), r(nv), v(nv);
177
178     // build vector b
179     for( int i=0; i<N-1; i++ ) {
180        double xc = (i+1)*h;
181        for( int j=0; j<N-1; j++ ) {
182           double yc = (j+1)*h;
183           int k = j*(N-1)+i;
```

```
184          b[k] = 0;
185          if( i==0 ) {
186            // left edge -- include -u(0,y)/h^2 in RHS
187            b[k] -= -yc*yc/hsq;
188          } else if( i==N-2 ) {
189            // right edge -- include -u(1,y)/h^2 in RHS
190            b[k] -= (1.-yc*yc)/hsq;
191          }
192          if( j==0 ) {
193            // bottom edge -- include -u(x,0)/h^2 in RHS
194            b[k] -= xc*xc/hsq;
195          } else if( j==N-2 ) {
196            // top edge -- include -u(x,1)/h^2 in RHS
197            b[k] -= (xc*xc-1)/hsq;
198          }
199        }
200      }
201
202      // ─────────────────────────────────────────────
203      // first -- Conjugate gradients alone
204
205      // zero x to start
206      for( int i=0; i<nv; i++ ) {
207        x[i] = 0;
208        r[i] = p[i] = b[i];
209      }
210
211      int iter=0;
212      do {
213        iter++;
214
215        // evaluate 2-norm of p to decide if convergence has been achieved
216        double dot = 0;
217        for( int i=0; i<nv; i++ ) dot += p[i]*p[i];
218        dot = sqrt(dot);
219
220        cout << iter << " CG L2(p) " << dot << endl;
221        if( dot < 0.001 ) break;
222
223        // v = A*p
224        Ax( v, p, N, hsq );
225
226        double oldrdr=0;
227        dot=0;
228        for( int i=0; i<nv; i++ ) {
229          oldrdr += r[i]*r[i];
230          dot += p[i]*v[i];
231        }
232
233        double alpha = oldrdr/dot;
234
235        for( int i=0; i<nv; i++ ) x[i] += alpha*p[i];
236
237        dot = 0;
238        for( int i=0; i<nv; i++ ) {
239          r[i] -= alpha*v[i];
240          dot += r[i]*r[i];
241        }
242
243        double gamma = - dot/oldrdr;
244
245        for( int i=0; i<nv; i++ ) p[i] = r[i] - gamma*p[i];
246
247      } while(1);
248
249      // ─────────────────────────────────────────────
250      // second -- Multigrid alone
251
252      for( int i=0; i<nv; i++ ) x[i] = 0;
253
```

```
254     iter = 0;
255     do {
256       iter++;
257       double r2 = Vcycle( 3, N, 1./hsq, b, x );
258       cout << iter << " MG L2(r) " << r2 << endl;
259       if( r2 < 0.001 ) break;
260     } while(1);
261
262     // ─────────────────────────────────────────────
263     // third — Multigrid preconditioned CG
264
265     int pcc = 1; // Vcycles per preconditioner
266
267     // zero x to start
268     for( int i=0; i<nv; i++ ) {
269       q[i] = x[i] = 0;
270       r[i] = b[i];
271     }
272
273     // p,q from approximate multigrid
274     for( int c=0; c<pcc; c++ ) Vcycle( 3, N, 1./hsq, r, q );
275     for( int i=0; i<nv; i++ ) p[i] = q[i];
276
277     iter=0;
278     do {
279       iter++;
280
281       // evaluate 2-norm of p to decide if convergence has been achieved
282       double dot = 0;
283       for( int i=0; i<nv; i++ ) {
284         dot += p[i]*p[i];
285       }
286       dot = sqrt(dot);
287       cout << iter << " MGCG L2(p) " << dot << endl;
288
289       if( dot < 0.001 ) break;
290
291       // v = A*p
292       Ax( v, p, N, hsq );
293
294       double oldrdq=0;
295       dot=0;
296       for( int i=0; i<nv; i++ ) {
297         oldrdq += r[i]*q[i];
298         dot    += p[i]*v[i];
299       }
300
301       double alpha = oldrdq/dot;
302
303       for( int i=0; i<nv; i++ ) x[i] += alpha*p[i];
304
305       for( int i=0; i<nv; i++ ) r[i] -= alpha*v[i];
306
307       // precondition with multigrid
308       for( int i=0; i<nv; i++ ) q[i] = 0;
309       for( int c=0; c<pcc; c++ ) Vcycle( 3, N, 1./hsq, r, q );
310
311       dot = 0;
312       for( int i=0; i<nv; i++ ) dot += r[i]*q[i];
313
314       double beta = dot/oldrdq;
315       for( int i=0; i<nv; i++ ) p[i] = q[i] + beta*p[i];
316     } while(1);
317   }
```

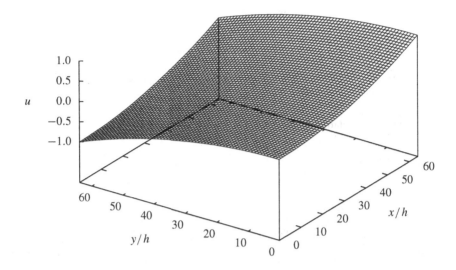

Figure S.20 Solution to Problems 4.4, 4.5, and 4.6.

Chapter 5

5.1 The Newton tableau for five equidistant points is:

$-2h$	y_{-2}				
		$\frac{y_{-1}-y_{-2}}{h}$			
$-h$	y_{-1}		$\frac{y_{-2}-2y_{-1}+y_0}{2h^2}$		
		$\frac{y_0-y_{-1}}{h}$		$\frac{-y_{-2}+3y_{-1}-3y_0+y_1}{6h^3}$	
0	y_0		$\frac{y_{-1}-2y_0+y_1}{2h^2}$		$\frac{y_{-2}-4y_{-1}+6y_0-4y_1+y_2}{24h^4}$
		$\frac{y_1-y_0}{h}$		$\frac{-y_{-1}+3y_0-3y_1+y_2}{6h^3}$	
h	y_1		$\frac{y_0-2y_1+y_2}{2h^2}$		
		$\frac{y_2-y_1}{h}$			
$2h$	y_2				

This gives the interpolation polynomial

$$p(x) = y_{-2} + (x + 2h)\left(\frac{y_{-1} - y_{-2}}{h} + (x + h)\left(\frac{y_{-2} - 2y_{-1} + y_0}{2h^2} + \right.\right.$$
$$x\left(\frac{-y_{-2} + 3y_{-1} - 3y_0 + y_1}{6h^3} + (x - h)\left(\frac{y_{-2} - 4y_{-1} + 6y_0 - 4y_1 + y_2}{24h^4}\right)\right)\right),$$

with error

$$\frac{y^{(5)}(\xi)}{5!}(x + 2h)(x + h)x(x - h)(x - 2h).$$

The derivative of $y(x)$ evaluated at $x = 0$ is

$$p'(0) = \frac{y_{-2} - 8y_{-1} + 8y_1 - y_2}{12h}$$

and the error at $x = 0$ is

$$\frac{y^{(5)}(\xi)\, h^4}{30}.$$

5.2 The following code computes the solution displayed in Figure S.21.

```
1    #include <iostream>
2    #include <iomanip>
3    #include <fstream>
4    #include "Matrix.H"
5    #include "FFT.H"
6    #include <math.h>
7    using namespace std;
8
9    int main()
10   {
11      int N=1024;
12      double sigma= 10;
13      double a1=10,  c1=100, w1=500;
14      double a2=50,  c2=400, w2=15;
15      Vector<double> fR(N), fI(N), gR(N), gI(N), convR(N), convI(N);
16
17      // create F and G data, write
18      fstream before;
19      before.open("conv.before", fstream::out );
20      before << scientific  << setprecision(16);
21      for( int i=0; i<N; i++ ) {
22         fR[i] = fI[i] = gR[i] = gI[i] = 0;
23         // fake stectrum
24         if( i < N/2 ) {
25           fR[i]  = 0.25*w1*w1*a1/((i-c1)*(i-c1)+0.25*w1*w1) +
26                    0.25*w2*w2*a2/((i-c2)*(i-c2)+0.25*w2*w2);
27         }
28         // fake filter
29         if( i <= 2*sigma )
30            gR[i] = exp(-i*i/(2*sigma*sigma))/(sigma*sqrt(2.*M_PI));
31         if( N-i <= 2*sigma )
32            gR[i] = exp(-(N-i)*(N-i)/(2*sigma*sigma))/(sigma*sqrt(2.*M_PI));
33
34         before << i << " " << fR[i] << " " << gR[i] << endl;
35      }
36      before.close();
37
38      // FFT convolution:
39      //    (1) make transform
40      fft( fR, fI, true );
41      fft( gR, gI, true );
42      //    (2) multiply, multiply by N
43      for( int i=0; i<N; i++ ) {
44         convR[i] = N*( fR[i]*gR[i] - fI[i]*gI[i] );
45         convI[i] = N*( fR[i]*gI[i] + fI[i]*gR[i] );
46      }
47      //    (3) back transform
48      fft( convR, convI, false );
49
50      // write
51      fstream after;
52      after.open("conv.after", fstream::out );
53      after << scientific  << setprecision(16);
54      for( int i=0; i<N; i++ ) {
55         after << i << " " << convR[i] << " " << convI[i] << endl;
56      }
57      after.close();
58   }
```

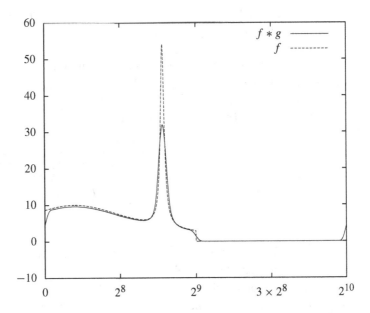

Figure S.21 Convolution solution; Problem 5.2.

5.3 Let $p_n(x)$ be an arbitrary polynomial of degree n, with leading coefficient 1, i.e.,

$$p_n(x) = a_0 + a_1 x + \cdots + a_{n-1} x^{n-1} + x^n.$$

If we expand $p_n(x)$ as a sum of Chebyshev polynomials, then

$$p_n(x) = \frac{1}{2^{n-1}} T_n(x) + b_{n-1} T_{n-1}(x) + \cdots + b_0 T_0(x).$$

Since $T_n(x) = \cos(n \cos^{-1} x)$, with the change of variable $x = \cos\theta$ we can express p_n as a Fourier cosine series,

$$g(\theta) = p_n(\cos^{-1} x) = \frac{1}{2^{n-1}} \cos(n\theta) + b_{n-1} \cos((n-1)\theta) + \cdots + b_0,$$

with a $1:1$ correspondence between the cosine terms and the Chebyshev terms. For any Fourier cosine series,

$$s_n(\theta) = \sum_{i=0}^{n} b_i \cos(i\theta),$$

one has

$$\max |s_n(\theta)| \geq |b_n|.$$

(Consider $t_n(\theta) = b_n \cos(n\theta)$, suppose $\max |s_n| < |b_n|$, and show that this leads to a contradiction. For $\theta_k = k\pi/n$,

$$s_n(\theta_k) - t_n(\theta_k) = s_n(k\pi/n) - b_n(-1)^n.$$

If max $|s_n| < |b_n|$, then this function changes sign n times in the interval $0 \le \theta \le \pi$, which means it has n zeros. However, $s_n - t_n$ is the cosine series $b_0 + \cdots + b_{n-1}\cos((n-1)\theta)$ equivalent to a degree $n-1$ polynomial, and therefore can have only $n-1$ zeros, so there is a contradiction.) Applied to p_n, this means max $|p_n| \ge 2^{1-n}$. The smallest maximum occurs with equality, max $|p_n| = 2^{1-n}$, which is the case $p_n(x) = \pm 2^{1-n} T_n(x)$.

Now, consider the construction of a polynomial of degree n with support points taken as the roots of T_{n+1}. The error formula for this polynomial will be

$$P_n(x) - f(x) = -\frac{y^{(n+1)}(\xi)}{(n+1)!} \underbrace{(x-x_0)\cdots(x-x_n)}_{\omega(x)}$$

$$= -\frac{y^{(n+1)}(\xi)}{(n+1)!} \underbrace{\frac{T_{n+1}(x)}{2^n}}_{\omega(x)},$$

where the degree $n+1$ term $\omega(x)$ on the right-hand side has the smallest possible value on $[-1, 1]$ of all degree $n+1$ polynomials.

The interpolation problem is solved in the following code listing. The results are displayed in Figure S.22, where it can be seen that Chebyshev support points give a much smoother result than interpolation from regularly spaced support points.

```
1   #include <iostream>
2   #include <iomanip>
3   #include "Matrix.H"
4   #include <math.h>
5   using namespace std;
6
7   // function to be interpolated
8   double f(double x)
9   {
10      return 1./(1.+12.*x*x);
11  }
12
13  // given support points, find the coefficients of a polynomial
14  // by Newton interpolation, for evaluation by Horner scheme.
15
16  // tableau:  packed into a vector, increasing index moves up and to right.
17  // This is an order that permits addition of a new support point at little cost.
18  // e.g.,
19  //
20  //   0
21  //       2
22  //   1       5
23  //       4       9
24  //   3       8
25  //       7
26  //   6
27
28  void Newton( Vector<double>& abscissas ,
29                Vector<double>& ordinates ,
30                Vector<double>& tableau )
31  {
32      int n = abscissas.size ();
33      assert( ordinates.size () == n );
34      assert( tableau.size () == (n*(n+1))/2 );
35
36      int j = 0;
37      for( int i=0; i<n; i++ ) {
38          tableau[j] = ordinates[i];
39          for( int k=j+1; k<=j+i; k++ ) {
40              tableau[k] = (tableau[k-1]-tableau[k-i-1])/(abscissas[i]-abscissas[i+j-k]);
```

```
41          }
42        j += i+1;
43      }
44   }
45
46   // evaluate a polynomial constructed by Newton tableau
47   // using Horner's scheme.  This implementation assumes x0<x1<...<xn
48   // and uses implicit permutation.
49
50   double Horner( Vector<double>& abscissas ,
51                  Vector<double>& tableau ,
52                  double x )
53   {
54     int n = abscissas.size();
55     Vector<double>xval(n-1);
56     Vector<double>fval(n);
57
58     // find the index of the abscissa greater closest to x
59     int r0=0;
60     double d0 = fabs(abscissas[r0]-x);
61     for( int r=1; r<n; r++ )  {
62       double d1 = fabs(abscissas[r]-x);
63       if( d1 < d0 ) {
64         d0 = d1;
65         r0 = r;
66       }
67     }
68     xval[0] = abscissas[r0];
69     int indx = (r0*(r0+1))/2;
70     fval[0] = tableau[indx];
71
72     // find the rest of the control points, assume x0<x1<x2...
73     int rm = r0-1;
74     int rp = r0+1;
75     for( int c=1; c<n-1; c++ ) {
76       if( rp < n ) {
77         if( rm >= 0 ) {
78           if( fabs(abscissas[rp]-x) < fabs(abscissas[rm]-x) ) {
79             xval[c] = abscissas[rp++];
80             indx += rp;
81             fval[c] = tableau[indx];
82           } else {
83             xval[c] = abscissas[rm--];
84             indx ++;
85             fval[c] = tableau[indx];
86           }
87         } else {
88           xval[c] = abscissas[rp++];
89           indx += rp;
90           fval[c] = tableau[indx];
91         }
92       } else {
93         xval[c] = abscissas[rm--];
94         indx ++;
95         fval[c] = tableau[indx];
96       }
97     }
98     fval[n-1] = tableau[(n*(n+1))/2-1];
99
100    double p = fval[n-1];
101    for( int c=n-2; c>=0; c-- ) {
102      p *= (x-xval[c]);
103      p += fval[c];
104    }
105
106    return p;
107  }
108
109  int main()
110  {
```

```
111     // interpolation function for regularly space points.
112     int n = 11; // number of points, = poly order + 1
113     Vector<double> abs_reg(n), ord_reg(n), tab_reg((n*(n+1))/2);
114     for( int i=0; i<n; i++ ) {
115       abs_reg[i] = -1 + 0.2*i;
116       ord_reg[i] = f(abs_reg[i]);
117     }
118     Newton( abs_reg, ord_reg, tab_reg );
119
120     // same for Chebyshev spacing
121     Vector<double> abs_cbs(n), ord_cbs(n), tab_cbs((n*(n+1))/2);
122     for( int i=0; i<n; i++ ) {
123       abs_cbs[i] = cos(M_PI*(2*i+1)/(2.*n));
124       ord_cbs[i] = f(abs_cbs[i]);
125     }
126     Newton( abs_cbs, ord_cbs, tab_cbs );
127
128     for( int i=0; i<=1000; i++ ) {
129       double x = -1 + 0.002*i;
130       double f1 = Horner( abs_reg, tab_reg, x );
131       double f2 = Horner( abs_cbs, tab_cbs, x );
132       cout << x << " " << f1 << " " << f2 << endl;
133     }
134   }
```

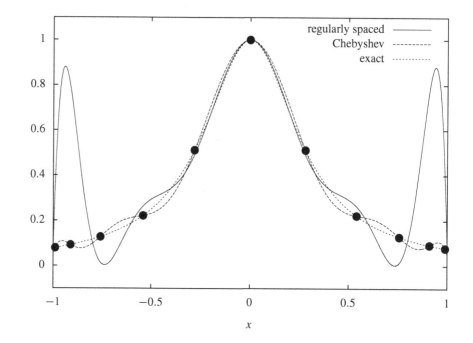

Figure S.22 Chebyshev interpolation; Problem 5.3.

An alternative scheme for generating quadrature points in some interval I was advocated by Leja [145], based on the following algorithm:

1. Start with an arbitrary point $x_0 \in I$. It is customary to select an end point of the interval.

2. Select $x_1 \in I$ to maximize $|x_1 - x_0|$. If x_0 is one end point, then x_1 will be the opposite end point.

3. Select x_{j+1} so that

$$\prod_{k=0}^{j} |x_{j+1} - x_k| = \max_{x \in I} \prod_{k=0}^{j} |x - x_k|.$$

To understand why this works well, consider the error associated with P_n:

$$-\frac{y^{(n+1)}(\xi)}{(n+1)!}(x - x_0)(x - x_1)\cdots(x - x_n).$$

If $y^{(n+1)}$ were constant, the greatest interpolation error for P_n would occur when $\prod_{j=0}^{n} |x - x_j|$ is a maximum. If we place a a new support point x_{n+1} at this location, P_{n+1} will now have zero error where P_n had its greatest error. Reichel [190] showed that Newton interpolation using Leja points has the lowest *condition number* – in the usual matrix sense.

For example, consider an interpolation with three support abscissas x_0, x_1, and x_2 specified. Then the finite differences used in the default Horner evaluation are given by a matrix operator

$$\mathbf{T}_0\mathbf{f} = \begin{pmatrix} 1 & 0 & 0 \\ -\frac{1}{x_1-x_0} & \frac{1}{x_1-x_0} & 0 \\ -\frac{1}{(x_2-x_0)(x_1-x_0)} & \frac{1}{(x_2-x_0)}\left[\frac{1}{x_1-x_0} - \frac{1}{x_2-x_1}\right] & \frac{1}{(x_2-x_0)(x_2-x_1)} \end{pmatrix} \begin{pmatrix} f_0 \\ f_1 \\ f_2 \end{pmatrix}$$

$$= \begin{pmatrix} f[x_0] \\ f[x_0, x_1] \\ f[x_0, x_1, x_2] \end{pmatrix}.$$

Then, another matrix operator $\mathbf{T}_1(x)$ performs interpolation:

$$P(x) = \mathbf{T}_1(x)\mathbf{T}_0\mathbf{f} = \begin{pmatrix} 1, & (x - x_0), & (x - x_0)(x - x_1) \end{pmatrix} \begin{pmatrix} f[x_0] \\ f[x_0, x_1] \\ f[x_0, x_1, x_2] \end{pmatrix}.$$

For fixed support abscissas, a mapping $\mathbf{T} = \mathbf{T}_1\mathbf{T}_0$ is established between the vector space of support ordinates and the space of degree two polynomials. The inverse operation $\mathbf{T}^{-1}$ consists of evaluating the polynomials at the support abscissas. Measuring the vector space with the L^∞ norm, and the polynomial space with the norm

$$\|P\| = \max_{x \in I} |P(x)|,$$

one can define the condition number $\mathrm{cond}(\mathbf{T}) = \|\mathbf{T}\| \|\mathbf{T}^{-1}\|$. This is the quantity Reichel shows to be minimized by the the choice of Leja support points. This condition number characterizes the propagation of errors, as in the case (2.20) and (2.23), and this inherent error is independent of algorithm. In particular, the order (permutation) of the chosen support points does not affect $\mathrm{cond}(\mathbf{T})$ – with the Newton method the best ordering is the ordering described beginning on p. 133, whether the points come from the Leja algorithm, Chebyshev polynomials, or anything else.

So, Chebyshev support abscissas minimize the approximation error. The correct ordering of these points, explicitly or implicitly, minimizes rounding errors. Leja points minimize inherent error. In most circumstances the approximation errors are the dominant source of error, so properly ordered Chebyshev abscissas are the best choice.

5.4 Write f_n for $f[x_0, \dots, x_n]$, etc. Then, the algorithm is

$$\eta_0 = f_n$$
$$\eta_1 = (x - x_{n-1})\eta_0 + f_{n-1}$$
$$\vdots$$
$$\eta_i = (x - x_{n-i})\eta_{i-1} + f_{n-i} \quad i = 2, \dots, n.$$

Including those numerical errors associated with intermediate steps (not input)

$$\bar{\eta}_0 = f_n$$
$$\bar{\eta}_1 = (((x - x_{n-1})(1 + \epsilon_1)\bar{\eta}_0)(1 + \epsilon_2) + f_{n-1})(1 + \epsilon_3)$$
$$= (x - x_{n-1})f_n(1 + \epsilon_1)(1 + \epsilon_2)(1 + \epsilon_3) + f_{n-1}(1 + \epsilon_3)$$
$$\vdots$$
$$\bar{\eta}_i = (((x - x_{n-i})(1 + \epsilon_{3i-2})\bar{\eta}_{i-1})(1 + \epsilon_{3i-1}) + f_{n-i})(1 + \epsilon_{3i}) \quad i = \dots, n$$
$$= f_{n-i}(1 + \epsilon_{3i}) + \prod_{j=n-i}^{n-1}(x - x_j)f_n \prod_{k=1}^{3i}(1 + \epsilon_k)$$
$$+ \sum_{\ell=n-i+1}^{n-1} \prod_{j=n-i}^{\ell-1}(x - x_j)f_\ell \prod_{k=3(n-\ell)}^{3i}(1 + \epsilon_k)$$
$$\vdots$$
$$\bar{\eta}_n = \prod_{j=0}^{n-1}(x - x_j)f_n \prod_{k=1}^{3n}(1 + \epsilon_k) + \sum_{\ell=1}^{n-1} \prod_{j=0}^{\ell-1}(x - x_j)f_\ell \prod_{k=3(n-\ell)}^{3n}(1 + \epsilon_k).$$

The Horner method gives an exact solution to

$$P(x) = \tilde{f}_0 + (x - x_0)\left(\tilde{f}_1 + (x - x_1)\left(\tilde{f}_2 + \cdots\right)\right),$$

where

$$\frac{|\tilde{f}_n - f_n|}{|f_n|} \le (3n)\epsilon$$
$$\frac{|\tilde{f}_i - f_i|}{|f_i|} \le (3i + 1)\epsilon \quad i = 1, \dots, n - 1.$$

Since coefficient error is a few times roundoff, Horner's scheme is considered stable.

Subtracting the exact solution η_n from the numerical solution $\bar{\eta}_n$ we get the error

$$\bar{\eta}_n - \eta_n = \sum_{\ell=1}^{n-1}\prod_{j=0}^{\ell-1}(x-x_j)f_\ell\left[\prod_{k=3(n-\ell)}^{3n}(1+\epsilon_k)-1\right]+\prod_{j=0}^{n-1}(x-x_j)f_n\left[\prod_{k=1}^{3n}(1+\epsilon_k)-1\right].$$

5.5 (i) Let $\bar{x}$ be the $(n+1-k)^{\text{th}}$ zero of $G(x) = P_n^{(k)}(x) - f^{(k)}(x) + K(x-\zeta_0)\cdots(x-\zeta_{n-k})$. By Rolle's theorem,

$$G^{(n+1-k)}(x) = -f^{(n+1)}(x) + (n+1-k)!K$$

has one zero in the interval $I(\bar{x}, \zeta_0, ..., \zeta_{n-k})$. Call this zero ξ. Then,

$$K = -\frac{f^{(n+1)}(\xi)}{(n+1-k)!},$$

and therefore

$$P_n^{(k)}(\bar{x}) - f^{(k)}(\bar{x}) = -\frac{f^{(n+1)}(\xi)}{(n+1-k)!}(\bar{x}-\zeta_0)\cdots(\bar{x}-\zeta_{n-k}),$$

for some ξ in the interval $I(\bar{x}, \zeta_0, \cdots, \zeta_{n-k})$.
(ii) Let x lie between neighboring support abscissas x_α and x_β. With $|x_\alpha - x_\beta| < h$, $|(x-x_\alpha)(x-x_\beta)| < h^2/4$. If x_γ is the next nearest neighbor to x, then $|x - x_\gamma| \leq 2h$. If x_δ is the next nearest neighbor, then $|x - x_\delta| \leq 3h$. Thus,

$$|P_n(x) - f(x)| \leq \frac{\max|f^{(n+1)}|}{(n+1)!}\frac{n!h^{n+1}}{4} = \frac{\max|f^{(n+1)}|}{n+1}\frac{h^{n+1}}{4},$$

for x in the interval $I(x_0, ..., x_n)$.
$x \in I(x_0, ..., x_n)$ is not necessarily bracketed by zeros of $P' - f'$. The greatest product occurs when x is outside the interval $I(\zeta_0, ..., \zeta_{n-1})$, and $\max|(x-\zeta_0)...(z-\zeta_{n-1})| \leq n!h^n$. Therefore,

$$|P_n'(x) - f'(x)| \leq \frac{\max|f^{(n+1)}|}{(n)!}n!h^n = \max|f^{(n+1)}|h^n,$$

for x in the interval $I(x_0, ..., x_n)$.
If x_0 is the smallest support abscissa, and if ζ_0 is now the smallest zero of $P'' - f''$, then $\max|x_0 - \eta_0| < 2h$. The next distance is bounded by $3h$, etc. This leads to

$$|P_n''(x) - f''(x)| \leq \frac{\max|f^{(n+1)}|}{(n-1)!}n!h^{n-1}.$$

For the third derivative, the sequence of maximal distances is $3h$, $4h$, ..., nh, and

$$|P_n^{(3)}(x) - f^{(3)}(x)| \leq \frac{\max|f^{(n+1)}|}{(n-2)!}\frac{n!h^{n-2}}{2!}.$$

Finally,

$$|P_n^{(k)}(x) - f^{(k)}(x)| \leq \frac{\max|f^{(n+1)}|}{(n+1-k)!}\frac{n!h^{n+1-k}}{(k-1)!}.$$

5.6 For this interpolation function to interpolate the degree zero function $f = 1$, it is necessary that

$$\sum_{i=1}^{n} c_i(x) = 1.$$

Then,

$$y(x) = \sum_i c_i y_i = y(\eta) \sum_i c_i = y(\eta)$$

is the discrete version of the mean value theorem for integration. This states that the interpolation result $y(x)$ will be some mean value of $y(\eta)$ of the support ordinates. In particular, $\min_i y_i \le y(\eta) \le \max_i y_i$. A quadratic function is not bounded, therefore this positive interpolation function cannot model a degree two polynomial exactly.

This is an application of Godunov's theorem [86], which famously shows that there is no second-order accurate scheme for partial differential equations that preserves monotonicity. There, as with interpolation and techniques built on interpolation, one risks instability when striving for higher-order accuracy.

5.7 Hermite interpolation using two points and nth derivatives will have

$$|P_{2n+1} - f| \le \frac{\max_{[a,b]} |f^{(2n+2)}|}{(2n+2)!} \left(\frac{b-a}{2} \right)^{2n+2}.$$

Cramér's formula gives

$$|\overset{(n)}{\mathrm{erf}}(x)| \le \frac{K 2^{(n-1)/2} \sqrt{(n-1)!}}{\sqrt{\pi}}$$

$$|\overset{(2n+2)}{\mathrm{erf}}(x)| \le \frac{K 2^{n+3/2} \sqrt{(2n+1)!}}{\sqrt{\pi}},$$

and with $b = 2$ and $a = 0$ one has $|\omega(x)| \le 1$. Therefore,

$$|P_{2n+1} - f| \le \frac{K 2^{n+1/2}}{(n+1)\sqrt{\pi (2n+1)!}}.$$

With $n = 3$ the upper bound is 2.45×10^{-2}, and with $n = 4$ it is 4.62×10^{-3}. The desired accuracy is therefore attained with $n = 4$. With the shorthand $s = \sqrt{2\pi}$ and $y = \mathrm{erf}(2)$, the tableau follows:

$$
\begin{array}{cccccccc}
0 & 0 \\
 & & \frac{2}{s} \\
0 & 0 & & 0 \\
 & & \frac{2}{s} & & \frac{-2}{3s} \\
0 & 0 & & 0 & & 0 \\
 & & \frac{2}{s} & & \frac{-2}{3s} & & \frac{4+3ys}{96s} \\
0 & 0 & & 0 & & \frac{4+3ys}{48s} & & \cdots \\
 & & \frac{2}{s} & & \frac{-4+ys}{8s} & & \frac{3e^{-4}+5-3ys}{24s} \\
0 & 0 & & \frac{-4+ys}{4s} & & \frac{4e^{-4}+8-3ys}{16s} & & \cdots \\
 & & \frac{y}{2} & & \frac{2e^{-4}+2-ys}{4s} & & \frac{-14e^{-4}-6+3ys}{16s} \\
2 & y & & \frac{4e^{-4}-ys}{4s} & & \frac{-24e^{-4}-4+3ys}{16s} & & \cdots \\
 & & 2\frac{e^{-4}}{s} & & \frac{-20e^{-4}+ys}{8s} & & \frac{61e^{-4}+3-3ys}{24s} \\
2 & y & & \frac{-4e^{-4}}{s} & & \frac{172e^{-4}-3ys}{48s} & & \cdots \\
 & & \frac{2e^{-4}}{s} & & \frac{14e^{-4}}{2s} & & \frac{-332e^{-4}+3ys}{96s} \\
2 & y & & \frac{-4e^{-4}}{s} & & \frac{-10e^{-4}}{3s} \\
 & & \frac{2e^{-4}}{s} & & \frac{14e^{-4}}{3s} \\
2 & y & & \frac{-4e^{-4}}{s} \\
 & & \frac{2e^{-4}}{s} \\
2 & y
\end{array}
$$

$$\frac{12e^{-4}+16-15ys}{192s}$$
$$\frac{-36e^{-4}-24+15ys}{128s}$$
$$\frac{-48e^{-4}-28+15ys}{96s}$$
$$\frac{532e^{-4}+176-105ys}{768s}$$
$$\frac{106e^{-4}+26-15ys}{96s}$$
$$\frac{-930e^{-4}-170+105ys}{768s}$$
$$\frac{164e^{-4}+24-15ys}{96s}$$
$$\frac{-1328e^{-4}-164+105ys}{768s}$$
$$\frac{-904e^{-4}-60+45ys}{384s}$$
$$\frac{-192e^{-4}-4+5ys}{64s}$$

Expanded out the interpolation polynomial is

$$
P_9 = \frac{2}{s}x - \frac{2}{3s}x^3 + \frac{-1256e^{-4}-300+189ys}{48s}x^5
$$
$$
+ \frac{2316e^{-4}+512-315ys}{48s}x^6 + \frac{-1072e^{-4}-220+135ys}{32s}x^7
$$
$$
+ \frac{7972e^{-4}+1536-945ys}{768s}x^8 + \frac{-930e^{-4}-170+105ys}{768s}x^9.
$$

A plot of the error calculated with numerical evaluation of $erf(x)$ is displayed in Figure S.23.

5.8 With everything real, g^* is g and the imaginary part of the autocorrelation is zero. The autocorrelation is symmetric about zero, so roughly half of it is displayed in Figure S.24. The 2000-day view does reveal a delta-like response. But, there is considerable structure at the month time scale.

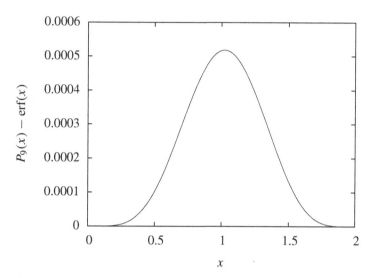

Figure S.23: Error in Hermite interpolation of erf(x); Problem 5.7.

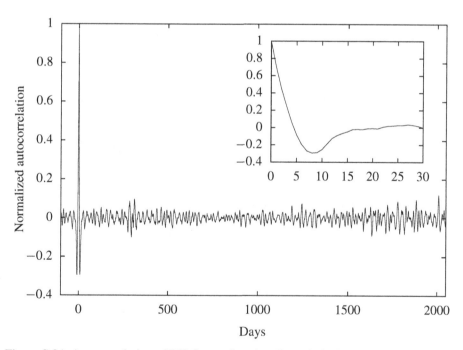

Figure S.24: Autocorrelation of DJI fluctuations in adjusted closing price for 4096 consecutive trading days ending 12/31/09; Problem 5.8.

The results will vary with the data used and the sampling interval, but the qualitative character of such analyses is robust.

The analysis procedure outlined for this problem applies to periodic signals, though times series stock market data is obviously not periodic. The treatment of aperiodic data with FFTs distorts the frequency response. One way to think of it is that there will be a jump at $t = 0$, and the Fourier transform of a discontinuity has amplitude at all frequencies (the Gibbs effect). This type of artifact can be attenuated by *windowing*, which amounts to the multiplication of the aperiodic data by a function which approaches 0 the 0 and $N - 1$ ends of the sequence. A detailed description of the phenomenon and a number of window functions is given by Harris [96].

```
1    #include <iostream>
2    #include <iomanip>
3    #include <fstream>
4    #include "Matrix.H"
5    #include "FFT.H"
6    #include <math.h>
7    using namespace std;
8
9    #define WINDOW
10
11   int main()
12   {
13     int N     = 4096;
14     int halfw = 10;
15
16     Vector<double> rR(N), rI(N), cR(N), cI(N), fR(N), fI(N);
17
18     // input stock data, save rate/price,  make a copy, create filter
19     fstream data;
20     data.open("dji", fstream::in );
21     double price;
22     data >> price;
23     for( int i=0; i<N; i++ ) {
24       double newprice;
25       data >> newprice;
26       rR[i] = (newprice-price)/price;
27   #ifdef WINDOW
28       double w = 1. - fabs((2*i)/N-1.); // Bartlet window
29       rR[i] *= w;
30   #endif
31       cR[i] = rR[i];
32       rI[i] = cI[i] = fI[i] = fR[i] = 0;
33       if( i <= halfw || N-i <= halfw ) {
34         fR[i] = 1./(2.*halfw+1);
35       }
36     }
37     data.close();
38
39     // smooth data:  convolution
40     fft( cR, cI, true );
41     fft( fR, fI, true );
42     for( int i=0; i<N; i++ ) {
43       double r1 = cR[i], r2 = fR[i];
44       double i1 = cI[i], i2 = fI[i];
45       cR[i] = N*( r1*r2 - i1*i2 );
46       cI[i] = N*( r1*i2 + i1*r2 );
47     }
48     fft( cR, cI, false );
49
50     // obtain noise by subtracting smoothed data
51     double sum = 0;
52     for( int i=0; i<N; i++ ) {
53       rR[i] -= cR[i];
54       sum += rR[i];
55     }
56
57     // find autocorrelation
58     fft( rR, rI, true );
59     for( int i=0; i<N; i++ ) {
```

```
60        rR[i] = N*( rR[i]*rR[i] + rI[i]*rI[i] );
61        rI[i] = 0;
62      }
63      // and complete autocorrelation
64      fft( rR, rI, false );
65
66      // variance is proportional to
67      double variance = rR[0];
68
69      // write in weird order for plotting
70      fstream result;
71      result.open( "autocorr.out", fstream::out );
72      result << scientific << setprecision(15);
73      for( int i=0; i<N; i++ ) {
74          int j = i+N/2;
75          j %= N;
76          result << ( (j>=N/2) ? (j-N) : j ) << " " << rR[j]/variance << endl;
77      }
78      result.close();
79  }
```

Chapter 6

6.1 Smale's formula (6.11) suggests evaluation of

$$\alpha(x) = \left| \frac{f(x)}{f'(x)} \right| \sup_{k>1} \left| \frac{f^{(k)}(x)}{k! f'(x)} \right|^{1/(k-1)}$$

$$= \left| \frac{x^2 - a}{2x} \right| \left| \frac{2}{2!(2x)} \right|^1 = \left| \frac{x^2 - a}{4x^2} \right|,$$

to find the x_0 that satisfies $\alpha(x_0) < \alpha_0 \approx 0.13071694$. This gives

$$\frac{a}{1 + 4\alpha_0} \le x^2 \le \frac{a}{1 - 4\alpha_0},$$

or

$$0.81034304\sqrt{a} \le x \le 1.4477068\sqrt{a},$$

and

$$-1.4477068\sqrt{a} \le x \le -0.81034304\sqrt{a}$$

for root $\xi = -\sqrt{a}$.

6.2

$$f(\xi) = 0 = f(x^{(0)}) + (\xi - x^{(0)}) f'(x^{(0)}) + \frac{1}{2}(\xi - x^{(0)})^2 f''(\eta)$$

$$x^{(1)} = x^{(0)} - f(x^{(0)})/f'(x^{(0)})$$

$$0 = (\xi - x^{(1)}) f'(x^{(0)}) + \frac{1}{2}(\xi - x^{(0)})^2 f''(\eta)$$

$$(\xi - x^{(1)}) = -(\xi - x^{(0)})^2 \frac{f''(\eta)}{2 f'(x^{(0)})},$$

for some η in the interval $I[\xi, x^{(0)}]$.

Now, for $f = x^2 - a$, $f'' = 2$ is constant, and

$$x^{(1)} = \frac{a + (x^{(0)})^2}{2x^{(0)}},$$

for Newton–Raphson. Note that the sign of x will not change during the Newton–Raphson procedure. We can therefore limit consideration to $\xi = +\sqrt{a}$ and $x^{(k)} > 0$, and

$$(x^{(1)} - \xi) = (x^{(0)} - \xi)^2 \frac{1}{2x}.$$

With the right-hand side positive, we must have $x > \xi$ for $(x^{(k)} - \xi)$ to have uniform sign. And, for the method to converge, we need

$$0 < \frac{x^{(0)} - \xi}{2x} < 1.$$

Both of these requirements are met by $x^{(0)} \geq \xi$. By symmetry, to converge monotonically to $-\sqrt{a}$, begin with $x^{(0)} < -\sqrt{a}$.

6.3 The complexity of the given formulas recommends the secant method over Newton's method. The following short program computes the vapor–liquid equilibrium phase diagram. The azeotrope is found with another secant method.

```
1   #include <iostream>
2   #include <iomanip>
3   #include <math.h>
4   using namespace std;
5
6   // ambient
7   double Pt = 101.325;
8
9   // gas constant
10  double R = 8.314472;        // J/K.mol
11
12  // saturation pressure parameters
13  double A1 = 7.16879, B1 = 1552.601, C1 = -50.731;
14  double A2 = 7.23255, B2 = 1750.286, C2 = -38.15;
15
16  // NRTL parapeters
17  double alpha12 = -0.351, alpha21 = -0.351;
18  double dg12 = 3159.86,   dg21 = -618.37;
19
20  void model( double T, double x1, double& y1, double& y2 )
21  {
22     double x2 = 1. - x1;
23
24     double tau12 = dg12/(R*T);
25     double tau21 = dg21/(R*T);
26     double G12 = exp(-alpha12*tau12);
27     double G21 = exp(-alpha21*tau21);
28
29     double d1 = x1 + x2*G21;
30     double d2 = x2 + x1*G12;
31     // activity coef ethanol
32     double gammal = exp( x2*x2*( tau21*G21*G21/(d1*d1) + tau12*G12/(d2*d2) ) );
33     // activity coef water
34     double gamma2 = exp( x1*x1*( tau12*G12*G12/(d2*d2) + tau21*G21/(d1*d1) ) );
35
36     // saturation pressure
37     double Psat1 = pow(10.,A1 - B1/(T+C1));
38     double Psat2 = pow(10.,A2 - B2/(T+C2));
```

```
39
40      y1 = gamma1*x1*Psat1/Pt;
41      y2 = gamma2*x2*Psat2/Pt;
42    }
43
44    void solve( double x1, double& T, double& y1)
45    {
46      double T0,T1,e0,e1,e2,y2;
47      double Tmin = -C1 + B1/(A1 - log10(Pt));
48      double Tmax = -C2 + B2/(A2 - log10(Pt));
49
50      T0 = Tmin;
51      model( T0, x1, y1, y2 );
52      e0 = y1+y2-1.;
53
54      T1 = Tmax;
55      model( T1, x1, y1, y2 );
56      e1 = y1+y2-1.;
57
58      int iter=0;
59      do {
60        // secant
61        T = (T0*e1-T1*e0)/(e1-e0);
62        model( T, x1, y1, y2 );
63        e2 = y1+y2-1.;
64
65        T0=T1;    T1=T;
66        e0=e1;    e1=e2;
67      } while( (fabs(e2)>1.e-12) && (++iter<1000));
68    }
69
70    int main()
71    {
72      cout << scientific << setprecision(15);
73      double Tmin = -C1 + B1/(A1 - log10(Pt));
74      double Tmax = -C2 + B2/(A2 - log10(Pt));
75      double x1, y1, T;
76
77      // data for plot
78      cout << Tmax << " " << 0. << " " << 0. << endl;
79      for( int ix=1; ix<1000; ix++ ) {
80        x1 = 0.001*ix;
81        solve( x1, T, y1);
82        cout << T << " " << x1 << " " << y1 << endl;
83      }
84      cout << Tmin << " " << 1. << " " << 1. << endl;
85
86      // find azeotrope. error = y1-x1;
87      double x10 = 0.1, x11=0.9, x12;
88      double e0, e1, e2;
89
90      solve( x10, T, y1 );
91      e0 = x10-y1;
92
93      solve( x11, T, y1 );
94      e1 = x11-y1;
95
96      int iter=0;
97      do {
98        // secant
99        x12 = (x10*e1-x11*e0)/(e1-e0);
100       solve( x12, T, y1 );
101       e2 = x12-y1;
102
103       e0=e1;    e1=e2;
104       x10=x11;  x11=x12;
105     } while( (fabs(e2)>1.e-12) && (++iter<1000));
106     cout << "azeotrope T=" << T << " x1=" << x12 << endl;
107   }
```

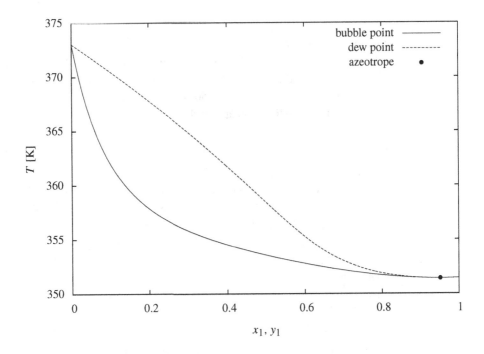

Figure S.25 Water–ethanol azeotrope; Problem 6.3.

6.4 The problem seeks the minimum of

$$L^2 = (\bar{x} - x)^2 + (\bar{y} - \tan(x))^2,$$

so we could seek the extremum where $f(x) = d/dx\, L^2$ is zero:

$$f(x) = x - \bar{x} + \sec(x)^2(\tan(x) - \bar{y})$$
$$f'(x) = 1 + \sec(x)^4 + 2\sec(x)^2 \tan(x)(\tan(x) - \bar{y}).$$

Therefore, the Newton–Raphson method to find $f(x) = 0$ is

$$x^{(k+1)} = x^{(k)} - \frac{x^{(k)} - \bar{x} + \sec(x^{(k)})^2(\tan(x^{(k)}) - \bar{y})}{1 + \sec(x^{(k)})^4 + 2\sec(x^{(k)})^2 \tan(x^{(k)})(\tan(x^{(k)}) - \bar{y})}.$$

With starting guess 0 it converges to 0.274 999.

6.5 A Newton–Raphson formulation of this problem is

$$\mathbf{f} = \begin{pmatrix} \frac{1}{R_{ab}} - \frac{1}{x_3} - \frac{1}{x_1+x_2} \\ \frac{1}{R_{bc}} - \frac{1}{x_1} - \frac{1}{x_2+x_3} \\ \frac{1}{R_{ac}} - \frac{1}{x_2} - \frac{1}{x_1+x_3} \end{pmatrix}$$

$$\nabla \mathbf{f} = \begin{pmatrix} \frac{1}{(x_1+x_2)^2} & \frac{1}{(x_1+x_2)^2} & \frac{1}{x_3^2} \\ \frac{1}{x_1^2} & \frac{1}{(x_2+x_3)^2} & \frac{1}{(x_2+x_3)^2} \\ \frac{1}{(x_1+x_3)^2} & \frac{1}{x_2^2} & \frac{1}{(x_1+x_3)^2} \end{pmatrix}$$

$$\mathbf{x}^{(k+1)} = \left[\mathbf{x} - (\nabla\mathbf{f})^{-1}\mathbf{f}\right]^{(k)}.$$

Depending on the starting value, this converges rapidly to the solution 149.99, 199.99, 175.03 to two decimal places. This formulation converges best when the starting vector $\mathbf{x}^{(0)}$ is small compared to the solution vector, e.g., (50, 50, 50). If it is large compared to the solution vector, e.g., (500, 500, 500) this formulation diverges.

A modified Newton–Raphson approach,

$$\mathbf{x}^{(k+1)} = \left[\mathbf{x} - \lambda(\nabla\mathbf{f})^{-1}\mathbf{f}\right]^{(k)},$$

will stabilize the method for a broader choice of starting values. Choose $\lambda = 2^{-K}$ with the smallest integer $K \geq 0$ such that each component of $\mathbf{x}^{(k+1)}(\lambda)$ is positive, which comes from consideration of the physical problem, and such that $(\mathbf{f}^{\mathrm{T}}\mathbf{f})^{(k+1)}$ is smaller than $(\mathbf{f}^{\mathrm{T}}\mathbf{f})^{(k)}$. This stabilizes the problem for starting value (500, 500, 500).

```
1    #include <iostream>
2    #include <stdlib.h>
3    #include <iomanip>
4    #include <math.h>
5    #include <Matrix.H>
6    #include <GE.H>
7    using namespace std;
8
9    #define MAX(A,B)  (((A)>(B))?(A):(B))
10
11   // evaluate the function and return the square of its 2-norm
12
13   double func( Vector<double>& f, const Vector<double>& x, const Vector<double> params )
14   {
15       double Rab = params[0];
16       double Rbc = params[1];
17       double Rac = params[2];
18       double sum = x[0] + x[1] + x[2];
19       f[0] = 1./Rab - 1./x[2] - 1./(x[0]+x[1]);
20       f[1] = 1./Rbc - 1./x[0] - 1./(x[1]+x[2]);
21       f[2] = 1./Rac - 1./x[1] - 1./(x[0]+x[2]);
22       double fdf = f[0]*f[0] + f[1]*f[1] + f[2]*f[2];
23       return fdf;
24   }
25
26   int main()
27   {
28       cout << scientific << setprecision(15);
29       Vector<double> x(3), y(3);
30       Vector<double> f(3), g(3), params(3);
31       Matrix<double> Df(3,3);
32       Vector<int> Pl(3), Pr(3);
33
34       // a starting value
35       x[0] = x[1] = x[2] = 500.;
36
37       double Rab = params[0] = 116.68;
38       double Rbc = params[1] = 107.14;
39       double Rac = params[2] = 123.81;
40
41       double infnorm;   // infinity norm of f(x)
42
43       int k = 0;
44       do {
45           Df[0][0] = 1./((x[0]+x[1])*(x[0]+x[1]));
46           Df[0][1] = 1./((x[0]+x[1])*(x[0]+x[1]));
47           Df[0][2] = 1./(x[2]*x[2]);
48
49           Df[1][0] = 1./(x[0]*x[0]);
```

```
50      Df[1][1] = 1./((x[1]+x[2])*(x[1]+x[2]));
51      Df[1][2] = 1./((x[1]+x[2])*(x[1]+x[2]));
52
53      Df[2][0] = 1./((x[0]+x[2])*(x[0]+x[2]));
54      Df[2][1] = 1./(x[1]*x[1]);
55      Df[2][2] = 1./((x[0]+x[2])*(x[0]+x[2]));
56
57      double fdf = func( f, x, params );
58
59      infnorm = MAX(fabs(f[0]),fabs(f[1]));
60      infnorm = MAX(infnorm,fabs(f[2]));
61      // stop if converged
62      if( infnorm < 1.e-9 ) break;
63
64      // replace f with Df^-1 f
65      ge( Df, Pl, Pr, COMPLETE );
66      solveLR( Df, Pl, Pr, f );
67
68      double lambda = 1;
69      do {
70        // a trial solution of modified Newton-Raphson
71        for( int i=0; i<3; i++ ) y[i] = x[i] - lambda*f[i];
72        // if any value is negative, then we need a different lambda.
73        // if all are positive, check size of f^T f
74        if( y[0] > 0 && y[1] > 0 && y[2] > 0 ) {
75          double fdf2 = func( g, y, params );
76          if( fdf2 < fdf ) {
77            // f^T f has diminished, so we have new x iterate.
78            for( int i=0; i<3; i++ ) x[i] = y[i];
79            break;
80          }
81        }
82        // try a smaller lambda
83        lambda /= 2;
84        // if lambda is too small, we are probably not converging
85        if( lambda < 1./1024 ) {
86          cout << "Stop iteration because lambda too small " << lambda << endl;
87          exit(1);
88        }
89
90      } while ( 1 );
91
92      ++k;
93    } while( infnorm > 1.e-9 );
94
95    cout << "after " << k << " iterations, the solution is" << endl << x << endl;
96  }
```

Chapter 7

7.1 First, write the problem in standard form:

$$\text{maximize} \quad z$$

$$4.1x_A + 4.3x_B + 5.8x_C + 6.0x_D + 7.6x_E + 7.5x_F + 7.3x_G + 6.9x_H + 7.3x_I + z = 0$$

$$10x_A + 10x_B + 40x_C + 60x_D + 30x_E + 30x_F + 30x_G + 50x_H + 20x_I = 30$$

$$10x_A + 30x_B + 50x_C + 30x_D + 30x_E + 40x_F + 20x_G + 40x_H + 30x_I = 30$$

$$80x_A + 60x_B + 10x_C + 10x_D + 40x_E + 30x_F + 50x_G + 10x_H + 50x_I = 40$$

$$x_A, x_B, x_C, x_D, x_E, x_F, x_G, x_H, x_I \geq 0.$$

For this problem, a feasible basis consists of x_E, z, and any two other independent variables, e.g., $J = \{z, x_A, x_B, x_E\}$. Then, $K = \{x_C, x_D, x_F, x_G, x_H, x_I\}$. The final solution is $J = \{z, x_D, x_B, x_A\}$, with $\mathbf{x}_J = (-4.98, 0.4, 0.6, 0)$, i.e., the lowest cost solution is a 3:2 mix of alloys B and D with a cost of $-z = \$4.98$.

7.2 Redundancy in the equation set causes phase I to create bases J containing artificial variables with zero value. For this problem, phase I constructs the following problem in standard form

$$\text{maximize} \quad x_3$$

$$x_0 - x_1 + x_4 = 0$$

$$x_1 - x_0 + x_5 = 0$$

$$x_0 + 2x_1 + x_2 + x_3 = 3$$

$$x_4 + x_5 + x_6 = 0$$

$$x_0, x_1, x_2, x_4, x_5, x_6 \geq 0,$$

with $J = \{x_3, x_4, x_5, x_6\}$ and all basic variables have zero value.

7.3 The exact solution $(1, 0, 0)$ of this helical function is approximated in ≈ 24 steps using strict thresholds 10^{-9} for termination and golden section search.

```
1   #include <iostream>
2   #include <iomanip>
3   #include <math.h>
4   #include "Matrix.H"
5   #include "VMM.H"
6   using namespace std;
7
8   double func( const Vector<double>& x)
9   {
10      // the Fletcher-Powell Helix
11      double theta = atan(x[1]/x[0])/(2*M_PI) + ((x[0]<0)?0.5:0);
12      double r = sqrt( x[0]*x[0] + x[1]*x[1] );
13      double f = 100*( (x[2]-10*theta)*(x[2]-10*theta) + (r-1)*(r-1) ) + x[2]*x[2];
14      return f;
15  }
16
17  void grad( const Vector<double>& x, Vector<double>& g )
18  {
19      // grad of Fletcher-Powell Helix
20      double r = sqrt(x[0]*x[0] + x[1]*x[1]);
21      double drdx0 = x[0]/r;
22      double drdx1 = x[1]/r;
23      double theta = atan(x[1]/x[0])/(2*M_PI) + ((x[0]<0)?.5:0);
24      double dtdx0 = -x[1]/(2*M_PI*r*r);
25      double dtdx1 = x[0]/(2*M_PI*r*r);
26      g[0] = 200*(r-1)*drdx0 + 2000*(10*theta-x[2])*dtdx0;
27      g[1] = 200*(r-1)*drdx1 + 2000*(10*theta-x[2])*dtdx1;
28      g[2] = 202*x[2] - 2000*theta;
29  }
30
31  int main()
32  {
33      Vector<double> x0(3);
34      int iteration;
35      cout << scientific << setprecision(4);
36
37      x0[0]=-1; x0[1]=0; x0[2]=0; VMM( func, grad, x0, DFP, iteration, 1.e-9, 1.e-9 );
38      cout << "DFP:   iterations=" << iteration << " solution=" << endl << x0 << endl;
39
40      x0[0]=-1; x0[1]=0; x0[2]=0; VMM( func, grad, x0, BFGS, iteration, 1.e-9, 1.e-9 );
```

```
41    cout << "BFGS:   iterations=" << iteration << " solution=" << endl << x0 << endl;
42
43    x0[0]=-1; x0[1]=0; x0[2]=0; VMM( func, grad, x0, OS,   iteration, 1.e-9, 1.e-9 );
44    cout << "OS:     iterations=" << iteration << " solution=" << endl << x0 << endl;
45  }
```

7.4 If this could be represented as a symmetric positive definite quadratic programming problem then the algorithm of Goldfarb and Idnani could be used. The fact that objective function f consists of a sum of squares is encouraging, and suggests the decomposition

$$f = \mathbf{x}^{T} \begin{pmatrix} 1 & 0 & 0 & 0 & 0 \\ -1 & 1 & 0 & 0 & 0 \\ 0 & 1 & 0 & 0 & 0 \\ 0 & 0 & 1 & 0 & 0 \\ 0 & 0 & 0 & 1 & 0 \end{pmatrix} \begin{pmatrix} 1 & -1 & 0 & 0 & 0 \\ 0 & 1 & 1 & 0 & 0 \\ 0 & 0 & 0 & 1 & 0 \\ 0 & 0 & 0 & 0 & 1 \\ 0 & 0 & 0 & 0 & 0 \end{pmatrix} \mathbf{x} + \mathbf{x}^{T} \begin{pmatrix} 0 \\ -4 \\ -4 \\ -2 \\ -2 \end{pmatrix} + 6.$$

But, this matrix is not positive because it is rank deficient. This could be remedied by adding $(x_1 + 3x_2)^2$ to the function. This new factor is zero on the constraint manifold, so the constrained optimum of f should be equal to the constrained optimum of g:

$$g = (x_1 + x_2)^2 + (x_2 + x_3 - 2)^2 + (x_4 - 1)^2 + (x_5 - 1)^2 + (x_1 + 3x_2)^2$$

$$= \mathbf{x}^{T} \begin{pmatrix} 1 & 0 & 0 & 0 & 1 \\ -1 & 1 & 0 & 0 & 3 \\ 0 & 1 & 0 & 0 & 0 \\ 0 & 0 & 1 & 0 & 0 \\ 0 & 0 & 0 & 1 & 0 \end{pmatrix} \begin{pmatrix} 1 & -1 & 0 & 0 & 0 \\ 0 & 1 & 1 & 0 & 0 \\ 0 & 0 & 0 & 1 & 0 \\ 0 & 0 & 0 & 0 & 1 \\ 1 & 3 & 0 & 0 & 0 \end{pmatrix} \mathbf{x} + \mathbf{x}^{T} \begin{pmatrix} 0 \\ -4 \\ -4 \\ -2 \\ -2 \end{pmatrix} + 6$$

$$= \mathbf{x}^{T} \mathbf{B}^{T} \mathbf{B} \mathbf{x} + \mathbf{x}^{T} \mathbf{b} + 6.$$

The unconstrained optimum occurs when $\nabla g = 0$, or $2\mathbf{B}^{T}\mathbf{B}\mathbf{x} = -\mathbf{b}$, giving $\mathbf{x} = (0, 0, 2, 1, 1)$ with $g = 0$. The Goldfarb–Idnani algorithm finds the constrained optimum:

$$x_1 = -0.767\,441\,860\,465\,12$$
$$x_2 = 0.255\,813\,953\,488\,37$$
$$x_3 = 0.627\,906\,976\,744\,19$$
$$x_4 = -0.116\,279\,069\,767\,44$$
$$x_5 = 0.255\,813\,953\,488\,37$$
$$f = 4.093\,023\,255\,814.$$

```
1   #include <iostream>
2   #include <iomanip>
3   #include <sstream>
4   #include "Matrix.H"
5   #include "Cholesky.H"
6   #include "ConvexQP.H"
7   using namespace std;
8
9   int main()
10  {
11      stringstream ss( stringstream::in | stringstream::out );
12      Matrix<double> G(5,5), Ci, Ce(5,3);
```

```
13      Vector<double> x(5), ci, ce(3);
14      cout << setprecision(14);
15
16      ss << "4 4 0 0 0   4 22 2 0 0   0 2 2 0 0   0 0 0 2 0   0 0 0 0 2";
17      ss >> G;
18
19      ss.clear();
20      ss << "1 0 0   3 0 1   0 1 0   0 1 0   0 -2 -1 0 0 0";
21      ss >> Ce;
22      ss >> ce;
23
24      ss.clear();
25      ss << "0 0 2 1 1";
26      ss >> x;
27
28      // convert G to L
29      Cholesky( G );
30
31      // impact of constraints
32      double df = 0;
33      int code = ConvexQP( G, x, Ce, ce, Ci, ci, df, 1.e-16 );
34      cout << "code=" << code << " Convex Quadratic Solution x=" << endl << x << endl;
35      cout << "change to residual " << df << endl;
36
37      cout << "function  " << (x[0]-x[1])*(x[0]-x[1]) + (x[1]+x[2]-2)*(x[1]+x[2]-2) +
38                              (x[3]-1)*(x[3]-1) + (x[4]-1)*(x[4]-1) << endl;
39      cout << "constraint1 " << x[0] + 3*x[1] << endl;
40      cout << "constraint2 " << x[2] + x[3] - 2*x[4] << endl;
41      cout << "constraint3 " << x[1] - x[4] << endl;
42  }
```

7.5 In standard form,

$$\text{maximize}\quad 121x_0 + 160x_1 + 135x_2 + 825x_3,$$

subject to

$$4.5x_0 + 2.5x_1 + 3.5x_2 + 3.25x_3 \le 1800$$
$$6.0x_0 + 4.2x_1 + 5.6x_2 + 14.0x_3 \le 5000$$
$$33.25x_3 \le 6000$$
$$x_0 + x_1 + x_2 + x_3 \le 600,$$

and x_0, x_1, x_2, x_3 are restricted.

Add objective x_4 and restricted slack variables x_5, x_6, x_7, x_8 to give matrix and vector

$$\begin{pmatrix} 121 & 160 & 135 & 825 & -1 & 0 & 0 & 0 & 0 \\ 4.5 & 2.5 & 3.5 & 3.25 & 0 & 1 & 0 & 0 & 0 \\ 6.0 & 4.2 & 5.6 & 14.0 & 0 & 0 & 1 & 0 & 0 \\ 0 & 0 & 0 & 33.25 & 0 & 0 & 0 & 1 & 0 \\ 1 & 1 & 1 & 1 & 0 & 0 & 0 & 0 & 1 \end{pmatrix}, \quad \begin{pmatrix} 0 \\ 1800 \\ 5000 \\ 6000 \\ 600 \end{pmatrix}.$$

On solution objective x_4 is \$216,000 with 419.55 acres of wheat and 180.45 acres of tomato.

Chapter 8

8.1 See Figure S.26.

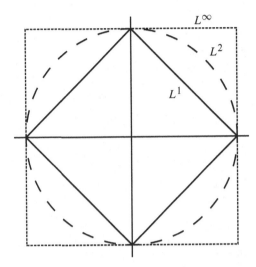

Figure S.26 L^p circles; Problem 8.1.

8.2 The point of this exercise is that

$$\mathbf{A}^\mathsf{T}\mathbf{A} = \begin{pmatrix} 1+\mu^2 & 1 & 1 & 1 & 1 \\ 1 & 1+\mu^2 & 1 & 1 & 1 \\ 1 & 1 & 1+\mu^2 & 1 & 1 \\ 1 & 1 & 1 & 1+\mu^2 & 1 \\ 1 & 1 & 1 & 1 & 1+\mu^2 \end{pmatrix},$$

and if $\mu^2 < 2^{1-t}$ then this will round to a matrix full of ones, which has no inverse, so the problem is not solvable by the normal equations.

But, symbolically the normal equations have a solution

$$\mathbf{x} = \mu^{-1}\begin{pmatrix} 1 \\ -4 \\ 6 \\ -4 \\ 1 \end{pmatrix},$$

and this is achievable by any of the numerically stable pseudoinverse type methods. The code below gives the following output:

```
Householder:
x0 +1.0000e+09
x1 -4.0000e+09
```

```
        x2 +6.0000e+09
        x3 -4.0000e+09
        x4 +1.0000e+09
        SVD:
        x0 +1.0000e+09
        x1 -4.0000e+09
        x2 +6.0000e+09
        x3 -4.0000e+09
        x4 +1.0000e+09
```

```
1   #include <iostream>
2   #include <iomanip>
3   #include <math.h>
4   #include "Matrix.H"
5   #include "Householder.H"
6   #include "SVD.H"
7   using namespace std;
8
9   #define MAX(A,B)  (((A)>(B))?(A):(B))
10
11  int main()
12  {
13    cout << scientific << setprecision(4) << showpos;
14
15    Matrix<double>A(6,5);
16    Vector<double> b(6),  diag(5);
17    Vector<int> Pl(6), Pr(5);
18    double mu = 1.e-9;
19
20    for( int i=0; i<6; i++ ) {
21      for( int j=0; j<5; j++ ) A[i][j] = ((i==0) ? 1 : 0);
22      if( i>0 ) A[i][i-1] = mu;
23    }
24    b[0]=mu; b[1]=b[5]=0; b[2]=b[4]=-5; b[3]=5;
25
26    Householder( A, Pl, Pr, diag, POWELLREID );
27    solveQR( A, Pl, Pr, diag, b );
28
29    cout << "Householder:" << endl;
30    for( int i=0; i<A.columns(); i++ )
31      cout << "x" << noshowpos << i << " " << showpos << b[i] << endl;
32
33    // rebuild A,b for SVD solve
34    for( int i=0; i<6; i++ ) {
35      for( int j=0; j<5; j++ ) A[i][j] = ((i==0) ? 1 : 0);
36      if( i>0 ) A[i][i-1] = mu;
37    }
38    b[0]=mu; b[1]=b[5]=0; b[2]=b[4]=-5; b[3]=5;
39
40    Matrix<double> U(6,6);
41    Matrix<double> VT(5,5);
42    Vector<double> sigma(5);
43    SVD( A, U, VT, sigma );
44
45    double smax = sigma[0];
46    for( int i=1; i<A.columns(); i++ ) smax = MAX(smax,sigma[i]);
47
48    // U sigma VT x = b
49    // x = V 1/sigma UT b
50    Vector<double> UTb(6);
51    for( int i=0; i<6; i++ ) {
52      UTb[i] = 0;
53      for( int j=0; j<6; j++ ) UTb[i] += U[j][i]*b[j];
54    }
55    // divide by sigma where sigma not zero
```

```
56      for( int i=0; i<5; i++ ) UTb[i] *= ((sigma[i]>=1e-10*smax) ? (1/sigma[i]) : 0);
57      // and multiply by V, store result in b
58      for( int i=0; i<5; i++ ) {
59        b[i] = 0;
60        for( int j=0; j<5; j++ ) b[i] += VT[j][i]*UTb[j];
61      }
62
63      cout << "SVD:" << endl;
64      for( int i=0; i<A.columns(); i++ )
65        cout << "x" << noshowpos << i << " " << showpos << b[i] << endl;
66    }
```

8.3 The Goldfarb–Idnani method is designed to find the minimum of quadratic function with constraints

$$f(\mathbf{x}) = \frac{1}{2}\mathbf{x}^T\mathbf{G}\mathbf{x} + \mathbf{b}^T\mathbf{x} + c,$$

where $\mathbf{G}$ is symmetric positive definite. The implementation on p. 420 uses only the $\mathbf{L}\mathbf{L}^T$ representation of $\mathbf{G}$, in particular just the lower triangular $\mathbf{L}$.

For the problem at hand, the least squares solution of $\mathbf{A}\mathbf{x} = \mathbf{b}$ can be found from $\mathbf{R}\mathbf{x} = \mathbf{Q}^T\mathbf{b}$, and $\mathbf{x}$ is the minimum of

$$\frac{1}{2}\mathbf{x}^T\mathbf{R}^T\mathbf{R}\mathbf{x} - \mathbf{b}^T\mathbf{A}\mathbf{x} + \frac{1}{2}\mathbf{b}^T\mathbf{b}.$$

The significance of this representation is that a starting point for the constrained least squares problem is the solution of the unconstrained problem, and the matrix $\mathbf{R}$ created in the course of making the starting point is $\mathbf{L}^T$: the transpose of the matrix needed for the Goldfarb–Idnani algorithm.

The unconstrained problem gives solution

$$a = -0.014\,857\,142\,857\,14$$
$$b = 0.873\,126\,984\,126\,98$$
$$c = 1.163\,714\,285\,714\,3$$

with residual $0.359\,526\,805\,411\,67$ in L^2. Upon enforcement of the constraints,

$$a = 0$$
$$b = 1.047\,637\,681\,159\,4$$
$$c = 1.047\,637\,681\,159\,4$$

and the residual grows by an additional $0.249\,559\,680\,124\,22$.

```
1   #include <iostream>
2   #include <sstream>
3   #include <iomanip>
4   #include "Matrix.H"
5   #include "Householder.H"
6   #include "ConvexQP.H"
7   using namespace std;
8
9   int main()
10  {
11    cout << setprecision(14);
12    Matrix<double> A(4,3), Ce, Ci(3,4);
13    Vector<double> b(4), diag(3), ce, ci(4);
14    Vector<int> Pl(4), Pr(3);
15
```

```
16      stringstream ss( stringstream :: in | stringstream :: out );
17      ss << "1 1 1     1 4 2     1 1 6     1 3 4     1.985 5.990 7.970 6.982";
18      ss >> A >> b;
19
20      // 4 inequality constraints in form Ci^T x + ci >= 0
21      ss.clear();
22      ss << "1 0 0 0     0 1 0 1     0 0 1 -1     0 0 0 0";
23      ss >> Ci >> ci;
24
25      // least squares solution without constraints
26      Householder( A, Pl, Pr, diag, POWELLREID );
27      double r2 = solveQR( A, Pl, Pr, diag, b );
28      // at this point, no longer overspecified.  Resize so L is square.
29      A.resize(3,3);
30      b.resize(3);
31      cout << "initial ls solution x=" << endl << b << endl;
32      cout << "with L2 residual " << r2 << endl;
33
34      // A contains part of R, and representation of Q.  Convert A to L = R^T
35      for( int i=0; i<3; i++ ) {
36        for( int j=i+1; j<3; j++ ) {
37          A[j][i] = A[i][j];
38          A[i][j] = 0;
39        }
40        A[i][i] = diag[i];
41      }
42
43      // However, depending on the Householder pivoting scheme,
44      // R may also include a right permutation Pr.  To keep the form of R,
45      // we must permute solution "b", and constraints.
46
47      for( int i=b.size()-1; i>=0; i— ) {
48        int k = Pr[i];
49        while( k > i ) k = Pr[k];
50        if( k != i ) {
51          double d=b[k]; b[k]=b[i]; b[i]=d;
52          for( int j=0; j<Ce.columns(); j++ ) {
53            d = Ce[k][j]; Ce[k][j]=Ce[i][j]; Ce[i][j]=d;
54          }
55          for( int j=0; j<Ci.columns(); j++ ) {
56            d = Ci[k][j]; Ci[k][j]=Ci[i][j]; Ci[i][j]=d;
57          }
58        }
59      }
60
61      // compute impact of constraints with Goldfarb Idnani method
62      double df = 0;
63      ConvexQP( A, b, Ce, ce, Ci, ci, df, 1.e-16 );
64
65      // and, undo permutation on solution vector to express in natural order
66
67      for( int i=b.size()-1; i>=0; i— ) {
68        int k = Pr[i];
69        while( k < i ) k = Pr[k];
70        if( k != i ) {
71          double d=b[k]; b[k]=b[i]; b[i]=d;
72        }
73      }
74
75      cout << "Convex Quadratic Solution x=" << endl << b << endl;
76      cout << "with additional L2 residual " << df << endl;
77  }
```

8.4 A fairly general program to implement line fitting with errors in all quantities is given below.

```
1   #include <iostream>
2   #include <iomanip>
3   #include <math.h>
```

```
4    #include <Matrix.H>
5    #include <Householder.H>
6    #include <QR.H>
7    using namespace std;
8
9    int main()
10   {
11     cout << setprecision(10) << scientific;
12
13     // read inputs:
14     int npts;
15     cin >> npts;
16     cout << "number of data points " << npts << endl;
17
18     Vector<double> x(npts), y(npts), X(npts);
19     for( int i=0; i<npts; i++ ) {
20       cin >> x[i];
21       X[i] = x[i];
22       cin >> y[i];
23     }
24     cout << "x input" << endl << x << endl;
25     cout << "y input" << endl << y << endl;
26
27     Matrix<double> Sigma(2*npts,2*npts);
28     for( int i=0; i<2*npts; i++ ) {
29       for( int j=0; j<=i; j++ ) {
30         cin >> Sigma[i][j];
31         Sigma[j][i] = Sigma[i][j];
32       }
33     }
34
35     double beta; // nonlinear term mixing
36     double zeta; // damping
37     cin >> beta;
38     cin >> zeta;
39
40     // replace Sigma with Sigma^(-1/2)
41
42     Matrix<double> P(2*npts,2*npts);
43     Vector<double> lambda(2*npts);
44     QReig( Sigma, P, lambda );
45
46     for( int i=0; i<2*npts; i++ ) {
47       for( int j=0; j<2*npts; j++ ) {
48         Sigma[i][j] = 0;
49         for( int k=0; k<2*npts; k++ ) Sigma[i][j] += P[i][k]*P[j][k]/sqrt(lambda[k]);
50       }
51     }
52
53     Matrix<double> A(2*npts,npts+2);
54     Vector<double> b(2*npts), diag(npts+2);
55     Vector<int> Pl(2*npts), Pr(npts+2);
56
57     double dbi;
58     int iteration=0;
59
60     double slope = 0;
61     double intercept = 0;
62
63     Vector<double> oldb(npts+2);
64     for( int i=0; i<npts; i++ ) oldb[i] = x[i];
65     oldb[npts] = slope;
66     oldb[npts+1] = intercept;
67
68     do {
69       ++iteration;
70
71       // solve weighted least squares problem
72       for( int i=0; i<npts; i++ ) {
73         A[i    ][npts] = A[i    ][npts+1] = 0;
```

```
74        A[i+npts][npts] = A[i+npts][npts+1] = 0;
75        b[i] = b[i+npts] = 0;
76        for( int j=0; j<npts; j++ ) {
77          A[i      ][j] = Sigma[i      ][j] + slope*beta*Sigma[i      ][j+npts];
78          A[i+npts][j] = Sigma[i+npts][j] + slope*beta*Sigma[i+npts][j+npts];
79          A[i      ][npts  ] += (1.-beta)*Sigma[i      ][j+npts]*X[j];
80          A[i+npts][npts  ] += (1.-beta)*Sigma[i+npts][j+npts]*X[j];
81          A[i      ][npts+1] += Sigma[i      ][j+npts];
82          A[i+npts][npts+1] += Sigma[i+npts][j+npts];
83          b[i      ] += Sigma[i      ][j]*x[j] + Sigma[i      ][j+npts]*y[j];
84          b[i+npts] += Sigma[i+npts][j]*x[j] + Sigma[i+npts][j+npts]*y[j];
85        }
86      }
87
88      Householder( A, Pl, Pr, diag, BJORCK );
89      solveQR( A, Pl, Pr, diag, b );
90
91      // measure change (L inf) before relaxation
92      dbi = 0;
93      for( int i=0; i<npts+2; i++ ) {
94        double t = fabs(b[i] - oldb[i]);
95        dbi = max(dbi,t);
96      }
97
98      // relax
99      for( int i=0; i<npts+2; i++ ) {
100       b[i] = (1.-zeta)*b[i] + zeta*oldb[i];
101       oldb[i] = b[i];
102     }
103
104     // replace X with new value
105
106     slope = b[npts];
107     intercept = b[npts+1];
108     for( int i=0; i<npts; i++ ) {
109       X[i] = b[i];
110     }
111
112     cout << iteration << " change dbi=" << dbi
113                       << " m=" << slope
114                       << " b=" << intercept << endl;
115   } while( dbi > 1.e-6 && iteration <100 );
116
117   cout << "after " << iteration << " iterations " << endl;
118
119   // now find covariance
120
121   cout << "slope m      " << slope << endl;
122   cout << "intercept b   " << intercept << endl;
123
124   for( int i=0; i<npts; i++ ) {
125     A[i      ][npts] = A[i      ][npts+1] = 0;
126     A[i+npts][npts] = A[i+npts][npts+1] = 0;
127     for( int j=0; j<npts; j++ ) {
128       A[i      ][j] = Sigma[i      ][j] + slope*Sigma[i      ][j+npts];
129       A[i+npts][j] = Sigma[i+npts][j] + slope*Sigma[i+npts][j+npts];
130       A[i      ][npts  ] += Sigma[i      ][j+npts]*X[j];
131       A[i      ][npts+1] += Sigma[i      ][j+npts];
132       A[i+npts][npts  ] += Sigma[i+npts][j+npts]*X[j];
133       A[i+npts][npts+1] += Sigma[i+npts][j+npts];
134     }
135   }
136   Householder( A, Pl, Pr, diag, NOTHING );
137   // make Q
138   Matrix<double> Q(2*npts,2*npts);
139   for( int i=0; i<2*npts; i++ ) {
140     Vector<double>& v = Q[i];
141     for( int j=0; j<2*npts; j++ ) v[j] = 0;
142     v[i] = 1;
143     QTb(A,v);
```

```
144     }
145     Matrix<double> Bdag(npts+2,2*npts);
146     // solve R Bdag = QT for Bdag
147     for( int i=0; i<2*npts; i++ ) {
148       for( int j=npts+1; j>=0; j— ) {
149         double t = Q[i][j];
150         for( int k=j+1; k<npts+2; k++ )
151           t -= A[j][k]*Bdag[k][i];
152         Bdag[j][i] = t/diag[j];
153       }
154     }
155
156     double varm=0, varb=0, covmb=0;
157
158     for( int i=0; i<2*npts; i++ ) {
159       varm  += Bdag[npts   ][i]*Bdag[npts   ][i];
160       varb  += Bdag[npts+1][i]*Bdag[npts+1][i];
161       covmb += Bdag[npts   ][i]*Bdag[npts+1][i];
162     }
163     cout << "var m " << varm << endl;
164     cout << "sig m " << sqrt(varm) << endl;
165     cout << "var b " << varb << endl;
166     cout << "sig b " << sqrt(varb) << endl;
167     cout << "covar mb " << covmb << endl;
168   }
```

For this particular problem the data input is

```
10
0.0 5.9
0.9 5.4
1.8 4.4
2.6 4.6
3.3 3.5
4.4 3.7
5.2 2.8
6.1 2.8
6.5 2.4
7.4 1.5
1.000e-3
0 1.000e-3
0 0 2.000e-3
0 0 0 1.250e-3
0 0 0 0 5.000e-3
0 0 0 0 0 1.250e-2
0 0 0 0 0 0 1.667e-2
0 0 0 0 0 0 0 5.000e-2
0 0 0 0 0 0 0 0 5.556e-1
0 0 0 0 0 0 0 0 0 1
0 0 0 0 0 0 0 0 0 0 1
0 0 0 0 0 0 0 0 0 0 0 5.556e-1
0 0 0 0 0 0 0 0 0 0 0 0 2.500c-1
0 0 0 0 0 0 0 0 0 0 0 0 0 1.250e-1
0 0 0 0 0 0 0 0 0 0 0 0 0 0 5.000e-2
```

```
0 0 0 0 0 0 0 0 0 0 0 0 0 0 0 0 5.000e-2
0 0 0 0 0 0 0 0 0 0 0 0 0 0 0 0 1.429e-2
0 0 0 0 0 0 0 0 0 0 0 0 0 0 0 0 1.429e-2
0 0 0 0 0 0 0 0 0 0 0 0 0 0 0 0 0 1.000e-2
0 0 0 0 0 0 0 0 0 0 0 0 0 0 0 0 0 2.000e-3
0.5
0.5
```

The result is

m	4.849×10^{-1}
b	5.503
σ_m^2	3.334×10^{-3}
σ_b^2	8.645×10^{-2}
$\mathrm{cov}(m, b)$	-1.634×10^{-2}

8.5 First consider the divided differences of a product $h(t) = f(t)g(t)$:

$$h[t_i, t_{i+1}] = \frac{f(t_{i+1})g(t_{i+1}) - f(t_i)g(t_i)}{t_{i+1} - t_i}$$

$$h[t_i, t_{i+1}] = \frac{f(t_{i+1})g(t_{i+1}) - f(t_i)g(t_{i+1}) + f(t_i)g(t_{i_1}) - f(t_i)g(t_i)}{t_{i+1} - t_i}$$

$$= f[t_i, t_{i_1}]g[t_{i+1}] + f[t_i]g[t_i, t_{i+1}],$$

which generalizes to a Leibniz formula

$$h[t_i, \ldots, t_{i+d}] = \sum_{k=0}^{d} f[t_i, \ldots, t_{i+k}]g[t_{i_k}, \ldots, t_{i+d}].$$

Let us apply this to $h = (t - x)_+^{d-1}(t - x) = (t - x)_+^d$, i.e., $h = (t - x)f_x^{d-1} = f_x^d$:

$$f_x^d[t_i, \ldots, t_{i+d+1}] = f_x^{d-1}[t_i, \ldots, t_{i+d+1}](t_{i+d} - x) + f_x^{d-1}[t_i, \ldots, t_{i+d}]$$

$$= \frac{f_x^{d-1}[t_{i+1}, \ldots, t_{i+d+1}] - f_x^{d-1}[t_i+, \ldots, t_{i+d}]}{t_{i+d+1} - t_i}(t_{i+d} - x)$$

$$- f_x^{d-1}[t_i, \ldots, t_{i+d}]$$

$$(t_{i+d+1} - t_i)f_x^d[t_i, \ldots, t_{i+d}] = f_x^{d-1}[t_{i+1}, \ldots, t_{i+d}](t_{i+1} - x)$$

$$- f_x^{d-1}[t_i+, \ldots, t_{i+d-1}](t_i - x)$$

$$= (t_{i+d+1} - t_{i+1})f_x^{d-1}[t_{i+1}, \ldots, t_{i+d}]\frac{t_{i+d+1} - x}{t_{i+d+1} - t_{i+1}}$$

$$- (t_{i+d} - t_i)f_x^{d-1}[t_i+, \ldots, t_{i+d}]\frac{t_i - x}{t_{i+d} - t_i}$$

$$B_{i,d}(x) = \frac{t_{i+d+1} - x}{t_{i+d+1} - t_{i+1}}B_{i+1,d-1}(x) + \frac{x - t_i}{t_{i+d} - t_i}B_{i,d-1}(x).$$

B-splines were first defined by these divided differences [212], and didn't find much practical use until the recursive form (8.24) was discovered by de Boor [53].

Chapter 9

9.1

$$f(x) = f(a) + (x - a)f'(a) + \int_a^x (x - t)f''(t)dt$$

$$\mathcal{L}f = \int_a^b P_2(x)dx - \int_a^b f(x)dx$$

$$= \mathcal{L}\int_a^x (x - t)f''(t)dt = \int_a^b \mathcal{L}(x - t)_+ f''(t)dt$$

$$\mathcal{L}(x - t)_+ = \frac{h}{2}(a - t)_+ + \frac{h}{2}(b - t)_+ - \int_a^b (x - t)_+ dx$$

$$= \frac{1}{2}(b - t)(t - a).$$

Since the kernel has uniform sign on interval $[a, b]$ the mean value theorem may be used

$$\tilde{I} - I = f''(\xi) \int_a^b \frac{(b - t)(t - a)}{2} dt = \frac{h^2}{12}(b - a)f''(\xi).$$

9.2 A short C++ program calculates these two integrals with the trapezoidal sum rule, and measures their convergence with Richardson extrapolation (9.22b).

```
1    #include <iostream>
2    #include <iomanip>
3    #include <math.h>
4    using namespace std;
5
6    double TSR_bad(int N)
7    {
8       double x = 0,   h = 1./N,   I = 0;
9       for( int i=0; i<=N; i++ ) {
10         double f = (i==0) ? 0 : expml(x)/sqrt(x);
11         I += f*h*( i%N ? 1 : 0.5 );
12         x += h;
13      }
14      return I;
15   }
16
17   double TSR_good(int N)
18   {
19      double u = 0,   h = 1./N,   I = 0;
20      for( int i=0; i<=N; i++ ) {
21         double f = 2*expml(u*u);
22         I += f*h*( i%N ? 1 : 0.5 );
23         u += h;
24      }
25      return I;
26   }
27
28   int main()
29   {
30      cout << scientific << setprecision(8);
31
32      cout << "bad" << endl;
33      int N = 100;
34      double I0=0, I1=0, I2=0;
35      for( int i=0; i<6; i++ ) {
36         I2=I1;  I1=I0;  I0=TSR_bad(N);
37         cout << 1./N << " " << I0;
```

```
38      if( i >= 2 )
39        cout << " " << log(fabs(I2-I1)/fabs(I1-I0))/log(2.);
40      cout << endl;
41      N *= 2;
42    }
43
44    cout << "good" << endl;
45    N = 100;
46    I0=0, I1=0, I2=0;
47    for( int i=0; i<6; i++ ) {
48      I2=I1; I1=I0; I0=TSR_good(N);
49      cout << 1./N << " " << I0;
50      if( i >= 2 )
51        cout << " " << log(fabs(I2-I1)/fabs(I1-I0))/log(2.);
52      cout << endl;
53      N *= 2;
54    }
55
56    return 0;
57  }
```

The results are shown in the following tables. The form (9.39) that does not possess derivatives shows an order of convergence of ≈ 1.5, significantly less than the theoretical $\mathcal{O}(h^2)$, which is obtained by the form (9.40) that does possess derivatives.

h	I (9.39)	order
1.000-e02	9.25110971-e01	
5.000-e03	9.25233844-e01	
2.500-e03	9.25278470-e01	1.46118220e+00
1.250-e03	9.25294546-e01	1.47304724e+00
6.250-e04	9.25300304-e01	1.48118090e+00
3.125-e04	9.25302359-e01	1.48680972e+00

h	I (9.40)	order
1.000-e02	9.25394100-e01	
5.000-e03	9.25326144-e01	
2.500-e03	9.25326144-e01	1.99997746e+00
1.250-e03	9.25304908-e01	1.99999436e+00
6.250-e04	9.25303846-e01	1.99999858e+00
3.125-e04	9.25303580-e01	2.00000048e+00

The non-integer order of convergence is a consequence of the non-existence of the derivative at the limit $x \to 0$ [29, 71]. One way to understand it is to begin with the Euler–Maclaurin expression (9.21), then express the non-existent derivatives $f^{(k)}(a)$ in terms of the function at h where all derivatives exist.

Consider the representative "difficult" function $f = x^\alpha$, for $\alpha > 0$ not an integer. Let $a = \lfloor \alpha \rfloor$ be the greatest integer less than α. Then, $f^{(a+1)}(0)$ and higher derivatives do not exist. However, these derivatives may be expressed in a series centered at h:

$$f(x) = \sum_{k=0}^{\infty} \frac{(x-h)^k}{k!} f^{(k)}(h)$$

$$= \sum_{k=0}^{\infty} \frac{(x-h)^k}{k!} \alpha(\alpha-1)\cdots(\alpha+1-k)h^{\alpha-k}$$

$$= \sum_{k=0}^{\infty} \frac{(x-h)^k}{k!} \frac{\Gamma(\alpha+1)}{\Gamma(\alpha+1-k)} h^{\alpha-k}$$

$$f^{(p)}(x) = \sum_{k=0}^{\infty} \frac{(x-h)^{k-p}}{(k-p)!} \frac{\Gamma(\alpha+1)}{\Gamma(\alpha+1-k)} h^{\alpha-k}$$

$$f^{(p)}(0) = h^{\alpha-p} \sum_{k=0}^{\infty} \frac{(-1)^{k-p}}{(k-p)!} \frac{\Gamma(\alpha+1)}{\Gamma(\alpha+1-k)}.$$

Then, terms involving $f^{(p)}(0)$ in the series (9.21) become

$$-\sum_{p=1}^{q} \frac{h^{2p} B_{2p}}{(2p)!} f^{(2p-1)}(0) = -\sum_{p=1}^{q} \frac{h^{2p} B_{2p}}{(2p)!} h^{\alpha+1-2p} \sum_{k=0}^{\infty} \frac{(-1)^{k+1-2p}}{(k+1-2p)!} \frac{\Gamma(\alpha+1)}{\Gamma(\alpha+1-k)}$$

$$= \beta h^{\alpha+1}.$$

Now, the specific integrand of this problem can be written $f(x) = x^{1/2} g(x)$ where $g(x)$ has a Taylor series convergent at the origin. This means $f(x)$ has a series in powers $i + 1/2$, $i = 0, 1, \ldots$, and this in turn implies that the $f^{(p)}(0)$ contribution to (9.21) will look like

$$\sum_{i=1} \beta_i h^{\alpha+i},$$

and, therefore, equation (9.21) specialized to this problem becomes

$$I - T(h) = \beta_1 h^{1+1/2} + \beta_2 h^{2+1/2} + \beta_3 h^{3+1/2} + \cdots$$
$$+ \alpha_2 h^2 + \alpha_4 h^4 + \alpha_6 h^6 + \cdots.$$

The lowest power of h is $3/2$, which is what we see in our convergence results.

Bulirsch and Stoer [31] show how the Romberg extrapolation idea can be extended to functions whose series has the arbitrary form

$$T(h) = a_0 + a_1 h^{\gamma_1} + a_2 h^{\gamma_2} + \cdots.$$

Notationally, they label numbers in the Neville tableau as follows:

$$
\begin{array}{ccccccc}
T_0^{(0)} & & & & \\
& T_1^{(0)} & & \\
T_0^{(1)} & & T_2^{(0)} \\
& T_1^{(1)} & \\
T_0^{(2)} & &
\end{array}
$$

Now

$$T_0^{(i)} = a_0 + a_1 h_i^{\gamma_1} + a_2 h_i^{\gamma_2} + \cdots$$
$$T_0^{(i+1)} = a_0 + a_1 h_{i+1}^{\gamma_1} + a_2 h_{i+1}^{\gamma_2} + \cdots,$$

and to eliminate the a_1 term in the construction of $T_1^{(i)}$ we must have

$$T_1^{(i)} = \frac{h_{i+1}^{\gamma_1} T_0^{(i)} - h_i^{\gamma_1} T_0^{(i+1)}}{h_{i+1}^{\gamma_1} - h_i^{\gamma_1}}$$

$$= a_0 + a_2 \left[\frac{h_{i+1}^{\gamma_1} h_i^{\gamma_2} - h_i^{\gamma_1} h_{i+1}^{\gamma_2}}{h_{i+1}^{\gamma_1} - h_i^{\gamma_1}} \right] + \cdots,$$

and similarly,

$$T_1^{(i+1)} = a_0 + a_2 \left[\frac{h_{i+2}^{\gamma_1} h_{i+1}^{\gamma_2} - h_{i+1}^{\gamma_1} h_{i+2}^{\gamma_2}}{h_{i+2}^{\gamma_1} - h_{i+1}^{\gamma_1}} \right] + \cdots.$$

The construction of the column T_1 of this new tableau follows the normal Romberg procedure, but using $h_i^{\gamma_1}$ in place of h_i^2, etc. We essentially create a weighted average of terms in the T_0 column, where the weighting factor depends on h_i and h_{i+1}:

$$T_1^{(i)} = \alpha T_0^{(i)} + (1 - \alpha) T_0^{(i+1)}$$

$$\alpha = \frac{h_{i+1}^{\gamma_1}}{h_{i+1}^{\gamma_1} - h_i^{\gamma_1}}.$$

In constructing the next column, we should aim to eliminate the a_2 terms. However, examination of the coefficients of a_2 now shows that $T_2^{(i)} = \alpha T_1^{(i)} + (1 - \alpha) T_1^{(i+1)}$ involves an α that is a complicated function of h_i, h_{i+1}, h_{i+2}, γ_1, and γ_2. A simple Neville-like interpolation rule is not possible for arbitrary h_i and γ_i.

If, however, one chooses $h_i = h_0 b^i$, where $0 < b < 1$ is a constant, then

$$T_1^{(i)} = \frac{b^{\gamma_1} T_0^{(i)} - T_0^{(i+1)}}{b^{\gamma_1} - 1}$$

$$= a_0 + a_2 h_i^{\gamma_2} \frac{b^{\gamma_1} - b^{\gamma_2}}{1 - b^{\gamma_1}} + \cdots$$

$$T_1^{(i+1)} = a_0 + a_2 h_i^{\gamma_2} b^{\gamma_2} \frac{b^{\gamma_1} - b^{\gamma_2}}{1 - b^{\gamma_1}} + \cdots,$$

so

$$T_2^{(i)} = \frac{b^{\gamma_2} T_1^{(i)} - T_1^{(i+1)}}{b^{\gamma_2} - 1}$$

eliminates coefficient a_2 throughout the T_2 column.

The general rule, then, is choose $h_i = h_0 b^i$, with $0 < b < 1$, and compute

$$T_m^{(i)} = \frac{b^{\gamma_m} T_{m-1}^{(i)} - T_{m-1}^{(i+1)}}{b^{\gamma_m} - 1}.$$

To demonstrate this method, consider the integral

$$I = \frac{3}{2} \int_0^1 \sqrt{x}\,dx = 1$$

with the power series already deduced: $\gamma_1 = 3/2$, $\gamma_i = 2(i-1)$ for $i \geq 2$. With $h_0 = 1$, and $b = 1/2$, compare the unmodified Romberg result with the Bulirsch and Stoer variant. The values $T_m^{(0)}$ are each the best interpolation using $m+1$ support points $0, \ldots, m$.

	Romberg	Bulirsch and Stoer
$T_0^{(0)}$	0.7500000000000000	0.7500000000000000
$T_1^{(0)}$	0.9571067811865476	0.9902829307627806
$T_2^{(0)}$	0.9866349049223435	0.9999295226155950
$T_3^{(0)}$	0.9954113536684386	0.9999992208746942
$T_4^{(0)}$	0.9983892976941983	0.9999999946926856
$T_5^{(0)}$	0.9994315485507613	0.9999999999831570

Clearly the interpolation method should respect the underlying polynomial representation.

9.3 With the three given resolutions one has

$$I(h_0) = \frac{h_0}{2}(f(0) + f(h_0))$$

$$I(h_1) = \frac{h_0}{4}(f(0) + 2f(h_0/2) + f(h_0))$$

$$I(h_2) = \frac{h_0}{8}(f(0) + 2f(h_0/4) + 2f(h_0/2) + 2f(3h_0/4) + f(h_0)).$$

The Neville tableau for this problem is

$$
\begin{array}{ccccc}
h_0^2 & P_0 = I(h_0) & & & \\
 & & P_{01} & & \\
h_0^2/4 & P_1 = I(h_1) & & P_{012} & \\
 & & P_{12} & & \\
h_0^2/16 & P_2 = I(h_2) & & &
\end{array}
$$

with

$$P_{01} = \frac{(0 - h_0^2)P_1 - (0 - h_0^2/4)P_0}{h_0^2/4 - h_0^2}$$

$$= \frac{h_0}{6}(f(0) + 4f(h_0/2) + f(h_0))$$

$$= \frac{h_1}{3}(f(0) + 4f(h_1) + f(2h_1))$$

$$P_{12} = \frac{(0 - h_0^2/4)P_2 - (0 - h_0^2/16)P_1}{h_0^2/16 - h_0^2/4}$$

$$= \frac{h_0}{12}(f(0) + 4f(h_0/4) + 2f(h_0/2) + 4f(3h_0/4) + f(h_0))$$

$$P_{012} = \frac{(0 - h_0^2)P_{12} - (0 - h_0^2/16)P_{01}}{h_0^2/16 - h_0^2}$$

$$= \frac{h_0}{90}(7f(0) + 32f(h_0/4) + 12f(h_0/2) + 32f(3h_0/4) + 7f(h_0))$$

$$= \frac{2h_2}{45}(7f(0) + 32f(h_2) + 12f(2h_2) + 32f(3h_2) + 7f(4h_2)).$$

Here, P_{01} is the Romberg result with two resolutions, and the resulting formula is Simpson's rule. P_{012} is the result of Romberg integration with three resolutions, and it corresponds to Boole's rule.

9.4 To analyze this question, consider that the Romberg/Neville tableau applied to algebraic formulas as in Problem 9.3 performs polynomial extrapolation on each coefficient of a given support abscissa f_i in even powers of h.

If $h_i = bh_{i-1}$, and $1/b$ is not an integer, then one could have a circumstance as follows. For $h_1 = b - a/2$,

$$T_1 = T(h_1) = \frac{h_1}{2}\left(f(a) + 2f(\frac{a+b}{2}) + f(b)\right)$$

$$T_2 = T(\frac{2h_1}{3}) = \frac{h_1}{3}\left(f(a) + 2f(\frac{2a+b}{3}) + 2f(\frac{a+2b}{3}) + f(b)\right).$$

Romberg extrapolation

$$\frac{T_2 - \frac{4}{9}T_1}{1 - \frac{4}{9}}$$

results in the coefficient of $f((a+b)/2)$ being negative:

$$\frac{-\frac{4}{9}h_1}{1 - \frac{4}{9}} = -\frac{4}{5}h_1.$$

This is why we need $1/b$ to be an integer.

Now limit consideration to the case where $1/b \geq 2$ is a positive integer. With this choice, once an ordinate f_i appears in a calculation $T_j = T(h_j)$, it will appear in all subsequent calculations $T_{j+1}, ..., T_n$. The weight of f_i in each calculation is known: relative to a weight of 1 in T_j, the weight is b in T_{j+1}, and b^{n-j} in T_n. Finally, polynomials of given degree are unique, so we could perform this polynomial extrapolation by any method. Rather than using the Neville tableau, a simpler analysis follows from the Lagrange form. Let f_i be absent in T_k, $k = 0, ..., j-1$, and present in T_k for $k = j, ..., n$.

The weight of f_j in the extrapolated value, relative to its weight in T_j, is

$$\sum_{k=j}^{n} b^{k-j} \prod_{\ell=0,\ell\neq k}^{n} \frac{-b^{2\ell}}{b^{2k}-b^{2\ell}}\cdot$$

$$= \sum_{k=j}^{n} b^{k-j} \underbrace{\left(\prod_{m=1}^{n-k} \frac{b^{2m}}{b^{2m}-1}\right)\left(\prod_{m=1}^{k} \frac{1}{1-b^{2m}}\right)}_{c_k},$$

using the Lagrange interpolation function. This a series with alternating signs, and with the coefficient of the $k=n$ term positive. That is, the sum has the form

$$\underbrace{c_n}_{\text{pos.}} + \overbrace{c_{n-1}}^{\text{neg.}} + \underbrace{c_{n-2}}_{\text{pos.}} + \overbrace{c_{n-3}}^{\text{neg.}} + \cdots + c_j.$$

If the relative magnitude of successive terms in the alternating sign series is less than or equal to 1 (in absolute value), then the series will be positive: we will either have

$$(c_n + c_{n-1}) + (c_{n-2} + c_{n-3}) + \cdots + (c_{j-2} + c_{j-1}) + \underbrace{c_j}_{\text{pos.}}$$

or

$$(c_n + c_{n-1}) + (c_{n-2} + c_{n-3}) + \cdots + (c_{j-1} + c_j)$$

where all pairs in parenthesis are positive.

With some preliminary algebra,

$$c_k = b^{k-j} \left(\prod_{m=1}^{n-k} \frac{b^{2m}}{b^{2m}-1}\right)\left(\prod_{m=1}^{k} \frac{1}{1-b^{2m}}\right)$$

$$= b^{k-j} \left(\prod_{m=1}^{n-k} \frac{b^{2m}}{b^{2m}-1}\right)\left(\prod_{m=1}^{k-1} \frac{1}{1-b^{2m}}\right)\frac{1}{1-b^{2k}}$$

$$c_{k-1} = b^{k-1-j} \left(\prod_{m=1}^{n+1-k} \frac{b^{2m}}{b^{2m}-1}\right)\left(\prod_{m=1}^{k-1} \frac{1}{1-b^{2m}}\right)$$

$$= b^{k-1-j} \left(\prod_{m=1}^{n-k} \frac{b^{2m}}{b^{2m}-1}\right)\frac{b^{2(n+1-k)}}{b^{2(n+1-k)}-1}\left(\prod_{m=1}^{k-1} \frac{1}{1-b^{2m}}\right),$$

the relative magnitude of the successive terms in the series is

$$\alpha_{k-1} \equiv \frac{c_{k-1}}{c_k} = -\frac{b^{2(n+1-k)}(1-b^{2k})}{b(1-b^{2(n+1-k)})}.$$

We now show that $0 < -\alpha_{k-1} \leq 1$, or

$$b^{2(n+1-k)} - b^{2(n+1)} \leq b - b^{1+2(n+1-k)}$$

$$b^{2(n+1)}\left(1-b^{2k}\right) \leq b\left(b^{2k}-b^{2(n+1)}\right).$$

If $k = n$,

$$b^{2(n+1)} \left(1 - b^{2n}\right) \le bb^{2n} \left(1 - b^2\right)$$

$$b \left(1 - b^{2n}\right) < \frac{1}{2} < \frac{3}{4} \le \left(1 - b^2\right).$$

And, if $k < n$,

$$b^{2(n+1)} \left(1 - b^{2k}\right) < b^{2n+2} < b^{2n+1} < b \left(b^{2k} - b^{2(n+1)}\right).$$

Therefore, the series is positive and the Romberg procedure is stable. The fact that $1/b$ is the basis for knowing that when a term f_j appears in the tableau for a given h, it will appear in all subsequent entries of the tableau.

9.5 One has

$$f'(a) = \frac{f(a + h) - f(a)}{h} + \mathcal{O}(h)$$

$$f'(b) = \frac{f(b) - f(b - h)}{h} + \mathcal{O}(h),$$

and therefore

$$\frac{h^2 B_2}{2}[f'(b) - f'(a)] = \frac{h}{12}(f(a) - f(a + h) - f(b - h) + f(b)) + \mathcal{O}(h^3),$$

so

$$\hat{T}(h) = T(h) - \frac{h}{12}(f(a) - f(a + h) - f(b - h) + f(b))$$

$$= \int_a^b f(x)dx + \mathcal{O}(h^3) + \sum_{p=2} \frac{h^{2p} B_{2p}}{(2p)!}[f^{(2p-1)}(b) - f^{(2p-1)}(a)]$$

$$\hat{T}(h) = \int_a^b f(x)dx + \mathcal{O}(h^3).$$

To improve the order of accuracy, we need a higher-order estimate of the derivatives. Using Taylor series, one can show

$$f'(a) = -\frac{3}{2h} f(a) + \frac{2}{h} f(a + h) - \frac{1}{2h} f(a + 2h) + \mathcal{O}(h^2)$$

$$f'(b) = \frac{3}{2h} f(b) - \frac{2}{h} f(b - h) + \frac{1}{2h} f(b - 2h) + \mathcal{O}(h^2),$$

so

$$T(h) - \frac{h}{12}\left(\frac{3}{2}f(a) - 2f(a + h) + \frac{1}{2}f(a + 2h) + \frac{1}{2}f(b - 2h) - 2f(b - h)\right.$$

$$\left. + \frac{3}{2}f(b)\right) = \int_a^b f(x)dx + \mathcal{O}(h^4).$$

9.6 The orthogonal polynomials associated with interval $[0, 1]$ and weight

$$w(x) = \frac{e^{-x/2}}{\sqrt{x}}$$

are called Rys polynomials. They are used extensively to evaluate electron repulsion integrals in *ab initio* computational chemistry applications.

The first order of business in solving this problem is change variables to eliminate the singularity

$$\int_0^1 2e^{-u^2/2} f(u^2)\,du.$$

Table S.1: Rys quadrature points, using Stieltjes' method; Problem 9.6.

	x_i	w_i
$n=1$	2.911250947727932e-01	1.711248783784297e+00
$n=2$	1.049211205285722e-01	1.192693723011120e+00
	7.194003698220316e-01	5.185550607731769e-01
$n=3$	5.301084180676952e-02	8.828846890483311e-01
	4.184224471238822e-01	5.912890071315668e-01
	8.603873672007244e-01	2.370750876043990e-01
$n=4$	3.181151534070814e-02	6.954455345110818e-01
	2.646578547010797e-01	5.464411730597030e-01
	6.208815065780603e-01	3.345442860199524e-01
	9.177718978199244e-01	1.348177901935604e-01
$n=5$	2.116574114799685e-02	5.720800341035382e-01
	1.807274494384905e-01	4.867501416799652e-01
	4.497992896450101e-01	3.528926975504390e-01
	7.392073296074377e-01	2.126789297231564e-01
	9.460909953356571e-01	8.684698072719967e-02
$n=10$	5.716834451403585e-03	3.010270231139561e-01
	5.070594182374957e-02	2.879031506794258e-01
	1.367640057198072e-01	2.635192554682825e-01
	2.563220540902696e-01	2.310526684070460e-01
	3.987098547602415e-01	1.941340152021735e-01
	5.509810226892831e-01	1.560144806231198e-01
	6.989981330081722e-01	1.190161899965165e-01
	8.287235957404431e-01	8.434532551014380e-02
	9.276126805665695e-01	5.219623696984956e-02
	9.859689954973575e-01	2.204043781378351e-02

The program listing below uses this modified integral to compute the quadrature data using two methods. The first method is the Stieltjes algorithm: the integrals computed are (p_i, p_i) and (xp_i, p_i), where in both cases the numerical evaluation of $p_i(x)$ comes from recursive definition using calculated parameters δ_l, γ_l. The matrix $\mathbf{J}$ is then found, and is evaluated using the QR method. The second method first computes moments $\int x^i w(x)\,dx$, then performs the Cholesky decomposition of the moment matrix.

The parameters δ, γ, are gleaned from this Cholesky decomposition, then used to compose the Jacobi matrix **J**. Again, **J** is analyzed by the QR method. The output of these algorithms are listed in Tables S.1 and S.2, with the digits that are in error underlined.

In general, the use of moments is not advisable: the condition number of the mapping from moments to δ, γ data can be enormous, and is especially troubling as the number of quadrature points increases. This trend can be seen in the computed tables. This error is not yet disastrous for $n \leq 5$, but 2/3 of the significant figures are lost when $n = 10$, which is included in the tables to emphasize this point.

Table S.2: Rys quadrature points, using moments; Problem 9.6.

	x_i	w_i
$n = 1$	2.911250947727932e−01	1.711248783784297e+00
$n = 2$	1.049211205285743e−01	1.192693723011128e+00
	7.194003698220353e−01	5.185550607731695e−01
$n = 3$	5.301084180678939e−02	8.828846890484487e−01
	4.184224471239451e−01	5.912890071314877e−01
	8.603873672007446e−01	2.370750876043608e−01
$n = 4$	3.181151534079829e−02	6.954455345118166e−01
	2.646578547014173e−01	5.464411730593394e−01
	6.208815065782929e−01	3.345442860196775e−01
	9.177718978199800e−01	1.348177901934638e−01
$n = 5$	2.116574114868275e−02	5.720800341114821e−01
	1.807274494423524e−01	4.867501416797156e−01
	4.497992896501173e−01	3.528926975470472e−01
	7.392073296107780e−01	2.126789297201568e−01
	9.460909953364443e−01	8.684698072589590e−02
$n = 10$	5.716674921404596e−03	3.010229544999735e−01
	5.070466175452440e−02	2.879005173491594e−01
	1.367611484322787e−01	2.635185267865969e−01
	2.563179353362912e−01	2.310533991588744e−01
	3.987052328396876e−01	1.941354905982277e−01
	5.509767218114061e−01	1.560161237741822e−01
	6.989947807174006e−01	1.190176497586023e−01
	8.287214825627176e−01	8.434643603395910e−02
	9.276117267890770e−01	5.219694542735107e−02
	9.859688040106883e−01	2.204074039737141e−02

```
1   #include <iostream>
2   #include <iomanip>
3   #include <math.h>
4   #include "Matrix.H"
5   #include "QR.H"
6   #include "Cholesky.H"
```

```
 7   #include "AdaptiveSimpson.H"
 8   using namespace std;
 9
10   // Stieltjes values are global for benefit of integrator
11   Vector<double> s_delta;
12   Vector<double> s_gammasq;
13
14   // more global info for integrator
15   int  s_pwr;
16   bool s_extra_x;
17
18   // evaluate polynomial degree "deg" with Stieltjes recursion
19   double polyfn( double x, int deg )
20   {
21     double pm2=0, pm1=0, p=1;
22     for( int i=1; i<=deg; i++ ) {
23       pm2 = pm1;
24       pm1 = p;
25       p *= x-s_delta[i-1];
26       if( i!=1 ) p -= s_gammasq[i-2]*pm2;
27     }
28     return p;
29   }
30
31   // the function to be evaluated for Stieltjes form
32   double orth_rys_f( double x )
33   {
34     double pi = polyfn(x*x,s_pwr);
35     double retval = 2.*pi*pi*exp(-x*x/2);
36     if( s_extra_x ) retval *= x*x;
37     return retval;
38   }
39
40   void coefficients_from_Stieltjes( int maxorder, double& p0p0 )
41   {
42     s_delta.resize(maxorder);        // in case
43     s_gammasq.resize(maxorder-1);
44
45     double oldIpipi;
46     for( int i=0; i<maxorder; i++ ) {
47       s_pwr    = i;      // global variable passes degree to orth_rys_f
48       s_extra_x = false;
49       double Ipipi = adaptiveSimpson( 0., 1., 1.e-15, orth_rys_f );
50       s_extra_x = true;
51       double Ixpipi = adaptiveSimpson( 0., 1., 1.e-15, orth_rys_f );
52
53       if( i == 0 ) p0p0 = Ipipi;
54       s_delta[i] = Ixpipi/Ipipi;
55       if( i != 0 ) s_gammasq[i-1] = Ipipi/oldIpipi;
56       oldIpipi = Ipipi;
57     }
58   }
59
60   // global power for benefit of moment integrator
61   int f_pwr;
62
63   // integrand for moment calculation
64   double moment_rys_f( double x )
65   {
66     return 2*exp(-x*x/2)*pow(x,2.*f_pwr);
67   }
68
69   void coefficients_from_moments( int maxorder, double& p0p0,
70                                   Vector<double> delta,
71                                   Vector<double> gamma  )
72   {
73     Matrix<double> M( maxorder+1, maxorder+1);
74
75     for( int i=0; i<maxorder+1; i++ ) {
76       for( int j=i; j<maxorder+1; j++ ) {
```

```
77            f_pwr = i+j;  // global variable passes degree to moment_rys_f
78            M[j][i] = M[i][j] = adaptiveSimpson( 0., 1., 1.e-15, moment_rys_f );
79          }
80        }
81
82        // save (p0,p0) for later
83        p0p0 = M[0][0];
84
85        // replace M with tildeL
86        Cholesky(M);
87
88        delta.resize(maxorder); // in case
89        delta[0] = M[1][0]/M[0][0];
90        for( int j=1; j<maxorder; j++ )
91          delta[j] = M[j+1][j]/M[j][j] - M[j][j-1]/M[j-1][j-1];
92
93        gamma.resize(maxorder-1); // in case
94        for( int j=0; j<maxorder-1; j++ )
95          gamma[j] = M[j+1][j+1]/M[j][j];
96      }
97
98      int main()
99      {
100        cout << scientific << setprecision(15);
101
102        int maxorder = 5;
103        double p0p0;
104        Vector<double> delta(maxorder), gamma(maxorder-1);
105
106        for( int method=0; method<2; method++ ) {
107          switch(method) {
108            case 0:
109              coefficients_from_moments( maxorder, p0p0, delta, gamma );
110              cout << "compute Jacobi matrix coefs with moments" << endl;
111              break;
112            default:
113              coefficients_from_Stieltjes( maxorder, p0p0 );
114              cout << "compute Jacobi matrix coefs with Stieltjes method" << endl;
115              // copy from global array
116              for( int i=0; i<maxorder-1; i++ ) {
117                delta[i] = s_delta[i];
118                gamma[i] = sqrt(s_gammasq[i]);
119              }
120              delta[maxorder-1] = s_delta[maxorder-1];
121              break;
122          }
123          cout << "delta" << endl << delta << endl;
124          cout << "gamma" << endl << gamma << endl;
125
126          for( int order=1; order<=maxorder; order++ ) {
127
128            // J
129            Matrix<double> J(order,order);
130            for( int i=0; i<order; i++ ) {
131              for( int j=0; j<order; j++ ) J[i][j] = 0;
132              J[i][i] = delta[i];
133            }
134            for( int i=0; i<order-1; i++ )
135              J[i][i+1] = J[i+1][i] = gamma[i];
136
137            // eigenvalues and vectors
138            Matrix<double> V(order,order);
139            Vector<double> x(order);
140
141            // make eigensystem
142            QReig( J, V, x );
143
144            // weights
145            Vector<double> w(order);
146            for( int i=0; i<order; i++ ) {
```

```
147            double len = 0;
148             for( int j=0; j<order; j++ ) len += V[j][i]*V[j][i];
149            w[i] = (p0p0/len)*V[0][i]*V[0][i];
150            }
151
152            cout << "Rys quadrature method with " <<  order << " abscissas:" << endl;
153            cout << "   quadrature points" << endl << x << endl;
154            cout << "   quadrature weights" << endl << w << endl;
155            }
156        }
157    }
```

Since the coefficients of a Gaussian quadrature method are calculated once but presumably used many times, the extra expense of using higher-precision calculations to obtain greater accuracy is easily justified.

9.7 The derivative of the integrand $f^{(n)}$ is bounded by Cramér's formula as

$$|f^{(n)}(x)| \leq \frac{2}{\sqrt{\pi}} K 2^{n/2} \sqrt{n!}.$$

With the trapezoidal sum rule,

$$\frac{b-a}{12} \frac{2}{\sqrt{\pi}} K 2\sqrt{2!} h^2 \leq 10^{-6} \quad \rightarrow \quad h = 0.001\,313,$$

so a composite of 1523 steps is required; $\text{erf}(2) \approx 0.995\,322\,253\,138\,966$ is calculated, which carries an error of 1.2×10^{-8}.

With Simpson's rule,

$$\frac{b-a}{180} \frac{2}{\sqrt{\pi}} K 2^2 \sqrt{4!} h^4 \leq 10^{-6} \quad \rightarrow \quad h = 0.043\,96;$$

a composite of 23 steps is required; $\text{erf}(2) \approx 0.995\,322\,248\,605\,844$ is calculated, which carries an error of 1.6×10^{-8}.

With Boole's rule,

$$\frac{2(b-a)}{945} \frac{2}{\sqrt{\pi}} K 2^3 \sqrt{6!} h^6 \leq 10^{-6} \quad \rightarrow \quad h = 0.098\,16;$$

a composite of six steps is required; $\text{erf}(2) \approx 0.995\,322\,265\,080\,178$ is calculated, which carries an error of 6.1×10^{-11}.

9.8 To integrate the given data the first order of business is to construct a smooth function that is integrable. It is important to observe that the integral is unbounded unless $Z(0) = 1$. This number is included in the tabulated data, but it may not be respected by a fit to the data. This would make the fit unsuitable to its intended purpose. There are two ways to make sure the fit respects this constraint: (1) do not use an overdetermined system. If the system is not overdetermined, then the given data will be reproduced exactly (with exact math); (2) use an overdetermined system, but impose constraints (e.g., the approach of Goldfarb and Idnani).

The Newton–Cotes approach uses polynomial fits, so one might try that. However, high-degree polynomials are extremely unstable and susceptible to wild oscillations (Figure S.27). Instead, since the data are actually quite smooth, one should use a data

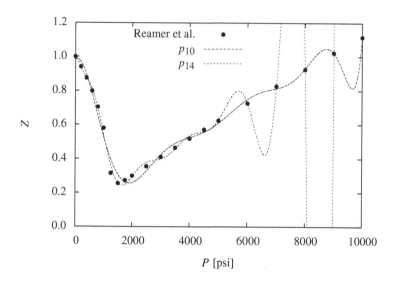

Figure S.27 Experimental data of Reamer *et al.* [189] and some polynomial fits; Problem 9.8.

fitting method that retains this smoothness: splines. To maximize smoothness, the "natural" assumption will be made (zero second derivative at end points). The spline construction requires knots outside the domain of the data. Assuming knots at negative pressure are spaced at 200 psi intervals and knots beyond 10 000 psi are spaced at 1000 psi intervals, the resulting fit is shown in Figure S.28.

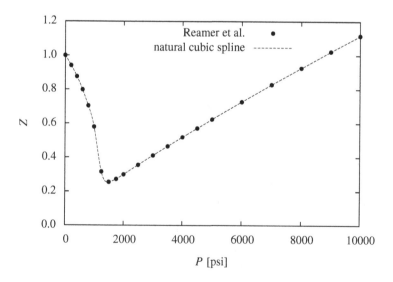

Figure S.28 Natural cubic spline fit to experimental data of Reamer *et al.* [189]; Problem 9.8.

As for integration, it should be noted that one must integrate $(Z(P) - 1)/P$, which is no longer a polynomial function. Two approaches are possible: (1) use a Newton–Cotes quadrature scheme to, in effect, fit $(Z - 1)/P$ to a polynomial then integrate the polynomial, or (2) devise a way to integrate exactly $(Z - 1)/P$ for the case that Z is piecewise cubic. Approach (1) is perfectly acceptable, and is by far the easier.

To integrate the spline fit exactly, refer to the finite difference definition of the Z/P part of the integrand

$$\frac{Z}{P} = \sum_i W_i (t_{i+4} - t_i) \frac{(t - P)^3_+ [t_i, \ldots, t_{i+4}]}{P},$$

where W_i are the weights determined by fitting the data. Introduce the Heaviside function

$$H(t - P) = \begin{cases} 0 & P > t \\ 1 & P \leq t \end{cases}$$

and use it to explicitly enforce the condition implied by the subscript $+$. Then,

$$\frac{Z}{P} = \sum_i W_i (t_{i+4} - t_i) \frac{H(t - P)(t^3 - 3t^2 P + 3t P^2 - P^3)[t_i, \ldots, t_{i+4}]}{P},$$

integrate

$$\int_0^P \frac{Z}{s} ds = \sum_i W_i (t_{i+4} - t_i) \left(t^3 \ln s - 3t^2 s + \frac{3}{2} t s^2 - \frac{1}{3} s^3 \right) \Big|_0^{\max(\min(t, P), 0)} [t_i, \ldots, t_{i+4}].$$

The integrand also contains $-1/P$, which can be written using the partition of unity property of B-splines

$$1 = \sum_i B_{i,3}(P).$$

The integral of $-1/P$ thus resembles the integral of Z/P with -1 in place of W_i. So,

$$\ln \phi = \sum_i (W_i - 1)(t_{i+4} - t_i) \left(t^3 \ln s - 3t^2 s + \frac{3}{2} t s^2 - \frac{1}{3} s^3 \right) \Big|_0^{\max(\min(t, P), 0)} [t_i, \ldots, t_{i+4}].$$

This still looks dangerous, since the logarithm will be undefined at the lower limit of integration. This is not really a problem, however. First, with $Z - 1 \to 0$ as a polynomial, we know $(Z - 1)/P$ is bounded. Thus, any apparent singularities occurring from $\ln(0)$ are artifacts. Simply omit those terms. Also, a fourth divided difference of a cubic function is zero. Whenever the argument is constant from the $s = 0$ limit, all terms are zero. In short, we evaluate this finite difference expression as implied, but omit $\ln(s)$ whenever $s \leq 0$. The final computed fugacity ϕ is plotted in Figure S.29.

Chapter 10

10.1 When Problem 3.3 was solved as an eigenvalue problem, there were multiple solutions. In fact, the number of eigenvalues depends on the arbitrary size N of the dis-

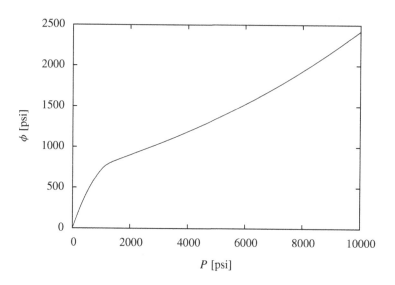

Figure S.29 Fugacity of pure CO_2 at $100°F$ computed by integrating a spline fit of $Z(P)$ as a sum of finite differences; Problem 9.8.

cretization, which means that there are an infinite number of possible solutions. A good initial guess is therefore necessary to achieve the desired answer.

The other lesson to learn from the eigenvalue problem is that the solution $U(x)$ corresponds to the eigenvector, and eigenvectors are not unique: they can be scaled by any nonzero constant and still be eigenvectors. Therefore, the scale of U is not important. With this insensitivity to scale, consider the BVP:

$$c(x)^2 U_{xx} = -\omega^2 U$$
$$U(0) = 0$$
$$U'(0) = 1$$
$$U(1) = 0$$

with ω^2 the parameter to be varied. As a system of first-order equations we have

$$\frac{d}{dx} \underbrace{\begin{pmatrix} U \\ U_x \end{pmatrix}}_{\mathbf{Y}} = \begin{pmatrix} U_x \\ -\omega^2 U/c(x)^2 \end{pmatrix} = \begin{pmatrix} Y_2 \\ -\omega^2 Y_1/c(x)^2 \end{pmatrix}$$

$$\mathbf{Y}(0) = \begin{pmatrix} 0 \\ 1 \end{pmatrix}.$$

The derivative with respect to ω^2 is simple

$$\mathbf{W} = \begin{pmatrix} W_1 \\ W_2 \end{pmatrix} = \frac{d}{d\omega^2}\mathbf{Y}$$

$$\frac{d}{dx}\mathbf{W} = \begin{pmatrix} W_2 \\ -(Y_1 + \omega^2 W_1)/c(x)^2 \end{pmatrix}$$

$$\mathbf{W}(0) = \begin{pmatrix} 0 \\ 0 \end{pmatrix}.$$

The simple shooting method implementation of these equations is listed below. The solution converges rapidly to $\omega \approx 10.247$ using the fourth-order Runge–Kutta integrator in 100 steps.

```
1    #include <math.h>
2    #include <iostream>
3    #include <iomanip>
4    #include "Matrix.H"
5    using namespace std;
6
7    // global parameters of problem
8    double omegasq;
9
10   // 4th order Runge Kutta
11
12   void rk4( double h, double x, Vector<double>& y,
13             void (*f)(double, Vector<double>&, Vector<double>&) )
14   {
15     // make static to avoid re-allocating each call
16     static Vector<double> k1, k2, k3, k4, tmp;
17
18     int n = y.size();
19     if( k1.size() != n ) {
20       k1.resize( n );
21       k2.resize( n );
22       k3.resize( n );
23       k4.resize( n );
24       tmp.resize( n );
25     }
26
27     (*f)(x,y,k1);
28     for( int i=0; i<n; i++ ) {
29       k1[i] *= h;
30       tmp[i] = y[i] + k1[i]/2.;
31     }
32
33     (*f)(x+h/2., tmp, k2);
34     for( int i=0; i<n; i++ ) {
35       k2[i] *= h;
36       tmp[i] = y[i] + k2[i]/2.;
37     }
38
39     (*f)(x+h/2., tmp, k3);
40     for( int i=0; i<n; i++ ) {
41       k3[i] *= h;
42       tmp[i] = y[i] + k3[i];
43     }
44
45     (*f)(x+h, tmp, k4);
46     for( int i=0; i<n; i++ ) {
47       k4[i] *= h;
48       y[i] += (k1[i]+2.*(k2[i]+k3[i])+k4[i])/6.;
49     }
50   }
51
52   // the derivative function called by rk4
53   // problem-specific
54   void deriv( double x, Vector<double>&y, Vector<double>&dy )
55   {
56     extern double omegasq;
57     double c = 3 + x*x;
58     double csq = c*c;
```

```
59
60      dy[0] = y[1];
61      dy[1] = -omegasq*y[0]/csq;
62      dy[2] = y[3];
63      dy[3] = -(y[0]+omegasq*y[2])/csq;
64    }
65
66   int main()
67   {
68      cout << scientific << setprecision(15);
69
70      extern double omegasq;
71      Vector<double> y(4);
72      double x;
73
74      int N = 100;
75      double h = 1./N;
76
77      omegasq = 100;
78      int iterations =0;
79      do {
80        // initial value
81        y[0] = 0;
82        y[1] = 1;
83        y[2] = 0;
84        y[3] = 0;
85        x = 0;
86
87        // integrate to x=1
88        for( int i=0; i<N; i++ ) {
89          rk4( h, x,  y, deriv );
90          x += h;
91        }
92
93        // Newton-Raphson adjustment of omegasq
94        omegasq -= y[0]/y[2];
95        ++iterations;
96
97        cout << "after iteration " << iterations << " omega = " << sqrt(omegasq) << endl;
98      }while( fabs(y[0])>1.e-15 && iterations <10 );
99      if( iterations >= 10 ) {
100         cout << "shooting method failed to converge" << endl;
101      }
102      cout << "omega " << sqrt(omegasq) << endl;
103   }
```

10.2 As a linear difference equation, we are interested in the roots $\lambda(x)$ of

$$\lambda^2 - 2x\lambda + 1 = 0$$

which comes from the substitution $T_k = \lambda^k$. The roots are a complex conjugate pair with magnitude 1 (see (6.20)). For $x \in (-1, 1)$ the roots are different, but at $x = -1$ and $x = +1$ they are degenerate: $\lambda(-1) = -1, -1$ and $\lambda(+1) = 1, 1$. As a linear difference equation, we conclude that the Chebyshev recurrence formula is stable for x in the open interval $(-1, 1)$, and that it is not numerically stable at -1 and $+1$. Since $T_0(x) = 1$, and $T_1(x) = x$, for this unstable case one can use $T_k(\pm 1) = (\pm 1)^k$.

For this recurrence, having a conservative polynomial means that the recurrence could be applied stably for either ascending index k, or descending k. For example, let $x = 0.1$. In the forward direction we calculate

$$T_0 = 1$$
$$T_1 = 0.1$$

$$\vdots$$

$$T_{999} = 0.447\,270\,449\,285\,074\,6$$

$$T_{1000} = 0.934\,642\,576\,731\,592\,7.$$

Then, reversing the direction with T_{999} and T_{1000} as the starting values,

$$T_{1000} = 0.934\,642\,576\,731\,592\,7$$

$$T_{999} = 0.447\,270\,449\,285\,074\,6$$

$$\vdots$$

$$T_1 = 0.100\,000\,000\,000\,000\,6$$

$$T_0 = 1.000\,000\,000\,000\,002.$$

Only $\approx 20\epsilon$ relative error crept into this calculation.

10.3 Let $c = x/\sqrt{1-x^2}$, and consider a limit in which $0 \ll m \ll \ell$. Then,

$$\frac{P_\ell^{m+1}}{m^2} + \frac{2c\,P_\ell^m}{m} + \frac{\ell^2 P_\ell^{m-1}}{m^2} = 0.$$

If $P_\ell^m \approx (m\lambda)^m$ then

$$m^{m-1}\lambda^{m+1} + 2cm^{m-1}\lambda^m + \ell^2 m^{m-3}\lambda^{m-1} = 0,$$

with m positive and ℓ larger than m,

$$\lambda^2 + 2c\lambda < -1.$$

Clearly for some c this inequality will possess solutions $|\lambda| > 1$, which is bad, but the main point is that the solutions are independent of m. This means that for any solution $\lambda \neq 0$ the recurrence will have parasitic growth like $(m\lambda)^m$, which is unstable.

For the other recurrence consider $\ell \gg m, 1$, so

$$(\ell+1)P_{\ell+1}^m - 2\ell x\,P_\ell^m + (\ell-1)P_{\ell-1}^m \approx 0.$$

If $P_\ell^m = (\lambda^\ell/\ell)$ then

$$\lambda^2 - 2x\lambda + 1 = 0$$

this difference equation has roots of magnitude 1 and they are equal if $x = \pm 1$, like the Chebyshev Problem 10.2. If they are different, then the parasitic solution for P_ℓ^m will grow like $\pm 1/\ell$, which is stable. If the roots are the same, then $\ell(\lambda^\ell/\ell) = \pm 1$ is a solution to the P_ℓ^m difference equation, which is still stable.

Recurrence calculations are are started with

$$P_m^m(x) = (-1)^m (2m-1)!!(1-x^2)^{m/2},$$

where $(2m-1)!!$ is the product of odd integers smaller than or equal to $(2m-1)$. With

$\ell = m$ and $P_{m-1}^m = 0$ (recall $0 \le |m| \le \ell$) the ℓ recurrence gives a second starting value

$$P_{m+1}^m = (2m + 1)x P_m^m.$$

Now, with these two values the ℓ recurrence can find all P_ℓ^m. Another useful relation is

$$P_\ell^{-m}(x) = (-1)^m \frac{(\ell - m)!}{(\ell + m)!} P_\ell^m(x).$$

Table S.3 Unstable associated Legendre recurrence: $P_{10}^m(0.99)$; Problem 10.3.

m	ascending m ⇑	descending m ⇓
10	5̌.710983878583193e+00	2.043275884884458̌e+00
9	-1.436̌866755672054e+01	-1.433955472726954̌e+01
8	4.7615̌18163175970e+01	4.761492107358553̌e+01
7	-9.87442̌0173708131e+01	-9.874419906282803̌e+01
6	1.41971984797̌9832e+02	1.4197198476578̌15e+02
5	-1.48217985894̌9003e+02	-1.482179858944301̌e+02
4	1.139983723918385̌e+02	1.13998372391829̌7e+02
3	-6.379629855553310̌e+01	-6.37962985555328̌6e+01
2	2.473372162613616̌e+01	2.473372162613614̌e+01
1	-5.838159341807792̌e+00	-5.838159341807789̌e+00
0	5.200890424821921̌e-01	5.2008904248219̌18e-01
-1	5.307417583461629̌e-02	5.307417583461627̌e-02
-2	2.0819630998431̌10e-03	2.0819630998431̌10e-03
-3	5.163518077856520̌e-05	5.163518077856546̌e-05
-4	9.415053563485933̌e-07	9.41505356348̌7090e-07
-5	1.360136647362321̌e-08	1.36013664736̌9250e-08
-6	1.628524519253040̌e-10	1.628524519846395̌e-10
-7	1.665690568678356̌e-12	1.665690̌641143890e-12
-8	1.487414614082461̌e-14	1.48742̌7689028596e-14
-9	1.178802490118905̌e-16	1.182646956130474̌e-16
-10	8.398512878929346̌e-19	3.261529116540806e-18

As a test, let $x = 0.99$ and compute P_{10}^0 with the ℓ recurrence, beginning with $P_0^0 = 1$ and $P_1^0 = x$ ($m = 0$ corresponds to Legendre polynomials). Then,

$$P_0^0 = 1$$
$$P_1^0 = 0.99$$
$$\vdots$$
$$P_9^0 = 0.597\,245\,524\,499\,472\,3̌$$
$$P_{10}^0 = 0.520\,089\,042\,482\,19\,1̌\,7.$$

Here, check marks indicate the first decimal digit where the calculated result differs from a very high-precision calculation: the calculation is stable. Reversing this calculation with the computed P_9^0 and P_{10}^0 as starting values,

$$P_{10}^0 = 0.520\,089\,042\,482\,191\,7$$
$$P_9^0 = 0.597\,245\,524\,499\,472\,3$$

$$\vdots$$

$$P_1^0 = 0.989\,999\,999\,999\,998\,0$$
$$P_0^0 = 0.999\,999\,999\,999\,996\,9$$

only a few tens of ϵ relative error accumulated over this calculation.

Alternatively, one can calculate P_{10}^{10}, and P_{10}^9 (after P_9^9). This provides starting values for a descending m recurrence. Or, P_{10}^{-10} can be calculated from P_{10}^{10}, and P_{10}^{-9} can be calculated from P_{10}^9 with the m parity equation, and these values permit use of the ascending m recurrence. The calculated sequences are shown in Table S.3 with the underlined variables being the trustworthy starting values. A check mark indicates the first decimal digit that disagrees with a very high-precision calculation. Both ascending and descending sequences do well while $|m|$ is descending. Once the sequences pass $m = 0$, the errors grow rapidly (unstably).

10.4 Introduce the variable $v(t) = \dot{x}(t)$ to make a coupled system of first-order equations

$$\frac{d}{dt}\begin{pmatrix} x \\ v \end{pmatrix} = \begin{pmatrix} 0 & 1 \\ -k & 0 \end{pmatrix}\begin{pmatrix} x \\ v \end{pmatrix},$$

with initial condition $(x_0,\ v_0)^{\mathrm{T}}$.

The forward Euler equation applied to this system is

$$\begin{pmatrix} x \\ v \end{pmatrix}^{n+1} = \begin{pmatrix} x \\ v \end{pmatrix}^n + h\begin{pmatrix} 0 & 1 \\ -k & 0 \end{pmatrix}\begin{pmatrix} x \\ v \end{pmatrix}^n$$
$$= \begin{pmatrix} 1 & h \\ -kh & 1 \end{pmatrix}\begin{pmatrix} x \\ v \end{pmatrix}^n$$
$$\begin{pmatrix} x \\ v \end{pmatrix}^n = \begin{pmatrix} 1 & h \\ -kh & 1 \end{pmatrix}^n \begin{pmatrix} x \\ v \end{pmatrix}^0.$$

The matrix appearing in the last equation can be diagonalized:

$$\frac{i}{2\sqrt{k}}\begin{pmatrix} 1 & 1 \\ i\sqrt{k} & -i\sqrt{k} \end{pmatrix}\begin{pmatrix} 1+ih\sqrt{k} & \\ & 1-ih\sqrt{k} \end{pmatrix}\begin{pmatrix} -i\sqrt{k} & -1 \\ -i\sqrt{k} & 1 \end{pmatrix},$$

and with the diagonal form $\mathbf{A} = \mathbf{X}\mathbf{\Lambda}\mathbf{X}^{-1}$ the power $\mathbf{A}^n = \mathbf{X}\mathbf{\Lambda}^n\mathbf{X}^{-1}$ follows. To facilitate taking the power of the eigenvalues, express them in the polar form

$$1 \pm ih\sqrt{k} = \sqrt{1+kh^2}\,e^{i\theta}$$
$$\theta = \arctan h\sqrt{k}.$$

Then,

$$\begin{pmatrix} x \\ v \end{pmatrix}^n = (1 + kh^2)^{n/2} \begin{pmatrix} \cos n\theta & \frac{1}{\sqrt{k}} \sin n\theta \\ -\sqrt{k} \sin n\theta & \cos n\theta \end{pmatrix} \begin{pmatrix} x_0 \\ v_0 \end{pmatrix}.$$

For the coordinate equation in particular, this gives

$$x(hn) = (1 + kh^2)^{n/2} \left[x_0 \cos n\theta + \frac{v_0}{\sqrt{k}} \sin n\theta \right].$$

The forward Euler approximation error can be characterized as an *amplitude error* and a *phase error*. The former is the factor of

$$(1 + kh^2)^{n/2} \approx \left(1 + \frac{kh^2}{2} - \frac{(kh^2)^2}{8} + \frac{(kh^2)^3}{16} - \cdots \right)^n,$$

which clearly has no analog in the exact solution. The phase error is the discrepancy between θ and $h\sqrt{k}$:

$$\theta = \arctan h\sqrt{k} \approx (h\sqrt{k}) - \frac{(h\sqrt{k})^3}{3} + \frac{(h\sqrt{k})^5}{5} - \frac{(h\sqrt{k})^7}{7} + \cdots .$$

Both the phase and amplitude errors go to zero as $h\sqrt{k} \to 0$.

For the backward Euler method one has

$$\begin{pmatrix} x \\ v \end{pmatrix}^{n+1} = \begin{pmatrix} x \\ v \end{pmatrix}^n + h \begin{pmatrix} 0 & 1 \\ -k & 0 \end{pmatrix} \begin{pmatrix} x \\ v \end{pmatrix}^{n+1}$$

$$\begin{pmatrix} 1 & -h \\ kh & 1 \end{pmatrix} \begin{pmatrix} x \\ v \end{pmatrix}^{n+1} = \begin{pmatrix} x \\ v \end{pmatrix}^n$$

$$\begin{pmatrix} 1 & -h \\ kh & 1 \end{pmatrix}^n \begin{pmatrix} x \\ v \end{pmatrix}^n = \begin{pmatrix} x \\ v \end{pmatrix}^0.$$

The matrix appearing in the last equation can be diagonalized:

$$\frac{-i}{2\sqrt{k}} \begin{pmatrix} -1 & -1 \\ i\sqrt{k} & -i\sqrt{k} \end{pmatrix} \begin{pmatrix} 1 + ih\sqrt{k} & \\ & 1 - ih\sqrt{k} \end{pmatrix} \begin{pmatrix} -i\sqrt{k} & 1 \\ -i\sqrt{k} & -1 \end{pmatrix},$$

with eigenvalues identical to the forward Euler case. Simplifying,

$$\begin{pmatrix} x \\ v \end{pmatrix}^n = (1 + kh^2)^{-n/2} \begin{pmatrix} \cos n\theta & \frac{1}{\sqrt{k}} \sin n\theta \\ -\sqrt{k} \sin n\theta & \cos n\theta \end{pmatrix} \begin{pmatrix} x_0 \\ v_0 \end{pmatrix}$$

$$x(hn) = (1 + kh^2)^{-n/2} \left[x_0 \cos n\theta + \frac{v_0}{\sqrt{k}} \sin n\theta \right].$$

For this problem, the backward Euler and forward Euler solutions have the same phase error. The amplitude errors are different, with backward Euler damping the solution.

The modified Euler method is

$$
\begin{pmatrix} x \\ v \end{pmatrix}^{n+1} = \begin{pmatrix} x \\ v \end{pmatrix}^{n} + h \begin{pmatrix} 0 & 1 \\ -k & 0 \end{pmatrix} \left[\begin{pmatrix} x \\ v \end{pmatrix}^{n} + \frac{h}{2} \begin{pmatrix} 0 & 1 \\ -k & 0 \end{pmatrix} \begin{pmatrix} x \\ v \end{pmatrix}^{n} \right]
$$

$$
= \begin{pmatrix} 1 - \frac{kh^2}{2} & h \\ -kh & 1 - \frac{kh^2}{2} \end{pmatrix} \begin{pmatrix} x \\ v \end{pmatrix}^{n}
$$

$$
\begin{pmatrix} x \\ v \end{pmatrix}^{n} = \begin{pmatrix} 1 - \frac{kh^2}{2} & h \\ -kh & 1 - \frac{kh^2}{2} \end{pmatrix}^{n} \begin{pmatrix} x \\ v \end{pmatrix}^{0}.
$$

For this IVP, Heun's method is identical to the modified Euler method. The matrix appearing in the last equation can be diagonalized:

$$
\frac{i}{2\sqrt{k}} \begin{pmatrix} 1 & 1 \\ i\sqrt{k} & -i\sqrt{k} \end{pmatrix} \begin{pmatrix} 1 - \frac{kh^2}{2} + ih\sqrt{k} & \\ & 1 - \frac{kh^2}{2} - ih\sqrt{k} \end{pmatrix} \begin{pmatrix} -i\sqrt{k} & -1 \\ -i\sqrt{k} & 1 \end{pmatrix},
$$

so

$$
\begin{pmatrix} x \\ v \end{pmatrix}^{n} = \left(1 + \frac{(kh^2)^2}{4} \right)^{n/2} \begin{pmatrix} \cos n\phi & \frac{1}{\sqrt{k}} \sin n\phi \\ -\sqrt{k} \sin n\phi & \cos n\phi \end{pmatrix} \begin{pmatrix} x_0 \\ v_0 \end{pmatrix}
$$

$$
x(hn) = \left(1 + \frac{(kh^2)^2}{4} \right)^{n/2} \left[x_0 \cos n\phi + \frac{v_0}{\sqrt{k}} \sin n\phi \right]
$$

$$
\phi = \arctan \frac{h\sqrt{k}}{1 - \frac{kh^2}{2}} \approx h\sqrt{k} + \frac{(h\sqrt{k})^3}{6} - \frac{(h\sqrt{k})^5}{20} - \frac{(h\sqrt{k})^7}{56} + \cdots .
$$

The phase error is similar in magnitude to the phase error of the forward and backward Euler methods, but the amplitude error is an order of h improved. These methods are compared in Figure S.30 with parameters $k = 4\pi^2$, $h = 0.02$, $x_0 = 0$, and $v_0 = -2$.

10.5 To evaluate the order of accuracy of this method using the available treatment of first-order systems, one must first find an equivalent first-order representation. One solution is the implicit time-staggered system

$$
\begin{pmatrix} x^{n+1} \\ v^{n+1/2} \end{pmatrix} = \begin{pmatrix} x^{n} \\ v^{n-1/2} \end{pmatrix} + h \begin{pmatrix} v^{n+1/2} \\ f(x^n) \end{pmatrix}.
$$

It is easily shown that this gives the Verlet equation, and that it is second order.
To model the harmonic oscillator, consider

$$
\begin{pmatrix} x^{n+1} \\ x^{n} \end{pmatrix} = \begin{pmatrix} 2 - kh^2 & -1 \\ 1 & 0 \end{pmatrix} \begin{pmatrix} x^{n} \\ x^{n-1} \end{pmatrix}
$$

$$
\begin{pmatrix} x^{n} \\ x^{n-1} \end{pmatrix} = \begin{pmatrix} 2 - kh^2 & -1 \\ 1 & 0 \end{pmatrix}^{n} \begin{pmatrix} x^{0} \\ x^{-1} \end{pmatrix}.
$$

The matrix has eigenvalue decomposition

$$
\frac{1}{\lambda_+ - \lambda_-} \begin{pmatrix} \lambda_+ & \lambda_- \\ 1 & 1 \end{pmatrix} \begin{pmatrix} \lambda_+ & \\ & \lambda_- \end{pmatrix} \begin{pmatrix} 1 & -\lambda_- \\ -1 & \lambda_+ \end{pmatrix},
$$

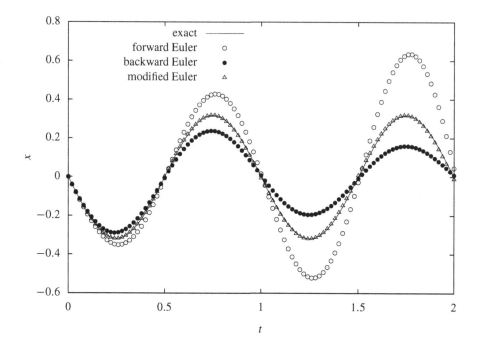

Figure S.30 Harmonic oscillator; Problem 10.4.

where $\lambda_\pm$ are the eigenvalues

$$\lambda_\pm = 1 - \frac{kh^2}{2} \pm i\sqrt{kh^2 - \frac{k^2h^4}{4}}$$

$$= e^{\pm i\theta}$$

$$\theta = \arctan \frac{\sqrt{kh^2 - \frac{k^2h^4}{4}}}{1 - \frac{kh^2}{2}} \approx \sqrt{k}h + \frac{(\sqrt{k}h)^3}{24} + \frac{3(\sqrt{k}h)^5}{640} + \cdots .$$

Note that the magnitude of the eigenvalues is 1, and this means that there is no amplitude error. After some simplification,

$$x^n = (\cot\theta \sin n\theta + \cos n\theta) x^0 - (\csc\theta \sin n\theta) x^{-1}.$$

The absence of amplitude error makes this method very popular for the solution of Newton's laws of motion when the force is a function of coordinates only, as is commonly the case in molecular dynamics.

The Verlet method displays the periodicity of the true solution only when $h < 2/\sqrt{k}$. If h is larger, the eigenvalues are real, so θ is imaginary, and the solution will display exponential growth. The forward Euler, backward Euler, and modified Euler schemes of Problem 10.4 have periodic behavior for any step size h.

10.6 This is a boundary value problem, where k is to be chosen so $S(b; k) - 100Z(b; k) =$

0, where $b = 365$, and we will begin from $a = 0$. If k is to be updated by Newton–Raphson, then

$$k^{(n+1)} = k^{(n)} - \frac{S(b;\ k^{(n)}) - 100Z(b;\ k^{(n)})}{\frac{d}{dk}S(b;\ k^{(n)}) - 100\frac{d}{dk}Z(b;\ k^{(n)})},$$

so one must augment the system of ODEs with equations for S_k and Z_k.

If

$$\frac{d}{dt}\begin{pmatrix} S \\ Z \\ R \end{pmatrix} = \begin{pmatrix} B - \beta SZ - \delta S \\ (\beta - \alpha)SZ + \zeta R \\ \delta S - \zeta R \end{pmatrix}$$

then, differentiating with respect to k,

$$\frac{d}{dt}\begin{pmatrix} S_k \\ Z_k \\ R_k \end{pmatrix} = \begin{pmatrix} -\beta(S_kZ + SZ_k) - \delta S_k \\ (\beta - \alpha)(S_kZ + SZ_k) + \zeta R_k \\ \delta S_k - \zeta R_k \end{pmatrix},$$

and since none of $S(a)$, $Z(a)$, or $R(a)$ depend on k, the initial values are zero:

$$S_k(a) = Z_k(b) = R_k(a) = 0.$$

At each impulsive eradication event (time $t = 5j$, $j = 1, ..., 73$) one has

$$S(t_+) = S(t_-)$$
$$R(t_+) = R(t_-)$$
$$Z(t_+) = (1 - k)Z(t_-),$$

and differentiating with respect to k,

$$S_k(t_+) = S_k(t_-)$$
$$R_k(t_+) = R_k(t_-)$$
$$Z_k(t_+) = (1 - k)Z_k(t_-) - Z(t_-).$$

The solution strategy is to compute the six quantities at time $365+$. If convergence has not been achieved, update k with the Newton–Raphson equation and re-evaluate. Each evaluation will consist of 73 cycles of (i) integrating the ODE systems for a period of 5 days, followed by (ii) application of the impulsive jump equations.

The solution occurs with $k = 4.468 \times 10^{-2}$, but the population is decimated: $S = 3.647$ (Figure S.31).

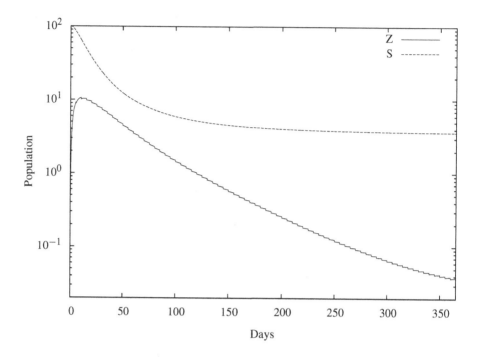

Figure S.31 Infectious disease model; Problem 10.6.

Chapter 11

11.1 Forward, as a derivation:

$$dv = -\gamma v dt + \sigma dW$$
$$d(ve^{\gamma t}) = \sigma e^{\gamma t} dW$$
$$ve^{\gamma t} - v_0 = \sigma \int_0^t dW_s e^{\gamma s}$$
$$v(t) = v_0 e^{-\gamma t} + \sigma \int_0^t dW_s e^{-\gamma(t-s)}.$$

Backward, by differentiation.

$$v(t) = v_0 e^{-\gamma t} + \sigma \int_0^t dW_s e^{-\gamma(t-s)}.$$
$$dv = -\gamma v_0 e^{-\gamma t} dt + \gamma \sigma \int_0^t dW_s e^{-\gamma(t-s)} dt + \sigma dW$$
$$dv = -\gamma v dt + \sigma dW.$$

Here, Leibniz' integral rule is used to differentiate the integral.

11.2 Forward, as a derivation:

$$dS = \mu S dt + \sigma S dW$$
$$dS/S = d \ln S = \mu dt + \sigma dW.$$

This looks like $d \ln(S)$ on the left, but Itô's formula for $d \ln(S)$ gives

$$d \ln(S) = \left(\frac{1}{S} f - \frac{1}{2S^2} g^2 \right) dt + \frac{1}{S} g dW$$

$$= \left(\mu - \frac{\sigma^2}{2} \right) dt + \sigma dW$$

$$\int_0^t d \ln S = \ln \frac{S}{S_0} = \left(\mu - \frac{\sigma^2}{2} \right) t + \sigma W$$

$$S(t) = S_0 \exp \left[\left(\mu - \frac{\sigma^2}{2} \right) t + \sigma W \right].$$

Backwards,

$$S(t) = S_0 \exp[(\mu - \sigma^2/2)t + \sigma W(t)].$$

Apply Itô's formula

$$dS = S(\mu - \sigma^2/2)dt + \frac{1}{2} S \sigma^2 dt + S \sigma dW$$

$$dS = \mu S dt + S \sigma dW.$$

11.3

$$R(t, s) = \mathbb{E} \frac{(W(t + h) - W(t))(W(s + h) - W(s))}{h^2}$$

$$= \frac{\min(t + h, s + h) + \min(t, s) - \min(t + h, s) - \min(t, s + h)}{h^2}$$

$$= \begin{cases} 0 & s + h < t \\ \frac{h+s-t}{h^2} & s < t < s + h < t + h \\ \frac{h+t-s}{h^2} & t < s < t + h < s + h \\ 0 & t + h < s \end{cases}$$

$$= \frac{1}{h} \max \left(0, 1 - \frac{|t - s|}{h} \right),$$

so it is wide sense stationary.

$$S(\omega) = \frac{1}{\sqrt{2\pi}} \int_{-\infty}^{+\infty} e^{i\omega\tau} \frac{\max\left(0, 1 - \frac{|\tau|}{h}\right)}{h} d\tau$$

$$= \sqrt{\frac{2}{\pi}} \int_{0}^{\infty} \cos\omega\tau \frac{\max\left(0, 1 - \frac{|\tau|}{h}\right)}{h} d\tau$$

$$= \sqrt{\frac{2}{\pi}} \int_{0}^{h} \cos\omega\tau \frac{h - \tau}{h^2} d\tau$$

$$= \sqrt{\frac{2}{\pi}} \frac{1 - \cos h\omega}{h^2\omega^2}$$

$$= \frac{1}{\sqrt{2\pi}} \left(\frac{\sin\frac{h\omega}{2}}{\frac{h\omega}{2}}\right)^2.$$

As $h \to 0$, the factor in parenthesis approaches 1, and this spectral density approaches that of the white noise. And, as $h \to \infty$ this density approaches zero.

11.4 (i) We have $\mathbb{E}d W_i^n = 0$ if n is odd, and $(n - 1)!!\Delta t_i^{n/2}$ if n is even (11.29). Since dW occurs only once in problem (i), the expectation of this product is zero.
(ii) $\mathbb{E}d W_i d W_j = dt_i\delta(t_i - t_j)$, so dW_0 and dW_1 cannot combine because $t_1 < t_0$ always, but dW_3 can combine with either dW_0 or dW_1, depending on the time interval.

$$\mathbb{E}\int_0^t dW_0 f(s_0) \int_0^{s_0} dW_1 g(s_1) \int_0^t ds_2 p(s_2) \int_0^{s_2} dW_3 q(s_3)$$

$$= \mathbb{E}\int_0^t ds_2 p(s_2) \int_0^{s_2} dW_3 q(s_3) \int_0^t dW_0 f(s_0) \int_0^{s_0} dW_1 g(s_1)$$

$$= \mathbb{E}\int_0^t ds_2 p(s_2) \int_0^{s_2} dW_3 q(s_3) \Big[\int_{s_2}^t dW_0 f(s_0) \int_0^{s_2} dW_1 g(s_1)$$

$$+ \int_{s_2}^t dW_0 f(s_0) \int_{s_2}^{s_0} dW_1 g(s_1) + \int_0^{s_2} dW_0 f(s_0) \int_0^{s_0} dW_1 g(s_1)\Big]$$

$$= \mathbb{E}\int_0^t ds_2 p(s_2) \int_0^{s_2} dt_3 q(s_3) g(s_3) \int_{s_2}^t dW_0 f(s_0)$$

$$+ \mathbb{E}\int_0^t ds_2 p(s_2) \int_0^{s_2} ds_3 q(s_3) f(s_3) \int_0^{s_3} dW_1 g(s_1),$$

and this is zero because $\mathbb{E}d W^1 = 0$.
(iii)

$$\mathbb{E}\int_0^t dW_0 f(s_0) \int_0^t dW_1 p(s_1) \int_0^{s_1} dW_2 q(s_2) = \mathbb{E}\int_0^t ds_2 f(s_2)q(s_2) \int_0^t dW_1 p(s_1),$$

and this is zero because $\mathbb{E}d W^1 = 0$.

(iv) Use $\mathbb{E}\, dW_0 dW_3 = dt_0 \delta(t_0 - t_3)$ and change the order of integration.

$$\mathbb{E}\int_0^t dW_0 f(s_0)\int_0^{s_0} ds_1 g(s_1)\int_0^t ds_2 p(s_2)\int_0^{s_2} dW_3 q(s_3)$$

$$= \int_0^t ds_2 p(s_2)\int_0^{s_2} ds_3 q(s_3) f(s_3)\int_0^{s_3} ds_1 g(s_1),$$

and this is not necessarily zero.

11.5

$$S(t) = S(0)\exp\left[(\mu - \sigma^2/2)t + \sigma W(t)\right]$$

$$\mathbb{E}S(t) = S(0)\exp\left[(\mu - \sigma^2/2)t\right]\left(\sum_{n=0}^{\infty}\mathbb{E}\frac{(\sigma W(t))^n}{n!}\right)$$

$$= S(0)\exp\left[(\mu - \sigma^2/2)t\right]\left(\sum_{n=0,\text{even}}^{\infty}\frac{\sigma^n t^{n/2}(n-1)!!}{n!}\right)$$

$$= S(0)\exp\left[(\mu - \sigma^2/2)t\right]\left(\sum_{n=0,\text{even}}^{\infty}\frac{\sigma^n t^{n/2}}{n!!}\right)$$

$$= S(0)\exp\left[(\mu - \sigma^2/2)t\right]e^{t\sigma^2/2}$$

$$= S(0)\exp[\mu t]$$

$$\mathbb{E}(S(t) - \mathbb{E}S(t))^2 = \mathbb{E}S(0)^2\exp[(2\mu - \sigma^2)t + 2\sigma W(t)] - 2\mathbb{E}S(0)^2 e^{2\mu t}$$

$$= S(0)^2\exp[(2\mu - \sigma^2)t]e^{2t\sigma^2} - S(0)^2 e^{2\mu t}$$

$$= S(0)^2\exp[(2\mu + \sigma^2)t] - S(0)^2 e^{2\mu t}$$

$$= S(0)^2 e^{2\mu t}(e^{\sigma^2 t} - 1).$$

With $S(0) = 1$, $\mu = 4$, and $\sigma = 1$, the mean at $t = 1$ is $e^4 \approx 55$ and the standard deviation is $e^4\sqrt{e-1} \approx 71.5$. About 68% of the paths should have an answer within one standard deviation of the mean, and about 95% within two standard deviations of the mean. The spread of values observed is on the order of two standard deviations, which is roughly consistent. This analysis assumes that S is Gaussian distributed, although it is not.

Based on the calculated standard deviation, to achieve 1% error,

$$\frac{1}{100}S(0)e^{\mu t} = \frac{S(0)e^{\mu t}\sqrt{e^{\sigma^2 t} - 1}}{\sqrt{n}}$$

$$n = 10^4(e^{\sigma^2 t} - 1),$$

so about 17 200 paths would have to be sampled to achieve the indicated error. Again, this assumes a Gaussian distribution.

References

[1] IEEE standard for floating-point arithmetic. *IEEE Std 754-2008*, pages 1–58, 2008.

[2] M. Abramowitz and I. A. Stegun. *Handbook of Mathematical Functions*. Dover, New York, 1972.

[3] A. Aitken. On Bernoulli's numerical solution of algebraic equations. *Proc. Roy. Soc. Edinburgh*, **46**:289–305, 1926.

[4] A. Arce, J. Martínez-Argeitos, and A. Soto. VLE for water + ethanol + 1-octanol mixtures. Experimental measurements and correlations. *Fluid Phase Equil.*, **122**:117–129, 1996.

[5] L. Bairstow. *Applied Aerodynamics*. Longmans, Green, and Co., New York, 2nd edition, 1939.

[6] F. Bashforth and J. C. Adams. *An Attempt to Test the Theories of Capillary Action by Comparing the Theoretical and Measured Forms of Drops of Fluid*. Cambridge University Press, 1883.

[7] H. Bateman. On a set of Kernels whose determinants form a Sturmian sequence. *Bull. Am. Math. Soc.*, **18**(4):175–179, 1912.

[8] F. L. Bauer. Optimally scaled matrices. *Numer. Math.*, **5**:73–87, 1963.

[9] F. L. Bauer. Computational graphs and rounding error. *SIAM J. Num. Anal.*, **11**(1):87–96, 1974.

[10] G. H. Behforooz. A comparison of the E(3) and not-a-knot cubic splines. *Appl. Math. Comput.*, **72**:219–223, 1995.

[11] G. H. Behforooz and N. Papamichael. End conditions for cubic spline interpolation. *J. Inst. Maths. Applics.*, **23**:355–366, 1979.

[12] E. Beltrami. On bilinear functions. *Giornale di Matematiche*, **11**, 1893. English translation by D. Boley, 1995, published in *3rd International Workshop on SVD and Signal Processing, Aug. 96, Leuven, Belgium*, Elsevier, 1996.

[13] C. Benoit. Note sur une méthode de résolution des équations normales provenant de l'application de la méthode des moindres carrés à un système d'équations linéaires en nombre inférieur à celui des inconnues. Application de la méthode à la résolution d'un system défini d'équations linéaires (procédé du Commandant Cholesky). *Bull. Géodésique*, **2**:67–77, 1924.

[14] N. Berglund and B. Gentz. Beyond the Fokker-Planck equation: pathwise control of noisy bistable systems. *J. Phys. A: Math. Gen.*, **35**:2057–2091, 2002.

[15] J.-P. Berrut and L. N. Trefethen. Barycentric Lagrange interpolation. *SIAM Rev.*, **46**(3):501–517, 2004.

[16] M. Bhatti and P. Bracken. The calculation of integrals involving B-splines by means of recursion relations. *Appl. Math. Comput.*, **172**:91–100, 2006.

[17] A. Björck. *Numerical Methods for Least Squares Problems*. SIAM, Philadelphia, 1996.

[18] F. Black and M. Scholes. The pricing of options and corporate liabilities. *J. Pol. Econ.*, **81**:637–654, 1973.

[19] G. Boole. *A Treatise on the Calculus of Finite Differences*. Macmillan and Co., London, 3rd edition, 1880.

[20] J. Boothroyd. Algorithm 27: Rearrange the elements of an array section according to a permutation of the subscripts. *Comp. J.*, **10**(2):310, 1967.

[21] M. Born and R. Oppenheimer. Zur Quantentheorie der Molekeln. *Ann. Phys.*, **84**(20):457–484, 1927.

[22] G. E. P. Box and M. E. Muller. A note on the generation of random normal deviates. *Ann. Math. Stat.*, **29**(2):610–611, 1958.

[23] S. F. Boys. Electronic wave functions. I. A general method of calculation for the stationary states of any molecular system. *Proc. Roy. Soc. London.*, **200A**(1063):542–554, 1950.

[24] A. Brandt. Multi-level adaptive technique (MLAT) for fast numerical solution of boundary-value problems. In H. Cabannes and R. Temam, editors, *Proceedings of the Third International Conference on Numerical Methods in Fluid Mechanics, Paris, 1972*. Lecture notes in Physics 18, pages 82–89. Springer, New York, 1973.

[25] A. Brandt. Multi-level adaptive solutions to boundary-value problems. *Math. Comp.*, **31**(138):333–390, 1977.

[26] W. L. Briggs, V. E. Henson, and S. F. McCormick. *A Multigrid Tutorial*. SIAM, 2nd edition, 2000.

[27] R. Brown. A brief account of microscopical observations made in the months of June, July, and August 1827, on the particles contained in the pollen of plants; and on the general existence of active molecules in organic and inorganic bodies. *Phil. Magazine*, **4**(21):161–173, 1828.

[28] C. G. Broyden. The convergence of a class of double-rank minimization algorithms 1. General considerations. *J. Inst. Math. Applics.*, **6**:76–90, 1970.

[29] R. Bulirsch. Bemerkungen zur Romberg-integration. *Numer. Math.*, **6**:6–16, 1964.

[30] R. Bulirsch. Die Mehrzielmethode zur numerischen Lösung von nichtlinearen Randwertproblemen und Aufgaben der optimalen Steuerung. Report of the Carl-Cranz-Gesellschaft, Oberpfaffenhofen, Germany, 1971.

[31] R. Bulirsch and J. Stoer. Fehlerabschätzungen und extrapolation mit rationalen funktionen bie verfahren vom Richardson-typus. *Numer. Math.*, **6**:413–427, 1964.

[32] C. A. Cantrell. Review of methods for linear least-squares fitting of data and application to atmospheric chemistry problems. *Atmos. Chem. Phys.*, **8**:5477–5487, 2008.

[33] G. Cardan. *Artis Magnae, Sive de Regulis Algebraicis Liber Unus*. J. Petreius, Nuremberg, 1545.

[34] A. L. Cauchy. Cours d'analyse de l'école royale polytechnique, 1821. In *Œuvres Complètes d'Augustin Cauchy. series 2*, Vol. 3. Gauthier-Villars, Paris, 1897.

[35] A. L. Cauchy. Résumé des leçons données a l'école royale polytechnique sur le calcul infinitésimal. In *Œuvres Complètes d'Augustin Cauchy series. 2*, Vol. 4. Gauthier-Villars, Paris, 1899.

[36] P. Chebyshev. Théorie des mécanismes connus sous le nom parallelogrammes. *Mémoires des Savants Étrangers Présentes à l'Academie de Saint-Pétersbourg*, **7**:539–586, 1854.

[37] A. J. Chorin. Hermite expansions in Monte-Carlo computation. *J. Comput. Phys.*, **8**:472–482, 1971.

[38] E. B. Christoffel. Sur une classe particulière de fonctions entières et de fractions continues. *Ann. Mat. Pura Appl.*, **8**(2):1–10, 1877.

[39] C. W. Clenshaw. Chebyshev series for mathematical functions. In *National Physical Laboratory Mathematical Tables, Vol. 5*. Her Majesty's Stationery Office, London, 1962.

[40] P. Colella. A direct Eulerian MUSCL scheme for gas dynamics. *SIAM J. Sci. Stat. Comput.*, **6**(1):104–117, 1985.

[41] E. U. Condon. The theory of complex spectra. *Phys. Rev.*, **36**(7):1121–1133, 1930.

[42] J. W. Cooley and J. W. Tukey. An algorithm for the machine calculation of complex Fourier series. *Math. Comp.*, **19**(90):297–301, 1965.

[43] R. Cotes. *Harmonia Mensurarum*. Cambridge, 1722.

[44] A. J. Cox and N. J. Higham. Stability of Householder QR factorization for weighted least squares problems. In D. F. Griffiths, D. J. Higham, and G. A. Watson, editors, *Numerical Analysis 1977, Proceedings of the 17th Dundee Biennial Conference*, pages 57–73. Addison Wesley Longman, Harlow, 1998.

[45] H. Cramér. On some classes of series used in mathematical statistics. In *Sixth Scandinavian Mathematical Congress*, pages 399–425. Copenhagen, 1925.

[46] J. H. Curry, L. Garnett, and D. Sullivan. On the iteration of a rational function: Computer experiments with Newton's method. *Comm. Math. Phys.*, **91**(2):267–277, 1983.

[47] C. F. Curtiss and J. O. Hirschfelder. Integration of stiff equations. *Proc. Natl. Acad. Sci.*, **38**:235–243, 1952.

[48] G. Dahlquist. Convergence and stability in the numerical integration of ordinary differential equations. *Math. Scand.*, **4**:33–53, 1956.

[49] G. B. Dantzig. Programming of interdependent activities: II mathematical model. *Econometrica*, **17**(3/4):200–211, 1949.

[50] G. B. Dantzig. *Linear Programming and Extensions*. Princeton University Press, Princeton, New Jersey, 1963.

[51] W. C. Davidon. Variable metric method for minimization. Technical report, Argonne National Laboratory, 1959. report ANL-5990 (reprinted as *SIAM J. Opt.*, **1**(1):1–17, 1991).

[52] C. H. Davis. *Theory of the Motion of the Heavenly Bodies Moving about the Sun in Conic Sections: A Translation of Gauss's "Theoria Motus"*. Little, Brown and Co., Boston, 1857.

[53] C. de Boor. On calculating with B-splines. *J. Approx. Theory*, **6**:50–62, 1972.

[54] C. de Boor. *A Practical Guide to Splines*. Springer-Verlag, New York, 1978.

[55] C. de Boor. Convergence of the cubic spline interpolation problem with the not-a-knot condition. Technical report, Mathematics Research Center, University of Wisconsin, Madison, 1984. Technical Summary Report #2876.

[56] J. Demmel, M. Gu, S. Eisenstat, I. Slapničar, K. Verelić, and Z. Drmač. Computing the singular value decomposition with high relative accuracy. *Linear Alg. Appl.*, **299**(1):21–80, 1999.

[57] J. Demmel and K. Veselić. Jacobi's method is more accurate than QR. *SIAM J. Matrix Anal. Appl.*, **13**(4):1204–1245, 1992.

[58] C. Eckart and G. Young. The approximation of one matrix by another of lower rank. *Psychometrika*, **1**(3):211–218, 1936.

[59] L. Euler. Methodus generalis summandi progressiones. *Comment. Acad. Sci. Imp. Petrop.*, **6**:68–97, 1738.

[60] L. Euler. *Methodus Inveniendi Lineas Curvas Maximi Minimive Proprietate Gaudentes, Sive Solutio Problematis Isoperimetrici Latissimo Sensu Accepti*. Marc-Michel Bousquet, Lausanne & Geneva, 1744.

[61] R. P. Fedorenko. A relaxation method for solving elliptic difference equations. *USSR Comp. Math. and Math. Phys.*, **1**(4):1092–1096, 1962. First published in Russian in *Zh. Vych. Mat.* **1**(5):922–927, 1961.

[62] R. P. Fedorenko. The speed of convergence of one iterative process. *USSR Comp. Math. and Math. Phys.*, **4**(3):227–235, 1964. First published in Russian in *Zh. Vych. Mat.* **4**(3):559–564, 1964.

[63] L. Feigenbaum. Brook Taylor and the method of increments. *Arch. Hist. Ex. Sci.*, **34**(1-2):1–140, 1985.

[64] N. Ferguson. Wall Street lays another egg. *Vanity Fair*, page 190, December 2008.

[65] R. Fletcher. A new approach to variable metric algorithms. *Comput. J.*, **13**(3):317–322, 1970.

[66] R. Fletcher and M. J. D. Powell. A rapidly convergent descent method for minimization. *Comput. J.*, **6**(2):163–168, 1963.

[67] V. Fock. Näherungsmethode zur Lösung des quantenmechanischen Mehrkörperproblems. *Zeit. Phys. A*, **61**(1-2):126–148, 1930.

[68] G. E. Forsythe. Gauss to Gerling on relaxation. *Math. Tables Aids Comput.*, **5**(35):255–258, 1951.

[69] G. E. Forsythe and C. B. Moler. *Computer Solution of Linear Algebraic Systems*. Prentice-Hall, Englewood Cliffs, NJ, 1967.

[70] J. B. J. Fourier. Solution d'une question particulière au calcul des inégalités, second extrait. *Histoire de l'Académie des Sciences pour 1824*, pages xlvii–lv. Reprinted in *Œuvres de Fourier*, ed. G. Darboux, Vol. 2, pp. 325–328. Gauthier-Villars, Paris, 1890.

[71] L. Fox. Romberg integration for a class of singular integrands. *Comput. J.*, **10**(1):87–93, 1967.

[72] J. G. F. Francis. The QR transformation: a unitary analogue to the LR transformation – Part 1. *Comput. J.*, **4**(3):265–271, 1961.

[73] J. G. F. Francis. The QR transformation – Part 2. *Comput. J.*, **4**(4):332–345, 1962.

[74] C. G. Fraser. Isoperimetric problems in the variational calculus of Euler and Lagrange. *Historia Mathematica*, **19**:4–23, 1992.

[75] D. C. Fraser. *Newton's Interpolation Formulas*. C. & E. Layton, London, 1927.

[76] A. Galántai and C. J. Hegedüs. Hyman's method revisited. *J. Comp. Appl. Math.*, **226**(2):246–258, 2009.

[77] W. Gander and W. Gautschi. Adaptive quadrature – revisited. *BIT*, **40**(1):84–101, 2000.

[78] C. F. Gauss. Methodus nova integralium valores per approximationem inveniendi. *Comment. Soc. Regiae Sci. Gottingensis Recentiores*, **3**:39–76, 1814. reprinted in *Werke*, vol 3, pp. 163–196.

[79] C. F. Gauss. *Briefwechsel zwischen Carl Friedrich Gauss und Christian Ludwig Gerling*. Otto Elsner, Berlin, 1927.

[80] W. Gautschi. Construction of Gauss–Christoffel quadrature formulas. *Math. Comp.*, **22**(102):251–270, 1968.

[81] W. Gautschi. On the construction of Gaussian quadrature rules from modified moments. *Math. Comp.*, **24**(110):245–260, 1970.

[82] H. Geiringer. On the solution of linear equations by certain iteration methods. In Staff of the Department of Aeronautical Engineering and Applied Mechanics of the Polytechnic Institute of Brooklyn, editor, *Reissner Anniversary Volume. Contributions to Applied Mechanics*. Edwards Brothers, Ann Arbor, Michigan, 1949.

[83] W. M. Gentleman and G. Sande. Fast Fourier transforms: for fun and profit. In *AIEE-IRE Proceedings of the November 7-10, 1966, Fall Joint Computer Conference*, pages 563–578, 1966.

[84] S. Gerschgorin. Über die abgrenzung der eigenwerte einer matrix. *Izv. Akad. Nauk. SSSR Otd. Fiz.-Mat. Nauk.*, **7**:749–754, 1931.

[85] W. Givens. Computation of plane unitary rotations transforming a general matrix to triangular form. *J. SIAM*, **6**(1):26–50, 1958.

[86] S. K. Godunov. Finite difference method for numerical computation of discontinuous solutions of the equations of fluid dynamics. *Math. Sbornik*, **47**(3):271–306, 1959. Translated from Russian by I. Bohachevsky.

[87] D. Goldfarb. A family of variable-metric methods derived by variational means. *Math. Comp.*, **24**(109):23–26, 1970.

[88] D. Goldfarb and A. Idnani. Dual and primal-dual methods for solving strictly convex quadratic programs. In J. Hennart, editor, *Numerical Analysis: Proceedings of the Third IIMAS Workshop held at Cocoyoc, Mexico, Jan. 19–23, 1981*, pages 226–239. Springer-Verlag, New York, 1982.

[89] D. Goldfarb and A. Idnani. A numerically stable dual method for solving strictly convex quadratic programs. *Math. Programming*, **27**:1–33, 1983.

[90] G. Golub. Numerical methods for solving linear least squares problems. *Numer. Math.*, **7**(3):206–216, 1965.

[91] G. Golub and W. Kahan. Calculating the singular values and pseudoinverse of a matrix. *J. SIAM B, Num. Anal.*, **2**(2):205–224, 1965.

[92] G. H. Golub and C. Reinsch. Singular value decomposition and least squares solution. *Numer. Math.*, **14**(5):403–420, 1970.

[93] G. H. Golub and C. F. van Loan. *Matrix Computations*. John's Hopkins University Press, Baltimore, 1983.

[94] G. H. Golub and J. H. Welsch. Calculation of Gauss quadrature rules. *Math. Comp.*, **23**(106):221–221, 1969.

[95] J. P. Gram. Ueber die entwickelung reeller functionen in reihen mittelst der methode der kleinsten quadrate. *J. Reine Angew. Math.*, **94**:41–73, 1883.

[96] F. J. Harris. On the use of windows for harmonic analysis with the discrete Fourier transform. *Proc. IEEE*, **66**:51–83, 1978.

[97] H. L. Harter. Method of least-squares and some alternatives. *Int. Stat. Rev.*, **42**(2):147–174, 1974. continued in 42(3):235–264, 1974; 43(1):1–44, 1975; 43(2):125–190, 1975; 43(3):269–278, 1975; and 44(1):113-159, 1976.

[98] D. R. Hayes and L. Rubin. A proof of the Newton–Cotes quadratures formulas with error term. *Amer. Math. Mon.*, **77**(10):1065–1072, 1970.

[99] W. J. Hehre, R. F. Stewart, and J. A. Pople. Self-consistent molecular-orbital methods. I. Use of Gaussian expansions of Slater-type atomic orbitals. *J. Chem. Phys.*, **51**(6):2657–2664, 1969.

[100] M. T. Heideman, D. H. Johnson, and C. S. Burrus. Gauss and the history of the Fast Fourier Transform. *Arch. Hist. Exact Sci.*, **34**(3):265–277, 1985.

[101] C. Hermite. Sur la formulè d'interpolation de Lagrange. *J. Reine Angew. Math.*, **84**:70–79, 1878.

[102] G. Herzberg. The dissociation energy of the hydrogen molecule. *J. Molec. Spectry.*, **33**(1):147–168, 1970.

[103] M. R. Hestenes and E. Stiefel. Methods of conjugate gradients for solving linear systems. *J. Res. Nat. Bur. Standards*, **49**(6):409–436, 1952.

[104] N. J. Higham. *Accuracy and Stability of Numerical Algorithms*. SIAM, Philadelphia, 2nd edition, 2002.

[105] N. J. Higham. The numerical stability of barycentric Lagrange interpolation. *IMA J. Numer. Anal.*, **24**(4):547–556, 2004.

[106] N. J. Higham and D. J. Higham. Large growth factors in Gaussian elimination with pivoting. *SIAM J. Matrix Anal. Appl.*, **10**(2):155–164, 1989.

[107] E. Hille. A class of reciprocal functions. *Ann. Math.*, **27**(4):427–464, 1926.

[108] F. S. Hillier and G. J. Lieberman. *Operations Research*. Holden-Day, San Francisco, 2nd edition, 1974.

[109] C. A. R. Hoare. Algorithm 64: Quicksort. *Comm. ACM*, **4**(7):321, 1961.

[110] C. A. R. Hoare. Quicksort. *Comp. J.*, **5**(1):10–15, 1962.

[111] W. G. Horner. A new method of solving numerical equations of all orders, by continuous approximation. *Phil. Trans. Roy. Soc. London*, **109**:308–335, 1819.

[112] A. S. Householder. Unitary triangularization of a nonsymmetric matrix. *J. ACM*, **5**(4):339–342, 1958.

[113] R. E. Howitt. Agricultural and environmental policy models: calibration, estimation, and optimization. online at `http://agecon.ucdavis.edu/people/faculty/richard-howitt/docs/master.pdf`, 2005.

[114] J. Hubbard, D. Schleicher, and S. Sutherland. How to find all roots of complex polynomials by Newton's method. *Invent. Math.*, **146**:1–33, 2001.

[115] T. E. Hull, T. F. Fairgrieve, and P.-T. P. Tang. Implementing complex elementary functions using exception handling. *ACM Trans. Math. Softw.*, **20**(2):215–244, 1994.

[116] M. A. Hyman. Eigenvalues and eigenvectors of general matrices. In *12th National Meeting of the Association for Computing Machinery, Houston, Texas*, 1957. (presentation).

[117] K. Itô. On stochastic differential equations. *Memoirs Amer. Math. Soc.*, **4**:1–51, 1951.

[118] C. G. J. Jacobi. Über eine neue Auflösungsart der bei der Methode der kleinsten Quadrate vorkommenden lineären Gleichungen. *Astron. Nach.*, **22**(20):297–306, 1845. Translated by G. W. Stewart, *On a new way of solving the linear equations that arise in the method of least squares*, IMA Preprint Series #951, 1992.

[119] C. G. J. Jacobi. Über ein leichtes Verfahren die in der Theorie der Säcularstörungen verkommenden Gleichungen numerisch aufzolösen. *J. Reine Angew. Math.*, pages 51–94, 1846.

[120] M. Jankowski and H. Woźniakowski. Iterative refinement implies numerical stability. *BIT*, **17**(3):303–311, 1977.

[121] C. Jordan. Mémoire sur la réduction et la transformation des systèmes quadratiques. *J. Math. Pure. Appl.*, **19**:397–422, 1874.

[122] W. Kahan. Interval arithmetic options in the proposed IEEE floating point arithmetic standard. In K. L. E. Nickel, editor, *Interval Mathematics 1980*, pages 99–128. Academic Press, New York, 1980.

[123] W. Kahan. Mathematics written in sand. *Proceedings of the Joint Statistical Meeting of the American Statistical Association*, pages 12–26, 1983. Version 22 is available online as `http://www.cs.berkeley.edu/~wkahan/MathSand.pdf`.

[124] B. Kallemov and G. H. Miller. A second-order strong method for the Langevin equations with holonomic constraints. *SIAM J. Sci. Comput.*, **33**(2):653–676, 2011.

[125] L. V. Kantorovich. On an efficient method of solving some classes of extremal problems. *Dokl. Akad. Nauk SSSR*, **28**:212–215, 1940.

[126] L. V. Kantorovich. Functional analysis and applied mathematics. *Usp. Mat. Nauk*, 3(6):89–185, 1948. (in Russian). Translated by C. D. Benster as National Bureau of Standards report 1509, 1952.

[127] W. Karush. Minima of functions of several variables with inequalities as side conditions. Master's thesis, The University of Chicago, 1939.

[128] A. Khintchine. Korrelationstheorie der stationären stochsachen Prozesse. *Math. Ann.*, **109**(1):604–615, 1934.

[129] J. Kiefer. Sequential mimimax search for a maximum. *Proc. Amer. Math. Soc.*, 4(3):502–506, 1953.

[130] H. F. King and M. Dupuis. Numerical integration using Rys polynomials. *J. Comp. Phys.*, **21**(2):144–165, 1976.

[131] P. E. Kloeden and E. Platen. *Numerical Solution of Stochastic Differential Equations.* Springer, New York, 3rd edition, 1999.

[132] T. Kojima. On the limits of the roots of an algebraic equation. *Tohoku Math J.*, **11**:119–127, 1917.

[133] W. Kołos and L. Wolniewicz. Improved theoretical ground-state energy of the hydrogen molecule. *J. Chem. Phys.*, **49**(1):404–410, 1968.

[134] D. König. Vonalrendszerek és determinánsok. *Math. és Term. tud. Ért.*, **33**:433–444, 1915.

[135] H. W. Kuhn and A. W. Tucker. Nonlinear programming. In *Proceedings of the Second Berkeley Symposium on Mathematics, Statistics, and Probability*, pages 481–492. University of California Press, 1981.

[136] R. Kupferman, G. A. Pavliotis, and A. M. Stuart. Itô versus Stratonovich white-noise limits for systems with inertia and colored multiplicative noise. *Phys. Rev. E*, **70**(037120):1–9, 2004.

[137] W. Kutta. Beitrag zur naherungsweisen Integration totaler Differentialgleichungen. *Z. Math. und Phys.*, **46**:435–453, 1901.

[138] J. L. Lagrange. Leçon sur le calcul des fonctions, 1806. In J.-A. Serret, editor, *Œuvres de Lagrange*, Vol. 10. Gauthier-Villars, Paris, 1867.

[139] J. L. Lagrange. Théorie des fonctions analytiques, contenant les principles du calcul différentiel, degagés de tout considération d'infiniment petits, d'évanouissans, de limites et de fluxions, et réduits à l'analyse algébrique des quantités finies, 1797. In J.-A. Serret, editor, *Œuvres de Lagrange*, Vol. 9. Gauthier-Villars, Paris, 1867.

[140] J. L. Lagrange. Leçons élémentaires sur les mathématiques données a l'École Normale, 1795. In J.-A. Serret, editor, *Œuvres de Lagrange*, Vol. 7. Gauthier-Villars, Paris, 1877.

[141] P. Langevin. Sur la thèorie du mouvement Brownien. *C. R. Hebdomadaires des Séances de l'Académie des Sciences*, **146**:530–533, 1908.

[142] P. Läuchli. Jordan-Elimination und Ausgleichung nach kleinsten Quadraten. *Numer. Math.*, **3**:226–240, 1961.

[143] C. L. Lawson and R. J. Hanson. *Solving Least Squares Problems*. SIAM, Philadelphia, 1995.

[144] G. W. Leibniz. *Mathematische Schriften*, Vol. 2. Georg Olms Verlag, Hildesheim, Germany, 1849. reprinted in 1962. Letter to the Marquis de l'Hôpital dated April 28, 1693, pages 236–241.

[145] F. Leja. Sur certaines suites liées aux ensembles plans et leur application à la représentation conforme. *Ann. Polon. Math.*, **4**:8–13, 1957.

[146] R. J. LeVeque. *Finite Volume Methods for Hyperbolic Problems*. Cambridge University Press, 2002.

[147] P. Lévy. Wiener's random function, and other Laplacian random functions. In *Proceedings of the Second Berkeley Symposium on Mathematical Statistics and Probability*, pages 171–187. University of California Press, 1951.

[148] R. Lipschitz. De explicatione per series trigonometricas instituenda functionum unius variablis arbitrariarum, et praecipue earum, quae per variablis spatium finitum valorum maximorum et minimorum numerum habent infintum disquisitio. *Z. Angew. Math.*, **63**:296–308, 1864.

[149] P.-O. Löwdin. On the non-orthogonality problem connected with the use of atomic wave functions in the theory of molecules and crystals. *J. Chem. Phys.*, **18**(3), 1950.

[150] J. N. Lyness. Notes on the adaptive Simpson quadrature routine. *J. ACM*, **16**(3):483–495, 1969.

[151] J. K. L. MacDonald. Successive approximations by the Rayleigh–Ritz variation method. *Phys. Rev.*, **43**(10):830–833, 1933.

[152] H. Maehly. Zur iterativen Auflösung algebraischer Gleichungen. *Z. Agnew. Math. Physik*, **5**:260–263, 1954.

[153] P. C. Mahalanobis. On the generalized distance in statistics. *Proc. Nat. Inst. Sci. India*, **2**:49–55, 1936.

[154] G. Marsaglia and T. A. Bray. A convenient method for generating normal variables. *SIAM Rev.*, **6**(3):260–264, 1964.

[155] S. F. McCormick. An algebraic interpretation of multigrid methods. *SIAM J. Num. Anal.*, **19**(3):558–560, 1982.

[156] W. M. McKeeman. Algorithm 145: Adaptive numerical integration by Simpson's rule. *Comm. ACM*, **5**(12):604, 1962.

[157] W. M. McKeeman. Certification of algorithm 145 adaptive numerical integration by Simpson's rule. *Comm. ACM*, **6**(4):167–168, 1963.

[158] L. E. McMurchie and E. R. Davidson. One- and two-electron integrals over Cartesian Gaussian functions. *J. Comp. Phys.*, **26**(2):218–231, 1978.

[159] D. McQuarrie. *Quantum Chemistry*. University Science Books, Mill Valley, California, 1983.

[160] E. Meijering. A chronology of interpolation: from ancient astronomy to modern signal and image processing. *Proc. IEEE*, **90**(3), 2002.

[161] A. Melman. Modified Gershgorin disks for companion matrices. *SIAM Rev.*, **54**:355–373, 2012.

[162] A. Miele, E. E. Cragg, R. R. Iyer, and A. V. Levy. Use of the augmented penalty function in mathematical programming problems, part 1. *J. Opt. Theory. Appl.*, **8**(2):115–130, 1971.

[163] G. N. Mil'shtein. A theorem on the order of convergence of mean-square approximations of solutions of system of stochastic differential equations. *Theory Prob. Appl.*, **32**:738–741, 1988.

[164] G. N. Milstein and M. V. Tretyakov. *Stochastic Numerics for Mathematical Physics*. Springer, New York, 2004.

[165] E. H. Moore. On the reciprocal of the general algebraic matrix. *Bull. Am. Math. Soc.*, **26**(9):394–395, 1920.

[166] G. E. Moore. Cramming more components onto integrated circuits. *Electronics*, **38**(8):114–117, 1965.

[167] D. D. Morrison, J. D. Riley, and J. F. Zancarano. Multiple shooting method for two-point boundary value problems. *Comm. ACM*, **5**(12):613–614, 1962.

[168] F. R. Moulton. *New Methods in Exterior Ballistics*. The University of Chicago Press, Chicago, 1926.

[169] P. Munz, I. Hudea, J. Imad, and R. J. Smith. When zombies attack! Mathematical modeling of an outbreak of zombie infection. In T. M. Tchuenche and C. Chiyaka, editors, *Infectious Disease Modelling Research Progress*, pages 133–150. Nova Science, Hauppauge, NY, 2010.

[170] L. Neal and G. Poole. A geometric analysis of Gaussian elimination. II. *Linear Alg. Appl.*, **173**:239–264, 1992.

[171] E. H. Neville. Iterative interpolation. *J. Indian Math. Soc.*, **20**:87–120, 1933.

[172] S. S. Oren. Self-scaling variable metric (SSVM) algorithms. Part II: Implementation and experiments. *Management Sci.*, **20**(5):863–874, 1974.

[173] S. S. Oren and D. G. Luenberger. Self-scaling variable metric (SSVM) algorithms. Part I: Criteria and sufficient conditions for scaling a class of algorithms. *Management Sci.*, **20**(5):845–862, 1974.

[174] S. S. Oren and E. Spedicato. Optimal conditioning of self-scaling variable metric algorithms. *Math. Programming*, **10**:70–90, 1976.

[175] E. E. Osborne. On the least squares solutions of linear equations. *J. ACM*, **8**(4):628–636, 1961.

[176] Parameśvara. *Siddhāntadīpikā*. ca. 1380/1460. Reprinted in T. S. Kuppanna Sastri, *Mahābhāskarīya of Bhāskarācārya with the Bhāsya of Godindasvāmin and the supercommentary Siddhāntadīpikā of Parameśvara*, Government Oriental Library, Madras, 1957.

[177] W. Pauli. Über den Zusammenhang des Abschlusses der Elektronengruppen im Atom mit der Komplexstruktur der Spektren. *Zeit. Phys.*, **31**(1):765–783, 1925.

[178] G. Peano. Residuo in formulas de quadratura. *Mathesis, Quatrième Série*, **4**:5–10, 1914.

[179] K. Pearson. On lines and planes of closest fit to systems of points in space. *Phil. Mag. Series 6*, **2**:559–572, 1901.

[180] R. Penrose. A generalized inverse for matrices. *Math. Proc. Cambridge Phil. Soc.*, **51**(3):406–413, 1955.

[181] G. Peters and J. H. Wilkinson. Practical problems arising in the solution of polynomial equations. *J. Inst. Math. Applics.*, **8**(1):16–35, 1971.

[182] E. Platen and W. Wagner. On a Taylor formula for a class of Itô processes. *Prob. Math. Stat.*, **3**(1):37–51, 1982.

[183] K. Plofker. The secant method of iterative approximation in a fifteenth-century Sanskrit text. *Historia Mathematica*, **23**:246–256, 1996.

[184] G. Poole and L. Neal. A geometric analysis of Gaussian elimination. I. *Linear Alg. Appl.*, **149**:249–272, 1991.

[185] G. Poole and L. Neal. The rook's pivoting strategy. *J. Comp. Appl. Math.*, **123**:353–369, 2000.

[186] M. J. D. Powell and J. K. Reid. On applying Householder transformations to linear least squares problems. Technical report, Mathematics Branch, Theoretical Physics Division, Atomic Energy Research Establishment, Hartwell, 1968. T.P. 322.

[187] J. Raphson. *Analysis Æquationum Universalis, seu, Ad Æquationes Algebraicas Resolvendas Methodus Generalis, et Expedita: ex Nova Infinitarum Serierum Doctrina Deducta ac Demonstrata*. Abel Swale, London, 1690.

[188] L. Rayleigh. On the calculation of Chladni's figures for a square plate. *Phil. Mag. Series 6*, **22**(128):225–229, 1911.

[189] H. H. Reamer, R. H. Olds, B. H. Sage, and W. N. Lacey. Phase equilibria in hydrocarbon systems: methane-carbon dioxide system in the gaseous region. *Ind. Eng. Chem.*, **36**:88–90, 1944.

[190] L. Reichel. Newton interpolation at Leja points. *BIT*, **30**:332–346, 1990.

[191] L. F. Richardson. The approximate arithmetical solution by finite differences of physical problems involving differential equations, with an application to the stresses in a masonry dam. *Phil. Trans. Roy. Soc. London*, **210**:307–357, 1911.

[192] L. F. Richardson. The deferred approach to the limit. Part I – Single lattice. *Phil. Trans. Roy. Soc. London A*, **226**:299–349, 1927.

[193] W. Ritz. Über eine neue Methode zur Lösung gewisser Variationsprobleme der mathematischen Physik. *J. Reine Angew. Math.*, **135**(1):1–61, 1909.

[194] J. J. Rodríguez. An improved bit-reversal algorithm for the fast Fourier transform. In *IEEE International Conference on Acoustics, Speech, and Signal Processing, 1988*, Vol. 3, pages 1407–1410, 1988.

[195] M. Rolle. *Démonstration d'une Méthode Pour Résoudre les Égalités de Tous les Degrés, Suivies de Deux Autres Méthodes, Dont la Première Donne les Moyens de Résoudre ces Mêmes Égalités par la Géométrie, et al Second pour Résoudre Plusieurs Questions de Diophante qui n'ont point été Résolues*. Paris, 1691.

[196] W. Romberg. Vereinfachte numerische Integration. *Det Kongelige Norske Videnskabers Selskab Forhandlinger (Trondheim)*, **28**(7):30–37, 1955.

[197] C. C. J. Roothaan. New developments in molecular orbital theory. *Rev. Mod. Phys.*, **23**(2):69–89, 1951.

[198] C. C. J. Roothaan and P. S. Bagus. Atomic self-consistent field calculations by the expansion method. In B. Alder, S. Fernbach, and M. Rotenberg, editors, *Methods in Computational Physics*, Vol. 2, pages 47–94. Academic Press, New York, 1963.

[199] B. J. Rosenberg, W. C. Ermler, and I. Shavitt. Ab initio SCF and CI studies on the ground state of the water molecule. II. Potential energy and property surfaces. *J. Chem. Phys.*, **65**(10):4072–4080, 1976.

[200] B. J. Rosenberg and I. Shavitt. Ab initio SCF and CI studies on the ground state of the water molecule. I. Comparison of the CGTO and STO basis sets near the Hartree-Fock limit. *J. Chem. Phys.*, **63**(5):2162–2174, 1975.

[201] H. H. Rosenbrock. An automatic method for finding the greatest or least value of a function. *Comput. J.*, **3**(3):175–184, 1960.

[202] E. Rouché. Mémoire sur la série de Lagrange. *J. de l'École Polytechnique, Paris*, **39**:217–219, 1862.

[203] C. Runge. Über die numerische Auflösung von Differentialgleichungen. *Math. Ann.*, **46**(2):167–178, 1895.

[204] C. Runge. Über empirische Funktionen und die Interpolation zwischen äquidistanten Ordinaten. *Z. Math. Phys.*, **46**:224–243, 1901.

[205] T. W. F. Russell and M. M. Denn. *Introduction to Chemical Engineering Analysis*. Wiley, New York, 1972.

[206] H. Rutishauser. Solution of eigenvalue problems with the LR-transformation. *Nat. Bur. Standards Appl. Math. Ser.*, **49**:47–81, 1958.

[207] J. Rys, M. Dupuis, and H. F. King. Computation of electron repulsion integrals using the Rys quadrature method. *J. Comp. Chem.*, **4**(2):154–157, 1983.

[208] R. A. Sack and A. F. Donovan. An algorithm for Gaussian quadrature given modified moments. *Numer. Math.*, **18**(5):465–478, 1972.

[209] H. B. Schlegel and J. J. W. McDouall. Do you have SCF stability and convergence problems? In C. Ögretir and I. G. Csizmadia, editors, *Computational Advances in Organic Chemistry*, pages 167–185. Kluwer Academic, Netherlands, 1991.

[210] E. Schmidt. Zur Theorie der linearen und nichtlinearen Integralgleichungen. *Math. Ann.*, **63**:433–476, 1907.

[211] H. Schnieder. The concepts of irreducibility and full indecomposability of a matrix in the works of Frobenius, König and Markov. *Linear Alg. Appl.*, **18**(2):139–162, 1977.

[212] I. J. Schoenberg. Contributions to the problem of approximation of equidistant data by analytic functions. *Quart. Appl. Math.*, **4**:45–99, 112–141, 1946.

[213] E. Schrödinger. An undulatory theory of the mechanics of atoms and molecules. *Phys. Rev.*, **28**(6):1029–1070, 1926.

[214] L. Seidel. Über ein Verfahren, die Gleichungen, auf welche die Methode der kleinsten Quadrate führt, sowie lineäre Gleichungen überhaupt, durch successive Annäherung aufzulösen. *Abh. Math.-Phys. Cl. Kön. Bayr. Akad. Wiss.*, **11**(3):81–108, 1873.

[215] D. F. Shanno. Conditioning of quasi-Newton methods for function minimization. *Math. Comp.*, **24**(111):647–656, 1970.

[216] I. Shavitt. The Gaussian function in calculations of statistical mechanics and quantum mechanics. In B. Alder, S. Fernbach, and M. Rotenberg, editors, *Methods in Computational Physics*, Vol. 2, pages 1–45. Academic Press, New York, 1963.

[217] T. Simpson. *Mathematical Dissertations on a Variety of Physical and Analytical Subjects*. T. Woodward, London, 1743.

[218] J. C. Slater. The theory of complex spectra. *Phys. Rev.*, **34**(10):1293–1322, 1929.

[219] J. C. Slater. Atomic shielding constants. *Phys. Rev.*, **36**(1):57–64, 1930.

[220] S. Smale. Newton's method estimates from data at one point. In R. Ewing, K. Gross, and C. Martin, editors, *The Merging of Disciplines: New Directions in Pure, Applied, and Computational Mathematics*, pages 185–196. Springer-Verlag, New York, 1986.

[221] D. E. Smith. *A Source Book in Mathematics*. Dover, New York, 1959.

[222] P. Stein and R. L. Rosenberg. On the solution of linear simultaneous equations by iteration. *J. London Math. Soc.*, **1**(2):111–118, 1948.

[223] G. W. Stewart. On the early history of the singular value decomposition. *SIAM Rev.*, **35**(4):551–566, 1993.

[224] T. J. Stieltjes. Quelques recherches sur la thèorie des quadratures dites méchaniques. *Ann. Sci. École Norm. Paris Sér. 3*, **1**:409–426, 1884. Reprinted in *Oeuvres*, vol I, pp. 377–396.

[225] J. Stoer and R. Bulirsch. *Introduction to Numerical Analysis*. Springer, New York, 2nd edition, 1993.

[226] C. Störmer. Méthode d'intégration numérique des équations différentielles ordinaires. *Comptes Rendus du Congres International des Mathématiciens, Strasbourg, 22–30 Step. 1920*, pages 243–257, 1921.

[227] G. Strang. *Linear Algebra and its Applications*. Academic Press, New York, 2nd edition, 1980.

[228] H. Taketa, S. Huzinaga, and K. O-Ohata. Gaussian-expansion method for molecular integrals. *J. Phys. Soc. Japan*, **21**(11):2313–2324, 1966.

[229] D. Talay and L. Tubaro. Expansion of the global error for numerical schemes solving stochastic differential equations. *Stochastic Anal. Appl.*, **8**(4):483–509, 1990.

[230] B. Taylor. *Methodus Incrementorum Directa et Inversa*. London, 1715.

[231] R. S. Varga. *Matrix Iterative Analysis*. Prentice-Hall, Englewood Cliffs, NJ, 1962.

[232] S. Venit. The convergence of Jacobi and Gauss-Seidel iteration. *Math. Mag.*, **48**(3):163–167, 1975.

[233] L. Verlet. Computer "experiments" on classical fluids. I. Thermodynamic properties of Lennard-Jones molecules. *Phys. Rev.*, **159**(1):98–103, 1967.

[234] R. von Mises and H. Pollaczek-Geiringer. Praktische Verfahren der Gleichungsauflösung. *Z. Angew. Math. Mech.*, **9**(1&2):58–77,152–164, 1929.

[235] W. Wagner and E. Platen. Approximation of Itô integral equations. Preprint ZIMM, Akad. Wissenschaften, DDR, Berlin, 1978.

[236] E. Waring. Problems concerning interpolations. *Phil. Trans. Roy. Soc. London*, **69**:59–67, 1779.

[237] T. Weddle. On a new and simple rule for approximating to the area of a figure by means of seven equidistant ordinates. *Cambridge and Dublin Math. J.*, **9**:79–80, 1854.

[238] H. Wielandt. Das Iterationsverfahren bei nicht selbstadjungierten linearen Eigenwertaufgaben. *Math. Z.*, **50**(1), 1944.

[239] N. Wiener. Generalized harmonic analysis. *Acta Math.*, **55**(1):117–258, 1930.

[240] N. Wiener. The homogeneous chaos. *Am. J. Math.*, **60**(4):897–936, 1938.

[241] H. S. Wilf. *Finite Sections of Some Classical Inequalities*. Springer-Verlag, New York, 1970.

[242] J. H. Wilkinson. *Rounding Errors in Algebraic Processes*. Prentice-Hall, Englewood Cliffs, NJ, 1963.

[243] J. H. Wilkinson. *The Algebraic Eigenvalue Problem*. Clarendon Press, Oxford, 1965.

[244] J. H. Wilkinson. The perfidious polynomial. In G. H. Golub, editor, *Studies in Numerical Analysis*, Vol. 24. Mathematics Association of America, Washington, DC, 1984.

[245] E. B. Wilson, J. C. Decius, and P. C. Cross. *Molecular Vibrations: the Theory of Infrared and Raman Vibrational Spectra*. McGraw-Hill, New York, 1955.

[246] D. York. Least-squares fitting of a straight line. *Can. J. Phys.*, **44**:1079–1086, 1966.

[247] T. J. Ypma. Historical development of the Newton–Raphson method. *SIAM Rev.*, **37**:531–551, 1995.

Index

A-norm, 101

absolute norm, 37
Adams–Bashforth methods, 284
Adams–Moulton methods, 284
adaptive
  hp, 291
  multistep methods, 289
  one-step methods, 288
  Simpson's rule integration, 257
additive noise, 302
adjoint, 116
Aitken's Δ^2 method, 179, 257
amplification factor, 6
amplitude error, 546
anticipatory, 307
  non-, 307
approximation error, 1, 13
Archimedes' method, 462
artificial variables
  linear programming, 195
associated Legendre polynomial, 300, 543
asymptotic regime, 161, 256, 266
autocorrelation, 152

B-spline, 235, 538
back substitution, 23, 44, 215
backward error, 17, 40, 49, 52, 81
backward Euler method, 284
Bairstow's method, 175
banana
  Rosenbrock's, 190
banded matrix, 228
basic variables
  linear programming, 192
basis vector, 45
Bernoulli
  numbers, 253
  polynomials, 253
bisection, 155
Björck pivoting, 49
Black–Scholes equation, 152, 302, 325
Boole's rule, 244
bottom solver, 122
Box–Muller transform, 305

bracketing, 182
Broyden–Fletcher–Goldfarb–Shanno, 188

Cauchy convergence criterion, 153
Cauchy sequence, 153
ceiling function, 167, 318
characteristic equation, 365
chasing the bulge, 77
Chebyshev interpolation, 347
Chebyshev polynomial, 101, 150, 300, 497
Cholesky decomposition, 49
Clenshaw's method, 347
code example
  `AdaptiveMS-test.cpp`, 434
  `AdaptiveMS.H`, 430
  `AdaptiveMS.cpp`, 431
  `AdaptiveSS-test.cpp`, 430
  `AdaptiveSS.H`, 428
  `AdaptiveSS.cpp`, 429
  `AdaptiveSimpson-test.cpp`, 428
  `AdaptiveSimpson.H`, 426
  `AdaptiveSimpson.cpp`, 427
  `CG-test.cpp`, 402
  `CG.H`, 400
  `CG.cpp`, 401
  `Cholesky-test.cpp`, 387
  `Cholesky.H`, 386
  `Cholesky.cpp`, 386
  `ConvexQP-test.cpp`, 426
  `ConvexQP.H`, 420
  `ConvexQP.cpp`, 421
  `FFT-test.cpp`, 407
  `FFT.H`, 405
  `FFT.cpp`, 406
  `GE-test.cpp`, 379
  `GE.H`, 375
  `GE.cpp`, 376
  `Givens.H`, 388
  `GoldenSection.H`, 408
  `GoldenSection.cpp`, 408
  `HFRopt.cpp`, 442
  `HQsort.H`, 384
  `HQsort.cpp`, 385
  `Householder-test.cpp`, 385

`Householder.H`, 380
`Householder.cpp`, 381
`Matrix.H`, 373
`QR-test.cpp`, 393
`QR.H`, 389
`QR.cpp`, 389
`RCP.H`, 371
`Rysfit.cpp`, 438
`SDEintegrate.cpp`, 434
`SVD-test.cpp`, 400
`SVD.H`, 393
`SVD.cpp`, 393
`Simplex-test.cpp`, 419
`Simplex.H`, 413
`Simplex.cpp`, 414
`VMM-test.cpp`, 413
`VMM.H`, 409
`VMM.cpp`, 410
`Vcycle-example.cpp`, 402
conjugate gradient, 490
data fitting in Mahalanobis norm, 520
FFT autocorrelation, 507
FFT convolution, 496
multigrid, 490
multigrid preconditioned conjugate gradient, 490
Newton interpolation, 498
cofactor, 364
colored noise, 303
column space, 72
companion matrix, 169, 367
complete pivoting, 30
complex numbers, 369
Euler's formula, 370
composite rule, 249
condition number, 6, 38, 501
conjugate, 94
conjugate gradients, 93
conservative polynomial, 282, 542
consistent
norm, 36
continuous
uniformly, 353
continuous function, 353
convergence
Cauchy, 153
convolution, 148
Cooley–Tukey algorithm, 141
Coulomb matrix, 332
covariance, 219, 303
matrix, 220

damped Jacobi, 118
damping, 226, 336
Davidon–Fletcher–Powell, 188
defective matrix, 57
deflation, 170
determinant, 30, 363

deviate, 305
diagonal matrix, 361
diagonalizable matrix, 56
discrete convolution, 148
discrete Fourier transform, 138
divided difference, 21, 131, 159, 242
divided differences, 132
do not use this, 24, 82, 95, 118, 125, 143, 213, 214, 230, 246, 261, 278
dot product, 361
dual, 202, 221

eigenvalue, 365
    Gerschgorin's theorem, 57
    inverse vector iteration, 62
    power method, 62
    QR algorithm, 70
    shifting, 63, 80
    spectrum, 57
    vector iteration, 62
eigenvector, 365
    inverse iteration, 62
    iteration, 62
    power method, 62
    QR algorithm, 70
elementary operation, 4, 8
equilibrate, 34
error
    amplitude error, 546
    approximation, 1, 13
    backward, 17, 40, 49, 52, 81
    experimental, 1, 9
    inherent, 9
    phase error, 546
    roundoff, 1
    sampling, 1, 312
error function, 151, 273, 339, 340
Euclidean norm, 37
Euler's formula, 102, 370
Euler–Maclaurin sum formula, 255
Euler–Maruyama
    strong analysis, 317
    weak analysis, 319
example
    Adams–Moulton, 285
    adaptive multistep method, 291
    adaptive Runge–Kutta, 288
    adaptive Simpson's method, 258
    Aitken's Δ^2 method, 179
    approximation versus algorithm error, 13
    Bairstow's method, 177
    bisection, 155
    Broyden–Fletcher–Goldfarb–Shanno variable metric method, 190
    Cholesky decomposition, 51
    conjugate gradient, 97
    cubic spline, 239

data fitting in norms L^1, L^2, and L^∞, 232
fast Fourier transform, 144
Gauss–Seidel iteration, 111
Gaussian elimination, 24, 28, 29
   complete pivoting, 31
   various pivoting choices, 35
Gaussian quadrature for a function with a
   convergent Taylor series, 266
Gaussian quadrature for a function without a
   convergent Taylor series, 265
Gerschgorin estimate, 59
Gerschgorin estimate with scaling, 61
Goldfarb–Idnani quadratic programming, 207
Hermite interpolation tableau, 137
Householder reduction, 47
integration with trapezoidal sum rule, 243
inverse vector iteration, 63
irreducibility, 110
iterative convergence, 154
iterative refinement, 53
Jacobi iteration, 106
least squares fit with experimental error in y, 223
multigrid, 123
multiple shooting method, 297
Neville interpolation tableau, 130
Newton interpolation tableau, 134
Newton–Raphson, 163
numerical trustworthiness, 6, 10
order of a convergence of a sequence, 155
order of convergence, 155
Peano kernel for integration, 250
Peano kernel for interpolation, 127
polynomial roots by complex Newton's method,
   174
propagation of errors from x and y to m and b,
   227
QR for eigenvalues, 77
quadratic formula, 16
regula falsi, 157
Romberg integration, 255
secant method, 160
simple shooting method, 293
simplex method phase I, 197
simplex method phase II, 192
simplex step with matrix modification, 201
singular value decomposition, 83
strong order of Euler–Maruyama, 317
truncated SVD, 87
vector iteration, 65
vector iteration for complex eigenvalues, 67
weak order of Euler–Maruyama, 319
exchange matrix, 332
expectation, 219, 303
experimental error, 1, 9

fast Fourier transform (FFT), 141
feasible basis

linear programming, 192
Fock matrix, 332
forward Euler method, 275, 306
Fourier transform
   discrete, 138
   fast, 141
Frobenius matrix, 24, 199, 369
Frobenius norm, 37
Frobenius normal matrix, 169, 281, 366
fundamental theorem of algebra, 102, 126, 262

Gauss–Seidel relaxation, 111
Gaussian basis functions, 337
Gaussian elimination, 23
Gaussian normal distribution, 219
   integral moments, 317
Gaussian quadrature, 259
   Gauss–Chebyshev, 263
   Gauss–Hermite, 264
   Gauss–Laguerre, 265
   Gauss–Legendre, 264
   Rys, 273, 345
geometric mean, 316
geometric series, 356
Gerschgorin's theorem, 57, 107, 169
Givens matrix, 75, 210
global discretization error, 276, 315
golden ratio, 160, 183, 185
golden section search, 184
Golub pivoting, 48
Gram–Schmidt orthogonalization, 64, 204, 216, 336
graph theory, 109
growth factor, 40

h-adaptive, 291
h.o.t., 168
Hartree–Fock–Roothaan equation, 332
Hermite interpolation, 136, 183, 249
Hermitian matrix, 362
Hessenberg matrix, 368
Hessian, 186
Heun's method, 277
Hilbert matrix, 87
Horner's scheme, 17, 132
Householder reduction, 43
hp-adaptive, 291
Hyman's method, 89

identity matrix, 362
improper integral, 271
induced norm, 36
infimum, 159
inherent error, 9
inner product, 361
integral remainder formula, 359
integration
   Boole's rule, 244
   Gaussian quadrature, 259
      Gauss–Chebyshev, 263

Gauss–Hermite, 264
Gauss–Laguerre, 265
Gauss–Legendre, 264
Newton–Cotes, 244
Romberg, 253
Simpson's 3/8 rule, 244
Simpson's rule, 244
trapezoidal sum rule, 244, 307
Weddle's rule, 244
interpolation
   Chebyshev, 347
   Hermite, 136, 183, 249
   Lagrange formula, 125
      barycentric form, 129
   modified Lagrange formula, 128
   Neville's method, 129, 255
   Newton's method, 131
irreducible, *see* reducible matrix
Itô calculus, 306
Itô's formula, 308
Itô–Taylor series, 309

Jacobi
   method for eigenvalues, 81
   relaxation, 105
Jacobi preconditioner, 103
Jordan block, 57
Jordan normal form, 57
Julia set, 172

Kantorovich's theorem, 164
Karush–Kuhn–Tucker, 202
knot, 236
Krylov space, 100

Löwdin decomposition, 222, 333
Lagrange interpolation formula, 125
   barycentric form, 129
   modified, 128
Lagrange multipliers, 201
Lagrange remainder formula, 169, 357
Langevin equation, 302
least squares, 22, 24, 46, 48, 49, 92, 105, 111, 203,
   213, 223, 230
least upper bound norm, 36
Lebesgue norm, 37, 241
left triangular matrix, 49, 368
Legendre polynomial
   associated, 300, 543
Legendre polynomials, 264
Leja points, 500
likelihood, 221
linear difference equation, 281
linear programming, 191
Lipschitz condition, 164, 277, 282, 315
Lipschitz continuous, 353
local discretization error, 276

machine precision, 4

Mahalanobis
   distance, 221
   norm, 221
matrix, 361
   banded, 228
   characteristic equation, 365
   cofactor, 364
   companion, 169, 367
   Coulomb, 332
   covariance, 220
   defective, 57
   determinant, 363
   diagonal, 361
   diagonalizable, 56
   eigenvalue, 365
   eigenvector, 365
   equilibrated, 34
   exchange, 332
   Fock, 332
   Frobenius, 24, 199, 369
   Frobenius normal, 169, 281, 366
   Givens, 75, 210
   Hermitian, 362
   Hessenberg, 368
   Hessian, 186
   Hilbert, 87
   identity, 362
   Jordan normal form, 57
   left triangular, 49, 368
   matrix, 43
   minor, 364
   modification, 199, 209
   norm, 36
   normal, 56
   overlap, 332
   permutation, 24
   positive, 113
   positive definite, 367
   pseudoinverse, 203, 216
   reducible, 79, 108
   right triangular, 23, 43, 368
   similar, 367
   skew Hermitian, 362
   symmetric, 362
   symmetric positive definite, 49, 81, 87, 93, 116,
      213, 220, 269, 333, 368, 519
   tridiagonal, 369
   Vandermonde, 54, 92, 219, 270
mean value theorem, 355
metric
   tensor, 186, 221
   variable, 185
minor matrix, 364
modification, 199, 209
modified Euler method, 277
modified Lagrange interpolation formula, 128

modified Newton–Raphson, 167
Moore–Penrose pseudoinverse, 216
multiindex notation, 318, 360
multiple shooting method, 295
multiplicative noise, 302
multivariate Taylor series, 318, 360

natural norm, 36
nested multiplication, 17
Neville's interpolation method, 129, 255
Newton map, 172
Newton's interpolation method, 131
Newton–Cotes methods, 244
Newton–Raphson, 162
    complex plane, 172
    Kantorovich's theorem, 164
    modified, 167
    Smale's theorem, 165
noise
    additive, 302
    colored, 303
    multiplicative, 302
    white Gaussian distributed, 302
nonanticipatory, 307
nonbasic variables
    linear programming, 192
norm, 36
    A, 101
    absolute, 37
    consistent, 36
    Euclidean, 37
    Frobenius, 37
    induced, 36
    least upper bound (lub), 36
    Lebesgue, 37, 241
    Mahalanobis, 221
    natural, 36
    Schur, 37
    submultiplicative, 36
    subordinate, 36
normal equations, 24, 49, 105, 111, 214
normal matrix, 56
numerically harmless, 10

ODE
    Adams–Bashforth methods, 284
    Adams–Moulton methods, 284
    backward Euler method, 284
    forward Euler method, 275, 306
    Heun's method, 277
    modified Euler method, 277
    multiple shooting method, 295
    Runge–Kutta method, 278
    simple shooting method, 292
order
    of accuracy, 245, 251, 276, 278, 311
        strong, 312
        weak, 312

of convergence, 154
of magnitude, 4
Oren–Sedicato, 189
orthogonal polynomials, 260
orthogonalization
    Gram–Schmidt, 64, 204, 216, 336
outer product, 361
overlap matrix, 332

paranoia, 454
Parseval's theorem, 144
partial pivoting, 30
partition of unity, 236
Peano kernel, 126, 251
permutation matrix, 24
phase error, 546
pivoting
    Björck, 49
    complete, 30
    Golub, 48
    partial, 30
    Powell–Reid, 48
    rook, 32
    scaled partial, 34
    trivial, 30
polar coordinates, 369
polynomial
    associated Legendre, 300
    characteristic, 365
    Chebyshev, 101, 150, 497
    conservative, 282, 542
    orthogonal, 260
    Wilkinson's, 19, 171
positive definite, 367
positive matrix, 113
Powell–Reid pivoting, 48
power method, 62
precondition, 103
    Jacobi, 103
principal vector, 57
probability density function, 219
prolongation, 115
pseudoinverse, 203, 216

QR algorithm for eigenvalues, 70
QR method, 43
quadratic
    equation, 16
    programming, 201

Rayleigh quotient, 62
Rayleigh's energy theorem, 144
reducible matrix, 79, 108
regula falsi, 157
residual, 41, 53, 93, 230
residual correction method, 52
restricted variables
    linear programming, 192
restriction, 115

Richardson extrapolation, 252, 286, 312
right triangular matrix, 23, 43, 368
Rolle's theorem, 126, 151, 168, 248, 354, 358, 472
Romberg integration, 253
rook pivoting, 32
Rosenbrock's banana, 190
Rouché's theorem, 124
roundoff error, 1
Runge–Kutta method, 278

sampling error, 1, 312
scaled partial pivoting, 34
Schur norm, 37
SDE
    Euler–Maruyama, 317–320
    Milstein, 318
    Milstein–Talay, 320
secant method, 159
series
    geometric, 356
    Taylor, 357
        common, 359
shifting
    eigenvalue, 63, 80
similarity transformation, 60, 367
simple shooting method, 292
simplex, 194
simplex method
    linear programming, 195
Simpson's 3/8 rule, 244
Simpson's rule, 244
singular value decomposition (SVD), 81
skew Hermitian matrix, 362
slack variables
    linear programming, 192
Slater determinant, 328
Smale's theorem, 165
smoother, 115
spectral radius, 56
spline
    basic (B), 235, 538
    natural cubic, 238
Störmer integration, 300
standard deviation, 6, 219, 304, 312
standard form
    linear programming, 192
steepest descents, 100
stencil, 217
Stieltjes' algorithm, 260
Stratonovich calculus, 307
strong row sum criterion, 107
submultiplicative, 36
subordinate norm, 36
subspace iteration, 72
supremum, 159
symmetric matrix, 362

symmetric positive definite, 49, 81, 87, 93, 116,
    213, 220, 269, 333, 368, 519
Taylor series, 357
    common, 359
    integral remainder formula, 359
    Lagrange remainder formula, 169, 357
    multivariate, 318
    multivariate with Lagrange remainder, 360
    proof of Euler's formula, 370
trapezoidal sum rule, 244, 307
triangle inequality, 58, 108, 154, 354
tridiagonal matrix, 369
trivial pivoting, 30

uniformly continuous, 353
unitary matrix, 43

Vandermonde matrix, 54, 92, 219, 270
variable metric methods, 185
    Broyden–Fletcher–Goldfarb–Shanno, 188
    Davidon–Fletcher–Powell, 188
    Oren–Spedicato, 189
variance, 6, 152, 219, 322, 325
variance reduction, 312
variational conditions, 116
variational principle, 334
vector, 361
    dot product, 361
    inner product, 361
    outer product, 361
    principal, 57
vector iteration, 62, 72
Verlet integration, 300

Wagner–Platen analysis, 309
weak order, 312
weak row sum criterion, 107
Weddle's rule, 244
white noise, 302
Wiener process, 304
Wilkinson's polynomial, 19, 171, 270

zero suppression, 170
zombies, 301

Printed in the United States
by Baker & Taylor Publisher Services